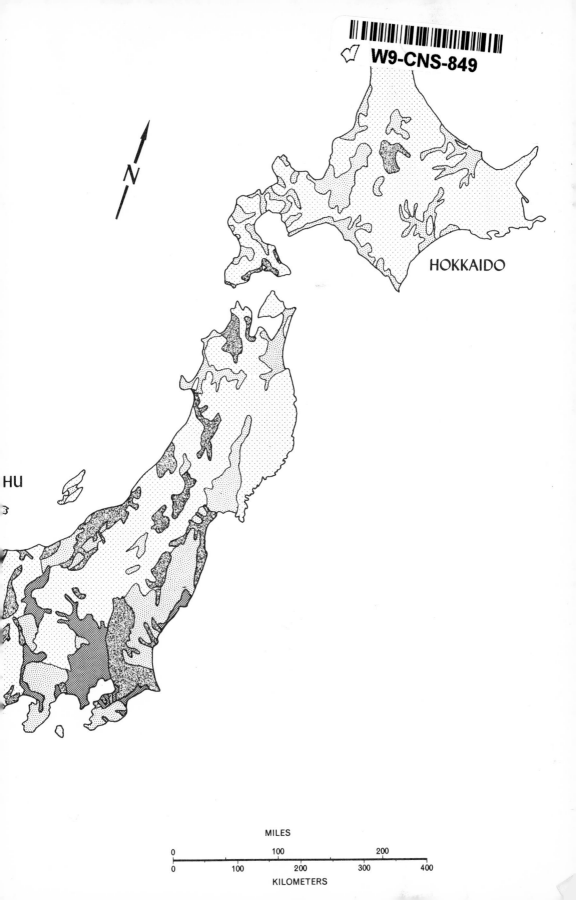

HOKKAIDO

HU

N

MILES

0   100   200

0   100   200   300   400

KILOMETERS

JAPAN

日本

JAPAN

A GEOGRAPHY

GLENN T. TREWARTHA

THE UNIVERSITY OF WISCONSIN PRESS, MADISON & MILWAUKEE
METHUEN & CO. LTD., LONDON

Published 1965
The University of Wisconsin Press
Box 1379, Madison, Wisconsin 53701

Published in Great Britain
Methuen & Co. Ltd.
11 New Fetter Lane, London EC4

Copyright © 1965
The Regents of the University of Wisconsin
All rights reserved

Printings 1965, 1970

Printed in the United States of America
ISBN 0-299-03440-2; LC 65-11200

PREFACE

The present book has two forebears, *A Reconnaissance Geography of Japan*, published in 1934 as Number 22 of the University of Wisconsin Studies in the Social Sciences and History, and *Japan: A Physical, Cultural, and Regional Geography*, published in 1945 by the University of Wisconsin Press. The grandparent volume, based upon fieldwork done in 1926–1927, and again in 1932, was written in the early 1930's before Japan was well launched upon those momentous developments of the thirties which eventually led to the war in the Pacific. To be sure, portents of the army-instigated expansionist policy in Asia were already observable in the 1931 invasion of Manchuria, and subsequently in the flouting of international censure by a haughty withdrawal from the League of Nations. But Japan had scarcely begun the revamping of her industrial and commercial structure required by her plans for domination in eastern Asia and the contingency of war which these plans entailed. That was a feature of the remaining years of the decade, following 1932. The first book, therefore, described Japan as of about 1930, before the period of transformation.

The second book treated the geography of Japan as of about 1940, on the eve of the Pacific War. By that date the country had already remarkably increased its efficiency to wage war through an unprecedented expansion in the heavy industries and a general conversion from consumer's to producer's goods. While a span of only about a decade separates

v

the datum planes of the two volumes, it was, nevertheless, a decade of great change.

Still more momentous changes, during and following the war, were to occur in the two decades that separate the temporal base level of the present book from that of its predecessor. Consequently, Japan of the early 1960's represents a quite different situation from that of either 1930 or 1940.

Not uncommonly offspring bear varying degrees of resemblance to their ancestors, and so it is with the present volume. In its preparation there has been some salvage, to be sure, from its parent and grandparent. Although the physical geography of Japan has of course not changed appreciably over a span of three decades, nevertheless the sections dealing with terrain, climate, soil, and minerals have had to be largely rewritten because of new materials that have become available and new concepts generated. Still, it is in the field of Japan's cultural-economic geography that the war and postwar periods have produced the most revolutionary changes, which in turn have been matched by a wealth of new materials. Consequently, the amount of salvage here has been small, necessitating a complete rewriting, which has the benefit of first-hand observation of Japan on two different occasions since the war—in 1947 when the country was still in the slough of despond, and in 1962 when it was riding the crest of the wave of prosperity.

It seems the part of wisdom to admit that the book has been written with at least two groups of readers in mind This dual focus has resulted in the inclusion of a greater variety of materials at a number of levels than would ordinarily be the case where only one audience is considered. Thus, while it is the professional geographers whose needs are most seriously regarded, I have not been unmindful of the large lay public who manifested their interest in Japan's geography by their wide use of the 1945 edition. The several hundred illustrations included, most of them new, are intended as an aid more especially to the non-professional readers. This has resulted in the inclusion of some illustrations which, though they fall short of photographic excellence, do portray with honesty a desired geographic feature or combination of features.

The fact that specific acknowledgements of assistance and courtesies received are so few only indicates that the donors are unusually numerous. An inclusive list would defeat by tedium the purposes of gratitude. Two groups of benefactors I feel constrained to single out as deserving special acknowledgement of my deep indebtedness: the university geographers of Japan and the government officials—national, prefectural, and municipal—both of which groups gave freely of their time in providing counsel and information as well as in collecting for me needed published and unpublished materials. It almost goes without saying that a book on the geography of Japan should be written by a Japanese geographer, and the fact that I have been so presumptuous as to undertake the task only makes me the more conscious that without the generous outpouring of assistance from these two groups the book could never have been completed. Certain of the chapters have benefited from a critical reading by Japanese specialists—those on terrain and climate by Tadashi Machida and Eiichiro Fukui of Tokyo Kyoiku University; that on agriculture by Yoshikatsu Ogasawara, Chief of the Land Survey Section of the Geographical Survey Institute; and the one on manufacturing by Kikukazu Doi of Shizuoka University. The entire manuscript has been read by Shoshichi Nomura of Yokohama University. But while I am eager to acknowledge the benefits received from the critiques of these five geographers, I hasten to exonerate them from blame for such errors as still remain, reserving that exclusively for myself.

G.T.T.

Madison, Wisconsin
June 1964

CONTENTS

JAPAN

INTRODUCTION

Japan is unique in the world—the first and only non-Western nation to achieve a high degree of industrialization, the single modern nation outside the West, the first and only genuine consumer market beyond the limits of the North Atlantic Basin. Japan has proved that the methods, tools, and policies of a free economy can generate rapid economic development in a non-occidental environment.

From the ashes of defeat in World War II, Japan rose again like the phoenix and now, less than two decades later, has achieved a level of production and consequent prosperity unheard-of before the war. With an economic growth rate in recent years unequalled elsewhere (averaging over 9 per cent for the period 1952–1962), she has forged ahead of several thriving nations in Europe to attain fourth rank in industrial production. This remarkable recovery has given the country new pride and self-respect. The old attitude of self-depreciation is giving way to a new and seemingly healthy, but heady, nationalism which at present pervades the whole country.

But if in some ways Japan is unique in the world, oddly enough she is more strikingly the exception within the less extensive Orient, of which she is a part and with which her contacts have been far more intimate and enduring. For all but a fraction of her two millenniums of history the greater world environment had almost no importance; it was only about a century ago that Japan emerged from her self-imposed seclusion to play a role in world affairs. Her roots are completely Asiatic, and the flavor of her culture derives chiefly from China, her vastly larger, older, and historically more prestigious continental neighbor. If Japan's nonconformity within Asia is to be emphasized, however, it is necessary first to establish the datum plane from which the exceptions are to be measured.

Eastern, southeastern, and southern Asia—an area sometimes called the Oriental Triangle, and perhaps less advisedly the Monsoon Realm—covers that corner of Asia reaching from maritime Siberia and Manchuria in the north and east to India–Pakistan–Ceylon, comprising South Asia, in the south and west. Between these pole areas lie Japan, Korea, Mainland China, and Formosa, in East Asia, and the numerous shatterzone countries of insular and peninsular Southeast Asia.

The prevailingly hilly Oriental Triangle, of which Japan is but a small fragment, faces upon the Pacific and Indian oceans. On its inner borders it is terminated by the great mountain masses and lofty plateaus which form the highland core of the continent, the 5000-foot contour being sometimes regarded as an approximate boundary. Slopes predominate, low level land being at a minimum. With the important exception of the Manchurian Plain, the lowlands are plains of river deposition—alluvial floodplains and deltas whose surfaces are essentially flat and hence easy to irrigate.

Partially offsetting the handicap of rough terrain is the area's relatively abundant rainfall, which greatly enhances its agricultural potential. The seasonal winds, known as monsoons, which are most perfectly developed in eastern and southern Asia, are an important unifying bond of the realm. In summer these monsoon winds blow from over warm tropical and subtropical seas into the heated continent, bringing with them an abundance of moisture which various kinds of weather disturbances cause to fall as copious rains. In winter, on the other hand, dry, cool to cold winds of land origin blow from the colder continent toward the warmer seas. Over most of southern and eastern Asia these winter monsoons yield scanty rain. The result is a region with a marked seasonal rhythm in precipitation; a land of winter drought and summer rains. On the average, summer rains penetrate the continent for several hundred up to a thousand miles. Beyond that are the dry lands of inner Asia, of pastoral nomads and oasis agriculture. The 20-inch rainfall line may also be taken as a fairly satisfactory boundary of the landward limits of the Monsoon Realm.

Because it is for the most part a humid realm, much of it was originally covered with forest, or with forest intermingled with woody shrubs and grass. Good natural grassland appears to have been confined largely to the Manchurian Plain. Here then is still another feature that sets apart the humid eastern and southern margins of Asia from the dry interior where sparse grasses and shrubs are prevalent. Because of the relatively abundant precipitation, the long growing season, and the forest cover characteristic of much of the region, residual soils are generally inferior, having been leached of important soluble minerals and being low in organic matter. Only in the subhumid north where good grassland prevailed, as in Manchuria, or where the parent soil material is basic lava, as it is in Java, are mature soils of good quality. To a great extent, however, it is the young alluvial soils of deltas and floodplains which provide the base for the intensive agriculture. Unfortunately, such fertile lowland soils occupy only a limited proportion of the realm's total area.

Eastern and southern Asia is not at present a large producer of important economic minerals. In part this reflects the prevailingly backward stage of economic development, for factory industry is only meagerly and spottily developed. In part, however, it reflects actual deficiencies or unfortunate location. By far the greatest coal reserves are in interior northern China, with much more modest ones in Japan and northeastern India. The only first class iron ore deposit is in northeastern India, although there are lesser ones in the Philippines and the Malay Peninsula. Petroleum is generally deficient, Indonesia possessing the most significant reserves. Southeast Asia is the world's greatest tin-producing region. What becomes clear is that while eastern and southern Asia possesses some important mineral deposits, they are, unfortunately, widely scattered, often poorly located, and not under the control of a single political unit which might integrate their use. It is dubious whether the mineral reserves of the Oriental Triangle are of such magnitude, quality, and geographic distribution as to be capable of supporting an industrial development comparable to that in the lands tributary to the North Atlantic Basin.

Culturally and economically the Oriental Triangle is one of the most distinctive of the earth's grand geographic subdivisions. It is one of the two Old World centers of ancient civilization and culture diffusion. And although the civilizations of China and India did not develop as early as those of Egypt and Mesopotamia in the eastern Mediterranean region, they have, on the other hand, endured down to the present time, whereas those earlier ones perished long since. Chinese civilization has flourished through forty centuries of time.

About 1.7 billion people are crowded into eastern, southeastern, and southern Asia, making it by far the most populous of the earth's subcontinental regions. Here on about 10 per cent of the earth's land area are concentrated 50 to 55 per cent of its people, a fact which by itself is sufficient to establish the high relative importance of the Oriental Triangle. The population density of about 115 per square kilometer (over 300 per square mile) is well above that of more highly urbanized and industrialized Europe. However, the figure gives no proper concept of true density since the huge total numbers are so highly concentrated on fertile plains of new alluvium while the more extensive hills and mountains are in many cases fairly empty. The consequence is a very patchy, fragmented pattern of population distribution, typifying a river-valley civilization.

Not only is the population already dense, but the high birth rates and declining death rates portend an increased density in the future. Birth rates may be about 40 per 1000 and death rates around 20 or a little more, so that the annual increase in population must be close to 2 per cent, or an addition each year of around 34 million to the present already superabundant population. It is this large and rapidly increasing population which makes it so difficult for the region to advance economically. Literacy is low, malnutrition is prevalent, and health is poor. Urban dwellers make up only a relatively small proportion of the total, for those functions which call cities into existence are poorly developed.

Probably upwards of 70 per cent of the 1.7 billion people are impoverished peasant farmers engaged in intensive subsistence agriculture. Their interest is primarily in the production of food crops for home consumption; to a vast majority of them the world market is not of great importance. Except in Japan, forests are inefficiently utilized or conserved. There is almost no good pasture land, a resource so essential in the economies of many world regions where livestock farming is important. Animals exist not in herds, but singly or in two's and three's, chiefly as work animals on farms. Animal-derived foods, which are very scarce, come mainly from scavenger animals such as swine and poultry, or from fish.

Rice is the principal food crop for two-thirds to three-quarters of the people, and 90 per cent of the world's production of rice comes from this part of Asia. In a few restricted areas of tropical southern and southeastern Asia, the raising of special commercial crops has become important, among them rubber, tea, Manila hemp, cinchona, and coconuts.

A corollary to the predominance of intensive subsistence agriculture is the relatively undeveloped state of modern factory industry. Though it is true that important progress has been made in manufacturing in Mainland China, India, and a few other areas since the war, nevertheless most of the comprehensive region has not been importantly touched by the Industrial Revolution. Much of the manufacturing is still in the workshop stage, with craft products being among the best known of oriental wares. Trade, like manufacturing, is meagerly developed, and transportation systems for large areas are inefficient. Clearly, the Oriental Triangle is to be numbered among the earth's underdeveloped regions, albeit something of an anomaly because of the immensity of its population, the intensity of its land use, and the antiquity of its culture.

Within the Orient whose basic characteristics have just been sketched, Japan is the exception, as was stated earlier. Unlike other parts of the region, Japan has never experienced outside control in the form of colonial rule and exploitation, with the variety of benefits and handicaps which colonialism involved. Thus, when in the second half of the nineteenth century she moved out of feudalism and undertook a rapid westernization, there were no ties with a foreign power that had to be broken before a modernization program could be launched. She could chart her course as she saw fit. Accordingly, she was able to modernize, and industrialize earlier than the rest of Asia, and to develop an efficient army and navy to bolster her standing among the world's nations. The modernization program was sparked by a strong central government which levied taxes to provide the capital for building railroads, merchant fleet, and factories, and for establishing a system of compulsory public education. It is difficult to overestimate the importance of the last of these in facilitating the modernization process, for by means of it Japan was early provided with a literate citizenry which could be relied upon to operate effectively in the transformation then taking place. Her efficient army and navy allowed her to be victorious in wars with China and Russia, out of which conflicts came the beginnings of a colonial empire. Maritime interests grew apace, so that foreign trade and a merchant marine became essential elements of her developing economy.

Thus, by the 1930's Japan already had attained a unique position in the Orient, for her superior military and economic strength provided her with an unrivaled opportunity to become the recognized leader in that part of the world. The treatment accorded the native peoples of Asia by the colonial powers was not such as to endear the masters to their subjects, so that if Japan had really had the welfare of the Asiatics at heart, and not so exclusively her own selfish interests, she might have elicited genuine cooperation from them in a non-violence program emphasizing "Asia for the Asiatics" and "The Greater Sphere of Co-prosperity in Eastern Asia." It soon became obvious, however, that such slogans were only a thin camouflage for a well-planned and systematic takeover in Asia. This led to the Pacific War.

Defeated in war and occupied by the enemy, shorn of her empire, and with her industrial structure in ruins, Japan has climbed back to a level of prosperity that makes her now more than ever the exception in Asia. But even so, her per capita income of $404 in 1961, high by Asiatic standards, was only about 15 per cent of that of the United States, 40 per cent of West Germany's, and 70 per cent of the Italian level.

Within the span of a decade after 1947 there occurred a vital revolution in Japan, scaling down both birth and death rates so that at present they rival the world's lowest, as does the rate of natural increase. Thus in her vital rates Japan is quite unlike the rest of the Orient. As cities grew in numbers and in size, percentage figures for urban population mounted; some 44 per cent of the nation's people now reside in compact settlements of more than 5000 population, a proportion two or three times higher than in most of the Orient. Corollary to this is the relative decline in rural population so that farm people make up a far smaller proportion of Japan's population than is the case in China, India, and their neighbors. Likewise in health, literacy, average years of formal schooling, quality of diet, and per capita real income the Japanese have far outdistanced the others.

In the economies, Japan's agriculture leads the Orient in efficiency and production per unit area of cropped land and also in the percentage of the output which goes on to the market. Thus, the agriculture is much more commercial than elsewhere. In manufacturing, the gap separating Japan from other oriental nations is even wider than in agriculture, and it continues to widen as modern factory industry maintains the greatest expansion rate of any sector of the economy. In overseas trade, also, it is Japan which provides the exception, there being not even a close rival within Asia. The same is true of merchant tonnage and other means of modern transportation.

Just why it should have been Japan that reared up so strikingly above the general low level of Asia is a question not easily answered. Undoubtedly it was an auspicious conjunction of a number of factors and events. Unquestionably the absence of colonial ties operated to stimulate the development of national initiative. A strong central government with farseeing men at the helm doubtless made a notable contribution as well. But by no means least of the contributing factors was Japan's intelligent, literate, and educated citizenry, the product of an early-established system of compulsory, universal public education.

In general organization the 1945 edition of this book has served as a model for the present edition. Parts I and II treat of Japan in its entirety.

In these sections those geographic characteristics belonging to the archipelago as a whole have been emphasized, while intraregional contrasts have received little attention. Part I concentrates on the physical aspects of Japan's geography and analyzes such resource elements and features of natural equipment as terrain, climate, soils, native vegetation, and economic minerals. Here there was the dual aim of supplying foundation material for an understanding of the subsequent treatment of the cultural geography, and providing the reader with a concrete picture of the physical face of Nippon. Part II does for the cultural elements of Japan's geography what Part I does for the physical. Population, settlements, agriculture, manufacturing, and transportation are the principal themes elaborated. Part III is concerned with the regional contrasts within Japan. To a notable extent its content is based upon personal observations made during four periods of investigation in Nippon. This regional treatment, it is hoped, will not only be intrinsically valuable, but will also provide a geographic framework for Japan into which past and future detailed studies of limited areas may fit.

For a regional treatment of the reconnaissance type, such as Part III contains, Japan presents peculiar difficulties. The country is composed of numerous small, isolated units which almost defy generalized synthesis. This individualism of units, resulting from the complicated relief, as well as from the country's long history and earlier feudal organization, is such a characteristic feature that a description employing broad strokes is less satisfactory than it would be in regions where the terrain and occupancy patterns are coarser and less complicated. In Part III this has led to a brief analysis of many more individual areas than would ordinarily be the case in what purports to be a regional study of reconnaissance type.

In the book as a whole consistency is lacking in the units of measurement employed, principally because the sources used presented their data in metric, English, and Japanese units, and I have not felt it necessary to convert them to a common denominator. Linear and area measurements are commonly given in metric units since most, but not all, Japanese statistical volumes present their data in this form. But, on the other hand, I have not hesitated to use English and Japanese units when these were employed in the original sources. Temperature data are always given in Fahrenheit degrees since, regrettably, a great majority of English-speaking people do not readily convert from Centigrade. Fortunately, a hectare and a Japanese cho are almost identical, each being approximately 2.5 acres. A conversion table is provided on the following page.

UNITS OF MEASUREMENT

Metric	American and English	Japanese
Length:		
Centimeter	0.3937 inch	3.3 bu
Meter	3.2808 feet	3.3 shaku
Kilometer	0.6214 mile	0.25 ri
Area:		
Square meter	10.7638 sq. feet / 1.1960 sq. yards	0.30 tsubo
Square kilometer	0.3861 sq. mile	0.065 sq. ri
Hectare	2.4710 acres	1.01 cho
Weight:		
Gram	0.03527 ounce	0.27 momme
Kilogram	2.2046 pounds	0.27 kan
Metric ton	0.9842 long ton / 1.1023 short ton	266.67 kan
Capacity:		
Liter	0.8799 imp. quart / 1.0567 U.S. quart	0.55 sho
Hectoliter	21.9973 imp. gallons / 26.4178 U.S. gallons / 2.7491 imp. bushels / 2.8378 U.S. bushels	0.55 koku

PRONUNCIATION OF JAPANESE NAMES

Unlike the English language, Japanese utilizes no variation in the amount of air from the lungs as the means of accent. So, to the ear of a native speaker of English, utterances in Japanese sound as if there were no accent. The Japanese accent consists of variations in pitch, the accented syllables being pronounced with a high pitch as in singing. The accented syllables may occur in the initial, medial, or final position of the word.

Vowels are pronounced nearly the same as the vowels of the musical scale: *a* as in *fa, e* as in *re, i* as in *mi, o* as in *do; u* has the sound of *oo* in boot. All five vowels have lengthened counterparts, such as the *o* in Osaka, which is held about twice as long as a long *o,* in English.

Diphthongs are pronounced as they are spelled. For example, *sai* is pronounced like *sigh* in English. Only *ei* is exceptional and is pronounced indiscriminately either like the vowel *a* in *fate* or as a lengthened Japanese *e*.

Consonants have much the same sound as in English. Double consonants are both pronounced. The letter *g* is always hard. In such names as *Tokyo, Kyoto,* and *Kyushu* in which a combination of a consonant + *y* + a vowel occurs, the *y* is pronounced as a consonant, so that *kyo* or *kyu* is rendered as one syllable. Thus *Kyushu* is pronounced *Kyoo-shoo.*

PART I

PHYSICAL EQUIPMENT AND
RESOURCES OF JAPAN

I · LAND-SURFACE FORM

Nature did not tailor the resource pattern of Japan to a scale befitting a great nation. Small in area, rugged of surface with a minimum of lowland, and sparingly supplied with most economic minerals, Japan in her climb to a position of world eminence has been handicapped by this scarcity of resources. Yet in recent years she has ranked fifth among the earth's nations in gross national product. That these natural handicaps did not prevent her from attaining high economic rank illustrates how fallacious is the doctrine of environmental determinism even as applied to nations. On the other hand, to belittle the effects of Japan's modest natural endowments upon the nation's economies and culture is as mistaken as to overemphasize and oversimplify their influence.

This same physical environment which has provided a resource base of only modest proportions has nevertheless blessed Japan with great natural beauty, including picturesque rugged coastlines and island-studded adjacent waters, verdant forested mountains, majestic volcanoes, and a total landscape softened and beautified by a pervading bluish moisture haze.

The early chapters of this book present an analysis of the land of Japan in its original and man-modified natural aspects. This is physical geography. Such an inventory of the natural endowments and resources provides a background against which to view and understand the development of a nation; for surface, climate, forests, soil, minerals, etc., provide the substance out of which a people fashions its civilization. These elements are more than merely a set of natural features combining to form the varied and attractive physical face of Nippon; along with other factors, they have influenced the very structure of anthropo-Japan.

LOCATION

Arc-shaped, like a bow tightly strung, Japan comprises a festoon of mountainous volcanic islands forming the outer façade of eastern Asia. Extending in a southwest-northeast direction for a distance of about 2200 kilometers, or close to 1400 miles, the main islands are separated from Korea by some 200 kilometers of ocean, while the shortest distance from Hokkaido to mainland U.S.S.R. approximates 300 kilometers. At its widest point, however, the Japan Sea measures over 900 kilometers, or about 550 miles. Physically, of course, this position remains unchanged, but its geographical significance has varied greatly through historic time.

Obviously, Japan's simultaneous proximity to Asia and separation from it by extensive bodies of water, have unusual climatic implications, as regards both temperature and precipitation. This location also strongly influenced the origins of the Japanese peoples, for they are predominantly of Mongoloid stock, while the country's culture shows a paramount Chinese influence.

Japan's insular position with respect to East Asia has been compared with that of Britain in Western Europe, although the breadth of the water barrier is significantly greater in the case of Japan. In each instance there occurred a development semidetached from that of the adjacent continent. Both countries are close enough to the mainland so that communication with it was relatively easy whenever it was deemed desirable, but at the same time the intervening seas were an important defense barrier on those occasions when withdrawal and seclusion fitted their needs. Japan's history is one of alternating periods of seclusion and communication with mainland East Asia. In both Britain and Japan maritime interests have loomed large, and each has emphasized sea power, an important merchant marine, and foreign trade. Each developed industry earlier than did the adjacent continent; and the resulting need for overseas industrial raw materials in turn stimulated the acquisition of a colonial empire. Japan, at the gateway to populous East Asia, has a strategic position both in an economic and in a military sense. Close ties with mainland East Asia appear to be a natural consequence, an association which at present is prevented by barriers of a political nature. And in spite of the fact that after World War II Japan was shorn of its insular possessions both to the north and south of the main islands, its strategic position with respect to Asia still persists. In any inventory of Japan's natural assets (and liabilities) this element of geographic location is a relatively enduring one, and as the country has capitalized on it in the past, no doubt it will continue to do so in the future. At present Japan is the westernmost Pacific bastion of the Free World, serving to counter the Communist thrust on the mainland, a position which is fraught with danger, and is not altogether relished.

ORIGIN AND PHYSICAL FRAMEWORK

Japanese legends depict the gods as a not very lively crowd, dwelling uneventfully in the "Plain of High Heaven." But the heavenly pace quickened with the arrival of the god-hero Izanagi and his female counterpart, Isanami. Among his numerous heroic gestures Izanagi once thrust his heavenly jeweled spear into the sea, and as he withdrew it the shower of drops that fell from the weapon was transformed into the dragon-shaped island group of Japan. In actual fact, the origin of the archipelago was not much less dramatic; only instead of showering down from above, the islands thrust upward from mighty ocean deeps.

Essentially the Japanese islands are the summits of gigantic submarine ridges which have risen from the floor of the Pacific Ocean, for the archipelago is but one sector of the unstable circum-Pacific orogenic zone. On its Pacific side the Japanese highlands rise abruptly from profound ocean deeps of 8,000–12,000 meters, and on the west likewise they are thrust upward steeply from the deep Japan Sea. Thus, Japan is a weak and unstable segment of the earth's crust where readjustments of deep-lying rock masses are constantly occurring, resulting in the hundreds of volcanoes that stud the archipelago, and the more than 1500 earthquake shocks that occur annually. Clearly the archipelago's most striking surface features, as expressed by altitude and by the arrangement of highlands, are the manifestation of orogenic processes resulting in differential movements of the earth's crust. Even some of the minor surface features—the ubiquitous marine and river terraces, fault scarps, submerged ria coasts, and emerged coastal plains—are of similar origin.

But while the larger primary surface features are a consequence of endogenic processes, with the result that areas of old worn-down mountains and erosional plains are small, at the same time gradational agents, chiefly running water, are responsible for many of the tertiary landform features. Abundant rainfall and steep slopes have resulted in a close network of vigorous streams whose erosive work is phenomenally rapid.

In area the Japanese Archipelago (369,766 sq. km., or 142,767 sq. mi.)

RYUKYU ARC

KYUSHU
NODE

SOUTHWESTERN ARC

NORTHEASTERN ARC

KARAFUTO ARC

MILES
0   200

0   300
KILOMETERS

BONIN ARC

CHUBU
NODE

HOKKAIDO
NODE

KURILE ARC

UNIVERSITY OF WISCONSIN CARTOGRAPHIC LABORATORY

Fig. 1-1.—Shape and terrain character of the Japanese archipelago are closely associated with structural arcs and nodes. Six main arcs are involved, and three principal nodal areas.

is slightly less than that of California and two-thirds that of France. As a result of territorial losses following World War II (chiefly Korea, Taiwan, the southern half of Sakhalin, and groups of Pacific islands) Japan's present territory has an area only 55 per cent of her prewar domain, a proportion which would be diminished further if less-permanently held Manchuria were considered. Four main islands—Hokkaido, Honshu, Kyushu, and Shikoku—comprise over 98 per cent of the total area, small adjacent islands making up the remainder. The principal islands are separated from one another by narrow straits resulting from local subsidence.

Arcuate in structure and convex toward the Pacific, the Japanese Archipelago comprises the Northeastern or Honshu Arc and the Southwestern Arc (Fig. 1-1). These two fuse in central Honshu where they are intersected by the Bonin (Shichito-Mariana) Arc pushing in from the south. In Hokkaido the Northeastern Arc is intersected by the Karafuto and Chishima (Kurile) arcs, and in Kyushu the Southwestern Arc is fused with the Ryukyu Arc, also intruding from the south. Terrain patterns within the two main Northeastern and Southwestern arcs are largely determined by the structural alignment of each arc, the grain of the relief being north-south in northern Honshu and west southwest–east northeast in southwestern Honshu and Shikoku. Where arcs intersect or coalesce, the terrain pattern is more confused, and extensive knots of high and rugged terrain result. These are called nodes, the most conspicuous being the Hokkaido Node in the north, the Chubu Node in central Honshu, and the Kyushu Node in the far south. Node areas characteristically are marked by numerous volcanoes, some of them active.

EARTH MATERIALS

Even a cursory examination of a geological map of Japan reveals that the archipelago is a minutely subdivided mosaic of various kinds of surface rocks and earth materials, differing in age, structure, and resistance, reflecting a long and complicated geological history. Still there is an observable arrangement and pattern in the distribution of the different rock formations, even though their patterns are complicated and the individual geologic units often of small size. The linear character of the islands and the approximate parallelism of the Pacific and Japan Sea coasts suggest a medial backbone of resistant rock, but such is not the prevailing structure. It is outside the scope of this book to trace the geological evolution and structure of the Japanese islands. But since lithic character and structure find expression in such geographically important features as terrain form, drainage patterns, soils, and economic minerals, brief comment will be made on surface rocks in connection with their prevalence, distribution, and relationship to slope and cultivation.

Since structural and erosional lowland plains are absent in Japan, most hard-rock regions (granite, older sedimentaries, volcanic, Tertiary) are predominantly regions of slope and so are to be classed as hill country and mountain. Table 1-1 shows that over 80 per cent of the country's area is composed of these four classes of surface rocks, with steep slopes prevalent. Some of the highest and most rugged terrain lies in the regions of the older sedimentary and metamorphic rocks. They comprise nearly one-quarter of the country's area and, while widely distributed, find their most exclusive development in the folded mountains of the Outer Zone of Southwestern Japan in southern Kyushu, southern Shikoku, and Kii Peninsula. They also form the high north-south mountain ranges

TABLE 1-1

Prevalence of different types of surface rocks in Japan and their relation to slope and cultivation

Surface rocks	% of country's total area	% of formation having slopes of 15° or less	% of formation under cultivation
Granite	12	16	10
Rocks older than Tertiary	24	4	2
Volcanic	26	25	8
Tertiary	20	46	11
Diluvium (older alluvium)	6	91	24
Alluvium	12	96	45

Source: Japanese Geological Survey.

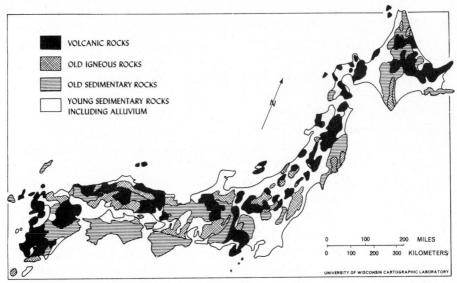

VOLCANIC ROCKS

OLD IGNEOUS ROCKS

OLD SEDIMENTARY ROCKS

YOUNG SEDIMENTARY ROCKS
INCLUDING ALLUVIUM

N

| 0 | | 100 | | 200 | MILES |
| 0 | 100 | 200 | 300 | | KILOMETERS |

UNIVERSITY OF WISCONSIN CARTOGRAPHIC LABORATORY

Fig. 1-2.—Terrain character in Japan is closely associated with the nature of the underlying bedrock. This map indicates a very complex surface geology, which in turn bespeaks complex terrain character. After map by Isida.

of Hokkaido. That slopes are steep in the older sedimentaries is borne out by the facts that only 4 per cent of their area has an inclination of less than 15°, and that only 2 per cent is under cultivation.

Granites are likewise widely distributed but are most extensively developed in those parts bordering the Inland Sea (extreme southwestern Honshu, northern Shikoku and northern Kyushu) where the terrain features are those of hills and low mountains. Milder slopes, with less than 15° inclination, are decidedly more widespread (16 per cent) in the granite areas than in the regions of older sedimentary rocks (4 per cent), with the consequence that the percentage of the total area under cultivation is distinctly higher in the granites (10 per cent as compared with 2 per cent).

Regions of volcanic rocks are likewise very prevalent (26 per cent of the country's area) and widely distributed, but their terrain character is highly variable. While some of the highest and wildest mountain country coincides with volcanic cones existing in groups, yet the lower slopes of many individual volcanoes are not so steep, a feature suggested by the fact that 25 per cent of the volcanic area has slopes of under 15°. Still, only 8 per cent of such areas is under cultivation, largely because of the low-grade soils and meager water supplies of the lower ash slopes and plateau-like uplands.

The younger Tertiary rocks (sandstone, shale, conglomerate, tuff) are often weak and poorly consolidated, so that where they exist at low elevations they characteristically form low, thoroughly dissected hill country, with moderate to steep slopes and considerable valley-floor areas. Such hill country usually is distinguished by fairly dense settlement, most of it concentrated on the alluvial floors of dendritic valley systems. Compared with granite and volcanics, Tertiary-rock areas are characterized by a smaller proportion of steep slopes (see Table 1-1), but still the percentage of land under cultivation is not markedly greater, indicating that Tertiaries are to be found at high as well as low elevations. Also of geographic significance is the fact that Tertiary rocks are the source of much of Japan's coal and petroleum.

Unconsolidated deposits, both older alluvium (diluvium) and newer alluvium, which constitute the plains of Japan together comprise only 18 per cent of the country's area. But it is on this restricted area that settlement and economic production are concentrated. Much of the old alluvium exists as benches and terraces, which rise in the form of low tablelands above the floodplains and deltas of new alluvium. The percentage of the alluvial surfaces having steep slopes is low in both the old and the new, but the discrepancy between them is striking in terms of the proportion of each under cultivation, for the older alluvium is more difficult to irrigate, has poorer soils, and is frequently deficient in water.

TECTONIC AND GEOMORPHIC SUBDIVISIONS

Since the gross features of terrain in Japan are closely associated with orogenic processes, which in turn have intimate connection with the genesis and evolution of the insular arcs, it is not unexpected that the geotectonic and geomorphic patterns show strong resemblances. Still they are not identical.

On the basis of geotectonic criteria, the country may be divided into two dissimilar parts, a north and a south, separated by the great depressed zone of the Fossa Magna which traverses the mid-part of the main island from the Pacific Ocean to the Sea of Japan, and is marked by a bold fault scarp on its western side (Fig. 1-3). Here it almost appears as though the Honshu mountain arc has been bent backward along this fracture zone and the resulting rift subsequently filled, in part, by younger strata and great volcanic piles. West of the Fossa Magna the country may be subdivided into distinct inner and outer zones by the Median Dislocation Line extending from central Honshu, through Shikoku and into Kyushu, marked by conspicuous fault scarps and tectonic depressions. While Japan north and east of the Fossa Magna is not so strikingly separated into inner and outer zones as is the southwest, still there is a contrast between the Pacific and the Japan Sea sides; the former is chiefly composed of older rocks and few recent volcanics, and the latter is abundantly volcanic. The line of demarcation separating the two lies along the eastern base of the central range and is denoted by modest fault and flexure scarps.

In general the Outer Zone of Japan, or the side facing the Pacific Ocean, is dominated by up- and down-warping instead of abrupt faulting; changes in coastal features are gradual; and geomorphic faults are relatively rare, with the epicenters of destructive earthquakes situated in the Pacific Ocean. By contrast, in the Inner Zone facing the Japan Sea the major features of terrain configuration are the result of intricate nets of geomorphic fault lines. As a result coastal features are of various types even within short distances. The earthquakes within this subdivision are mostly of minor intensity and their epicenters are located inland. Abrupt differential crustal movements are common, and earthquakes appear to be associated with the renewed activity of fault block formation.[1]

But with attention primarily to surface form, geographers are inclined to recognize a pattern of terrain subdivisions somewhat different from that previously sketched, consisting of four principal units (Fig. 1-4 inset).[2] It will be observed (Figs. 1-3, 1-4) that most of the main geotectonic boundaries continue to function also as landform boundaries. The principal contrast is that the Fossa Magna has been discarded as a major terrain boundary and the great central knot of Honshu, or the Chubu Node, has been recognized as one of the four chief subdivisions because of its distinctive landform character. Hokkaido proper has likewise been

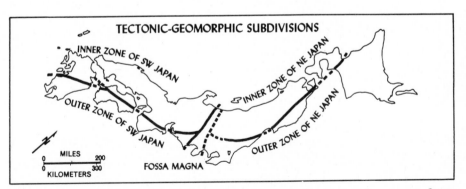

Fig. 1-3.—On the basis of structural and lithic contrasts, Japan may be separated into two great subdivisions, a northeastern and a southwestern, and each of these in turn may be further subdivided into an inner and an outer zone.

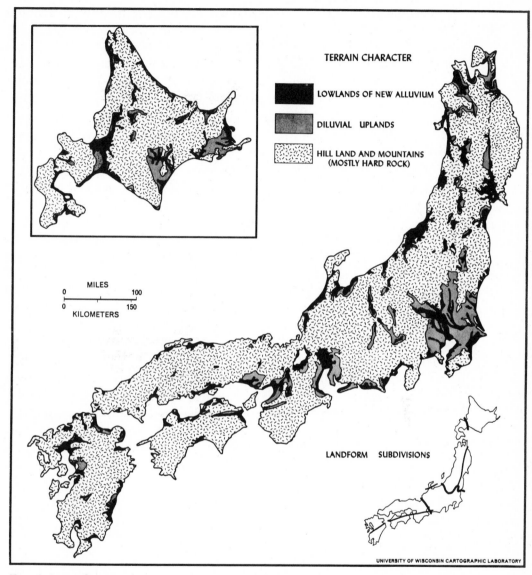

Fig. 1-4.—Predominantly Japan is a region of hill land and mountain where slopes prevail. Plains are minimal in extent.

raised to a first-order subdivision. Following are noted a few of the salient characteristics of each of the four primary landform subdivisions. Greater detail on surface features, including identification of the numerous subdivisions, will be reserved until a later section of this book dealing with the regional geography of Japan.

In the Hokkaido Proper Region the paramount surface features are a consequence of the coalescence of the Karafuto Arc from the north and the Chishima Arc from the east. Thus, the meridional backbone of the island, forming the two great north-south projections which terminate in Capes Soya and Erimo, is a southward continuation of the Karafuto Mountains. Similarly, the volcanic zone of Chishima and its basement range are continued westward in east-west trending volcanic mountains of central and eastern Hokkaido. Accordingly, the nearly right-angled intersection of these two mountain systems provides the skeletal framework for the island's terrain and its approximately rhomboid shape. Among the distinguishing terrain features of this region are subdued mountain forms and an extraordinary development of coastal and river terraces.

The Northeastern Region, including northern Honshu and the southwestern peninsula of Hokkaido, comprises the arc of northeastern Honshu. Here, corresponding to the structure of the arc, the trend of highlands and intervening lowlands is strongly meridional. Three parallel chains of north-south highland are separated from one another by lines of structural depressions. A major dislocation line marked by fault and flexure scarps follows the eastern margins of the central range and separates the region into eastern and western subdivisions. The eastern, composed chiefly of granites, gneisses, and older sedimentaries, is largely made up of the spindle-shaped Kitakami and Abukuma highlands where old erosion upland surfaces are conspicuous. The western subdivision includes a central backbone range, crowned with a row of volcanoes, and a lower, less continuous western range. Both of these are continued northward in peninsular Hokkaido.

The Central Region, where three major mountain arcs—the Northeastern and Southwestern of Honshu, and the Shichito-Mariana of the Pacific—coalesce and overlap, forms the broadest and highest part of the country. Terrain features are complicated. Lowlands include two of Japan's largest and most important plains—Kanto at the head of Tokyo Bay, and Nobi at the head of Ise Bay on the Pacific side—and a number of less extensive plains facing the Sea of Japan. In addition a series of mountain basins follows the depressed zone of the Fossa Magna. In the mountain lands lying west of the Fossa Magna the highest portions are found close to the western margins of that depressed zone which is marked by one of the boldest and most continuous fault scarps anywhere in the country. A chain of volcanoes, including Mt. Fuji, representing a northwestward projection of the volcanic chain of the Shichito-Mariana Arc, follows the great depressed zone east of this fault scarp and prevents it from being a continuous morphologic graben.

The Inner Region or Zone of Southwest Japan includes westernmost

Honshu (Chugoku), northern Kyushu, northern Shikoku, and the Inland Sea, this last subdivision representing a general zone of subsidence. Younger rock strata are not extensively developed, while granites are widespread. Volcanic materials are scarce except in Kyushu, where they are a result of the northward extension of the Ryukyu Arc and its intersection with the Southwestern Arc of Honshu. A regular and systematic arrangement of terrain elements is not conspicuous, as it is in northern Honshu and also in the Outer Zone of Southwest Japan. In this Inner Zone, terrain character is commonly associated with faulting which has produced great landform variety, including block mountains and associated basins, plateau-like upland surfaces, and within the Inland Sea itself, areas of open water separated by island groups.

The Outer Region or Zone of Southwest Japan composed of three fragments—Kii Peninsula, southern Shikoku, and southern Kyushu—is a region of young, folded mountains where lowlands are uncommon. It represents the longest, most conspicuous mountain range in the entire country, and actually extends eastward beyond Kii Peninsula into the Central Knot of Honshu where it becomes the Akaishi Mountains. The predominant rocks are crystalline schists and older sedimentaries. Local subsidence has produced the straits which divide the region into three separate parts. The appendage of volcanic materials in southernmost Kyushu is related to the intersection with the Ryukyu Arc. Along the northern margin of the Outer Zone is a northward-facing fault scarp coinciding in location with the great Median Dislocation Line.

TERRAIN CHARACTERISTICS

Since complexity and fineness of pattern are characteristic not only of the lithic features of Nippon but also of its terrain, broad generalization is difficult. Even within small areas the earth materials, their structures, and the resulting landforms are often of the greatest diversity. Lofty folded ranges have been altered by block movements, so that faulted and folded forms are much intermingled. Remnants of erosion surfaces at relatively high altitudes are widespread throughout the hill and mountain country, contrasting curiously with the steep slopes and great relief. Repeated volcanic eruptions and intrusions, extensive and wide-spread, have added further to the complexity of the terrain. Moreover, the recurring showers of volcanic ash, outpourings of lava, earthquakes, and changing strand lines prove that these tectonic forces are still active. Short, vigorous, steep-gradient streams, acting upon these complex structures and materials have sculptured a land surface whose lineaments are varied and intricate.

Hill and mountain lands

A core of moderately rugged hill land and mountain, containing a number of debris-choked depressions, with small discontinuous fragments of river- and wave-deposited plains fringing the sea margins of the mountain land—such is the gross geomorphic pattern of Japan. Along important stretches of the coast, plains are absent and the hill lands reach down to tidewater. Some three-quarters of the country's area is hill and mountain land whose average slopes exceed 15°, making it unfit for normal cultivation. Although 60 to 65 per cent of the land with a gradient of 15° or less is tilled, the area in farms is only 16 to 17 per cent of the country's total area, while but 14 per cent is cultivated. The highest elevations and the most mountainous terrain of Japan are to be found in the Chubu Node of central Honshu, where, in the so-called Japanese Alps, a dozen or more peaks rise to about 3000 meters, and Mt. Fuji, the highest of all, to 3776 meters. Just to the west and south of the Chubu Node, along the line of the Biwa Depression, is the most complete break in the Honshu highland barrier separating the Pacific and Japan Sea coasts.

In spite of the humid forest climate, much of it subtropical, a deep mantle of weathered rock material making for rounded terrain profiles is not characteristic. Instead sharp angular forms prevail, and steep slopes and narrow divides are characteristic, the conspicuous exception being

TABLE 1-2

Areas with slope of less than 15° (in thousands of cho*)

Region	Area with gradient below 15°		% of total area
Honshu		5602	24.1
Tohoku	1421		20.8
Hokuriku	631		25.1
Kanto	1509		30.1
Tokai	794		24.6
Kinki	515		24.7
Sanin	252		18.1
Sanyo	478		22.2
Shikoku		439	23.2
Kyushu		1102	26.0
Hokkaido		2158	26.6
ALL JAPAN		9303	24.8

* 1 cho = 2.45 acres, or about 1 hectare.
Source: Shiroshi Nasu, *Aspects of Japanese Agriculture* (New York, 1941).

Fig. 1-5.—Over large parts of Japan's mountain country sharp, angular terrain features prevail. Note the cultivated fields, both on the narrow floodplain and on the adjacent slopes. Settlements are on the slopes. Photograph by Setsutaro Murakami.

the areas of granitic rock. This seeming lack of harmony between climate and terrain form is probably the result of the recent and rapid uplift of many of Japan's mountain areas and the associated vigorous downcutting and removal by streams.[3] Youthful and imposing fault- and flexure-scarps, which commonly serve as the boundary zones between geomorphic subdivisions, are conspicuous features. Frequent landslides caused by earthquake tremors may have appreciably sharpened the contours of the terrain in some areas. Evidence of recent slides is to be observed in the numerous ruddy scars on certain mountain flanks, where the mantle of vegetation and a mass of regolith have recently been removed, exposing the raw bedrock beneath.[4] Upland remnants of old erosion surfaces are

Fig. 1-6.—A region of granite rock where rounded slopes are characteristic. The regolith mantle is thin, as is also the vegetation cover.

further evidence of relatively recent uplift. Especially where granite is the predominant rock type, rounded cupola features, covered with a thin mantle of weathered rock materials, are common. Southwesternmost Honshu (Chugoku) north of the Inland Sea is representative of this latter type of terrain (Figs. 1-5 and 1-6).

Another element giving variety and contrast to the terrain is the hundreds of volcanic cones, in various stages of activity as well as dissection, with their associated lava and ash plateaus. These cones provide some of the highest elevations of the archipelago. The symmetrical concave slopes of the young cones, and the radial patterns of their drainage lines and divides are distinguishing features. As noted earlier, volcanic rocks and ash deposits cover about one-quarter of the country's surface. Although they occur throughout the entire length of Japan, volcanoes are distributed according to a recognizable pattern, being numerous in some

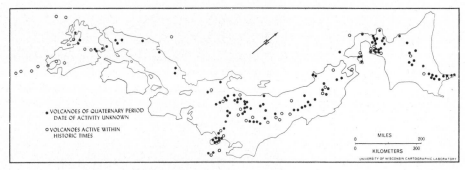

Fig. 1-7.—Distribution of volcanoes. After map by Isida.

areas and almost lacking in others. They are largely absent in the Outer Zone, or the Pacific side, of the country except in eastern Hokkaido, the vicinity of the Fossa Magna, and southern Kyushu. They are much more numerous in the Inner Zone and especially so in those areas where the mountain arcs coalesce.[5]

Lowlands

Japan lacks extensive lowlands, and the diminutive ones that lie within its borders are almost exclusively surfaces of river deposition—deltas, floodplains, and fans. There are no extensive structural plains underlain by relatively horizontal strata, such as exist in central and eastern North America and in northern and western Europe, and there are no erosional lowlands. Hard-rock areas in Japan are practically coextensive with hill and mountain land, almost the only exceptions being the fragments of wave-planed terrace along the coast, and certain upland erosion surfaces. The typical plain of Japan is a small isolated patch of river- and wave-worked sediments developed in a coastal indentation or in a mountain basin. Even the Kanto (Tokyo) Plain in east-central Honshu, which is by far the largest of these, has an area of only about 13,000 square kilometers, or 5,000 square miles. Being predominantly peripheral, most of the plains have frontage upon salt water. They seldom continue along the coast very far because of the frequent interruptions by spurs and larger masses of hard-rock hill land that extend down to the sea. Tunnels are numerous along those rail lines that attempt to follow the coast.

As the swift, turbulent mountain streams of Japan debouch from

Fig. 1-8.—The inland portion of a small coastal delta-plain wedged in between spurs of highland. The latter frequently reach to tidewater. Tea gardens largely cover the slopes in the foreground.

their highland valleys onto the plains, rapidly losing gradient, they deposit their coarser load near the base of the mountain in the form of conspicuous alluvial cones and fans. Some of what appear to be fans may actually have the structure of pediments. Most Japanese delta-fans steepen very perceptibly as the mountain hinterland is approached and the materials comprising the deposits become markedly coarser. Nevertheless, because of the small size of most Japanese plains, the mountain-fed rivers in time of flood are able to carry even to the sea margins much coarser materials than are ordinarily found in deltas of larger size.

Along their outer margins many of the delta-fans are bordered by a belt of beach ridges and dunes, the width and height of which depend on the strength of the winds and waves along any particular section of coast, as well as the type of recent vertical movement of the strand line. Where a low coast faces the open sea, dunes and beach ridges are usually prominent features, but along coasts fronting on quiet waters of deeply indented bays, especially the Inland Sea, they may be inconspicuous or lacking altogether. It is along the stormy Japan Sea littoral that dunes and multiple beach ridges are best developed. On the Niigata Plain, for example, the belt of parallel beach ridges and dunes attains a width of several miles, the ridges being separated by narrow, partially filled lagoons. Not infrequently the coastal dunes and beach ridges so obstruct the natural seaward movement of river waters that poorly drained land, or even lake and swamp, develops inland from them. Rivers are noticeably deflected from a direct seaward course and often are forced to flow parallel with the coast for some distance before they succeed in breaking through the dune and beach-ridge barriers to the sea. Many of the ridges have been planted with pine trees to prevent the sand from being blown inland over the fertile rice land, and to serve as windbreaks. These elevated beach ridges and dunes of coarse sand are obviously unsuited to a crop like rice, requiring inundation. Hence their use for orchards and unirrigated vegetable and cereal crops further distinguishes them from the plain proper to the rear, where paddy rice is likely to dominate.

Rivers on the lowlands are shallow and braided. They flow in broad gravel-choked beds, the several channels, except in flood periods, occupying only a small part of the total width of the bed. Characteristically, the rivers are *on* the lowlands rather than *in* them, for their channels and levees are the most elevated portions of the plains; that is, the land slopes downward away from the rivers. This is particularly apparent to anyone bicycling in the lowlands. Approach to a stream ordinarily is announced by harder pedaling, which eases up promptly once the bridge is crossed.

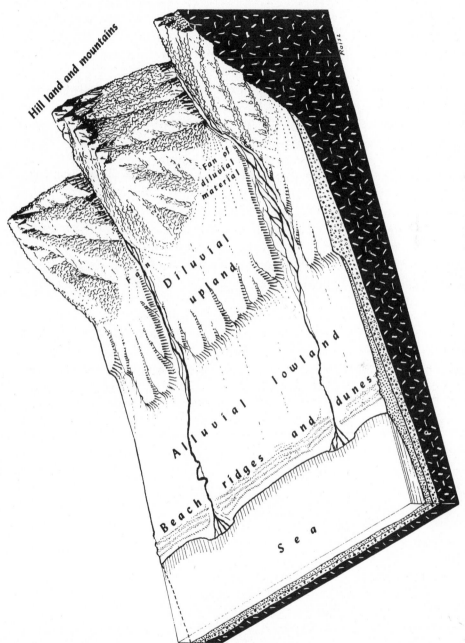

Hill land and mountains

Fan

Fan of diluvial material

Diluvial upland

Alluvial lowland

Beach ridges and dunes

Sea

Raisz

Fig. 1-9.—Representative arrangement of terrain features seaward from the mountains.

Fig. 1-10.—Cross section profile across an elevated stream channel.

In rare instances the streams are so much higher than the adjacent country level that roads and even railroads pass under them in tunnels instead of crossing over them. On the topographic maps, contours ordinarily bend downstream in the vicinity of river channels on the lowlands, indicating that they are above the general country level. Natural and man-made levees and dikes, often more than a single series of the artificial variety, endeavor to hold the lowland stream within its raised channel. Such elevated streams are both asset and liability—an asset in that a simple gravity system ordinarily serves to distribute their waters for irrigation purposes over the rice fields on the plain; a liability in that they may break their levees in time of flood and do an excessive amount of damage through inundation and even more through burying the precious rice fields under a layer of gravel and sand.

Common to many Japanese plains are riverine belts, where old and new levees, together with scars of old stream channels, result in terrain zones which not only are slightly more elevated, but whose surfaces are more uneven and whose materials are coarser than those of the plain proper. Such riverine zones, being more difficult to irrigate, are usually

Fig. 1-11.—An elevated natural-levee stream with man-made dikes. A road follows the crest of one levee. A village has also sought out the elevated dry site. Bamboo groves protect the dikes against washing.

planted to upland dry crops instead of paddy rice. The lower ends of some valley bottoms are poorly drained because of recent subsidence. In others, uplift has produced multiple alluvial terraces.

Diluvial uplands and terraces.—Not uncommonly, the inner margins of the delta-plains terminate abruptly at the base of the encircling hard-rock foothills. At the points where rivers debouch upon a plain the transition may be made less abrupt by the presence of steeply inclined alluvial fans or cones, which in certain areas are sufficiently continuous to form piedmont belts. But it is very common, also, for the descent from mountain to new alluvial plain to be broken by one or more intermediate steps in the form of sand and gravel terraces, or even broader uplands (Fig. 1-9). Such benches are of unconsolidated materials deposited by rivers and in some cases reworked by waves and currents. They are alluvium, preponderantly, but it is alluvium whose fertility has been sapped by long-continued leaching without being replenished by additions of new sediment. This older alluvium the Japanese geologists and geographers call *diluvium.* Most of the diluvial terraces and uplands are the result of recent uplift which has raised to their present positions coastal plains and delta-fans that were once near sea level. With the uplift the streams were rejuvenated and a portion of the elevated alluvium subsequently was cut away and carried seaward to help form the newer alluvial lowland. The presence of diluvial uplands in almost all parts of the country, from Hokkaido to Kyushu, has led one group of geomorphologists to conclude that rather general and contemporaneous uplift of the Japanese island group has taken place.

The usual form of a diluvial terrace in Japan is that of a low, flat to undulating upland plain which rises by fairly abrupt slopes above the adjacent lowland surface of new alluvium (Fig. 1-12). They vary markedly in elevation; some are so low as to be almost indistinguishable from the new alluvium, whereas others rise by precipitous slopes to heights of several hundred meters. Vertical downcutting by streams is rapid in these elevated unconsolidated sediments, creating a terrain characterized by shallow, steep-sided valleys and flat to gently sloping interfluves. This latter feature is a remnant of an earlier depositional surface; its preservation indicates how short a time, geologically speaking, has elapsed since the uplift. Within a relatively brief time, measured geologically, these diluvial crests are bound to disappear. As it is, the relatively even skylines of the diluvial uplands are one of the distinctive characteristics of their profiles as seen from a distance, distinguishing them from most hard-rock hills.

Less common are certain diluvial lands, ordinarily the older and higher ones, which have been so intricately carved by streams that flattish upland surfaces are mostly absent and the region has been reduced to a badland condition in which slopes predominate. The Tama Hills of the Kanto (Tokyo) Plain are of this type, as are certain other diluvial areas in the southern part of the Lake Biwa Basin north of Kyoto, along the margins of the Osaka Plain, and elsewhere. Morphologically, areas of dissected older diluvium are not easy to distinguish from low-level Tertiary areas, although usually the latter are composed of materials that are better consolidated, while their slopes are commonly steeper and the crest-lines of their ridges more uneven, so that the skyline is irregular.

Throughout Japan the diluvial uplands are much less intensively utilized than are the lower plains of new alluvium. Rice is not a common crop on the diluvial uplands; more characteristic are orchards and unirrigated fields of wheat, barley, vegetables, mulberry, and tea. Large areas are even kept in woodland and parts would be classed as virtual wasteland. This less intensive use of the diluvial uplands is partly a reflection of the difficulty of raising irrigation water from the incised valleys to the upland levels. And to the Japanese any land that cannot readily be inundated and therefore planted to paddy rice is, *ipso facto*, inferior land. Moreover, much diluvial upland is composed of such coarse materials that the soils tend to be droughty, while the water table may be so deep as to make an adequate water supply difficult and expensive to obtain. The soils are also notoriously infertile; leached of the soluble minerals and shy of humus, they are low in mineral plant foods and inferior in structure. In some parts of the country, the addition of acidic volcanic ash has done nothing to improve the soil condition.

Fig. 1-12.—A fragment of relatively high diluvial upland. Note the flattish crest and the deeply incised valley with steep slopes.

Where the veneer of ash is not present the soil may be quite sandy or even gravelly.

Interior basin lowlands.—Although a large majority of Japanese plains, including the most important ones, front upon the sea, a considerable number of isolated basins, a few of them fairly large, are located in the mountainous interior. A noteworthy feature of most inland basins is the small amount of genuinely level land characteristic of their alluvial and diluvial deposits. From the enclosing hill and mountain ramparts numerous short but vigorous streams debouch into a basin, building a series of coalescing alluvial cones and fans. The resulting terrain is one of smooth, moderately inclined depositional slopes in which fan or cone pattern is conspicuous. Multiple terraces are characteristic and coarse materials predominate. Typically, these interior basins are arranged in lines or rows, following important zones of crustal fracture and subsidence. Illustrating this linear arrangement are (1) the meridional basins in central Hokkaido; (2) the two parallel rows of meridional basins, one on either side of the central range, in northern Honshu; (3) the chain of basins across central Honshu, coinciding with the Fossa Magna; and (4) the series of east-west basins, corresponding with the great Median Dislocation Line, in Kii Peninsula and northern Shikoku.

THE COASTLINE[6]

Few regions of comparable area have either so long a coastline or such a variety of coastal features as does Japan proper. With its approximately 28,000 kilometers of sea frontage, the ratio is about one linear kilometer of coast for each 8.5 square kilometers of area.* The salutary influence of this unusually long line of contact between land and sea is accentuated by the fact that a great majority of the lowlands—the only areas capable of large-scale production, and consequently the areas of large population—have sea frontage. To an unusual degree, therefore, the people of Japan have a maritime outlook.

Coastal features of first magnitude in Nippon are due principally to faulting and warping, so that broad regional contrasts often represent differences in fault patterns and in the crustal movements that have occurred along the fault lines. Thus, the coastline of the Japan Sea littoral of Honshu is relatively smooth because it roughly parallels a fault system along which downwarping has occurred. Only occasionally, as at Wakasa Bay, does a fault system lie athwart the coastline, thereby pro-

* The ratio is about 1:13 for Great Britain.

ducing major irregularities. But along the Pacific coast of North Honshu the fault system cuts the north-south highlands at an oblique angle, resulting in an echelon type of development, with smoother stretches where the coast parallels the fault lines and other more irregular ones where it makes an angle with them. The Pacific side of southwestern Japan is indented by a series of channels and bays of fault origin ranging in a general north-south direction and consequently cutting across the trend of the coast. The result is a number of major irregularities, including Sagami-Tokyo, Suruga, Ise, and Osaka bays and Kii and Bungo straits. And finally, the Inland Sea and midwestern Kyushu owe their coastal complexity and irregularity to subsidence along a mosaic of fault lines crossing each other at several angles.

Other factors affecting the nature of Japan's coastline in various parts are the relative resistance of the rock and the nature of the crustal movements. Irregular and indented coasts are most commonly developed in resistant rocks, whereas the hinterlands of smooth coasts are usually composed of weak rocks.[7] It is believed that during the earliest alluvial period there was a general subsidence of the Japanese islands which created a coastline much more irregular than the present one.[8] Wave action subsequently has eroded the promontories and filled in the bays, thus gradually smoothing the coastline. On the basis of coastal movement, two types of coastline are recognized. The one, usually the smoother, has raised beaches and terraces which, with the incised rivers, are evidence of emergence. In the other type, which is more irregular, submergence is indicated.

The coastal configuration of Japan suggests a tripartite division of the country, as shown in Figure 1-13, with (1) a northeastern region where evidences of uplift in the form of coastal terraces, coastal plains, and dissected fans predominate; (2) an intermediate region where coastal forms resulting from elevation and depression are both well represented; and (3) a southwestern region, chiefly the Inland Sea borderlands and northwestern Kyushu, where forms due to subsidence predominate.[9]

The same Figure 1-13 also shows in more detailed fashion the nature of individual sections of the coastline. Clearly, the section facing Asia is less diversified than that along the Pacific, the ratio of their lengths being about 1 to 3.7. Most marked of the irregularities in the prevailingly regular coastline along the Asiatic side is the much-indented and island-studded coast of western Kyushu, where a mature land surface cut by a complicated system of faults has suffered subsidence. The smooth coasts of the plains along the Japan Sea are usually bordered by wide belts of

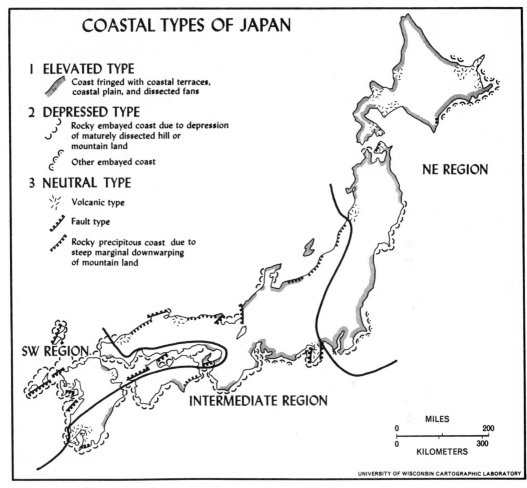

COASTAL TYPES OF JAPAN

1 ELEVATED TYPE
Coast fringed with coastal terraces, coastal plain, and dissected fans

2 DEPRESSED TYPE
Rocky embayed coast due to depression of maturely dissected hill or mountain land

Other embayed coast

3 NEUTRAL TYPE
Volcanic type

Fault type

Rocky precipitous coast due to steep marginal downwarping of mountain land

NE REGION

SW REGION

INTERMEDIATE REGION

MILES
0   200
0   300
KILOMETERS

UNIVERSITY OF WISCONSIN CARTOGRAPHIC LABORATORY

Fig. 1-13.—Coastal types and regions of Japan. After map by Akira Watanabe.

dunes and beach ridges. Relatively smooth and abrupt coasts are characteristic of the Pacific side of Japan north of the latitude of Tokyo, the single important exception being the irregular ria coast of the southern part of the Kitakami hill land in northern Honshu. Pacific southwestern Japan has many more large irregularities. The four great bays between Tokyo and Osaka have already been noted. Farther to the southwest is the island-studded Inland Sea with its ragged coastal margins, and the ria coasts of Kyushu, Shikoku, and Kii Peninsula bordering Bungo and Kii channels leading to that sea. The coastal outlines of the Pacific Folded Mountains, or the Outer Zone of Southwest Japan, in southern Kii Peninsula, southern Shikoku, and southern Kyushu, have few large indentations, although minor irregularities are numerous.

Unfortunately Japan's best natural harbors are situated in regions that in other respects are unsuited to the development of important ports. In general the hinterlands of the deeply indented coasts where good natural harbors are plentiful are regions of rugged terrain which are relatively unproductive and are not connected by easy natural routes with fruitful regions to the rear. Conversely, the productive plains regions are ordinarily paralleled by shallow silted waters, have smooth shorelines, and on their sea sides are bordered by belts of dunes and beach ridges or by abrupt terrace fronts. Where rivers enter the sea at the heads of large indentations, such as Tokyo, Ise, and Osaka bays, forming relatively extensive bay-head plains, so much sediment is deposited that artificially dredged channels are necessary to permit ships of even modest size to reach the river entrance. Moreover, the rivers are so small and shallow that seagoing vessels are unable to enter the plains.

The climates of Japan represent a composite of continental and marine elements, with the continental predominating. This is about what one would expect to find in a chain of mountainous islands stretched out in middle latitudes along the eastern side of the earth's greatest land mass. Continental control is chiefly to be observed in the temperature characteristics—winter and summer temperatures severe for the latitude, relatively large annual ranges of temperature, steep latitudinal temperature gradients in the colder months—but it is likewise evident in the prevalent summer maximum of precipitation. Marine control is more strongly evident in the humidity element—high relative and absolute humidity, abundant rainfall, moderately wet winters—but it appears also in the retarded temperature maximum of summer, for August is warmer than July. In many respects the climates of Japan resemble those of eastern North America in similar latitudes, or from Maine to northern Florida.

IMPORTANT CLIMATIC CONTROLS

Extending from about 31° to 45° N., Japan has a range of latitude which in itself tends to produce marked climatic contrasts between the northern and southern extremities. Fortunately, a relatively large part of the country lies in subtropical latitudes where climatic energy is abundant and the potentialities for plant growth high. Complexity of surface configuration and marked contrasts in elevation are factors making for striking local differences in climate.

Seasonal atmospheric circulations and air masses[1]

The contrasting seasonal circulation patterns over Japan are strongly influenced by the proximity of Asia. During the warm season three main air streams dominate eastern Asia (Fig. 2-1). North of about 40° N. are the zonal westerlies with which is associated a jet stream whose average position is approximately 40°–45° N., but which, nevertheless, undergoes wide non-periodic fluctuations in location. South of the zonal westerlies is a southwesterly current which has its origin in the equatorial latitudes of the Indian Ocean but reaches Japan via subtropical China. Commonly designated the summer southwest monsoon, this current is a poleward extension of the same Indian or equatorial southwesterlies which dominate most of tropical South Asia in summer. This deep and humid southwesterly current of maritime equatorial (*mE*) *air*, with a high rainfall potential, covers most of eastern continental Asia south of about 40° N., and probably supplies a large share of the moisture for the abundant summer rainfall of that region. Significantly the mean water vapor flux in summer is directed from west to east over Korea, much of China, and possibly Japan as well.[2] Separating the zonal westerlies from the equatorial southwesterlies is the polar-front convergence in whose disturbances is generated much of the summer rainfall of northern China and northern Japan. The summer monsoon is weaker but less variable than that of winter, with the highest velocities not more than 7–8 m/sec.

The third basic summer air current in eastern Asia is the tropical easterlies, which arrive in Japan as a southerly flow. Known as the summer southeast monsoon, this originates in the western margins of the Oga-

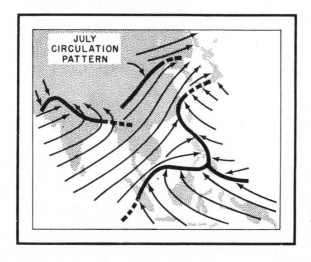

Fig. 2-1.—Principal elements of the surface atmospheric circulation over eastern and southern Asia during summer. Note that Japan is influenced both by equatorial southwesterlies and by tropical southeasterlies. After Thompson, Watts, Flohn, and others.

sawara subtropical high over the North Pacific Ocean. Probably more so than in China, this tropical maritime (*mT*) air mass from the Pacific plays an important role in supplying summer moisture to the Japan area, but there is debate over the comparative roles of the equatorial or Indian southwesterlies and the tropical southeasterlies in this respect. At about the 700 millibar level there is a moist tongue that coincides almost precisely with the axis of the Indian southwesterlies, which seems to indicate that this equatorial current plays a very important role in shaping the summer rainfall distribution over eastern Asia, including Japan.[3] Above the surface Indian southwesterlies is a marked belt of strong westerlies which Murakami has called the *Mai-u* jet. It is positioned over southern Japan in the warm season. Thus, Japan in summer experiences the effects of two jet streams, the polar jet in the north and the subtropical jet in the south. There is an upward motion at all levels, associated with the moist tongue of Indian equatorial air, and a rather weak upward motion in the tropical easterlies.[4] On the whole, the tropical easterlies appear to provide a less favorable atmospheric environment for precipitation processes than do the Indian or equatorial southwesterlies, for while the latter are deeply humidified and unstable, the Ogasawara air is warm and moist in its lower layers but drier aloft. Murakami is of the opinion that the southwesterly and southeasterly currents are both equally involved in the transport of moisture for the *baiu* ("plum season") rains, but that the latter is more important for the September rains.

In winter the *upper-level* zonal westerlies have shifted far enough south to be obstructed and bifurcated by the Tibetan Highlands and their flanking mountains, so that they flow around them both on the north and on the south (Fig. 2-2). Over eastern Asia there exist in winter, then, two jet streams, a fluctuating one to the north of Tibet and a positionally stable one to the south of it. The southern branch of the zonal westerlies and its jet cross southern China and are positioned over southern Japan. Thus, during the winter in Japan the air at about 10,000 feet is mainly in that branch of the zonal westerlies located north of Tibet, but is near to the convergence between these westerlies and those that arrive from south of Tibet.

But the winter *surface flow* is somewhat different from that described above, for the cold anticyclone over eastern Siberia results in a strong northerly surface circulation which floods eastern Asia with a succession of outpourings of dry, cold continental (*cP*) air (Fig. 2-3). This is the northwest, or winter, monsoon in Japan. The leading edge of each new surge of polar air forms a new front, either with old polar air or with tropical maritime air. In mean winter position the polar front lies eastward

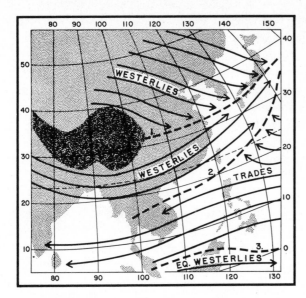

Fig. 2-2.—Characteristic features of the atmospheric circulation at about 10,000 feet elevation, November to March. (1) Tibetan lee convergence zone, (2) polar front, (3) intertropical convergence. Note that Japan is located in a zone of convergence. After Thompson.

of the continent and close to the southeast coast of Japan. It needs emphasizing that neither the winter nor the summer monsoon is a steady and continuous stream of air, for in both seasons the prevailing flow is interrupted frequently in direction and strength by passing atmospheric disturbances. On the average, the winter monsoon is stronger and more variable than that of summer. Average wind velocity of the winter monsoon is about 7–8 m/sec, but it does reach 15 m/sec and more. Consequently it is on the north and west sides that farmsteads are most commonly protected by windbreaks.

To what extent the winter monsoon (or the summer monsoon either)

TABLE 2-1

Air-mass calendar for Tokyo (in percentage frequencies)

Air mass types	J	F	M	A	M	J	J	A	S	O	N	D
cP	78	62	43	17	9	4	3	2	26	39	51	74
cNP	20	33	43	36	36	33	14	19	37	51	44	25
mP	0	0	0	0	0	1	1	4	1	0	0	0
mNP	0	0	2	4	5	13	12	13	9	2	0	0
cT	0	1	1	5	1	1	0	0	0	0	1	1
cNT	2	3	5	19	18	1	0	0	2	0	3	0
mT	0	0	0	3	3	26	58	48	13	1	0	0
mNT	0	1	5	16	27	21	12	13	11	6	1	0

Source: H. Arakawa and J. Tawara, "Frequency of Air-Mass Types in Japan," *Bulletin of the American Meteorological Society,* 30 (1949).

Fig. 2-3.—Principal elements of the surface atmospheric circulation over eastern and southern Asia in winter. Japan lies within the polar-front convergence. After Thompson, Watts, Flohn, and others.

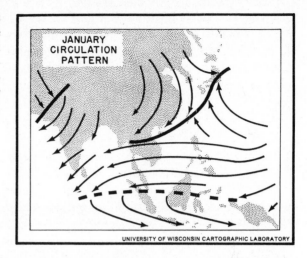

JANUARY CIRCULATION PATTERN

UNIVERSITY OF WISCONSIN CARTOGRAPHIC LABORATORY

of East Asia is thermally induced is debatable. It may be that in winter the well-developed Siberian High and intense Aleutian Low as climatic averages are to some extent associated with the cold land and the relatively warmer ocean, but the thermal hypothesis fails to clarify the changing synoptic patterns. It is unable to explain the intense outbreaks of cold Siberian air after a period of relatively mild weather, these outbreaks being associated with a clearly defined pressure pattern, composed of an intensified continental high and oceanic low, with both centers displaced abnormally far to the south. Known to the Japanese meteorologists as the "west-high–east-low" pressure pattern, it is this which brings cold spells to Japan and heavy snows to those parts facing the Japan Sea.[5] On such occasions dynamic anticyclogenesis occurs over the Arctic, the mean jet is intensified over East Asia and shifted southward, and an upper-level low forms to the north of Japan.

The thermal hypothesis fails also to explain why the west-high–east-low pattern, with its attendant cold weather, is of long duration in some winters and short in others. Cold outbreaks seemingly are favored by a special type of low-index situation with ridges positioned in the eastern Atlantic, west Siberia, and the eastern Pacific. Accordingly, the west-high–east-low pattern characteristic of the winter monsoon in East Asia probably is just one link in the general circulation.[6] Development of the Siberian anticyclone appears to be a result of dynamic anticyclogenesis over the Arctic and subsequent tropospheric cold advection directed southward on the western side of an upper-level cutoff low. The cold advection intensifies the frontal zone in East Asia, and a rapid deepening of cyclones follows. In the colder months temperatures in Japan are closely associated with the latitudinal position of the tropospheric frontal zone or the mean position of the jet stream. If the jet and the front are to the north of their average positions, relatively mild weather prevails, but if they are shifted to the south, the Siberian Anticyclone and Aleutian Low both intensify, establishing the west-high–east-low pressure pattern with cold outbreaks frequent and low temperatures prevalent in Japan.[7]

Weather disturbances

It is a misnomer to speak of the rainfall of eastern Asia, even that of summer, as being monsoonal in origin. Admittedly the flooding of the region with maritime equatorial and tropical air in summer does provide an atmospheric environment which has a high rainfall potential. But to an overwhelming degree the precipitation, except in highlands, originates in extensive traveling weather disturbances where convergence and lifting of the air occur. This is the case in all seasons.

In the colder months Japan's weather is greatly influenced by a succession of depressions moving eastward from the continent in the zonal westerlies. Two very general tracks are recognized, one in the north leading from Manchuria and Siberia, and another more southerly one in which the disturbances arrive from southern China. The two storm tracks appear to converge in the vicinity of Japan, making that area one of great cyclone frequency.[8] Some depressions on the southern route seem to originate along cold fronts as they begin to decelerate in their southward and eastward progress. Others are imbedded in the zonal westerly current which flows around Tibet on the south. Significantly, routes of cool-season depressions reaching Japan, from both the northwest and the southwest, tend to coincide with the two jet streams, one on either side of Tibet, and each jet in turn is related to a low-level frontal zone.[9]

Although most of the summer disturbances are relatively weak, they are nonetheless effective generators of precipitation since they are operating in mT and mE air. Some are of frontal origin, others are not. Among the latter are (1) surges or speed convergences in the equatorial or Indian southwesterlies; (2) frontal wave disturbances; (3) pressure waves, resembling easterly waves, in the tropical easterlies; and (4) tropical vortex storms of the typhoon variety, which are most frequent in Japan in late summer and fall, but which can strike at any time of year. Such violent storms not infrequently do great damage, especially since they are most likely to come roaring in just as the all-important rice crop is approaching maturity and is most vulnerable to the combination of furious wind and deluging rain. Strangely enough, Japan is not an area of numerous thunderstorms, even though it is dominated by equatorial and tropical air masses for half the year and is a region of rugged terrain. North of about latitude 38° days with thunderstorms are fewer than ten a year, while by far the largest part of subtropical Japan has fewer than twenty such days annually.

The surrounding seas and their ocean currents

Japan's winter climate is meliorated by the 300 to 900 kilometers of open sea that the polar air masses must cross before reaching Japan. As a consequence of the sea trajectory the lower levels of the continental polar air are warmed and humidified, so that winter temperatures in Japan are less severe than they are in similar latitudes on the continent, and winter precipitation is much heavier, especially on the side facing Asia. Here the amount of winter precipitation is somewhat dependent upon the width of the Japan Sea and consequently the amount of moisture which has been added at the base of the cold air, with the least precipitation along those parts of the coast opposite the narrowest parts of the sea.[10]

Two ocean currents—a cold one from the north, the *Oyashio* or Okhotsk Current, and a warm one from the south, the *Kuroshio* or Japan Current—have some modest influence upon Japan's climate (Fig. 2-4). The Japan Current bifurcates at the southern extremity of Japan Proper, the main stream flowing northward along the Pacific coast of Honshu to nearly the latitude of Tokyo (35°) before it turns northeastward into the Pacific. Surface temperatures of this warm current range from about 82° in late summer to 68° in late winter, but because it is on the lee side of Japan during the winter monsoon, its direct effects on the islands' temperatures are minimized. The western branch of the Kuroshio, the *Tsushima* Current, containing a much smaller amount of warm water, enters the Japan Sea through the Tsushima (Korea) Strait and flows northward in the eastern part of that sea as far as northern Hokkaido.

The cold Oyashio or Okhotsk Current, flowing southward from the Okhotsk Sea, hugs the Pacific side of Japan down to about latitude 36°, where it sinks below the waters of the Kuroshio. In the western part of the Japan Sea cold water is likewise present, but this can scarcely be regarded as an eastern branch of the Okhotsk Current, inasmuch as the strait between Sakhalin Island and the Asiatic mainland is only about 7 kilometers wide and 12 meters deep at its narrowest part. It is probably more the result of excessive cooling in winter, when seaward-moving *cP* air masses are prevalent. In the western part of the Japan Sea the sea surface temperature in summer is only about 67°, whereas in the

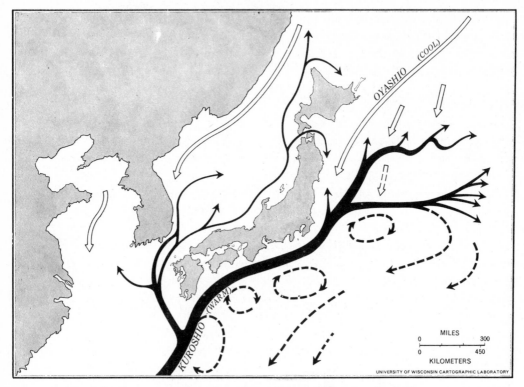

Fig. 2-4.—Ocean currents in the vicinity of Japan. After map by Isida.

eastern part it is approximately 81°. The cold Oyashio acts to reduce the summer temperatures and cause much summer fog along the littorals of northern and eastern Hokkaido and northeastern Honshu.

IMPORTANT CLIMATIC ELEMENTS

Air temperature

Japan's easterly location with respect to the land mass of Asia gives the islands a temperature regime that is more continental than marine, with relatively cold winters and warm to hot summers. Actually the Japan area in January is distinctly colder than the average for its latitude, with a negative temperature anomaly amounting to 10° to 20° F. Summers, although more normal for the latitude, are hot and sultry except in the north. Seasonal temperatures are comparable to those in the central and eastern United States, a region occupying the same latitudes and a similar eastern or leeward position with respect to a large continent. To be sure, Japan's insularity modifies somewhat the winter cold, but Siberian air is so intensely cold at its source that, despite the warming influence of the Japan Sea, it is still cold continental air when it arrives in Nippon. Thus, Asahigawa in Hokkaido at latitude 44° N. has winter and summer temperatures similar to those of St. Johnsbury, Vermont; Tokyo in mid-Japan at about 36° N. resembles Norfolk, Virginia; and Kagoshima in the extreme south at about 31° or 32° N. finds its homoclime in Jacksonville, Florida. In winter and spring, and especially in midwinter, air temperatures in fact are several degrees lower in insular Japan than they are along the continental American Atlantic Seaboard in similar latitudes.[11] In summer and fall, however, the differences are unappreciable.

Average January temperatures range from about 15° or 20° in northern and central Hokkaido, to 35° or 40° on the lowlands of central Japan, and 45° in the extreme south of Kyushu (Fig. 2-5). Thus the latitudinal temperature gradient, or rate of change of temperature, is very steep— approximately 2.6° for each degree of latitude, which is almost the same as on the American Atlantic Seaboard. The freezing isotherm for January is situated at about latitude 38° or 39°, in the general vicinity of Sendai and Niigata. In winter, isotherms tend to loop well southward over the islands, roughly paralleling the coasts and thus reflecting the effects both of altitude and of the colder land.

Despite the fact that the west coast faces the cold Asiatic continent,

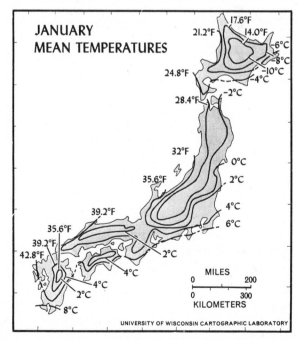

Fig. 2-5.

from which the polar air masses arrive in winter, there is little difference between the winter air temperatures of the Japan Sea side and
those of the Pacific coast at any given latitude. Actually the Japan Sea
side north of 36° (northern Honshu and Hokkaido) is slightly warmer
than the Pacific side. This is due to the rather constant cloudiness along
the Japan Sea coast in winter, which tends to reduce night cooling. The
clearer skies of the Pacific side, which accelerate earth radiation, ordinarily produce colder nights and warmer, sunnier days. But thermometer
recordings notwithstanding, there is no doubt that the windy, cloudy,
and more humid west side facing Asia *feels* colder in winter than the
quieter, brighter, and drier east side, where the winds are both descending and offshore.

High temperatures combined with high humidity make the summer
weather of all but northernmost Japan extremely sultry and oppressive.
July temperatures in central and southern parts range from 77° to 80°,
and August is slightly warmer than July at most stations (Fig. 2-6). Thus,
in midsummer, subtropical Japan has almost Amazonian heat. Tohoku,
or northern Honshu, is characterized by summer-month temperatures a
few degrees lower, 72° to 75° being normal. Hokkaido largely escapes uncomfortable heat, for there the July-August average is only 65° to 70°, resembling New England. Along the littorals in both eastern Hokkaido and

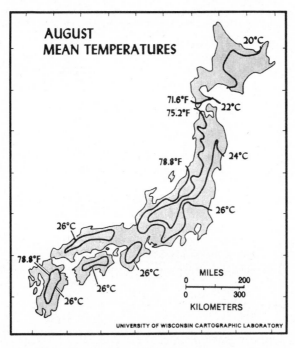

Fig. 2-6.

the Pacific side of North Honshu, where the cold Oyashio current prevails, summer temperatures are unusually cool, reaching the low 60°s in eastern Hokkaido and only to about 70° in northern Honshu. Here coastal fog is prevalent. The rate of change in temperature from north to south is much less in summer than in winter, the latitudinal temperature gradient being only 1.3° for each degree of latitude, or approximately half that of January.

The length of the frost-free, or growing, period ranges from about 120 or 130 days in central and eastern Hokkaido to 250 days or more along the extreme southern and eastern littoral (Fig. 2-7). In the United States northern New England and North Dakota are comparable to Hokkaido in length of frost-free season. Thus, St. Johnsbury, Vermont, has a frost-free period of 127 days, which is almost identical with that of Asahigawa in central Hokkaido; Grand Forks, North Dakota, is without frost for 132 days. The 250-day frostless season in extreme southern Japan is duplicated in southern Georgia, Alabama, and Mississippi. Tokyo in the central part, with a growing season of about 215 days, and Nagoya, with 207, compare well with the northern part of the American Cotton Belt. The fact that the isarithms of frost-free days tend to parallel the coasts shows that altitude and land-water controls are as influential as latitude in frost distribution. In northern Japan, especially in

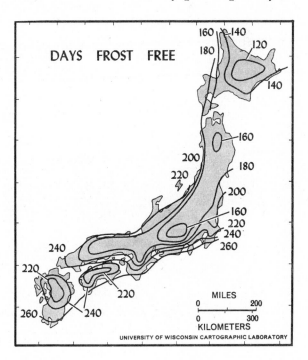

DAYS FROST FREE

Fig. 2-7.

MILES
0   200
0   300
KILOMETERS

UNIVERSITY OF WISCONSIN CARTOGRAPHIC LABORATORY

Hokkaido, early autumn frosts often do serious damage to the rice crop, and in the northern part of subtropical Japan late spring frosts not infrequently damage vegetables, mulberry, and tea.

PRECIPITATION

Unlike subhumid-semiarid North China and Manchuria in similar latitudes, Japan is a humid land, no section of which normally suffers from an annual or seasonal deficiency of rainfall. Annual amounts of precipitation on lowlands are two to three times as great as in comparable latitudes on the mainland, while the winter-dry climates of northern China find no counterpart in Japan. Thus, the summer concentration of precipitation is less marked. These contrasts as they apply to Japan are partly the result of her insular character, which assures a predominance either of genuinely maritime air masses or of continental air modified by a sea trajectory. Hilly and mountainous terrain also acts to increase the total precipitation, and in addition the country is well positioned with respect to fronts, jet streams, areas of cyclogenesis, and concentrated storm tracts, so that the perturbation element is both strong and seasonally persistent.

It is hard to generalize about the areal distribution of precipitation, for variable relief results in an exceedingly confused and patchy rainfall map, the larger patterns of which are obscured by the numerous closed isohyets and the very circuitous courses of others (Fig. 2-8). In general three regions of heavier-than-normal precipitation (80–120 inches) may be recognized: (1) the Pacific side south of about 35° N., which faces the southerly air flow of the warmer months, and which

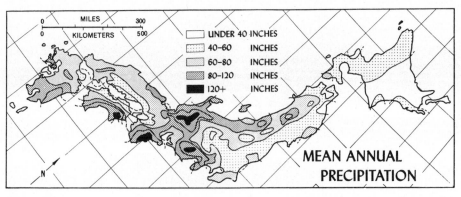

Fig. 2-8.

TABLE 2-2

Climatic data for selected stations

Station	Temperature (F°) Mean of warmest month (Aug.)	Mean of coldest month (Jan.)	Days of growing season	Precipitation (cm.) Jan.	Feb.	Mar.	Apr.	May	June	July	Aug.	Sept.	Oct.	Nov.	Dec.	Total
Southwest Japan																
Kagoshima	80	44	253	7.3	10.4	15.4	21.6	20.4	42.2	31.7	20.6	21.8	13.4	9.0	7.8	221.4
Kumamoto	80	40	211	5.6	7.3	12.0	16.1	15.8	35.3	29.7	16.5	17.6	11.0	6.9	6.3	180.1
Fukuoka	80	41	203	6.4	7.9	10.5	13.1	11.2	26.1	23.3	14.5	21.7	11.2	7.8	7.6	161.2
Hiroshima	80	39	221	4.6	6.0	10.6	15.4	14.0	25.7	21.1	11.3	20.1	11.4	6.1	5.2	151.3
Osaka	80	40	219	4.2	5.9	9.3	12.3	12.3	19.7	15.3	11.5	17.3	13.2	7.2	4.9	133.0
Kochi	79	41	241	6.1	9.4	18.9	24.9	25.8	34.2	32.3	33.4	36.6	21.1	10.7	7.4	260.7
Central Japan																
Nagoya	80	37	207	5.1	6.8	11.1	14.8	14.9	21.0	17.9	17.5	21.4	17.1	8.3	5.8	161.7
Hamamatu	79	41	281	2.3	4.6	9.1	11.1	9.0	14.0	14.6	14.1	15.9	15.0	6.6	3.9	118.4
Tokyo	78	38	215	4.8	7.6	10.8	13.4	14.5	17.4	14.6	16.4	24.6	22.2	9.2	5.7	161.0
Nagano	76	29	166	5.7	5.1	5.8	6.5	7.8	10.9	14.0	9.9	13.2	9.4	5.2	5.7	99.1
Fukui	79	36	205	30.7	20.5	16.1	14.9	13.0	17.4	18.3	13.7	20.8	18.1	22.3	33.4	233.9
Tohoku																
Sendai	75	31	181	3.3	3.2	5.9	15.3	13.5	10.0	9.1	11.3	15.2	14.4	6.8	5.2	112.9
Yamagata	75	29	168	10.0	7.9	7.6	7.8	7.8	9.3	14.0	14.0	13.9	10.6	8.7	11.9	123.6
Morioka	73	26	148	55.4	4.3	7.1	5.6	7.0	7.1	11.3	16.1	12.6	9.5	8.5	7.6	102.1
Akita	75	29	175	18.3	10.5	10.1	11.4	11.2	11.9	19.6	17.9	20.0	17.9	19.4	16.3	179.5
Hokkaido																
Sapporo	70	21	129	9.7	7.4	6.2	5.6	6.4	6.9	9.3	10.3	13.2	11.4	11.6	10.0	108.0
Asahigawa	69	14	127	7.3	5.6	5.3	5.2	6.7	7.9	11.7	12.5	14.5	10.9	11.3	10.4	109.3
Obihiro	67	13	121	3.9	3.5	5.8	6.3	8.5	9.1	9.8	12.9	15.2	9.5	6.8	4.3	95.7
Kushiro	64	20	141	5.2	3.6	6.7	9.0	9.3	11.1	11.5	14.1	15.3	11.6	7.4	5.0	109.8

Source: The Climatographic Atlas of Japan (Tokyo, 1948).

has high altitudes and a concentration of storm tracks, including those of typhoons; (2) the Japan Sea side north of about 35° N., where highlands obstruct the flow of the sea-modified winter monsoon; and (3) the highlands of central Honshu. There are at least four general areas which experience less than the country-wide average of precipitation, and where the annual totals are only about 40–45 inches: (1) most of Hokkaido, (2) large areas on the Pacific side of northern Honshu, (3) the central part of the Inland Sea basin, and (4) some of the interior mountain basins of central Honshu.

It bears re-emphasizing that, in contrast to the mainland, there are, on the average, no winter-dry climates (Cw-Dw) in Japan. This reflects the presence of more humid winter air masses in the islands, as well as a greater activity of atmospheric disturbances.

Over much the greater part of the country precipitation is heaviest in the warmer months (June through September), a reflection of the continental climate and well-developed system of monsoon winds (Fig. 2-9). Throughout much of subtropical Japan, except on the Japan Sea side, the precipitation of the wettest summer month is four to six times that of the driest winter month. The seasonal contrast is less striking in the north, in Hokkaido, where even the driest winter months usually have one to three inches of precipitation.

Throughout subtropical Japan, and even at some stations farther north, the general warm-season precipitation maximum has two peaks,

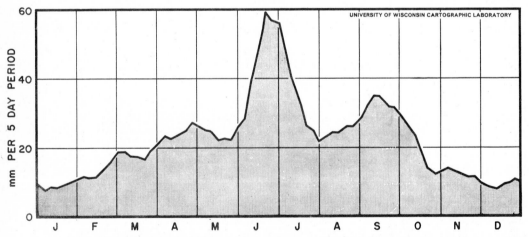

Fig. 2-9.—Composite annual profile of 5-day means of rainfall for 5 stations in subtropical Japan. Rainfall is concentrated in the warmer months, although winter is by no means dry. There are two peaks of warm-season rainfall, one in early summer (*baiu*) and the other in early fall (*shurin*), while mid and late summer are less wet.

one in June and early July and a second in September, separated by a July-August secondary minimum. The June peak, or so-called *baiu* maximum, is so coincident in time with certain changes in broad-scale pressure and circulation patterns that it is suggestive of a cause-and-effect relationship. Almost indisputably one element involved is the advance —northward with the sun, crossing China and reaching Japan in early summer—of the Indian or equatorial southwesterlies, whose temperature, humidity, and vertical-structure characteristics provide a favorable environment for abundant rainfall.[12] Still another indicator of the approaching baiu season is the intensification of the Okhotsk High to the north of Japan and a bifurcation of the westerly jet over the continent, with Japan lying in the confluence of the two jet branches. Branching of the summer jet is fairly concurrent with the development of the Okhotsk High and the northward penetration of the Indian southwesterlies. The northern branch of the jet lies in the vicinity of northern Sakhalin, while the southern branch is positioned at, or south of, 35° N. and hence in the latitude of southern Japan. Northerly air originating in the Okhotsk High (*mPK*) and over cold water, moves far southward and forms fronts with the equatorial-tropical air (*mT[E]W*) from the south; and along those fronts, positioned underneath the southerly jet, developing disturbances which bring the baiu rains pass at intervals of two or three days. The southern branch of the jet is always to be observed over southern Japan during the baiu season.

This early summer rainy period normally ends by about mid-July, as the North Pacific subtropical anticyclone advances northward to its highest latitudinal position, bringing to Japan more subsident and stable air. Simultaneously the southerly jet also retreats northward, the Okhotsk High weakens, more interruptions occur in the unstable equatorial flow over Japan, while intrusions of more stable maritime air increase, and rain-bringing synoptic patterns are weaker and occur less frequently. Pressure and temperature both increase over Japan in midsummer while, significantly, vapor pressure at the 700 millibar level declines. With such an interrelated series of events there is ushered in the less-rainy period of the July-August secondary minimum.

The second, or September, rainfall maximum is known as the *shurin* season. In many of its characteristics, and seemingly in its origin, also, shurin resembles baiu, for as the subtropical high retreats southward with the sun a less stable southerly air flow revives, the Okhotsk High reappears in the north, and a relatively durable front, between Okhotsk and southerly maritime air, again becomes positioned over southern Japan. Successive depressions developing along this front bring clouds and

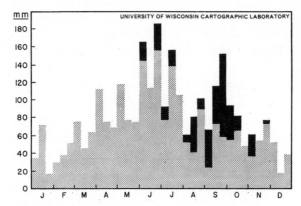

Fig. 2-10.—Average 10-day precipitation amounts for a station in subtropical southwestern Japan. Precipitation originating in typhoons is shown in black; that from other sources is shaded. Typhoon rainfall is at a maximum during the *shurin* season of early fall. After Saito.

rain which lower the air temperature a few degrees even though the humidity remains high. A considerable part of the shurin rain occurs in conjunction with tropical storms of the typhoon variety entering the Japanese area from the south (Fig. 2-10). On the whole the shurin maximum is a less dependable feature than is that of the baiu, and appears to be characteristic of fewer stations. According to Murakami, the moisture transport for shurin rainfall is chiefly accomplished by the southeasterly current.

Remaining unresolved is the question relating to the relative impor-

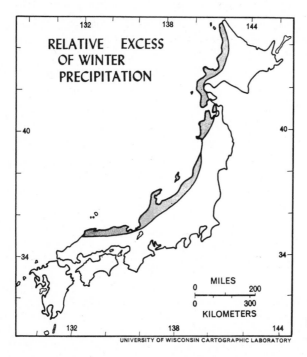

Fig. 2-11.—Much of the shaded area has a winter maximum of precipitation. After Okada.

tance of air mass vs. synoptic controls in contributing to the baiu and shurin maxima and the intervening midsummer secondary minimum. Actually the air-mass calendar for Tokyo and other subtropical stations shows a distinctly greater prevalence of low latitude maritime air in the less rainy months of July-August than in the rainier periods of June and September, which might seem to indicate that air masses are not an important cause. But the frailty of these surface data is that conditions aloft are omitted, and no distinction is made between mE and mT air.

The most marked regional exception to a general warm-season maximum of precipitation is the Japan Sea side of the country all the way from northern Hokkaido to southwestern Honshu. Thus, the contrast between the maps of January and July precipitation is striking. During the weak southerly monsoon of summer, strong windward-leeward rainfall effects are not pronounced, and from about the latitude of Tokyo northward the Japan Sea side has even more summer rain than does the Pacific side. But during the stronger winter monsoon, windward-leeward effects are much more striking and there is a strong concentration of precipitation, much of it in the form of snow, on the windward side facing the Japan Sea, where the sea-modified northwesterly air comes onshore and is forced upward by highlands. Across the highlands, on the Pacific side, where these same cold air masses are descending, the winter skies

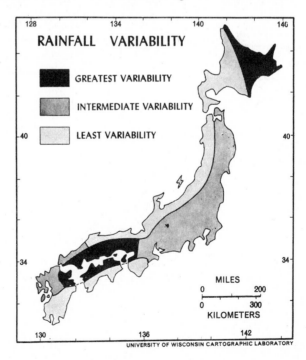

Fig. 2-12.—After Fukui.

RAINFALL VARIABILITY

GREATEST VARIABILITY

INTERMEDIATE VARIABILITY

LEAST VARIABILITY

MILES
0   200
0   300
KILOMETERS

UNIVERSITY OF WISCONSIN CARTOGRAPHIC LABORATORY

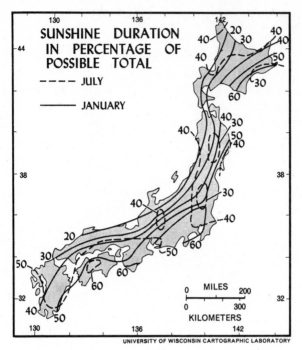

Fig. 2-13.

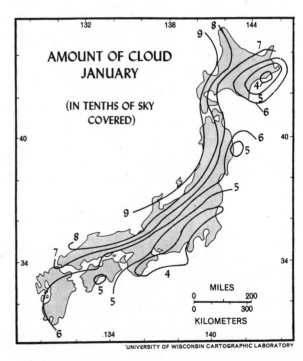

Fig. 2-14.

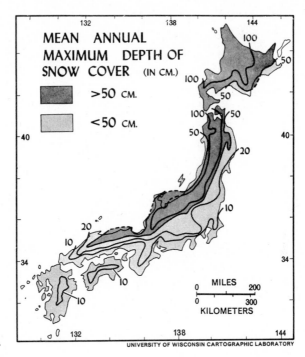

Fig. 2-15.

are clearer and precipitation less. As noted, summers are wet along the Japan Sea side, but in most parts winters are even wetter, with the result that snow lies deep on the ground throughout this section.

Snow falls over the whole of Japan, and a continuous snow cover of about three-months duration is characteristic of nearly all of Hokkaido and most of western highland Honshu south to about latitude 36°. Depth and duration of snow cover are strikingly less on the Pacific side than on the winter windward side facing the Sea of Japan (Fig. 2-15). It is noteworthy, however, that even on the snowy side facing Asia, the coastal lowlands are distinctly less snowy than are the slopes and highlands back of the plains. Throughout Hokkaido and western North Honshu snow offers serious problems to rail and highway traffic during several months of the year, so that snow sheds and snow barriers are common features along numerous stretches of rail line. A number of stations in western North Honshu and Hokkaido experience 25–30 days with snow during the month of January.

SUMMARY OF THE SEASONAL MARCH OF WEATHER[13]

In Japan's winter the prevailing large-scale weather situation is that of a west-high-(Siberian)–east-low-(Aleutian), out of which arise frequent non-periodic thrusts of northwesterly *cPK* air (Fig. 2-16). Structural modifications of this cold dry air over the Japan Sea result in masses of persistent cloud with heavy snow showers along the windward Asiatic side, but much fair weather on the leeward Pacific side. Between the windward and leeward sides the weather and climatic divides, as expressed in total precipitation, rainy days, and duration of sunshine, are well marked. Strong outflows of northwesterly air prevail for a week or so and then the anticyclone withdraws westward and the winter monsoon weakens or even halts. At such times the heavy cloud cover along the Japan Sea side briefly dissipates and blue sky may appear, while along the Pacific side warmer, quieter weather sets in. But after a lull of a few days the Siberian anticyclone again thrusts eastward and a new surge of polar air moves southeastward preceded by a sharply defined cold front. Along this cold front small cyclonic depressions develop over the China Sea and move over Japan from southwest to northeast bringing widespread rains. It is these disturbances which save the Pacific lee side from dry winters. Other cyclonic storms move on northern tracks via the Japan Sea in a northeasterly direction along the continental coast. Following the passage of these storms there is a fresh outbreak of polar air. This pattern of winter weather, repeated at intervals of about a week, persists until around the middle of March.

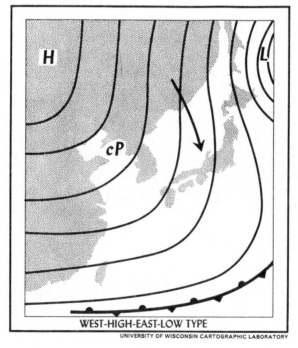

WEST-HIGH-EAST-LOW TYPE
UNIVERSITY OF WISCONSIN CARTOGRAPHIC LABORATORY

Fig. 2-16.—After Takahashi.

Spring is a season of variable wind strength and frequent cyclonic storms, so that the weather is changeable. Well-developed cyclones with sharp cold fronts between contrasting *cP* and *mT* air produce strong winds and increasing monthly precipitation. South winds associated with such depressions cause strong *föhn* effects, especially on the Japan Sea side. In this respect it is significant that 40 per cent of the country's large fires occur in the months of March and April. In May, as the Siberian anticyclone weakens and withdraws toward its core area, cyclones become fewer.

Toward the end of spring as the northward advancing *mT-mE* air masses begin to gain ascendancy over the weakening polar air, the rainy baiu season begins. An oscillating front over southwestern Japan separating *mE(T)* and Okhotsk modified polar maritime (*mPK*) air masses, generates much cloud and precipitation. As the warm air and its front push northward gradually, with frequent backtracks, the tropical southerly air increasingly dominates the Japan sector and summer is said to begin (Fig. 2-17). Unfortunately the advent, duration, and strength of baiu, and the arrival of drier midsummer are subject to many temporal and intensity variations, depending on the relative strengths of the Okhotsk and Ogasawara highs. If the baiu front is weak and lies abnormally far north the greatly anticipated baiu rains may be meager, while

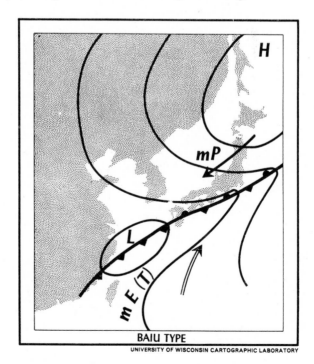

Fig. 2-17.—After Takahashi.

BAIU TYPE

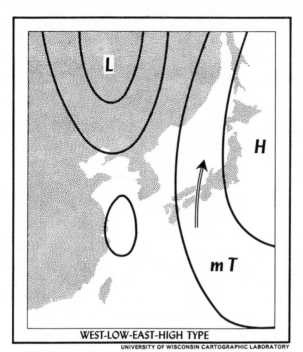

Fig. 2-18.—After Takahashi.

WEST-LOW-EAST-HIGH TYPE

UNIVERSITY OF WISCONSIN CARTOGRAPHIC LABORATORY

if it is located well south of its usual position a cool and rainy summer may be the consequence.

In July-August, the period of the southeasterly monsoon, the general pressure pattern is one of west-low–east-high, the reverse of winter's large-scale weather situation (Fig. 2-18). At this time the Ogasawara subtropical high is at its maximum northward and westward position, so that the southerly air from its western limb, while warm and moist in its lower layers, is drier and more stable aloft. In addition, summer has no well-developed and persistent front such as characterizes the baiu season. As a consequence there is much fair weather over Japan, both on the windward and leeward sides, and the annual rainfall curve shows a midsummer drop.

Fall has a more variable and complicated weather pattern than summer. Already in September as the Ogasawara High weakens and withdraws southward and the *mPK* and *cPK* air masses strengthen, a front resembling that of baiu is positioned over southern Japan. The result is a second and shorter warm-season rainy period, the shurin, when typhoons reach maximum frequency. In October-November invasions of polar air become more frequent and the eastward-moving highs bring beautiful fall days. Finally by December the west-high–east-low weather situation is in control again and winter sets in.

CLIMATIC TYPES AND REGIONS

When analyzing the climatic characteristics and subdivisions of a relatively small area like Japan, there is good reason for employing some recognized scheme of world classification of climates. By thus fitting the local climatic subdivisions into a world pattern, one is better able to compare them with climates elsewhere in the world. In the regional description of Japan's climates this book follows a modified form of the Köppen classification, in which the boundary between the mild mesothermal (*C*) and severe microthermal (*D*) climates is accepted as the 32° (0° C.) isotherm for the *coldest* month. Except for areas of high altitude, all of Japan falls within the *C* and *D* climatic groups (Fig. 2-19). Based upon monthly climatic averages, lowland northern Honshu (north of about 38° N.) and all of Hokkaido have microthermal climates (Köppen, *D*), in which the average temperature of the coldest month is below freezing. The remainder of lowland Japan, the more southerly subtropical part, is mesothermal (Köppen, *C*) in character, and, more specifically, belongs to the humid subtropical subdivision (Köppen, *Caf*) of that group. The microthermal north has two subdivisions, a *Daf* region in lowland northern Honshu where the average temperature of the warmest month is above 71.6° (22° C.) and a *Dbf* region in Hokkaido where it is below 71.6°.

If instead of employing climatic averages based on a period of twenty to thirty years for delimiting climatic regions, the same criteria are applied to individual years (year climates), it can be observed that the pattern of Japanese climates varies considerably from year to year, and in addition, certain new climatic types are introduced.[14] Thus, while summer-dry year climates (Köppen, *s*) are infrequent, they are to be observed in scattered areas as often as one year in ten. Winter-dry year climates (Köppen, *w*) are much more frequent and widespread. The *f* symbol, indicating no dry season, overwhelmingly predominates on the Japan Sea side of the country, but in a fair number of years it is replaced by *w* on the Pacific side. Core regions of most frequent *w* years are located as follows: eastern Kyushu, southeastern Shikoku, southern Chugoku, northern and western Kanto, and extreme northeastern Honshu. The fact that year climates in Japan show considerable annual variations indicates important annual shifts in intensity and position of the major climatic controls.

Since greater detail on regional climates is provided in Part III of this book, where Japan is analyzed regionally, only a brief summary is provided at this point.

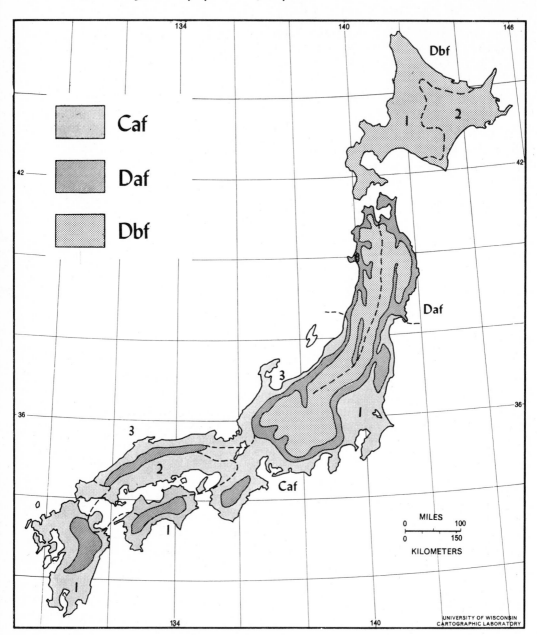

Fig. 2-19.

Cool-summer continental (Dbf) climates of Hokkaido.—Hokkaido, in the general latitude of the northern tier of American states, resembles climatically the northeastern part of New England and the Atlantic provinces of Canada. Summers tend to be short and also cool, the average for the warmest month usually being below 70°. These are delightful temperatures for human comfort but less desirable for the growing of paddy rice, Japan's staple food crop. The eastern and southeastern littorals, washed by the cool Okhotsk Current, are particularly cool in summer, warm-month averages there being below 65°. Summer fog is prevalent along these cool-water coasts. The normal frost-free season is between four and five months. Winters, on the other hand, are long, cold, and snowy. Except in the southwestern peninsula of Hokkaido, average temperatures of each of the four winter months are below freezing. Hokkaido is one of the driest parts of Japan; at most of its lowland stations precipitation totals only about 40 inches. There is no distinctly dry season, although stations close to the west coast have more precipitation in the winter half-year than in the summer half-year, whereas the converse is true of stations in the center and east.

The two climatic subdivisions in Hokkaido are separated by a boundary line approximately coinciding with the isotherm of 68° for the warmest month. In western subprovince 1, August temperatures are higher, and in the eastern subdivision lower, than 68°. Highland *Dbf* climate extends well to the south of Hokkaido, at least as far as central Honshu.

Warm-summer continental (Daf) climates of northern Honshu.—The chief distinction between *Daf* and *Dbf* climates is the difference in summer temperatures. It is therefore the warmer summers of northern Honshu that set it apart climatically from Hokkaido, although the winters also are less severe. Throughout all the lowland parts of the *Daf* region the average temperature of August, the warmest summer month, is above 71.6° F. (22° C.) and at most stations it is between 73° and 76°. These approximate the summer-month temperatures in the American Corn Belt. As a rule the summer temperatures at stations along the Pacific coast, which is paralleled by the cool Okhotsk Current, are a few degrees lower than at interior and west-coast stations. In fact, this whole eastern littoral of northern Honshu south to about 39° or 40° comes close to having a *Dbf* climate. The period between killing frosts ranges from 160 to 180 days.

Winters are considerably milder than in Hokkaido, average January temperatures being only a few degrees below freezing. At most stations in this climatic subdivision average temperatures fall below 32° in only one or two months. Such winter temperatures are comparable to those of southern New England, New Jersey, Pennsylvania, and the milder parts of the American Corn Belt.

Precipitation is heavier than in Hokkaido, virtually all the *Daf* region having over 50 inches and much of the western part over 60. Western Tohoku is one of the wetter sections of Japan, some stations recording 80 to 100 inches. On the Pacific side, precipitation is three to four times as heavy in the summer as in the winter. A seasonal reverse is true of the Japan Sea side, in many parts of which the winter precipitation equals or even exceeds that of summer.

Two subdivisions of the *Daf* climate, an east and a west, are shown in Figure 2-19, with the boundary between them approximately the line of 150 mm., or 6 inches, of January precipitation. On Figure 2-19 the boundary is made to coincide roughly with the central meridional mountain range. West of this line, in subprovince 1, the precipitation in January usually exceeds 6 inches, and it is not uncommon for the total winter precipitation to be greater than that of summer. Subprovince 2 is much drier in winter. Highland *Daf* climates extend far southward into subtropical Japan.

Humid subtropical (Caf) climates of central and southwestern Japan. —This is the part of Japan best known to Occidentals, who are likely to assume that it is typical of the entire country. Summer temperatures are high, the average for the warmest month being between 75° and 81°. At a majority of weather stations midsummer temperatures approximate those of the wet tropics. And since rainfall is abundant and humidity high, sensible temperatures as well as air temperatures are veritably equatorial. Winters are relatively mild, the coldest winter months having a mean temperature above freezing. During January, Tokyo has an average temperature of 37°, Osaka 40°, Kagoshima 45°. On sunny winter days midday temperatures are very pleasant, but when it is overcast and a strong wind is blowing, the humid cold is raw and penetrating. Many Japanese homes and public buildings are so ineffectively heated that indoor winter temperatures are distinctly uncomfortable; indeed, the foreigner in subtropical Japan is likely to have as unpleasant recollections of the winter cold as of the summer heat. Frosts are widespread throughout subtropical Japan in midwinter. Thus at Kumamoto, far to the south in central-western Kyushu, night temperatures drop below freezing on an average of 64 days a year.

Precipitation varies greatly in amount. Heavy average annual rainfalls of over 100 inches are recorded at some stations along the mountainous Pacific coast which faces the inflowing summer monsoon, and also along the Japan Sea coast, which is windward during the winter monsoon. By contrast the central section of the Inland Sea basin has only 40 to 50 inches of rainfall. Seasonal distribution of precipitation is no more uniform than the amounts. Over the larger part of the *Caf* region summer rainfall greatly predominates, but the Japan Sea side of

Honshu usually receives an excess of winter precipitation. Snow falls on occasions over the whole of subtropical Japan, Tokyo recording thirteen days with snow on an average and Osaka seventeen. Over much the larger part of this region snow remains on the ground for only a few hours, or days at the most. Only on the northwestern side, facing the Japan Sea, does snow lie deep on the ground throughout the winter.

There are three subprovinces of *Caf*. Subprovince 1, occupying the Pacific side of Japan, is so typical of the *Caf* region that it requires no further description. Subprovince 2 comprises chiefly the drier border-lands of the Inland Sea. The boundary as drawn follows closely the 1500 mm., or 60 inch, isohyet. The central portion of this subprovince, with an average annual rainfall of less than 45 inches, is one of the driest regions in Japan. Subprovince 3, comprising the Sanin and southern Hokuriku districts of Japan, is the milder subtropical southwestward extension of the snowy Japan Sea climates of northern Honshu (subprovince 1 of *Daf*). This region is unquestionably unique among the subtropical climates of the earth, for although it has the normal hot and rainy summers, with an average August temperature of 77° to 79°, its northeasterly part also has heavy snowfall in winter. No other region of the earth has such a combination of tropical heat and humidity in summer and a considerable depth of snow in winter.

FORESTS

Forest land comprises 68 per cent of the total area of Japan, one of the highest proportions for any of the earth's well-developed countries, rivaling even those for Sweden and Finland. Desert shrub is of course absent, and genuine natural grasslands of good grazing quality are negligible. But although over two-thirds of the country is classed as forest land, about 5 per cent of this is wasteland, not considered capable of producing usable wood, while another 7 per cent is *genya*, or wild grassland, much of it best described as intermingled coarse grasses, shrubs, and scattered low trees.

The fact that such a large proportion of the country still bears a forest cover, even though it has been occupied by agricultural peoples for upwards of two millenniums, is to be explained by the preponderance of hill and mountain terrain whose steep slopes are unsuited for growing cultivated crops. But although this rugged forest land has a much lower resource value than crop land, still it plays a significant role in the nation's economy, for not only is it a raw-material source for countless manufactured wood products, construction timber, woodpulp, charcoal, and fuel wood, but also it serves as a check against destructive erosion, while its runoff provides the main source of irrigation water and of hydroelectric power. The forest resource means more to the Japanese people than to many others, for 99 per cent of them live in wooden houses; and wood, either in its natural form or as charcoal, is a major domestic fuel, especially in rural areas. The connection between forest cover and other types of land and water resources is particularly close in Japan because of the predominance of steep slopes, the intensity of precipitation, the great relative importance of hydroelectric power, and the necessity for abundant irrigation water in the growing of paddy rice, which is by far the leading food crop.

Like mountains and hill land, the forest land is widely and rather evenly distributed over the country, occupying more than 60 per cent of the area of each of the four main islands. In 38 of the 46 prefectures of Japan, more than half the total land area is forested, and in 20 prefectures forests cover more than 70 per cent.

Forest zones

In a region having such wide ranges of latitude and of altitude, and hence of temperature, as Japan, different forest types naturally exist in the various latitudinal and altitudinal zones. Each zone is bounded by approximate temperature limits. The boundary planes of the altitudinal forest zones slope downward toward the north, intersecting Japan's sea-level surface at various latitudes, thereby forming the latitudinal belts. Three general latitudinal forest zones—the subtropical, the temperate, and the boreal—are usually recognized (Fig. 3-1).

The Subtropical Forest Zone descends to sea level at about latitude 37.5° or 38° and occupies the lowlands and lower slopes of southwestern Japan, where the mean annual temperatures are between about 55° F.

TABLE 3-1

Primary forest-land uses in Japan, 1960

Forest-land use	Area (1000 hectares)	Per cent
Saw timber	6,735	26.9
Coniferous	(3,655)	(14.6)
Broadleaf	(3,080)	(12.3)
Fuel wood	9,003	36.1
Inaccessible	2,533	10.2
Area in need of reforestation	2,749	11.2
Bamboo	98	0.4
Genya	1,829	7.3
Wasteland	1,270	5.1
Other	690	2.8
TOTAL	24,907	100.0

Source: Ministry of Agriculture and Forestry (Japan).

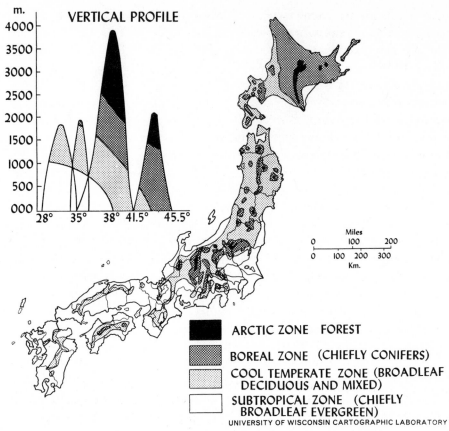

Fig. 3-1.—Original forest zones of Japan.

and 70° F. At increasingly higher altitudes to the south, above the iso-thermal surface of about 55° F., are deciduous and mixed forests, which are also characteristic of the lowlands in northern Japan. The original vegetation of this subtropical forest consisted of evergreen broadleaf trees, and remnants of these forests still survive in the isolated mountain districts of Kyushu and Shikoku. With long human occupancy, however, attended by careless cutting, fires, and partial reforestation, there has been an intrusion of broadleaf deciduous species and conifers. Oaks predominate and are the most widely used of the broadleaf trees. On the whole, however, this forest zone is not rich in good timber trees. Common species having special industrial uses are the Japanese tallow or wax tree (*Rhus succedanea*), the camphor tree (*Cinnamomum camphora*), and several varieties of bamboo, which grows in small groves rather than in the form of extensive woodland. In the extreme southern part of Kyushu are found numerous tropical elements, both among the trees and among the plants which compose the underwood. Here one sees palms and banana trees.

The Temperate Mixed-Forest Zone includes Honshu north of about latitude 37° or 38°, southwestern Hokkaido, and numerous highland districts south of latitude 37°—all regions where the mean annual temperatures are between about 43° F. and 55° F. Here the original cover was broadleaf deciduous forests and stands of mixed broadleaves and conifers. Deciduous trees still predominate, although the conifers are commercially more important and in the planted woodlands greatly outnumber all others. Mixed forests are most common—and, in autumn, most riotous in color, the reds, yellows, and browns of maple, birch, beech, poplar, and oak vivid against, the dark green of fir, pine, hemlock, and cedar. Among the conifers the most valuable trees are the Japanese cypress, arbor vitae, Japanese cedar, and fir. Of the broadleaf trees, elm, beech, ash, chestnut, poplar, and oak are of greatest commercial value. Within this temperate zone of deciduous and mixed woodland are included most of Japan's economically valuable forests.

In northern and eastern Hokkaido, where the average annual temperatures are below 43° F., is the Zone of Boreal Forests, where conifers predominate, principally fir and spruce. A great variety of broadleaf deciduous trees—birch, alder, aspen, willow—of little value for timber, are also present either in pure or in mixed stands. The undergrowth is thick, and dead timber clutters the forests, making penetration difficult. Many of the river valleys have a bog vegetation. This boreal forest type is also found in a small area of Honshu at elevations of 1800–2800 meters. Such

TABLE 3-2

Area of forest by composition of stand (unit: 1000 hectares)

Natural forest	17,018
Broadleaf	(13,528)
Coniferous	(3,490)
Planted forest	6,113
Coniferous	(5,800)
Broadleaf	(313)
Wild land	1,932
Cutover and damaged	150
Bamboo	168
Special forests	21

Source: *Japan Statistical Yearbook, 1962.*

highland forests are of limited commercial value because of their inaccessibility, but their effects on water supply and runoff are of great importance.

Nature and use of Japanese forests

A distinctive feature of the Japanese flora is the remarkable number of its tree species, which includes at any latitude elements belonging to warmer climates. This may reflect recent climatic oscillations. In general composition the forests of Japan show resemblances to those of North America. Originally the country was largely covered by dense forests in which broadleaf species predominated. But occupancy for two thousand years by a civilized agricultural people has so completely modified the character of all but the most inaccessible forests that they bear only slight resemblance to the original stands. The more accessible forests have been depleted through overcutting, poor management, and soil erosion, while further large-scale modification has resulted from the introduction of new species by artificial afforestation.

Broadleaf forests are most extensive, accounting for over one-half of the total forest area, while forests predominantly of conifers represent 30 per cent, and mixed broadleaf-conifers 16 per cent. But although it is the broadleaf forest which is most extensive, almost two-thirds of it, or nearly 27 per cent of the total forest area, is composed of small trees of coppice dimensions which are chiefly valuable for fuel wood.

Japanese forest land is about equally divided between public (national, prefectural, community) and private ownership. There are more than 5 million individual owners of forest land, of whom nearly 73 per cent have holdings of one hectare (2.47 acres) or less, and nearly 94 per cent hold less than four hectares. A great majority of these are farmers whose wood lots are considered adjuncts to farming. Thus, to an unusual

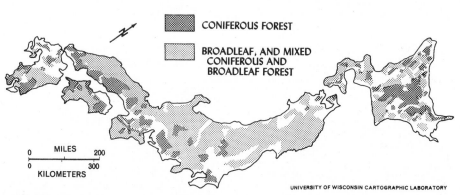

CONIFEROUS FOREST

BROADLEAF, AND MIXED CONIFEROUS AND BROADLEAF FOREST

MILES 0 200
KILOMETERS 0 300

UNIVERSITY OF WISCONSIN CARTOGRAPHIC LABORATORY

Fig. 3-2.—Present forest types of Japan.

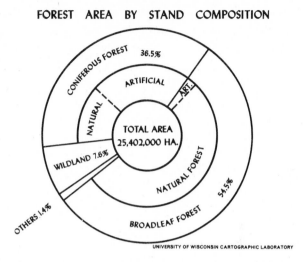

Fig. 3-3.—Fifty-five per cent of Japan's forest land is broadleaf forest, nearly all of it a natural growth. Of the 37 per cent which is conifers, about 70 per cent is planted forest.

FOREST AREA BY STAND COMPOSITION

CONIFEROUS FOREST 36.5%

ARTIFICIAL

ART.

NATURAL

TOTAL AREA 25,402,000 HA.

WILDLAND 7.6%

NATURAL FOREST

54.5%

OTHERS 1.4%

BROADLEAF FOREST

UNIVERSITY OF WISCONSIN CARTOGRAPHIC LABORATORY

degree, trees are looked upon as a crop in Japan, and this accounts for the patchy appearance of so many wooded slopes, on which there exist multitudes of woodland plats differing in size and shape and in the kinds and ages of trees being grown.

Planted forests, which are 90 per cent coniferous, amount to slightly less than one-quarter of the whole forest area. So while the broadleaf forest of Japan is overwhelmingly a natural growth, and the mixed forest is only somewhat less so, the reverse is true of the conifers, for in their case 70 per cent is planted forest. Because of their desirability as lumber trees, cedar, cypress, red pine, and black pine are the most important species in the planted forests; they require about forty years of growth to reach merchantable size.

Some 11 per cent of the total forest area is classified as in need of reforestation. Over the decade of the 1950's the area annually reforested varied between 500,000 and 650,000 hectares, one-half to three-quarters of which was artificial regeneration, nearly all of it conifer. Natural afforestation is more evenly divided between conifers and broadleaves. More than two-thirds of the afforestation is on privately owned lands.

Over a long period of time the annual area reforested by artificial and natural means has not kept pace with the rate of annual cut, so that the cumulative area in need of reforestation has steadily increased. As a consequence the area of farm lands damaged by floods has mounted rapidly as forest cutting has proceeded in the flood-source areas. The need for remedial action is urgent, especially when it is considered that less than 15 per cent of the forest area provides a preponderance of the timber for industrial and construction purposes.

Saw-timber forests comprise only about 27 per cent of the forest land

—14.6 per cent coniferous and 12.3 per cent broadleaf. But it is the less than 15 per cent which is in conifers that provides nearly 80 per cent of the cut of saw logs, while the 12.3 per cent which is in broadleaf species provides only 20 per cent of the timber harvest. This is because the most important broadleaf species (oak, beech, birch, maple) of saw-timber size are characteristically short-boled, crooked, and forked. The coniferous species providing most of the saw logs are cedar, pine, larch, and fir. Distribution of the saw logs cut from coniferous forest is very widespread but with some concentration in northernmost Honshu, central and eastern Hokkaido, the highlands of central Honshu, and the mountainous parts of Shikoku and Kyushu. The output of broadleaf logs is concentrated to a somewhat greater degree, with 37 per cent from Hokkaido alone.

To compensate for the loss of Karafuto, which before the war had been a main source for pulpwood, Japan has been obliged increasingly to turn to her home forests for this product. New techniques recently have made possible a greatly expanded use for pulp of red pine and certain broadleaf trees which grow in subtropical parts of the country. Consequently the pulpwood industry is gradually shifting from northern to southwestern Japan.[1] At present there are six main pulpwood supply areas, all of them highland in character: eastern Hokkaido, southern Tohoku, western Chubu, Chugoku, southwest Shikoku, and southwest Kyushu. All except the first two are in southwest Japan, with Chugoku being the single most important supply center.

In Japan more wood has always gone into fuel than into lumber; and fuel wood, including charcoal, is produced in every region of the country. The source of supply is principally coppice woodland in which stumps of various species of hardwood trees send out sprouts from the root crowns, the sprouts being harvested each year or at somewhat longer intervals. Overexploitation is characteristic of such woodlands, especially where they lie adjacent to villages and roads, with the result that they have the appearance of low brush wood on partially stocked hillsides. Of small value for soil protection, they yield only a fraction of the wood which they are capable of producing under efficient management.

Seven to eight per cent of the so-called forest land of Japan is *genya*, wild land primarily in coarse grass, either treeless or with shrubs and scattered low trees, It occupies different kinds of sites, mainly slope lands and diluvial uplands, although areas of marsh grass may be classified as genya. Scarcely a climatically induced type, genya is often a second growth following man's removal of the forest, but in some areas it appears to be an original natural growth related to soil character. Systematic utilization of genya is for the most part lacking. Cutting the wild grasses and raking up all litter for use as compost, fuel, fodder, and stock bedding are the common forms of utilization and are relatively haphazard, with low yields. Some genya is currently being converted to woodland.

Inaccessible forests are those lacking the transportation facilities needed for economical exploitation of timber resources. They may represent as much as 10 per cent of the total forest area, and they probably equal 20 per cent of the timber volume of the country. Such forests are chiefly located in rugged mountainous country, so that one of their most important values is that of watershed protection. It seems doubtful whether the timber resources of these inaccessible forests can ever be developed on a large scale.

Some 3 million hectares, or 13 to 14 per cent of the forest area, is designated as forest reserve or protection forest, whose value is social as well as economic. Besides providing scenic beauty, these areas protect headwaters, serve as windbreaks and fish shelters, and protect against floods and the spread of eroded sands and gravels over cropped fields. In terms of area, by far the most important functions are the protection of headwaters and the accompanying reduction of the flood menace. Some selective timber cutting is permitted even in the protection forests.

THE SOIL RESOURCE

Soil has importance as a resource chiefly in its relation to agricultural production. Such being the case, and bearing in mind that only about 14–15 per cent of Japan's total area is in cultivated crops, it is clear that the character and quality of soils over much the larger part of the country can have little direct economic significance. On the other hand, considering the high importance of agriculture in the Japanese economy, and the fact that about 36 per cent of the nation's households are engaged in agriculture, as a main or a subsidiary occupation, it is undeniable that soil is a major natural resource of Japan, even though its active use is restricted to a small part of the country's area. As a broad generalization it may be said that the quality of Japan's soils is not conducive to high agricultural productivity: Japanese farmers must rely on careful husbandry to achieve their high level of production.

Soil development and the resulting soil properties are influenced chiefly by four natural factors: climate, native vegetation, parent material, and slope of the land surface. In addition to these natural factors, man has also been influential in modifying the qualities of soils in a variety of ways—a fact particularly true in Japan, where intensive agricultural land use spans nearly two millenniums of time.

Two comparatively recent reconnaissance soil surveys have provided classifications, textual descriptions, and generalized maps of the soils of Japan.[2] While these two surveys are similar in many respects, still there are differences, some of them not easy to reconcile. A few examples of the significant differences may be noted:

1. The SCAP survey, a product of the Allied Occupation, classifies the new alluvial soils of Japan as azonal, or having slightly developed soil profiles. By contrast, Kamoshita, of Japan's Department of Soils, classifies these same soils as intrazonal, which signifies that they have fairly well developed profiles, and he describes the development of their horizons as complete.

2. The SCAP survey classifies all volcanic ash, or Ando, soils as belonging to the intrazonal group, while Kamoshita designates only the recent volcanic ash as intrazonal, while other older ash soils are classed as Brown Forest Soils within the zonal group.

3. On Kamoshita's soil map only 15 soil types are recognized, so that it is highly generalized. By contrast, the maps accompanying the SCAP reports identify over 60 soil types and subdivisions, with many of them differentiated in terms of type of location, texture of materials, and the nature of the relief.

Taken as a whole, the more recent Kamoshita soil classification does not appear to be a marked improvement over that contained in the SCAP publication. The latter, besides providing more detail on types and distribution of soils, supplies quantitative estimates of the areal extent of the several soil groups. In one significant respect, however, the Kamoshita classification does appear to be an improvement over the earlier one, in that it classifies the soils of deltas and floodplains ("Azonal Alluvial" soils of SCAP) as intrazonal. Probably important proportions of the soils designated by SCAP as "Azonal Alluvial" could more appropriately be classified as intrazonal soils, falling into what are now called the Low Humic Gley group.[3] The complex distribution pattern of each soil type and the limited extensiveness of individual soil areas make it impossible to construct a small-scale map which adequately represents soil distribution in Japan.

Young to immature soils, developed for the most part under a forest cover, predominate in Japan. Dark-colored, fertile grassland soils are absent. Virtually all Japanese soils are acid.

Azonal soils, with only slightly developed soil profiles, characterize

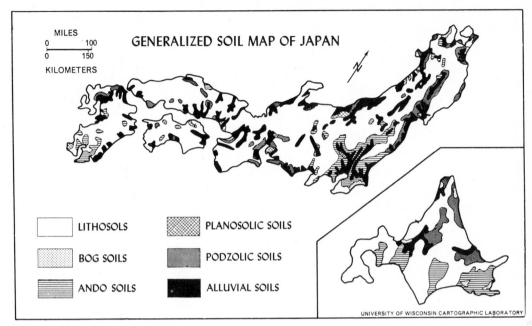

Fig. 3-4.—After map in "Reconnaissance Soil Survey of Japan," Report No. 110-I, "Summary," General Headquarters, Supreme Commander for the Allied Powers, Natural Resources Section (Tokyo, 1951).

nearly 82 per cent of the nation's area. The azonal order of soils in Japan comprises two soil groups—lithosols and new alluvial soils, with the former, unfortunately, much the more extensive. Lithosols almost exclusively are characteristic of hill and mountain lands with steep slopes, and they prevail over about two-thirds of the country. Here erosion keeps pace with weathering, so that the accumulation of soil materials is meager. As a result, most lithosols are shallow and stony, with properties strongly influenced by the nature of the underlying bedrock. Very restricted in extent are lithosols in the form of loose sands, found in beach-ridge and dune areas along coasts. These often are fairly well utilized, whereas only a small part of the slope lithosol area is under cultivation, most of it remaining in woodland.

The second azonal soil—the alluvial deposits of floodplains, deltas, and alluvial fans, covering 14 per cent of the country's area—is by far the most important agricultural soil of Japan. It comprises relatively immature soils with slightly-to-moderately developed profiles, the morphological characteristics depending mainly on the height of the ground water table and the drainage conditions. Texture varies widely, from coarse gravels in the upper parts of alluvial fans, to sands, sandy loams,

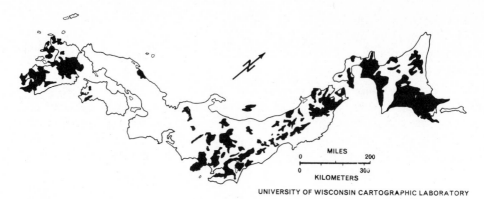

MILES
0   200
0   30ɔ
KILOMETERS

UNIVERSITY OF WISCONSIN CARTOGRAPHIC LABORATORY

Fig. 3-5.—Distribution of volcanic soils in Japan. After Toya.

and clays on the broader alluvial plains. This variety is related to distance from stream channel, size of stream, and the nature of the subsoils and earth materials on the upper watershed. Even on a small alluvial lowland there may be a considerable variety of soils arranged in very complex patterns of distribution. Alluvial soils are widely distributed, being found in all latitudes of Japan. Without doubt they are the most fertile of Japan's soils, and it is upon them that agricultural production, and especially paddy-rice production, is concentrated. The least productive of the alluvial soils are the bog, half-bog, and meadow varieties where poor drainage is the most serious handicap.

Intrazonal soils, having only fairly well developed profiles and strongly reflecting the influence of local factors such as poor drainage or recent additions of volcanic ash, cover some 10 per cent of the country. In general they are low in natural productivity. Three major types of intrazonal soils are recognized: Ando soils, Planosols, and Bog soils. Ando soils, derived largely from wind-deposited acidic volcanic ash, are the most extensively developed of the group (Fig. 3-5). They are dark in color, occur chiefly on uplands and terraces, are variable in texture, well drained, and of low fertility. But in spite of being infertile, they are exceeded in agricultural importance only by the soils of new alluvium. They are highly susceptible to erosion, both by wind and by water. Ando soils, where under cultivation, are mostly planted in dry crops rather than in irrigated rice. Although widely distributed, they are most extensively developed in southern and eastern Hokkaido, the Kanto Plain area in central Japan, and southern Kyushu. Planosols, of small extent, occur on slopes that have been bench-terraced for rice culture. As a result of long continued irrigation they are strongly leached and have developed a compact claypan subsoil. Bog soils develop in poorly drained areas and are composed of peat and muck formed from partially decayed swamp vegetation. They are not of high quality, but after expensive drainage and improvement they may be converted into fairly satisfactory crop-land.

Zonal soils, with well developed profiles and characteristics resulting from long exposure to the prevailing climatic and vegetative environments, are meagerly represented in Japan. This is not surprising since steeply sloping hill lands and depositional plains, neither conducive to mature soil development, make up a major part of the land surface. Predominantly, the zonal soils of Japan are of the podzolic group, gray-brown in the cooler climates of northern Honshu and Hokkaido, and yellow (or red) in the subtropical southwest. Characteristically they occur on rolling-to-hilly uplands of older alluvium or of Tertiary sands and clays, as well as on river terraces. Relatively low in mineral plant foods,

TABLE 3-3

Area of great soil groups in Japan

Order	Total area (hectares)	Extent (% of Japan)
Zonal soils		
Podzolic	2,678,615	7.27
Gray-brown	(1,704,631)	(4.63)
Yellow	(872,504)	(2.36)
Red	(101,480)	(0.28)
Reddish-brown lateritic	3,580	0.01
Total zonal soils	2,682,195	7.28
Intrazonal soils		
Bog	192,551	0.52
Half bog	30,615	0.08
Planosols	430,054	1.17
Ando (volcanic ash: brown and black)	3,099,607	8.42
Total intrazonal soils	3,752,827	10.19
Azonal soils		
Lithosols	24,995,786	67.81
Alluvial	5,195,900	14.09
Total azonal soils	30,191,686	81.90
Water	231,800	0.63
GRAND TOTAL	36,858,508	100.00

Source: "Reconnaissance Soil Survey of Japan, Summary" (1951).

they must be abundantly supplied with fertilizer if good yields are to be maintained. Where cultivated, they are usually in upland, or dry, crops.

It warrants re-emphasis that most of the agricultural soils of Japan have been altered from their original natural state through centuries of cultivation. While some of the soils have been tilled for nearly two thousand years, it is judged that their current productivity equals or exceeds what it was when cultivation first began. This is a credit to the high quality of Japanese husbandry, which has coaxed bountiful yields from originally unpromising land.

WATER RESOURCES

Compared with most parts of the earth, Japan has a generous water supply, thanks to the relatively abundant rainfall, which in turn produces a dense network of small rivers carrying surface runoff, and a relatively high ground-water table.

Rivers of Japan must of necessity be short in length, be steep in gradient, and have relatively small drainage basins. Even the Tone River, which is the largest, has a length of only 322 kilometers, and its watershed area is limited to 15,760 square kilometers. A further consequence of small size is that the annual variability of stream discharge is great, the ratio of maximum to minimum discharge being more than 200, with an extreme value of 500. This variability, and the susceptibility of a number of valleys and lowlands to floods, is a major handicap to Japan's agricultural system. The short, steep gradients of almost all rivers result in flash floods succeeding heavy rains, with an accompanying large discharge of sand, gravel, and even boulders. When this debris is spread over precious farm land it is more damaging than the floodwater itself. A discharge in a single flood equivalent to the total annual runoff of a particular stream has been recorded on numerous occasions. Since Japanese rivers carry relatively heavy loads of coarse materials all the way to their mouths, they tend to aggrade and so elevate their channels on the floodplain, as was mentioned in Chapter 1. Such a river is called "Tenjo-gawa" or, literally, a river flowing above the roofs. Steep stream gradients also make it difficult to store water for irrigation or waterpower purposes, while the few sites that are suitable for reservoirs likewise are valuable as agricultural land. Where reservoirs are built, serious trouble is encountered in the form of a rapid accumulation of sediment.

Stream runoff in Japan is at its maximum in the warmer months or growing season, when it is most useful for irrigating crops, more especially paddy rice. In general this reflects the season of maximum precipi-

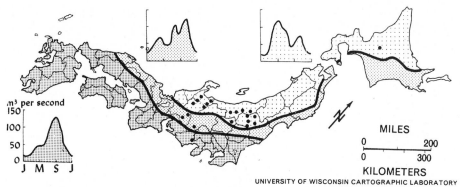

Fig. 3-6.—Three general regional types of annual river flow in Japan. Black dots represent large hydroelectric power plants. After *Regional Geography of Japan*, No. 7.

tation for most of the country. There are significant regional variations, however (Fig. 3-6). For example, on the Pacific side of southwestern Japan the maximum discharge is characteristically in late summer or early fall at the time of the heavy rains associated with tropical storms. In northern and northwestern Japan, by contrast, where the winter snow cover is heavy, the maximum runoff is often reached in April and May at the time of snow melt.

Ordinary stream diversion supplies about 68 per cent of the 30–60 million tons of irrigation water required to inundate the rice paddies. Since there is a general lack of storage capacity, the supply of irrigation water may be so reduced during drought years that rice yields are diminished. Adding to the problem is the fact that in many regions the rice area has been expanded to near the upper limits of the irrigation potential, making for even greater vulnerability to water supply shrinkage. Especially in Hokkaido and northern Honshu, diminished stream flow in middle and late summer may cause the rice crop to suffer, and no part of Japan is immune from such losses in particularly dry years. Supplementary irrigation to the amount of nearly one-third the total is provided by small artificial ponds which collect local runoff, and by springs, wells, and lakes.

Because of their small size, variable flow, and steep gradients Japanese rivers are little used for transportation. There is some rafting of logs and other forest products, and near their mouths the distributaries and canalized channels are employed for short barge hauls within the larger port cities. Industries tend to locate along these city waterways where they have the site advantage of being able to receive bulky imports of fuel and raw materials by cheap water transport.

On the other hand, the numerous small steep-gradient streams do pro-

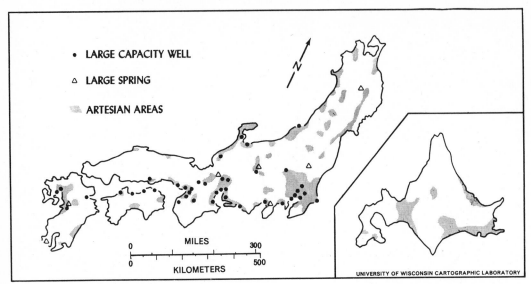

Fig. 3-7.—Distribution of underground water. After map by Geological Survey of Japan.

vide an abundance of potential waterpower, the developed part of which supplies close to a quarter of the nation's energy requirements. (See Chapter 4 for an amplification of the topic of electric power resources.)

In addition to the important supply of water from rivers, there would appear to be a large supplementary store of water available underground. Particularly abundant is the ground-water resource contained in the unconsolidated sediments of the alluvial plains and coastal plains, which in turn are the centers of population and hence of large water needs. The water capacities of some diluvial and Tertiary formations are likewise good, although lower than the capacity of the newer alluvium. Other than these it is the volcanic extrusives and the limestones that are most aquiferous and are the chief sources for large springs. Few productive aquifers are to be found in the igneous and the older sedimentary rocks which cover 63 per cent of the country's area.

Of the total amount of water stored in the unconsolidated sediments of the alluvial and coastal plains, estimated to be 500–1000 billion cubic meters, 2 to 2.5 per cent is annually available for pumping.[4] The most extensive of the artesian aquifers of the Quaternary age, with a thickness exceeding 200–300 meters, underlie the Kanto, Nobi (Nagoya), Osaka, and Ishikari plains (Fig. 3-7). Other smaller areas of high-pressure confined ground water are widely distributed. There is also much free ground water in the alluvial sediments, but as a rule it is not available in large quantities. In the diluvial uplands the water table usually

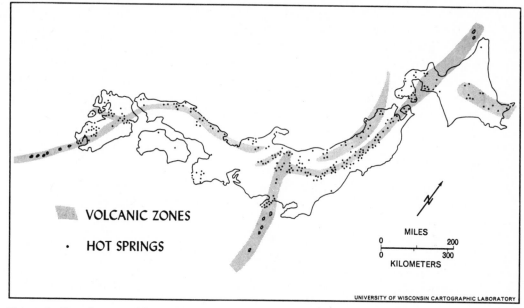

VOLCANIC ZONES

• HOT SPRINGS

MILES
0   200
0   300
KILOMETERS

UNIVERSITY OF WISCONSIN CARTOGRAPHIC LABORATORY

Fig. 3-8.–Distribution of hot springs. After map by Geological Survey of Japan.

is from 5 to 25 meters deep, and in a few alluvial cones it may be over 25 meters deep. Commonly springs are found along the lower margins of alluvial fans. Underground water in the form of wells and springs is a supplementary source of water for irrigation and is the most important source of water for domestic and industrial uses. Forty-two per cent of the water supplied by public waterworks is from ground-water sources, 34.5 per cent is from surface water, and 23.5 per cent is from springs.[5]

As late as the 1930's it was comparatively easy to obtain an adequate industrial supply of water from rivers or underground sources, but the phenomenal growth of industry in the postwar period, especially of such industries as iron and steel, metals, chemicals, and synthetic fibers, has resulted in an acute water shortage for industrial purposes. Much of the recent expansion of the industries named has taken place on the sea margins of alluvial plains, part of it on newly reclaimed lands, where fresh water from alluvial sources is relatively abundant and sea water is also available. As the supply of fresh water from surface sources has decreased and the dependence upon ground water has increased, overpumping has led to serious subsidence of the land in the industrial areas of Osaka and Tokyo, as well as acute shortages of water supply attended by a reduced artesian flow and an increase in the required pump lift.[6]

Mention should be made of mineral hot springs in Japan, which have stimulated the development of about 1000 spas, some of which attract tourists in large numbers. A great majority of the mineral hot springs are found in areas of young volcanic rocks, which are widely distributed, though less numerous on the Pacific side of the country (Fig. 3-8).[7] While such springs are used mainly for bathing and for their therapeutic benefits, recently some have been utilized for industrial purposes (evaporating salt, drying lacquer and plastics, fermentation processes, tannin extraction) and for heating purposes (hothouses, hotbeds, and irrigation water).

In parts of the country, especially on a few deltas, it is the excess of water that is a handicap. Some of this wet land which existed in the form of bogs and shallow lakes has already been reclaimed for agricultural land use, and additional reclamation is now in progress. Much more universal are those wet delta lands which, while satisfactory for a crop of summer rice, cannot be sufficiently rid of their surplus water after the rice harvest to permit the planting of unirrigated fall-sown cereals. Perhaps 20 per cent of the paddy area located in a climatic environment permitting winter crops remains unsown in fall because of poor drainage.

4 · MINERAL AND ENERGY RESOURCES

Considering Japan's area, her mineral resources are surprisingly diverse in character, but at the same time the country is relatively deficient in deposits of sufficient magnitude and quality to meet the needs of her present industrialization. Undoubtedly the great variety of metallic minerals is related to the complicated nature of the geologic structures and the widespread occurrence of igneous activity. Of thirty-three metals used in modern industry, twenty-two are either mined or known to exist in Japan.

On the basis of adequacy of domestic supply, three categories of minerals may be recognized: (1) those in which domestic production is adequate or nearly so—chromite, ordinary coal, gypsum, limestone, magnesium, pyrite, sulfur, lead, zinc, copper, gold, and silver; (2) those produced in insufficient quantity for domestic requirements so that a significant import is necessary—iron ore and other ferrous materials, coking coal, antimony, mercury, manganese, tin, titanium, tungsten, molybdenum, chromium, and vanadium; and (3) those which are lacking or so strongly deficient that a very large proportion of the nation's need must be met through import—nickel, cobalt, aluminum, nitrate, phosphate, potash, salt, rare-element minerals, and petroleum.

Characteristic of the Japanese exploitation of mineral resources is the multitude of small mines, many of which are operated by comparatively crude methods, a circumstance which reflects the small size of a majority of the widely scattered ore bodies. Significantly there are at present only seven ore mines in which the annual output of crude ore exceeds 500,000 tons. An additional feature of Japan's mineralization and mining operations is the association of diverse kinds of ores within a single deposit and, consequently, the multiform output of individual mines. As an example, at the Kamioka mine, one of the largest in Japan, zinc, lead, sulfur, silver, and gold are all included in the mineral output.

ENERGY MINERALS AND RESOURCES

Reflecting the remarkable postwar industrial expansion in Japan, the demand for energy has registered a striking increase. In terms of coal equivalent, energy consumption in 1960 was 152 million tons (188 million in 1962, or 2.1 times the prewar peak). While coal is still the primary source of energy, providing 38 per cent of the total, its position has declined (64 per cent in 1934 and 51 per cent in 1950) while that of petroleum has risen sharply (6 per cent in 1950, 24 per cent in 1960, 31 per cent in 1962). Like coal, hydroelectricity has also declined in relative importance as an energy resource, for while it accounted for 33 per cent of the power used in 1950, this had shrunk to 23 per cent in 1960

TABLE 4-1

Production of seven leading mines in 1961

Crude ore	Kamaishi	Kamioka	Hitachi	Yanahara	Besshi	Osarizawa	Matsuo
TOTAL (tons)	1,363,118	1,087,750	703,080	696,755	664,811	652,328	615,590
Metal content:							
Gold (kg)	42.4	54.4	337.2	....	144.9	157.3	....
	(0.1)	(0.1)	(0.5)		(0.3)	(0.3)	
Silver (kg)	3,683	28,184	3,258	....	4,288	5,512	....
	(8)	(26)	(5)		(6)	(9)	
Copper (tons)	4,481	....	6,430	94	7,105	7,149	....
	(0.3)		(0.9)		(1.1)	(1.1)	
Lead (tons)	....	8,664	....	....	....	2,283	
		(0.8)				(0.2)	
Zinc (tons)	....	53,694	4,734	.. ..	....	....	....
		(4.9)	(0.7)				
Iron (tons)	377,471	....	....	....	7,800	15,400	....
	(27.7)						
Sulfur (tons)	....	29,546	113,692	310,939	59,779	32,299	233,491
		(2.7)	(16.2)	(44.6)	(9.0)	(5.3)	(33.3)

Note: Numbers in parentheses indicate the percentage of total domestic production by an individual mine.

Source: Ministry of International Trade and Industry, *Statistical Year Book of Production at Individual Mines, 1962.*

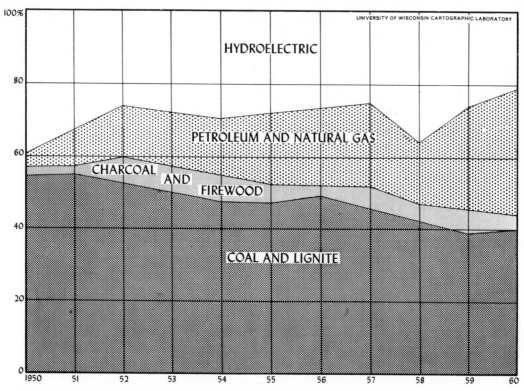

Fig. 4-1.—Ratio of the principal energy sources in Japan, 1950 to 1960. Thermoelectric power is included with petroleum or coal.

(Fig. 4-1). Japan appears to be shifting to new kinds of power resources, and more especially from solid to liquid fuels, as the limitations of her coal and hydroelectric resources become increasingly apparent. This is doubly significant when it is considered that the domestic supply of oil is completely inadequate, so that nearly the whole requirement must be imported, while ordinary coal and waterpower have important production domestically.

Coal

Coal, by far the most important of Japan's mineral resources, represented 47 per cent of the total value of the mineral output in 1960. Of some 55 million tons consumed annually, roughly 55 per cent is used in manufacturing and 35 per cent in public utilities (electric power, railways, etc.).

Japan is neither impoverished nor yet affluent as regards coal reserves. The official estimate of a theoretically recoverable reserve is 20.79 bil-

TABLE 4-2

Total practical coal reserves, 1956 estimate

Basis of estimate	% of estimated 3,226,853,000 T.
Location	
Hokkaido	48.5
Kyushu	38.4
Honshu and Shikoku	13.1
Certainty	
Proved	28.6
Possible	14.3
Estimated	54.1
Kind of coal	
Anthracite	2.7
Bituminous to high-rank lignite	94.5
Low-rank lignite	2.8

Source: Japanese Geological Survey, *Geology and Mineral Resources of Japan* (2d ed., 1960).

lion tons, but a more realistic figure is that of total practical coal reserves, or 3.23 billion tons, of which only 0.9 billion tons are proven. But even if one accepts the figure of 3.23 billion tons as reasonable, it becomes obvious that with an annual production of 50–55 million tons, this reserve would be depleted in about sixty-five years. Considering the natural difficulties under which coal is mined, it seems unlikely that Japan's annual output will exceed 50–60 million tons, so that if an expanding industry greatly increases its energy demands, new sources of power will have to be sought.

Moreover, there are certain basic weaknesses in Japan's coal reserve other than modest volume. Quality is not of the best, much of it being subbituminous with relatively low heat energy, while anthracite is meager (2.7 per cent) and high-grade coking coal nearly lacking. In addition, the coal seams are thin, steeply inclined, and made discontinuous by faults, so that the use of large-scale mining machinery is difficult and greater dependence upon cheap manual labor results. Also, a preponderance of the reserves and of the current production is located in the northern and southern extremities of the country (48.5 per cent in Hokkaido; 38.4 per cent in Kyushu), well removed therefore from the industrial heart of Japan (Fig. 4-2). The remaining 13 per cent is in scattered small deposits. There is practically no coal in Shikoku. Somewhat offsetting its smaller reserves, the quality of Kyushu coal is appreciably higher than that of Hokkaido, the latter having a larger percentage of subbituminous and brown coal.

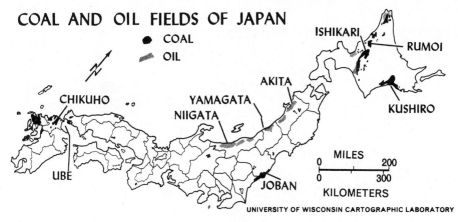

COAL AND OIL FIELDS OF JAPAN
● COAL
🔺 OIL

ISHIKARI
RUMOI
AKITA
CHIKUHO
YAMAGATA
NIIGATA
KUSHIRO
UBE
JOBAN

MILES
0   200
0   300
KILOMETERS

UNIVERSITY OF WISCONSIN CARTOGRAPHIC LABORATORY

Fig. 4-2.

Under the pressure of wartime needs Japan's coal production reached a maximum of about 53 million tons in 1944, but it slumped to 30 million tons in 1945. Since then it has risen again, to the 50–55 million tons of recent years. Some 12 million tons of coal were imported in 1962, most of it coking coal, chiefly from the United States.

Production distribution is not the same as the distribution of reserves, for the bulk of the output comes from four centers—northern Kyushu with about 50 per cent; Hokkaido, 36 per cent; eastern Honshu, 8 per cent; and western Honshu, 6–7 per cent (Fig. 4-3). Thus, Hokkaido and northern Kyushu reverse their positions as regards reserves and production. The fact that north Kyushu, with somewhat smaller reserves, has always been a larger producer than Hokkaido, reflects its easier accessibility with respect to the great industrial centers, as well as the fact that its geological structure is simpler and dips of its coal seams are less steep. Nevertheless, in spite of the handicap of more intricately folded coal structures, Hokkaido has gradually been increasing its pro-

TABLE 4-3

Coal production by regions
(in percentage of the country's total production)

Region	1930	1940	1955	1959	1960
Hokkaido	21.4	26.8	30.2	33.3	35.9
E. Honshu	8.1	6.7	8.6	8.2	8.1
W. Honshu	5.8	5.3	7.1	7.9	6.5
Kyushu	64.7	61.0	54.1	51.5	50.3

Source: Ryuziro Isida, *Geography of Japan* (Tokyo, 1961).

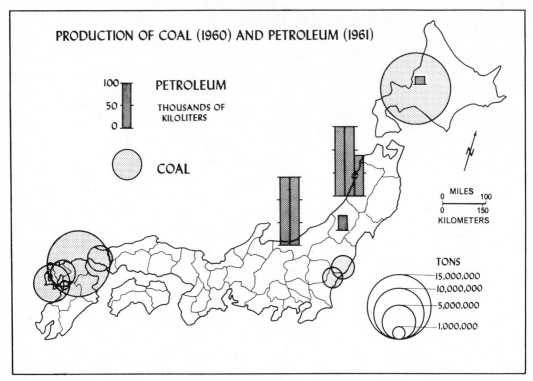

PRODUCTION OF COAL (1960) AND PETROLEUM (1961)

Fig. 4-3.

portion of the nation's coal production, while north Kyushu has passed its prime and has been declining relatively. In the three decades between 1930 and 1960, Hokkaido's proportion of the nation's coal production has risen from about 21 per cent to nearly 36 per cent, while that of Kyushu has declined from about 65 to 50 per cent.

Both in Hokkaido and in northern Kyushu coal is produced in a number of separate fields. Up to fourteen separate coal and lignite fields are designated by name for Hokkaido, but one field alone—the Ishikari in the central-western part, the largest coal field in Japan—contains 71 per cent of the island's exploitable coal reserve and accounts for about three-quarters of its production. Some eighteen separate coal fields are to be found in Kyushu, foremost being Chikuho in northernmost Kyushu, which contains 58 per cent of the island's reserve and supplies about 53 per cent of its annual output. After Ishikari it is the most important producing field in Japan, accounting for nearly 30 per cent of the nation's total. In Honshu, the two principal coal fields are Joban in the east, located on the coast about 200 kilometers north-northeast of Tokyo, and Ube in extreme southwest Honshu, near the western end of the Inland

Sea. Joban, with about 53 per cent of Honshu's exploitable coal reserve, produced over 4 million tons in 1960, while the comparable figures for Ube are 31 per cent and about 3 million tons. The former field has the distinct advantage of being favorably situated in close proximity to the great power-consuming Kanto industrial area.

Quality of the coal mined varies from field to field and even between different parts of the same field. The bulk of the Hokkaido product is bituminous or subbituminous, but nearly 22 per cent is classed as having weak coking quality. Kyushu coals are similar or somewhat better in quality, with nearly 29 per cent classed as weak coking coal. From both Joban and Ube the output is largely subbituminous, high in ash and moisture.

Japan's near lack of good coking coal causes imports of this special variety to run as high as 45–50 per cent of the total consumption. Because of the domestic shortage and the high cost of importing the product from eastern United States, great efforts have been made to reduce the coke consumption ratio in blast-furnace operations. Overwhelmingly the domestic reserves are of a semicoking type which must be mixed with imported heavy coking coal to make a satisfactory blast-furnace product. In 1961 the production of domestic coking coal amounted to 12 million tons, or about 22 per cent of the coal mined, a figure which seems unlikely to go much higher.[1] Nearly 55 per cent originated in the Kyushu fields and the remainder in Hokkaido. Heavy coking coal accounted for less than 4 per cent of the total, and all of it was from the Hokusho district in Kyushu.

Because of the size and shape of the country, all coal mining in Japan must of necessity be located not far from the sea, but the Kushiro field in Hokkaido, Joban and Ube in Honshu, and Miike in Kyushu are all

TABLE 4-4

Output of coal (except lignite) in 1961 (in tons)

| Kinds of coal | Hokkaido | Honshu | | Kyushu | Total |
		Eastern	Western		
Coking coal	5,416,404	....	....	.6,143,311	11,559,715
General	14,802,744	4,201,364	2,017,292	19,566,523	40,587,923
Anthracite	....	78,714	1,040,053	774,744	1,893,511
Natural coke	....	....	....	442,358	442,358
Total	20,219,148	4,280,078	3,057,345	26,926,936	54,483,507
Per cent	37.1	7.9	5.6	49.2	100

Source: Statistical Year Book of the Japanese Coal Industry, 1961.

coastal in location, and in most of these fields submarine seams are being worked. In recent years some 12–15 per cent of the nation's output has been taken from such submarine seams. This actual or near tidewater location of Japan's coal has the great advantage of permitting cheap water transport of such a bulky product.

The coal industry of Japan is in the doldrums, and its future looks dim in the light of high costs of production and declining demands. Technological changes are shifting the emphasis in the chemical industries away from coal-derived materials toward petrochemicals, and industry in general is turning from coal- to oil-generated power and heat.

Petroleum and natural gas

Of the fuel and power resources in Japan, petroleum is the weakest unit, and yet its use is being expanded more rapidly than that of either coal or hydroelectric power, both of which are present in greater abundance. At present rates of production the now-known oil fields may be

TABLE 4-5

Production of coal by coal fields, 1960
(thousands of metric tons)

Hokkaido		19,043
Ishikari	14,715	
Kushiro	2,393	
Rumoe	1,390	
Tenhoku	378	
Yoshinuma	107	
Tobu (NE. Honshu)		4,241
Joban	4,089	
Others	152	
Seibu (W. Honshu)		3,177
Yamaguchi (Ube)	3,140	
Others	37	
Kyushu		26,146
Chikuho	13,598	
Fukuoka	1,202	
Asakura	62	
Miike	1,207	
Karatsu	2,722	
Sasebo	4,068	
Saito-Takeshima	2,879	
Amakusa	408	
TOTAL JAPAN		52,607

Source: Statistical Year Book of the Japanese Coal Industry, 1961.

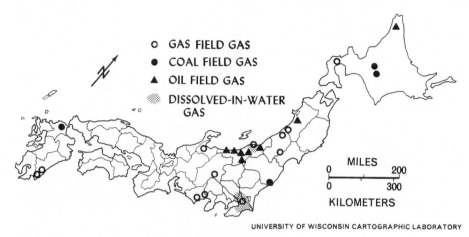

○ GAS FIELD GAS
● COAL FIELD GAS
▲ OIL FIELD GAS
▨ DISSOLVED-IN-WATER GAS

MILES
0　　　　200
0　　　　300
KILOMETERS

UNIVERSITY OF WISCONSIN CARTOGRAPHIC LABORATORY

Fig. 4-4.—After map by Geological Survey of Japan.

exhausted within a few decades, and there appears to be slight likelihood of important new oil discoveries. The reserves of oil are not only meager, but also widely scattered and associated with complicated geological structures.

The main oil-bearing belt extends from western Hokkaido southward along the Japan Sea side of northern Honshu to about the 37° parallel, with the wells located both on the alluvial plains and in the low foothills bordering them on the east. As a consequence of more efficient oil recovery methods and of modest new discoveries, production has trended upward recently and in 1962 reached 860,000 kiloliters (1 kl. = 6.29 barrels). About 57 per cent is from western Tohoku, and another 41 per cent is from adjacent Niigata Prefecture (1961). Japan is largely dependent upon imported petroleum, the domestic production representing less than 2 per cent of the amount imported. Accordingly, crude and partly refined petroleum normally is Japan's foremost import.

Although natural-gas production increased remarkably in the 1950's, it still remains a very minor element of the total national energy resource. About 52 per cent of the output is used as fuel in industries and in households; the remainder is consumed as a raw material in the chemical industries. Three-fifths of the production is from marine sediments of recent age where the gas is dissolved in water. Of this type, the largest producers are in Niigata Prefecture and south Kanto, the latter located in close proximity to the Tokyo market and the former connected with Tokyo by pipeline (Fig. 4-4). The remaining gas production is associated with coal fields and oil fields, 90 per cent of the production in the former being used in generators, boilers, etc., at the mines.

Hydroelectric power

For her size, Japan has large potential waterpower resources, thanks to the abundant rainfall and the generally rugged terrain which makes for steep-gradient rivers. Total hydroelectric resources are estimated at about 22.5 million kilowatts, of which 8.8 million kilowatts, or 39 per cent, are developed. More than 85 per cent of the electricity consumed in Japan is of hydro origin. In 1960 about 23 per cent of the nation's energy consumption was provided by hydroelectricity, so that waterpower is a very important element in the energy resource of the country. Because hydroelectric output is affected by the seasonality of the precipitation (storage capacity being insufficient to avoid reduction during low-stream stages) and because there is insufficient hydroelectric generating capacity within economic reach of certain large power-consuming centers, thermoelectric plants have been constructed to supplement the hydroelectric facilities. There is a large concentration of the thermoelectric power installations along the margins of the Inland Sea and in northern Kyushu where hydroelectric potentials are small and coal from the north Kyushu and Ube fields is readily available. Other large concentrations of thermoelectric power are in the Kanto, Nagoya, and Osaka-Kobe industrial areas.

While it is true that only 39 per cent of the potential hydroelectric power has been developed, on the other hand it is the more readily accessible and more economical sites that have been brought into use, and future developments are bound to be more costly, for they will be located in more inaccessible areas and will require the construction of expensive reservoirs. In recent years as more hydroelectric power has been produced, the cost of electric power has increased. This situation, which will doubtless continue to prevail, has led to the forecast that an increasingly larger percentage of Japan's electric power in the future will be of thermal origin. Significantly, in 1960, 2.38 million kilowatts of new generating capacity were added, 45 per cent of which was thermal. It follows, therefore, that the ratio of hydroelectric to thermoelectric generating capacity has been changing, in favor of the latter—79:21 in 1955 and 51:49 five years later.

Considering the small volume of most Japanese rivers, it is not surprising that the generating plants are prevailingly small. These inconspicuous units are scattered over scores of separate drainage basins, from the extreme north to the extreme south. By far the greater share of the plants use little or no impounded water, but depend upon natural

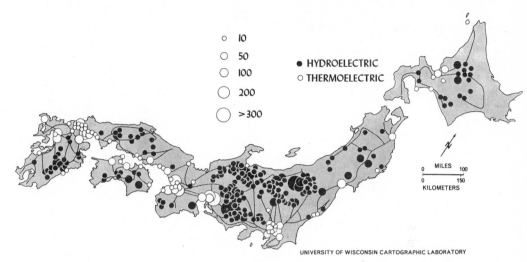

○ 10
○ 50
○ 100
○ 200
○ >300

● HYDROELECTRIC
○ THERMOELECTRIC

MILES 0 100
0 150 KILOMETERS

UNIVERSITY OF WISCONSIN CARTOGRAPHIC LABORATORY

Fig. 4-5.—Electric power plants and transmission lines, units in thousand kilowatts. After Isida.

stream flow, with a small diversion dam sending water into penstocks which feed directly to the turbines.

Hydroelectric generation, while widespread, is strongly concentrated in the broad central part of Japan where elevation is greatest, rivers are longest and have greatest volume, and precipitation is heavy (Fig. 4-5). This concentration includes the regional subdivisions of Tosan, Hokuriku, Tokai, Kanto, and southern Tohoku. Thus, fortunately, hydro-electric power is concentrated in those parts of the country where coal resources are meager, and also where it is readily accessible to the three great power-consuming industrial concentrations of the country.

METALLIC MINERALS

Iron ore

In 1961, Japan forged ahead of Britain to rank fourth among the nations in crude steel output; yet her domestic supplies of iron ore, as well as of coking coal, are completely inadequate to meet the requirements of her metallurgical industries. In 1962 the domestic output of iron ore was only 1.1 million tons, plus 1.4 million tons of iron-sand ore, while imports of iron ore amounted to some 15 million tons, obtained chiefly from Malaya, India, Philippines, and Canada.

Japan makes use of three kinds of domestic iron raw material in her steel industry: iron ore, iron sand, and iron pyrite cinder produced as a

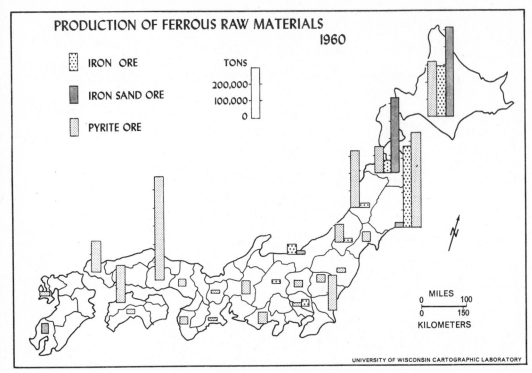

Fig. 4-6.—Distribution of domestic sources of ferrous materials, 1960.

by-product of sulfuric acid plants. The latter source is a relatively new one, but already it has become the most important domestic source of ferrous materials for the Japanese steel industry.

Iron ore reserves, which consist of scattered small deposits located almost exclusively in Hokkaido and northern Honshu, are not only limited in volume but are likewise of low grade, averaging about 35–36 per cent metallic ore. About fifty mines are in operation but only seven have annual outputs exceeding 50,000 tons. Largest of these, and producing about 28 per cent of the total, is the Kamaishi mine in eastern Iwate Prefecture near the coast, where the magnetite ore mined had a grade of 59 per cent after crushing and beneficiation (Fig. 4-6). Next comes the Gumma limonite mine with 9 per cent of the production, and an iron content of 21 per cent. These are followed in importance by four limonite mines, three in the volcanic region of southwestern Hokkaido, and one in Aomori Prefecture, each producing 4.5 to 6 per cent of the total.

It was not until World War II that the use of iron sand became common, but during the last ten years consumption has increased steadily so that it has become one of the most important sources of domestic ferrous materials. The amount of this product that can be used in blast-furnace sintering is limited, because of the high inclusion of titaniferous oxides. Over half the output is used in blast furnaces and a large part of the remainder in electric furnaces. Iron-sand deposits, characteristically located along sea coasts, are widely distributed and exist chiefly in the form of beach placers, or river placers in beach sands and terraces of Recent, or Pleistocene, age. There are eighty-five to one hundred working mines, with the chief producing districts located at Funkawan (along Volcano Bay) in Hokkaido, along the Pacific coast region of Aomori Prefecture in northernmost Honshu, and the Iioka district of Chiba Prefecture situated east of Tokyo. Together, these areas supply 80 per cent of the domestic production of iron sands. The iron mineral in the iron sand is magnetite which occurs in the form of grains, usually rounded and of fairly uniform size.

Other metallic minerals

Copper surpasses all other metallic minerals in value of production and in number of mines. Upwards of two hundred mines contribute to the copper output, although about half is from ten so-called large mines, most of them located in northeastern Honshu and in Shikoku. Ordinarily lead and zinc exist in close ore-body association, and both of them are not infrequently associated with copper. Mines are numerous and production is widely distributed, but one mine—Kamioka in Gifu Prefecture —normally accounts for nearly one-third of the country's zinc and one-sixth of its lead. Smelting plants for metallic ores characteristically have tidewater locations since they process imported as well as foreign ores.

Among the non-metallic minerals, iron sulfide is one of the more important, for Japan is a large producer of pyrite. After World War II, production increased remarkably in response to the increased demand

TABLE 4-6

Iron ore workable reserves

Reserves	Amount (metric tons)	Grade (% iron)
Proved	16,227,079	33.5
Probable	9,767,323	36.9
Possible	5,210,449	39.7
TOTAL AND AVERAGE	31,204,851	35.6
Iron sands	159,639,000	13.9

Source: Japanese Geological Survey, *Geology and Mineral Resources of Japan* (2d ed., 1960).

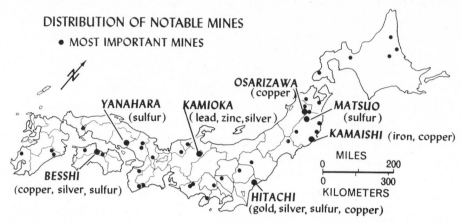

Fig. 4-7.—Distribution of the more important mines in Japan. The seven most important producers in 1961 are named. After map by Geological Survey of Japan.

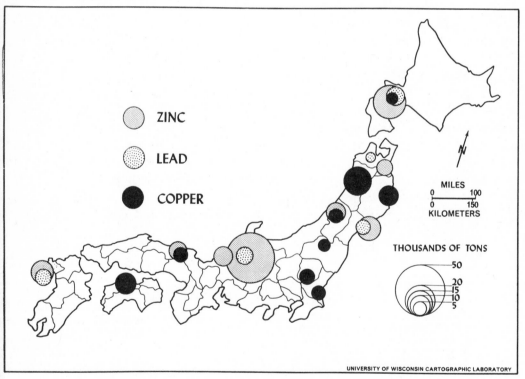

Fig. 4-8.—Production of copper, lead, and zinc in 1960, by prefectures.

for sulfuric acid, which is used chiefly in the manufacture of chemical fertilizers. Iron sulfide ores are used for the extraction of sulfurous acid gas, which is then processed into sulfuric acid and fertilizers. Iron sulfide cinder is used as a ferrous raw material. Two large mines, Matsuo in Iwate Prefecture in northeastern Honshu and Yanahara in Okayama Prefecture in southwestern Honshu, together produce about one-third of the total iron sulfide.

Native sulfur deposits, originating from recent volcanic activity, are scattered throughout the islands. Japan's supply of sulfur, including both pyrite and native deposits, is adequate for her needs.

PART II
CULTURAL FEATURES OF JAPAN

Occupied by a civilized people for nearly two millenniums, the land of Japan bears the indelible imprint of long and intensive tenure, not merely in its strictly man-made features, but in altered natural ones as well. "Japan is a country where the stones show human fingerprints; where the pressure of men on the earth has worn through to the iron rock" (*Fortune,* September, 1936). The crowding of many men on little land is a dominant feature: men and the works of men's hands are omnipresent. One of the most enduring adverse impressions Japan leaves with the traveler is the inconvenience and annoyance associated with a superabundance of people. Overpopulation seems not so much an objective fact of economics and living standard as it is a subjective feeling of suffocation and surfeit with mass humanity. Public conveyances and their terminals are jammed to overflowing; reservations on trains and planes are difficult to obtain and must be made far in advance; public parks and tourist spots are inundated by crowds; in the hearts of larger cities the rip and swirl of the surging human stream are a menace even where sidewalks exist, while the absence of sidewalks in congested side streets and alleys obliges the pedestrian to fight his way against unceasing vehicular traffic.

The present cultural scene in Japan is incongruous in many respects. Upon a basically feudal Oriental culture the elements of a modern, Occidental, machine-age civilization have been hastily and unconformably

101

superimposed. So recent has been Nippon's emergence from a state of Oriental feudalism that except in the downtown sections of the larger cities many of the ancient and indigenous features are still prevalent. Westernism is partly a façade, and a genuine amalgam is lacking. The West has assumed that Japan's new industrial and commercial wealth has meant a close approximation to Western standards and forms, but the West has been mistaken. The few broad boulevards of the metropolis feed directly into narrow, congested alleys flanked with indigenous combination shops and residences. Outside the city these same boulevards commonly terminate abruptly in narrow, pitted country roads where the traffic is a combination of motor vehicles and slow-moving man- and animal-powered conveyances. Even among the rural people the ways and standards of living are undergoing accelerated change, to be sure, but in some respects they are not as yet much different from those of their great-grandparents living under the feudal rule of the Tokugawa shoguns.

RACIAL AND CULTURAL ORIGINS

The exact origins of the Japanese people are still in dispute, but at least it is agreed that they are a blend of several racial strains. This mixture results in significant physical variations among individuals, but the representative male may be characterized as of short stature, averaging about 5 feet 4 inches,* with legs short in relation to trunk; he has little body hair, and his beard is light; his skin is tan, and his hair is black and usually straight, although occasionally wavy; his eyes are dark, and the epicanthic fold of the characteristic Mongoloid eye is common.

One element in the compounding of the present Japanese was the Neolithic peoples of Jomon culture who inhabited the archipelago for several millenniums prior to the beginning of the Christian Era. Jomon man was not Ainu, and he was only thinly blended with Mongoloid features, but on the other hand, these prehistoric people, like the present Japanese, were not of a single race.[1] Whatever the routes by which they entered Japan—by way of the southern islands, crossing from Korea, or entering from the continent via Sakhalin and Hokkaido—they were obliged to come by water, though in their small boats they were never for long out of sight of land.[2] Proto-Malayan and Polynesian strains were probably present, but what others is not known.

---

* The average stature of the Japanese male has increased from 156.7 cm. (60.7 in.) in 1898 to 164.3 cm. (63.7 in.) in 1958, an increase of 7.6 cm., or 3 in.

Still another element entering into the modern Japanese is the Ainu strain, a Caucasoid group with Mongoloid traces. An earlier theory that Jomon man was Ainu has recently lost favor, one current belief being that these Caucasoids arrived relatively late and entered Japan from the north at roughly the same time as the people of Yayoi culture, the "Japanese," moved up from the south.[3] Many place names in Hokkaido and northern Honshu are of Ainu origin, while farther south very few are. Historical records show that these people were entrenched in Tohoku for centuries during the Middle Ages and that they were able to slow up the northward-pushing Yamato or Japanese armies.

Originating from continental stimuli associated with the expansionist Hans, there began in the early part of the third century B.C. a large migration into Japan of Mongoloid peoples from the mainland. It was these newcomers who introduced the more advanced Yayoi culture, with its use of metals and its emphasis upon rice cultivation. Furthermore, this relatively late in-migration provided the Japanese people with their one common denominator, viz., their Mongoloid features, even though there exist important physical variations, in skull and facial structure and in skin color.[4]

Southwest Japan, around the borders of the Inland Sea, was the racial melting pot and the cultural hearth of the Japanese people. Yayoi culture probably centered first in North Kyushu, the closest point of contact between Japan and the mainland, but it spread eastward, first into the Inland-Sea and Chugoku areas, reaching their eastern parts, in Kinki, by around 200 B.C. It had spread to southern Tokai, in present Aichi and Shizuoka prefectures, a century later; to the Kanto area by the beginning of the Christian Era; and to central Tohoku by around 100 A.D.[5] Complete Mongolization of much of southwestern Japan resulted, with southeastern Kyushu and Tohoku resisting this imported culture longest. The consequence was a thorough transformation of the subtropical central and southwestern parts by the superposition of an agrarian pattern like that on the continent, in which rice was the staple crop and food. As continental pressures mounted, the Yayoi expansion gained momentum, with the result that by the third century A.D. the main cultural center had shifted from North Kyushu, the original focus, to the Yamato lowlands at the eastern end of the Inland Sea, in the vicinity of present-day Nara and Kyoto, where it remained for more than a millennium and a half.[6]

Thus, although the Japanese people are the product of a mixture of diverse strains, some of southern tropical origin and others from various parts of the colder mainland, the predominant physical characteristics doubtless have been provided by a late invasion of Mongoloid peoples, which apparently revolutionized the culture of the islands as well and laid the foundations for much that is considered to be typical of Japan even today. The establishment of paddy rice as the staple food crop was of prime importance, but in addition the Chinese-Korean influence is to be seen in numerous other elements of Japanese culture—written language, religion, land-survey system, government, and the arts and crafts. These borrowings from the mainland were not all contemporaneous with the early migrations. For many centuries after the large-scale inmigrations halted, China continued to be the font of culture for observers and students from Japan. It is understandable, then, why Japanese civilization bears a strong imprint of things Chinese. Nevertheless, it was not a slavish copying of China, for in being adapted to the new environment the older culture acquired something of an indigenous quality.

Much of what is known about Japan at the dawn of her written history comes from two ancient records, *Kojiki* ("Records of Ancient Matters") written in A.D. 712, and *Nihongi* ("Chronicles of Japan") of about A.D. 720.[7] Both are a mixture of mythology, Chinese legends, and factual history; and both were recorded in Chinese ideographs, an indication that Chinese influence was strong at the time. Japan at that period appears to have had a culture consisting of littoral and riverine communities engaged in agriculture, hunting, and fishing. The documents reveal also that by the close of the seventh century the armies from the south had driven the Ainu northeastward into Tohoku, beyond a line running roughly from latitude 39° on the west to 37° on the east. At nearly this same time the southern Yamato people moved their seat of government over 500 kilometers eastward from northern Kyushu to Nara, in the vicinity of the Biwa Depression, in order to be nearer the fighting frontier.

POPULATION IN THE PAST

Little is known with any degree of certainty concerning the size of the Japanese population during the formative centuries of the State. According to tradition, a "census" of A.D. 610 indicated a total population of nearly 5 million, and there are further estimates of 4.4 million by 990–1080 and of 9.75 million by 1185–1333.[8] If the latter figure has any validity, then already in medieval times, seven or eight centuries ago, an agricultural Japan supported 62 per cent as many people as highly urbanized, and somewhat larger, California in 1960. From the thirteenth through the sixteenth century population probably changed irregularly, but with a general upward trend.

Of particular importance, not only as related to population, but also with regard to almost all cultural features of modern Japan, are the two and one-half centuries of Tokugawa rule (1615–1867) which immediately preceded the restoration of the emperor in 1868 and the beginning of modern Japan. During this period society and government were organized along feudal lines with actual power in the hands of a Tokugawa military overlord, called a *shogun*, who ruled from Edo, now Tokyo. The emperor, largely without real power but ceremonially supreme, continued to reside in Kyoto. The country was divided into scores of small, isolated feudal fiefs, each of which was ruled over by a lesser military lord called a *daimyo*, who lived in feudal fashion in a moat-and-wall-encircled castle, and whose power was maintained by contingents of professional soldiers called *samurai*. A predominantly farming population paid taxes and owed allegiance to the daimyo and was under the control of his samurai. Among the virtues instilled in the commoners were filial piety, loyalty to one's superiors, frugality, industry, physical endurance, and willingness to sacrifice one's self for the State. Duties were emphasized; the concept of rights was unknown. These same Spartan virtues, derived from the feudal period, were inculcated in the modern Japanese as well, although there has been a notable weakening in their practice since World War II.

Feudal society under the Tokugawas was rigidly stratified, with the shogun and daimyos constituting the upper stratum, followed in descending order by the samurai, farmers and artisans, and finally the merchant class. This was also the period of government-enforced seclusion, so that foreign trade was negligible, the building of ocean-going ships was prohibited, and people were forbidden to leave the country. Railroads were lacking, industry was exclusively handicraft, large cities were few. Thus, Japan a century ago was a feudal country whose political, social, and economic organization resembled that of thirteenth-century France or England.

But decades before Commodore Perry, in 1853, obliged Japan to abandon her seclusion, the feudal system was crumbling. The increasing importance of trade and the merchant class spelled doom to Japan's self-sufficiency, which was based exclusively on her agriculture. In 1867 the Shogunate fell; shortly thereafter the Emperor Meiji was restored to power, and Japan was launched upon a period of rapid transformation and westernization.

The two and one-half centuries of Tokugawa rule saw important population changes. During the first half, as a strong central government throttled the devastating internal wars and instituted land-reclamation programs, population grew moderately, probably from less than 20 million at the end of the sixteenth century to about 26–27 million in the first quarter of the eighteenth. These added numbers of people were able to occupy newly reclaimed agricultural lands. But during the latter half of Tokugawa rule, as easily reclaimable good lands dwindled, growth of population slowed markedly. The actual, or presumed, stagnation in numbers at around 33 million for about a century and a half before the Meiji Restoration in 1868 is usually ascribed to two causes: the prevalence of infanticide and abortion, and the fact that population had increased to the maximum or saturaton point which the agricultural land could support at the technical levels of the closed feudalistic economy then prevailing.[9]

In the successive reports to the shogun concerning the number of commoners, the estimates provided are 26.5 million in 1726 and only 27.2 million in 1852. But although Japan as a whole showed a nearly static population during this century and a quarter, analysis of population change in different parts of the country reveals less consistent figures. There were sharp temporal as well as regional changes, with the regional ones sufficiently compensatory to provide what may have been a fictitious stability in population numbers for the country as a whole. The first half of the period from 1750 to 1850 was characterized by a small total decline in population and irregular regional and temporal changes. Regional variations in growth rates were partly a consequence of physical calamities which locally affected food production, but perhaps were also political and economic in origin, with decline in population chiefly evident in those areas most accessible to the tax-collecting officials of the shogun at Edo (northeast Honshu, Tokai, Kinki), and with growth more characteristic of those regions farther removed from Edo (the Japan Sea side, southwestern Honshu, Shikoku, Kyushu).[10]

During the second half of the century between 1750 and 1850 there were fewer regional variations in population growth, but there is evidence of a 6 per cent increase in the total number of commoners, suggesting that an expanding commercial economy during the late Tokugawa period had already initiated a cycle of slow population increase which was to accelerate in the subsequent modern period. The quickening of internal trade and population growth saw a commensurate expansion in the numbers and populations of cities and towns. It is impossible to measure the degree and distribution of urbanism as of this premodern period, but there were a number of places with populations of several hundred thousand. By far the largest number of Tokugawa cities were either feudal capitals of the daimyos or important traffic stations on the main highways.

Distribution of the 25–30 million people in Tokugawa Japan exhibited many of the present day patterns.[11] Density of population was already high, about 67 commoners per square kilometer (173 per square mile) of total land area in 1750, or 91 (235) if largely-unoccupied Hokkaido is excluded; and these figures must be multiplied up to ten times to get nutritional density, for the area under cultivation in mid-Tokugawa was probably not more than 3 million hectares, or one-half to three-fifths what it is today. Then as now the predominantly farming population was strongly concentrated on the small alluvial lowlands, so that a discontinuous, fragmented pattern was conspicuous. Already at this time settlement was densest in a belt along the Pacific side of the country extending from Edo and the Kanto Plain on the northeast, to Kinki at the eastern end of the Inland Sea, and thence along the borderlands of that waterway to northern Kyushu. Within this population belt the most conspicuous centers were those of Kinki, the lowlands around Ise Bay, and western Kanto. Most striking contrast between population distribution in feudal Japan of 1860 and that of a century later is that the former lacked the great increment of urban population which has accumulated chiefly during the last century. As at present, most of the mountain country was meagerly settled, and frontier Hokkaido had only about 65,000 inhabitants. The addition of some 60 million people since 1872 has not been primarily associated with an expansion of population into new frontier lands, but rather with a further piling up within the old ecumene, so that the distribution of population has become increasingly uneven.

POPULATION GROWTH AND REDISTRIBUTION IN MODERN JAPAN

Japan between 1868 and 1940 provides the perfect example of the classic assumption concerning the interaction of economic and demographic factors. As industrialization and urbanization proceeded, birth and death rates both declined. During this period economic expansion was paralleled by population growth, and while the latter was not rapid, never exceeding 1.5 per cent per year between 1872 and 1940, still over a period of seven decades a massive total increase resulted. For centuries the growth of Japan's population was a reflection of the country's economic evolution and was in step with it, so that it is only during the last decade that the biological increase has begun to lag well behind the soaring industrial expansion, a momentous fact having remarkable repercussions. Not the least of these are the rise in the standard of living and the greater mobility of the population.

Rates of population increase accelerated during the first five or six decades of the modern period, so that total population grew from about 35 million in 1872 to 55 million by 1920, the date of the first modern census. As of the earlier date, when over three-quarters of the employed population was engaged in primary industry, average density was already 91 per square kilometer (over 650 per sq. km. of cultivated land), high figures for a predominantly agricultural population. This addition of 20 million people between 1872 and 1920, in an already densely populated country, necessitated important regional redistributions. Between 1872 and 1885 the rate of growth was especially rapid in semi-frontier northern Honshu, and was lowest in Sanin and Kyushu in southwestern Japan. In the two decades between 1900 and 1920 when 11–12 million were added, the rate of increase was most rapid in frontier Hokkaido and in the six largest cities, with the prefectures containing the metropolises of Tokyo, Yokohama, Nagoya, Osaka, and Kobe absorbing about two-fifths of the total increment. In 1888 the six largest cities contained only 6 per cent of the nation's population; in 1918 this

TABLE 5-1

Index numbers of population and real national income per capita

Year	Population	Real income per capita
1934–36 average	100.0	100.0
1946	110.4	51.9
1947	113.8	53.3
1948	116.6	60.9
1949	119.1	68.6
1950	121.2	80.0
1951	123.2	86.7
1952	125.0	93.8
1953	126.7	98.1
1954	128.6	99.5
1955	130.1	109.5
1956	131.4	119.0
1957	132.4	126.2
1958	133.7	131.4
1959	134.9	151.0
1960	136.1	171.9

Source: Minoru Tachi, "Forecasting Manpower Resources," Institute of Population Problems, Tokyo, English Series, No. 55.

had risen to 11 per cent. The rapid growth in urban population was beginning. In 1888 only 12.9 per cent of the nation's people lived in communes of over 10,000 population, but this had risen to 20.6 per cent in 1903, 27.6 per cent in 1913, and 31.9 per cent in 1920.

The most recent period, 1920–1963, has witnessed far-reaching population changes. To the 55 million people in Japan in 1920 have been added an additional 41 million, the total effect of which is to further intensify the unevenness of distribution through swelling the population of a few great metropolitan areas. At present Japan is the world's sixth or seventh most populous country.

Trustworthy data on mortality rates are lacking for the period prior to 1920, but after that date, except for the war years, there is unmis-

TABLE 5-2

Total population and national income growth rates, 1880–1960

Year	Population (000 omitted)	Annual average increase rate (%)	Population density per sq. km.	Annual real national income growth rate (%)	Demographic elasticity
1880	38,166		101		
		0.6		4.3	7.2
1890	40,353		106		
		0.8		5.3	6.6
1900	43,785		115		
		1.1		2.9	2.6
1910	49,066		129		
		1.2		4.0	3.3
1920	55,391		146		
		1.3		5.1	3.9
1925	59,179		156		
		1.5		5.9	3.9
1930	63,872		168		
		1.4		3.8	2.7
1935	68,662		181		
		0.8		3.9	4.9
1940	71,400		188		
		0.2		. . .	. . .
1945	72,200		196		
		2.9		. . .	. . .
1950	83,200		226		
		1.4		8.7	6.2
1955	89,276		242		
		0.9		9.4	10.4
1960	93,419		253		

Source: Minoru Tachi, "Forecasting Manpower Resources," Institute of Population Problems, Tokyo, English Series, No. 55.

takable evidence of a consistent and rapid decline—from 24.4 per 1000 in 1920, to 14.6 in 1947, 7.6 in 1960, and 7.5 in 1962, the last figure representing one of the earth's lowest national rates. While the fall in death rates, no doubt, is associated with improvement in the general well-being and cultural attainments of the population, more specifically it is related to betterment of the public health services and to the success attained in greatly reducing the effects of some of the killing diseases. Birth rates likewise declined after about 1920 as the country's modernization progressed, falling off only gradually over the first two decades, but plummeting after the war, from a high of 34.3 per 1000 in 1947 to 17.2 in 1960 and 17.0 in 1962, the latter rate being well below that for the United States, and closely resembling the lowest to be found among the nations of Western Europe. Accordingly, the net reproduction rate in Japan dropped from 2.7 in 1920 to 0.99 in 1959. Between 1950 and

TABLE 5-3

Progress of urbanization

Year	Pop. in communes of over 10,000		Pop. in official shi, 1920–1960; in DID, 1960	
	Millions	% of total pop.	Millions	% of total pop.
1888	5.2	12.9	....	....
1893	6.7	16.0	....	....
1903	10.0	20.6	....	....
1913	15.2	27.6	....	....
1920	17.7	31.9	10.0	18.1
1930	25.9	40.6	15.4	24.1
1940	36.5	50.4	27.5	37.9
1950	44.9	54.0	31.2	37.6
1960	83.7	89.6	59.3	63.5
1960 (DID)	....	....	40.8	43.7

Note: Satisfactory data for showing the growth of urban population in Japan over the past half century are not available. The data by size of communes, or administrative units, while useful, provide only an approximate measure. Data by shi, or incorporated municipalities, generally with populations greater than 30,000, omit some unincorporated compact settlements with more than 30,000 inhabitants, and many more smaller compact settlements whose populations also are dependent on non-primary occupations. But at the same time the shi boundaries also include large areas whose populations are essentially rural. The 1960 data are scarcely comparable with those of previous decades because the Reorganization Act of 1953 so greatly increased the number and area of the shi while at the same time greatly reducing the number and area of the non-urban administrative units (machi and mura). Probably the most satisfactory measure of the present urban population is that for the Densely Inhabited Districts, the DID being defined as a contiguous agglomeration of population of 5000 or more inhabitants, in which the density is 4000 or more inhabitants per square kilometer. Unfortunately data for DID are not available for periods prior to 1960.

Sources: Compiled from *Japan Statistical Yearbook, 1961;* and Irene B. Taeuber, *The Population of Japan* (Princeton, 1958).

1957 births per 1000 women aged ten to forty-nine fell 41 per cent, probably the most rapid decline in birth rates ever experienced by any nation.

Synchronized declines in birth and death rates resulted in moderate rates of natural increase which have not changed drastically over the last four decades, except for a temporary rise associated with increased birth rates immediately following the war. Significantly, during the five-year period 1955–1960 the rate of natural increase dropped below 1 per cent per year, the first quinquennial period in which this has occurred, at least since 1920.

Clearly, what had been an evolutionary process down to about 1940, in the postwar period took on the attributes of a vital revolution. Stimulus for the sharp decline in births was the remarkable success in winning the fight against death, a success which brought a sustained rate of natural increase and a large population increment. Aided and

TABLE 5-4

Vital rates (per 1000 population)

Year	Birth rate	Death rate	Natural increase rate
1900–04	32.1	20.4	11.7
1905–09	32.2	21.0	11.2
1910–14	33.7	20.3	13.4
1915–19	32.5	22.6	9.9
1920–24	35.0	23.0	12.0
1925–29	34.0	19.8	14.3
1930–34	31.8	18.1	13.6
1935–39	29.2	17.4	11.9
1940–43	30.7	16.3	14.4
1947	34.3	14.6	19.7
1948	33.5	11.9	21.6
1949	33.0	11.6	21.4
1950	28.1	10.9	17.2
1951	25.3	9.9	15.4
1952	23.4	8.9	14.5
1953	21.5	8.9	12.6
1954	20.0	8.2	11.9
1955	19.4	7.8	11.6
1956	18.5	8.0	10.4
1957	17.2	8.3	9.0
1958	18.0	7.5	10.6
1959	17.6	7.5	10.1
1960	17.2	7.6	9.6
1961	16.8	7.4	9.5

Source: Minoru Tachi, "Forecasting Manpower Resources," Institute of Population Problems, Tokyo, English Series, No. 55.

abetted by the national government, the Japanese people responded to the dilemma by prompt and determined application of most of the then-known demographic checks on fertility—abortion, contraception, postponed marriage, and sterilization. The spur was not the threat of poverty, for economic growth had outdistanced that of population. Rather, given the situation of a sustained drop in mortality, the Japanese felt constrained to change their demographic behavior so as to better take advantage of the opportunities being provided by the developing economy.[12] The urge was the individual's desire for self-improvement. The phenomenal lowering of fertility rates was a major national accomplishment, the result of which was to make Japan completely unlike other Asian countries. It is only natural, therefore, that underdeveloped countries with serious population problems should turn to Japan for help and guidance.

The structural changes in population brought about by this recent vital revolution are already resulting in changes in the age distribution within the labor force and in the regional and occupational distributions of population as well. The Japanese economy is now confronted with a shortage of labor, especially of young males, and as this situation becomes aggravated it may even lead to propaganda favoring a return to higher fertility.[13] Although Japan's population is aging and its growth rate is declining, it is estimated that total numbers may reach 100 mil-

TABLE 5-5

Estimates of future population by age group

Year	Population (000 omitted)				% distribution		
	Total	0–14	15–64	65+	0–14	15–64	65+
1955	89,276	29,798	54,729	4,747	33.4	61.3	5.3
1960	93,419	28,028	60,512	5,360	29.9	64.4	5.7
1965	98,245	24,696	67,372	6,177	25.1	68.6	6.3
1970	102,216	23,197	71,920	7,099	22.7	70.4	7.0
1975	106,327	23,546	74,760	8,020	22.2	70.3	7.5
1980	109,688	23,713	76,975	9,001	21.6	70.2	8.2
1985	111,843	23,246	78,865	9,732	20.8	70.5	8.7
1990	112,943	21,745	80,342	10,856	19.3	71.1	9.6
1995	113,293	20,351	80,320	12,623	18.0	70.9	11.1
2000	113,053	19,687	78,956	14,409	17.4	69.8	12.8
2005	112,108	19,474	76,872	15,762	17.4	68.6	14.1
2010	110,247	19,141	73,943	17,162	17.4	67.1	15.6
2015	107,529	18,413	70,044	19,072	17.1	65.1	17.7

Source: Institute of Population Problems, *Selected Statistics Indicating the Demographic Situation in Japan* (Tokyo, 1961).

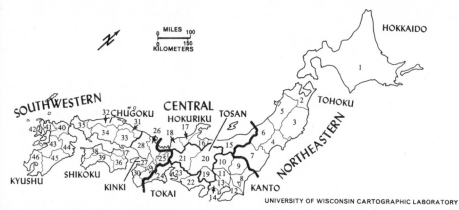

Fig. 5-1.—Index map of prefectures and main regional subdivisions of Japan.

1	Hokkaido	11	Saitama	20	Nagano	29	Nara	38	Ehime
2	Aomori	12	Chiba	21	Gifu	30	Wayakama	39	Kochi
3	Iwate	13	Tokyo	22	Shizuoka	31	Tottori	40	Fukuoka
4	Miyagi	14	Kanagawa	23	Aichi	32	Shimane	41	Saga
5	Akita	15	Niigata	24	Mie	33	Okayama	42	Nagasaki
6	Yamagata	16	Toyama	25	Shiga	34	Hiroshima	43	Kumamoto
7	Fukushima	17	Ishikawa	26	Kyoto	35	Yamaguchi	44	Oita
8	Ibaraki	18	Fukui	27	Osaka	36	Tokushima	45	Miyazaki
9	Tochigi	19	Yamanashi	28	Hyogo	37	Kagawa	46	Kagoshima
10	Gumma								

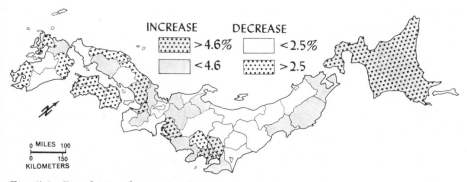

Fig. 5-2.—Population change for the 5-year period, 1955–1960, in per cent. National average: 4.6 per cent increase.

lion by about 1968 and 113 million by the last decade of this century, after which time they are expected to decline.

During the five-year period 1955–1960, when 4.1 million were added, twenty-six of the forty-six prefectures registered absolute losses in population and thirteen others showed modest gains below the country average of 4.6 per cent. A similar depopulation was experienced by approximately three-quarters of the 3500 minor civil divisions (*shi, machi, mura*) of the country. In only seven prefectures did the population gain at a rate exceeding the country average. Significantly, six of these seven embrace the metropolitan areas of the great industrial cities of Tokyo, Yokohama, Nagoya, Osaka, and Kobe (Fig. 5-2). Hokkaido was the only prefecture lying outside the great metropolitan areas that increased in population at a rate above the national average, and its rate of increase was the smallest of the seven. The six prefectures—Tokyo, Saitama, Kanagawa, Aichi, Osaka, and Hyogo—including or adjacent to the five industrial port cities noted above, accounted for over 95 per cent of the nation's quinquennial increase in population. Only Kyoto Prefecture, among those containing a great metropolis, increased at a rate below the country average. This reflects the fact that Kyoto city, although a metropolis, is neither a port nor an outstanding center of modern industry, so that its growth rate is more modest. Thus, in well over half the area of the country population is actually declining, and in most of the remainder it is increasing only slowly. Only in three restricted urban-industrial areas—Keihin, Hanshin, and Chukyo—is there a genuinely rapid multiplication of population. Internal migration is proceeding at a quickened pace stimulated by the disparity in incomes between rural and urban areas. Quite naturally, the industrial revolution in Japan was accompanied by an urban concentration of population, but only in recent years has this concentration, associated with an accelerated internal migration, reached a point where it is causing a depopulation of large parts of the country to a degree that the rural areas have begun to suffer from an acute shortage of young workers, while conversely the great metropolitan areas are experiencing an equally acute housing shortage and a near paralysis of their transportation systems.

General population losses in the period 1955–1960, characteristic of twenty-six prefectures, were widespread and conspicuous in all the principal islands except Hokkaido, although a well-defined regional pattern of distribution is difficult to discern. One of the most extensive areas of loss was in southwestern Japan, where of the sixteen prefectures in Chugoku, Kyushu, and Shikoku thirteen registered absolute losses and three showed gains below the country average. Another extensive and contiguous area of population loss lies in central and northern Honshu, and includes western and southern Tohoku, northern Kanto, and central and eastern Tosan. The whole pattern of loss distribution reflects the decline in number of people employed in rural occupations and, by contrast, the swelling number engaged in occupations associated with urban living—manufacturing, commerce, services.

The preceding generalization is further supported by an analysis of population change in *shi* (official incorporated cities) compared with *gun* (official counties, chiefly rural) during the period 1955–1960. As indicated earlier, twenty-six of the forty-six prefectures experienced total population losses, but a further refinement into the smaller shi and gun subdivisions shows that only three prefectures registered an absolute loss in total shi population, while forty experienced declines in total gun population. The six registering gains in gun or rural population were the same six urban prefectures noted earlier which recorded large total gains in population, all of which were influenced by great industrial cities. Here the gains in rural population appear to represent a suburban growth around the peripheries of the great cities. Only in semi-frontier Hokkaido is there an instance of a growth in gun population in a prefecture not directly influenced by the proximity of one of the metropolises.

It is of some geographical significance that the nation's center of population has been slowly shifting northeastward over the past several decades as the great Kanto metropolitan-industrial center has grown disproportionally. Thus, in 1920, the center was located on the east side of Lake Biwa in Shiga Prefecture, while in 1960 it was in Gumma Prefecture in Central Japan.

REGIONAL VARIATIONS IN POPULATION CHARACTERISTICS

Vital rates.—Although Japan represents a relatively homogeneous society, there have been in the past and there continue to be significant regional variations in fertility, mortality, and natural increase, as well as in certain other important population characteristics. As of 1960 the average national birth rate was 17.2 per 1000 (17.0 in 1962), with the highest for any prefecture being 20.9 (Aomori in northernmost Honshu), and the lowest 14.9 (Kyoto and Kochi in southwestern Japan). Higher-than-average birth rates are characteristic of the two extremities of the country, Hokkaido, Tohoku, and northern Kanto in the north, and southern and western Kyushu in the extreme south (Fig. 5-3). These peripheral areas are designated by Ogasawara as belonging to the frontier agricultural zone, regions somewhat isolated by distance from the nation's economic and cultural centrum, that have lagged behind in general culture, and are less urban in character. In most of these higher-fertility areas a relatively large percentage of the population is engaged in agriculture, and a low percentage in secondary industry. Below-aver-

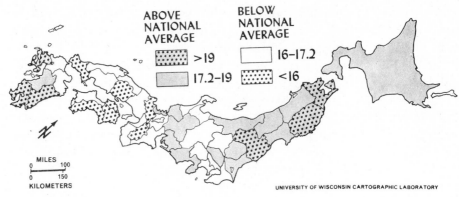

Fig. 5-3.—Birth rate, 1960. National average: 17.2 per thousand; high: 20.9; low: 14.9.

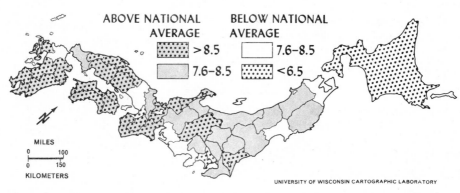

Fig. 5-4.—Death rate, 1960. National average: 7.6 per thousand; high: 9.6; low: 5.2.

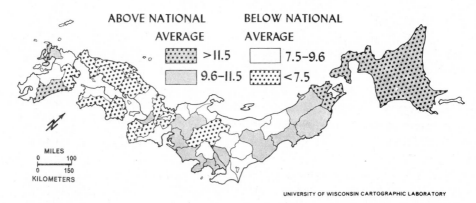

Fig. 5-5.—Natural increase in population, 1960. National average: 9.6 per thousand; high: 13.4; low: 5.3.

age fertility rates are generally characteristic of southwestern Japan lying west of the Biwa Depression (excluding southern Kyushu), including Kinki, Chugoku, northern Kyushu, and Shikoku, districts where settlement began early and rural crowding is great. Low birth rates are likewise characteristic of the more urbanized prefectures, especially those containing, and adjacent to, the great metropolises where per capita incomes are relatively high.

The degree of fall of the standardized birth rate over the five-year period 1955–1960 has been calculated for each prefecture.[14] In the country as a whole the decline of the standardized birth rate was about 15 per cent. Seventeen prefectures registered declines smaller than the national average (two of them, Osaka and Tokyo, even registering slight increases), while twenty-nine showed declines greater than the national average. Generally speaking, the degree of decline in birth rates was large in those districts where the birth rate was high in the past (Tohoku, Hokkaido, northern Kanto, Shikoku, Kyushu), and it was small where the birth rate was low in the past, especially in districts with large cities.

Throughout the country death rates have declined markedly and at present are low everywhere, but regional differences do exist, with the prefectural extremes being 9.6 per 1000 (Kochi) and 5.2 (Tokyo). Of the ten prefectures with mortality rates below the 1960 country average of 7.6 (7.5 in 1962), five contain great industrial metropolises, while a sixth includes northern Kyushu with the new industrial metropolis of Kitakyushu (Fig. 5-4). Of the five prefectures with the lowest death rates (below 7.0), all but Hokkaido are strongly urban and industrial.[15] This association of low death rates with urbanization probably reflects not only the better medical care and general living conditions in the larger cities, but also the effects of selective migration, with younger people leaving the farms for employment in the metropolitan areas. The standardized death rate over the 1955–1960 period showed a fall of only a little less than 3 per cent for the country as a whole, compared with 15 per cent for the birth rate. Regional differences in death rate decline were small.

Rates of natural increase (birth rate minus death rate) have never been high in modern Japan, and recently they have been even less than 1 per cent per year. In 1960 the country average was 9.6 per 1000 with a prefectural high of 13.4 (Aomori, in northernmost Honshu) and a low of 5.3 (Kochi, in southern Shikoku). Since regional variations in birth rates are larger than those of death rates, it is to be expected that regional variations in natural increase will follow, and be more consistent with, birth rates.[16] Thus, rates of natural increase in excess of the national

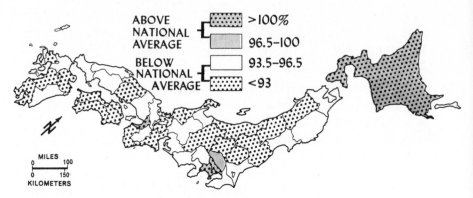

ABOVE NATIONAL AVERAGE
- [:::::] >100%
- [grey] 96.5–100

BELOW NATIONAL AVERAGE
- [white] 93.5–96.5
- [:::::] <93

MILES
0 ———— 100
0 ———— 150
KILOMETERS

Fig. 5-6.—Sex ratio [male/female], 1960. National average: 96.5.

average prevail in both the northern and southern extremities of the country, the so-called frontier zones, which are likewise regions of relatively high birth rates. The lowest rates of natural increase are strongly concentrated in the southwest (Kinki, Chugoku, and Shikoku), crowded regions of long occupance (Fig. 5-5). Admittedly there are a number of exceptions which do not fit the above-described pattern.

If the prefectures are further subdivided into the smaller administrative units of *shi* (city), *machi* (town), and *mura* (rural village), the relationship between natural increase and rural-urban living becomes still more obvious. Thus, for the country as a whole, and in all prefectures, the rates of natural increase are highest in the mura or rural villages and lowest in the shi or cities. Since this is consistently not the case in terms of total population change, the situation bespeaks a large rural-urban migration.

Sex ratios.—Unlike most of eastern and southern Asia, Japan has an excess of females, the male-female ratio being 96.5 to 100 in 1960. In only four prefectures do males predominate, and three of these (Tokyo, Kanagawa, Osaka) are strongly urbanized industrial prefectures, while the fourth is semifrontier Hokkaido (Fig. 5-6). In all of these a selective young-male in-migration has served to increase the number of men in proportion to women. In a large majority of the prefectures a selective out-migration of males has operated to reduce the proportion of men below the country average. In twenty-one prefectures the male-female ratio was below 93.5 or markedly below the country average. Seven of these male-scarce subdivisions are located in western Tohoku, northern Kanto, and Tosan; two, in Hokuriku; and twelve, in southwestern Japan (four in each of Shikoku and Kyushu). In all but one of these twenty-one prefectures, the real income per capita is far

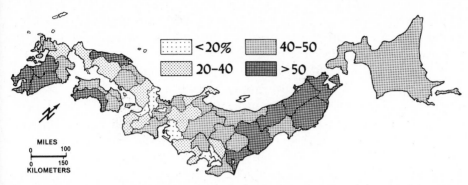

*Fig. 5-7.—*Percentage of Japan's employed population 15 years old and over engaged in the primary industries, 1960. National average: 34 per cent.

below the country average (under 65 per cent), while all of them showed varying potentials for out-migration.[17]

*Occupational status.—*Of Japan's employed population in 1960, about 34 per cent were engaged in primary industry, 28 per cent in secondary industry, and 38 per cent in tertiary industry. About 33 per cent were engaged in agriculture and forestry alone. This composition reflects a relatively advanced stage of economic development and is very unlike the occupational status of population in the other countries of eastern and southern Asia where agriculture supports a much larger proportion

TABLE 5-6

Change in population composition by industrial sectors

	Number (in millions) and per cent employed			
Year	Total	Primary industry	Secondary industry	Tertiary industry
1880*	19.5	16.1 82%	1.1 6%	2.4 12%
1890*	22.6	17.2 76	2.0 9	3.4 15
1900*	24.8	17.3 70	2.9 12	4.5 18
1910*	26.2	16.5 63	3.9 15	5.8 22
1920*	27.0	14.4 54	5.6 21	6.9 26
1930*	29.3	14.5 49	6.0 20	8.9 30
1940*	32.2	14.2 44	8.4 26	9.6 30
1947†	33.3	17.8 53	7.4 22	8.1 24
1950‡	35.6	17.2 48	7.8 22	10.6 30
1955§	39.3	16.1 41	9.2 24	13.9 35
1960§	43.7	14.9 34	12.2 28	16.6 38

* For total employed.   ‡ Employed 14 years old and over.
† Employed 10 years old and over.   § Employed 15 years old and over.
Source: Minoru Tachi, "Forecasting Manpower Resources," Institute of Population Problems, Tokyo, English Series, No. 55.

of the inhabitants and the other economic sectors a much smaller one. At the same time, Japan is unlike the United States and the more industrialized countries of Western Europe, where a larger proportion of employed persons is engaged in the secondary and tertiary industries. In Western countries agriculture commonly supported 65–70 per cent of the population before the Industrial Revolution and 20–25 per cent after it, while by contrast in Japan the comparable figures are 75–80 per cent and 45 per cent (33 per cent in 1960). Over the past three-quarters of a century as Japan modernized its economy, there has been a steady decrease in that part of the population engaged in primary industry, chiefly agriculture, and an increase in the proportion employed in secondary and tertiary industries (see Table 5-8). Between 1955 and 1960 employed persons in agriculture decreased by 11 per cent, while those in manufacturing increased by 38 per cent.

Of the forty-six prefectures, thirty-six show an employment status (1960) in which the percentage in primary industry is greater than the country average (34 per cent). The other ten prefectures are the more industrialized-urbanized ones, six of them containing metropolises of a million or more population. Employment in primary industries is unusually large (over 50 per cent) in Tohoku, northeastern Kanto, southern Shikoku, and central and southern Kyushu. This distribution pattern is approximately reversed for employment in secondary and tertiary industry (Figs. 5-7, 5-8, 5-9, 5-10, and 5-11).

Internal migration.—In an earlier section analyzing population change for the quinquennium 1955–1960, evidence given indicated a widespread out-migration from rural Japan and a strong in-migration to

TABLE 5-7

Employed persons 15 years old and over by industries, 1960
(in thousands of persons)

Primary industry	14,879	34.0%
Agriculture and forestry	(13,670)	(31.3)
Fishing	(676)	
Mining	(533)	
Secondary industry	12,198	28.0
Manufacturing	(9,495)	(21.7)
Construction	(2,703)	
Tertiary industry	16,604	38.0
TOTAL	43,681	

Source: Japan Statistical Yearbook, 1961.

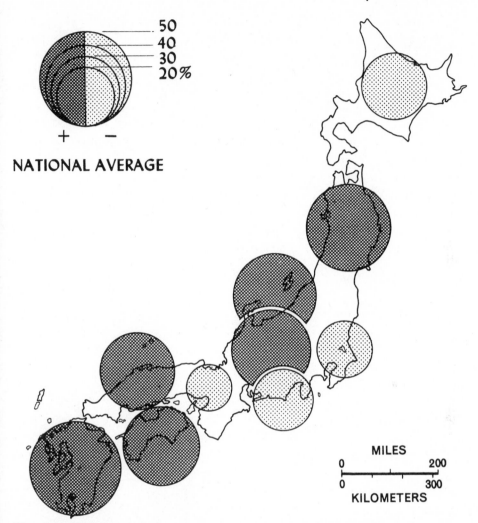

NATIONAL AVERAGE

Fig. 5-8.—Percentage of the labor force employed in agriculture, 1960, by large regional subdivisions. National average: 30.2 per cent.

TABLE 5-8

Employment of workers in Japan in different economies

Year	Agriculture	Commerce	Manufacturing
1872	78%	7%	4%
1920	53	12	20
1950	45	13	16
1960	31	17	28

Sources: International Geographical Union, Science Council of Japan, Regional Conference in Japan, *Regional Geography of Japan*, No. 7, *Geography of Japan;* and *Japan Statistical Yearbook, 1961.*

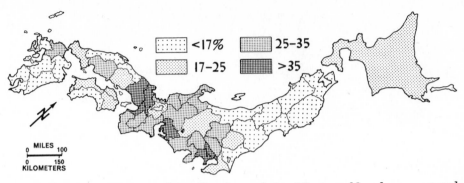

Fig. 5-9.—Percentage of Japan's employed population 15 years old and over engaged in the secondary industries, 1960. National average: 27.9 per cent.

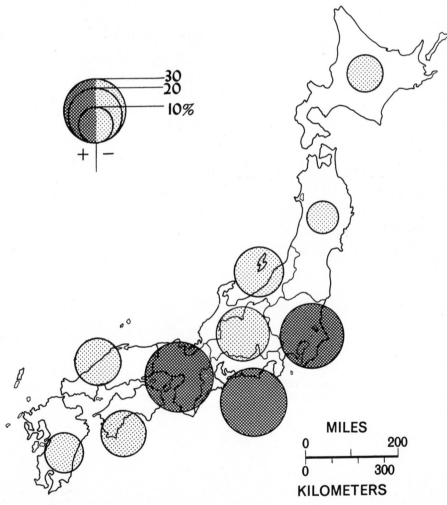

Fig. 5-10.—Percentage of the labor force employed in manufacturing, 1960, by large regional subdivisions. National average: 21.7 per cent.

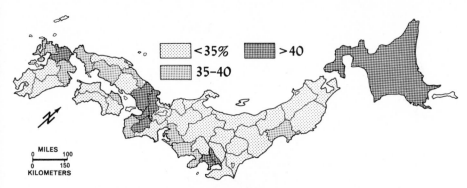

Fig. 5-11.—Percentage of Japan's employed population 15 years old and over engaged in the tertiary industries, 1960. National average: 38.0 per cent.

cities, particularly the larger ones. Altogether, thirty-five prefectures showed absolute losses in gun (rural) population and gains in shi (urban) population, with much the greater part of the gain concentrated in the Tokyo, Osaka, and Nagoya areas. This recent and large rural-urban migration in Japan is of two types: local, from rural villages to nearby cities, often within the same prefecture; and long-distance, chiefly from the more rural prefectures to the great metropolitan prefectures of Tokyo-Kanagawa-Saitama in Kanto, Aichi in Tokai, and Osaka-Hyogo in Kinki. The intraprefectural migrations show about equal numbers of men and women. In the longer interprefectural migrations about 60 per cent of the migrants are males, most often under thirty years of age.

Japan's population in recent years has become much more mobile than formerly, as indicated by change in place of residence. During the year

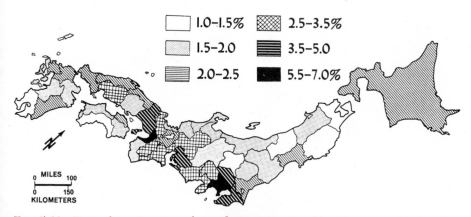

Fig. 5-12.—Rate of in-migration of population one year old and over (male), October 1, 1959 to September 30, 1960. After map by Yoichi Okazaki.

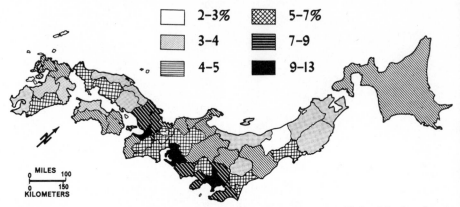

Fig. 5-13.—Rate of in-migration of population 15 to 29 years old (male), October 1, 1959 to September 30, 1960. After map by Yoichi Okazaki.

October 1, 1959 to September 30, 1960, 8 per cent of the total population in ages one year and over, or about 7.4 million persons, changed their place of residence. Some 2.6 millions, or 2.8 per cent of the population, removed their residence beyond the prefectural boundary.[18] Young in-migrants aged fifteen to twenty-nine from other prefectures comprised nearly two-thirds of the total in-migrants age one year and over, while in prefectures with large cities the proportion in this age group reached nearly three-quarters, indicating that it was preponderantly a youthful labor-force migration. The proportion of in-migrants (male) was very great in prefectures with large cities, reaching 6 per cent in Tokyo pre-

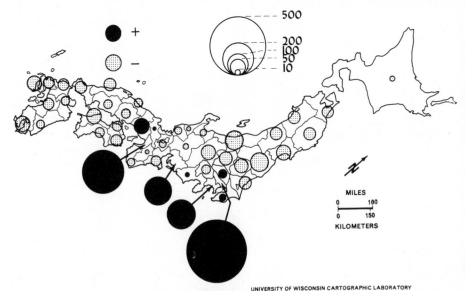

UNIVERSITY OF WISCONSIN CARTOGRAPHIC LABORATORY

Fig. 5-14.—Net migration volume of male labor force, 1955–1960. In thousands.

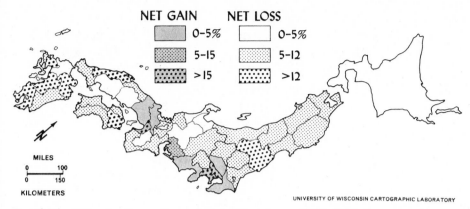

NET GAIN   NET LOSS

▨ 0-5%	☐ 0-5%	
▨ 5-15	⬚ 5-12	
▨ >15	⬚ >12	

MILES
0   100
0   150
KILOMETERS

UNIVERSITY OF WISCONSIN CARTOGRAPHIC LABORATORY

Fig. 5-15.—Net migration volume of male labor force, 1955–1960. In per cent.

fecture, over 5.5 per cent in Kanagawa and Osaka, and over 3.5 per cent in Saitama and Chiba in Kanto, Aichi, and Hyogo.[19] The rate of in-migration from other prefectures of employed persons fifteen years old and over was about 4 per cent for males and 3 per cent for females for the country as a whole, but the corresponding figures were far greater for the prefectures with large cities—in the case of Tokyo 8 per cent for males and 9 per cent for females. Of the employed persons who were in-migrants from other prefectures (1959–1960), agriculture employed only a small percentage. Manufacturing provided the greatest source of employment, followed by services, wholesale-retail trade, and construction. Construction employed chiefly males, and services largely females, while there is less disparity between the sexes in manufacturing

TABLE 5-9

Rate of in-migration of employed persons (15 years old and over) by industry and percentage by industry of employed non-migrants and in-migrants (Oct. 1, 1959 to Sept. 30, 1960)

Major industry group	Rate of in-migration from other prefectures		Percentage by industry			
			Non-migrants		In-migrants from other prefectures	
	Male	Female	Male	Female	Male	Female
TOTAL	4.2%	2.9%	100.0%	100.0%	100.0%	100.0%
Agriculture	0.3	0.4	25.1	44.7	1.8	6.0
Construction	9.7	4.1	8.2	1.6	21.2	2.4
Manufacturing	5.6	6.0	23.3	17.2	32.6	38.1
Wholesale-retail	4.2	3.6	14.7	16.4	15.1	21.5
Services	3.6	5.6	9.8	13.4	8.7	28.5

Sources: Masao Ueda, in *Annual Reports of the Institute of Population Problems.* No. 7 (1962).

and wholesale-retail (see Table 5-9).[20] Of the total male-employed involved in interregional migrations during this one year the strongest net gains (in-migrants minus out-migrants) were registered by Kanto and Kinki, with smaller ones by Chubu and Hokkaido. The other four regions experienced net losses, very large ones by Kyushu and Tohoku, and smaller ones by Shikoku and Chugoku. Thus, Japan southwest from Kinki is the country's major region of net loss by migration, with Tohoku next in order (Figs. 5-12, 5-13, 5-14, 5-15, and 5-16).

Scale of migration within Japan has increased during the decade of the 1950's, for the volume of the male labor force migration (15 years

T A B L E 5-10

Volume of male labor force and its net migration

Prefecture	Volume of labor force			Net migration volume		Net migration rate	
	1950	1955	1960	1950–55	1955–60	1950–55	1955–60
ALL JAPAN	22,310,219	24,435,282	27,081,800	....	....	......	......
Hokkaido	1,153,778	1,323,582	1,476,346	43,496	− 9,812	3.8	− 0.7
Aomori	337,404	359,874	372,671	−13,814	−31,671	− 4.1	− 8.8
Iwate	361,942	377,270	390,441	−22,432	−33,122	− 6.2	− 8.8
Miyagi	441,509	443,557	455,060	−40,851	−45,664	− 9.3	−10.3
Akita	348,914	356,648	367,623	−26,376	−28,339	− 7.6	− 7.9
Yamagata	362,502	359,144	357,021	−42,265	−34,101	−11.7	− 9.5
Fukushima	532,126	534,184	528,060	−56,160	−64,688	−10.6	−12.1
Ibaraki	534,001	545,241	581,775	−33,121	−35,541	− 6.2	− 6.5
Tochigi	393,971	403,480	405,485	−32,548	−43,496	− 8.3	−10.8
Gumma	420,916	428,405	441,733	−36,370	−38,881	− 8.6	− 9.1
Saitama	566,998	612,647	705,056	− 4,930	18,845	− 0.9	3.1
Chiba	552,793	584,775	661,842	−16,392	11,373	− 3.0	1.9
Tokyo	1,776,353	2,436,309	3,198,650	466,530	530,949	26.3	21.8
Kanagawa	692,482	840,452	1,078,333	81,875	135,742	11.8	16.2
Niigata	652,078	654,067	680,521	−64,524	−53,181	− 9.9	− 8.1
Toyama	271,543	276,715	294,623	−19,299	−12,788	− 7.1	− 4.6
Ishikawa	255,674	258,140	280,387	−13,974	− 6,294	− 5.5	− 2.4
Fukui	206,696	206,283	214,152	−13,974	−13,823	− 6.8	− 6.7
Yamanashi	210,880	214,010	217,080	−21,783	−20,720	−10.3	− 9.7
Nagano	553,496	560,230	574,101	−54,107	−42,601	− 9.8	− 7.6
Gifu	428,095	441,545	479,999	−19,286	−10,499	− 4.5	− 2.4
Shizuoka	653,357	728,059	830,659	8,078	6,252	1.2	0.9
Aichi	907,811	1,059,113	1,313,890	59,604	131,352	6.6	12.4
Mie	391,469	409,093	427,294	−19,302	−23,058	− 4.9	− 5.6

or over) was 34 per cent greater during the quinquennium 1955–1960 than during the preceding five years.[21] This reflects the increased employment opportunities in the cities as Japan's industrial capacity and productivity soared. Distribution of the net migration rate for the quinquennium 1955–1960 shows that only nine prefectures, all located within the country's three principal industrial nodes, four in Kanto, two in Tokai, and three in Kinki, had a plus net migration, while thirty-seven had a negative net migration (Table 5-10). When the forty-six prefectures are combined into thirteen regions, only South Kanto, Tokai, and Kinki, the three principal industrial areas, had a net in-migration; the ten others showed an excess of out-migration (Table 5-11). South Kanto, by all odds, is the most important region of net in-migration, followed by Kinki and Tokai. Most important of the net out-migration re-

TABLE 5-10—Continued

Prefecture	Volume of labor force			Net migration volume		Net migration rate	
	1950	1955	1960	1950–55	1955–60	1950–55	1955–60
Shiga	232,228	233,934	238,788	−14,258	−14,133	− 6.1	− 6.0
Kyoto	489,047	534,762	593,083	5,969	1,597	1.2	0.3
Osaka	1,081,401	1,353,722	1,768,445	177,817	268,207	16.4	19.8
Hyogo	910,473	1,015,159	1,181,219	39,323	49,390	4.3	4.9
Nara	207,428	216,548	232,427	− 4,237	− 4,484	− 2.0	− 2.1
Wakayama	266,462	283,190	289,272	− 2,853	−16,166	− 1.1	− 5.7
Tottori	158,261	164,157	166,092	− 8,074	−13,964	− 5.1	− 8.5
Shimane	245,626	255,715	245,957	− 8,998	−34,745	− 3.7	−13.6
Okayama	445,582	463,107	479,797	−22,789	−22,836	− 5.1	− 4.9
Hiroshima	569,053	592,504	635,893	−11,914	−20,184	− 2.1	− 3.8
Yamaguchi	420,375	442,479	448,498	−11,826	−31,578	− 2.8	− 7.1
Tokushima	226,411	228,924	222,229	−16,992	−29,230	− 7.5	−12.8
Kagawa	245,137	252,935	254,337	−15,749	−24,208	− 6.4	− 9.6
Ehime	398,506	402,414	399,023	−47,339	−47,748	−11.9	−11.9
Kochi	242,546	244,043	242,524	−10,734	−22,537	− 4.4	− 9.2
Fukuoka	947,247	1,016,333	1,081,463	−10,035	−29,881	− 1.1	− 2.9
Saga	241,950	245,665	239,091	−14,650	−33,700	− 6.1	−13.7
Nagasaki	437,655	449,188	454,354	−24,460	−40,481	− 5.6	− 9.0
Kumamoto	468,086	485,132	467,681	−25,624	−59,535	− 5.5	−12.3
Oita	322,930	330,175	321,985	−19,649	−40,158	− 6.1	−12.2
Miyazaki	286,316	298,332	305,124	−14,577	−32,564	− 5.1	−10.9
Kagoshima	460,711	514,041	481,716	−46,425	−87,296	−10.1	−17.0

Source: Yoichi Okazaki, in *Annual Reports of the Institute of Population Problems,* No. 7 (1962).

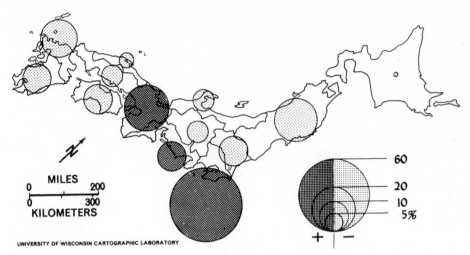

Fig. 5-16.—Net migration volume of male labor force, 1955–1960, by principal districts. In per cent of total net migration of 1,095,866 persons.

gions are southwest Japan (Kyushu, Shikoku, Sanyo, Sanin), Tohoku, North Kanto, Hokuriku, and Tosan.

Migration is motivated by various factors, but there is strong evidence that this phenomenon in Japan is fundamentally economic in its stimulation and that the migratory flows are led by movements of the labor force from areas of low real income to those of high real income.[22] Thus, Tsubouchi has studied migration in terms of what he calls population

TABLE 5-11

Net migration of male labor force, 1955–1960

District	Volume	Rate in per cent
ALL JAPAN	±1,095,866	±100
Hokkaido	− 9,812	− 0.7
Tohoku	− 237,585	− 21.7
North Kanto	− 117,918	− 10.8
South Kanto	+ 696,909	+ 63.6
Hokuriku	− 86,086	− 7.9
Tosan	− 73,820	− 6.7
Tokai	+ 114,546	+ 10.4
Kinki	+ 284,411	+ 26.0
Sanin	− 48,709	− 4.4
Sanyo	− 74,598	− 6.8
Shikoku	− 123,723	− 11.3
North Kyushu	− 203,755	− 18.6
South Kyushu	− 119,860	− 10.9

Source: Yoichi Okazaki, in *Annual Reports of the Institute of Population Problems*, No. 7 (1962).

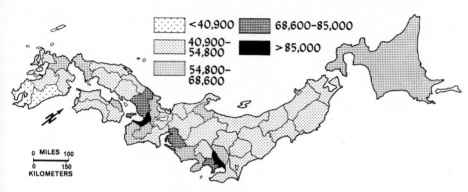

Fig. 5-17.—Real income per capita, 1957. In yen. After map by M. Tachi and M. Oyama.

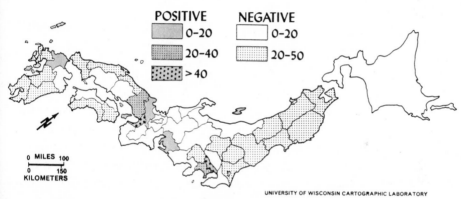

UNIVERSITY OF WISCONSIN CARTOGRAPHIC LABORATORY

Fig. 5-18.—Economic potential of internal population in-migration, 1955–1959 average, by prefectures. Positive indicates high per capita real income. Data from Minoru Tachi.

pressure, which is not just arithmetic density, but instead is population numbers in terms of the supporting capacity as indicated by regional income, which is expressed by the following formula:

$$\text{Population pressure} = \frac{\text{Total population of an area}}{\text{Total real income of an area}}$$

He finds population pressure to be low in a relatively few great metropolitan prefectures—within Kanto, Kinki, and Tokai. By contrast it is high in the non-industrial regions of Tohoku, Tosan, Sanin, Shikoku, and southern Kyushu. When his map showing distribution of population pressure by prefectures is compared with number and rate of population increase for 1950–1955, there is found to be a strong inverse correlation, i.e., the stronger the population pressure, the smaller the increase in population. A similar correlation is found between population pressure and (1) in-migration of population by prefectures and (2) excess of in-migrants over out-migrants (Figs. 5-17 and 5-18).[23]

On the assumption that population migration operates as a movement to balance the regional disparities of income distribution, Minoru Tachi has computed the hypothetical population of each prefecture if population were distributed according to income. By comparing this hypothetical population with the actual population he obtains what he calls the "potential of population migration" for each prefecture. He finds that a remarkable correlation exists (1) between the potential of population migration and the net migration rate (rate of net population increase minus the natural increase rate) and (2) between potential of population migration and the specific net migration rate for the male labor force.[24]

As a consequence of the selective migration taking place, in which the emphasis is upon young males, there has resulted a disproportionate marked aging of males in the primary industries, while the age structure of the male population engaged in manufacturing has been somewhat rejuvenated. Prewar out-migration in agricultural areas consisted chiefly of the younger sons of farmers who had no claim to the family farm. But since World War II, as the desire to rise in the social and economic scales has intensified and the migration rate has accelerated, the emigration stream from the rural areas has also included numerous heads of families and the favored eldest sons or heirs. This draining of young people, with the concomitant export of wealth, away from peripheral areas toward a few great urban industrial centers creates many problems, both economic and social. Conceivably the process may be economically sound, but how it should be viewed socially is still another question.

TABLE 5-12

Regional contrasts in income levels, 1960
(national average = 100)

High income metropolitan prefectures		Low income prefectures	
Tokyo	184.3	Aomori	70.3
Osaka	159.5	Iwate	65.0
Kanagawa	139.2	Miyazaki	67.5
Aichi	134.0	Kumamoto	65.9
Hyogo	121.5	Kagoshima	55.1

Source: *Oriental Economist*, *31*, No. 630 (1963).

Rural and urban residence.—One aspect of Japan's population which remains somewhat uncertain, because of the lack of definition of what constitutes an urban place, is the relative proportions of the total population residing in urban and in rural settlements. As of 1960, nearly 63 per cent of the population lived in shi, or within the boundaries of official cities, while about 37 per cent lived within the machi and mura of gun, and it is the latter proportion which is commonly classed as rural. Actually, however, this present division between shi and gun weights the urban population too heavily, for the boundaries of most shi include extensive non-urban areas, although admittedly the gun also contain some urban places.

A much closer approximation to the true proportions of urban and rural populations is provided by the new data for Densely Inhabited Districts (DID), as delimited by the 1960 Population Census, a DID being loosely defined as a compact settlement with 5000 or more inhabitants. Some 970 DID's were recognized, their total population in 1960 amounting to 40.8 million, or 43.7 per cent of the total population.[25] These figures are to be compared with 59.3 million contained within shi, or 63 per cent of the total. If the figure of 43.7 may be accepted as approximately correct for the proportion of population in Japan which is urban, then it may be noted that this is well below comparable figures for most of the industrial countries bordering the North Atlantic Basin, but, on the other hand, well above those for countries in eastern and southern Asia. Here again, as in so many other ways, Japan stands out as the exception in Asia.

Distribution of the urban (DID) population is most uneven. It is highly concentrated in the Manufactural Belt extending from the Kanto region at the Belt's northeastern extremity, along the Pacific side of Honshu and the Inland Sea borderlands to northern Kyushu at the southwestern extremity (Fig. 5-19). Nearly 54 per cent of the total urban population is included within the six prefectures (Tokyo, Kanagawa, Aichi, Osaka, Hyogo, Kyoto) containing the six metropolises of over a million inhabitants, and this figure rises to nearly 60 per cent if industrial Fukuoka Prefecture in northern Kyushu is added. Nearly one-third of the urban total is included within two prefectures, Tokyo and Osaka.

Seen in terms of the proportion of the total population of each prefecture contained within urban places (DID's), the figures range widely from highs of 92 per cent in Tokyo, and 81.4 per cent in Osaka prefecture, to a low of 15.3 per cent in Shimane Prefecture on the Japan Sea side of extreme southwestern Honshu. Twenty-one prefectures have urban percentages of 25 per cent or lower. These are concentrated in

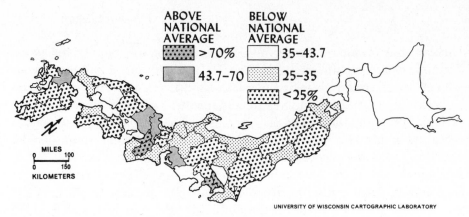

Fig. 5-19.—Distribution of urban population. Percentage of the total population in densely inhabited districts (DID's), 1960. National average: 43.7 per cent.

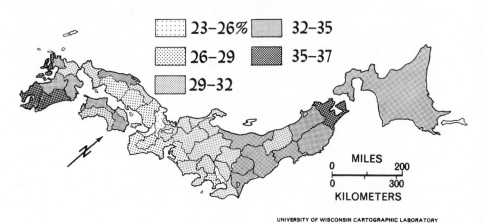

Fig. 5-20.—Percentage of the total population which is children (0–14 years), 1960.

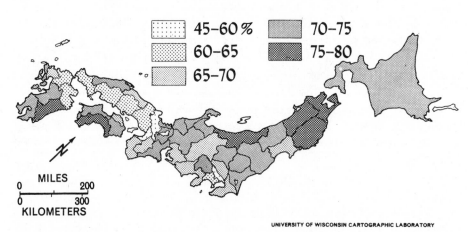

Fig. 5-21.—Percentage of population over 14 years of age with only a primary school education.

Tohoku, northern and eastern Kanto, mountainous Tosan, parts of Chugoku, most of Shikoku, and central and southern Kyushu. A large proportion of these less-urbanized prefectures are contained within Ogasawara's frontier and peripheral zones. There are a few, however, such as Okayama and Kagawa, bordering the Inland Sea and therefore included within the general Manufactural Belt, which are not highly urbanized. The six metropolitan prefectures, plus industrial Fukuoka in northern Kyushu, all have urban percentages in excess of 50 per cent. Hokkaido presents something of an anomaly in that its proportion of urban dwellers is relatively high—42.1 per cent—nearly equal to the country average, although in proportions of gainfully employed it is well below the country average for secondary industry and well above for agriculture and the other primary industries. Here the more commercial nature of the agriculture, together with the larger farms and dis-

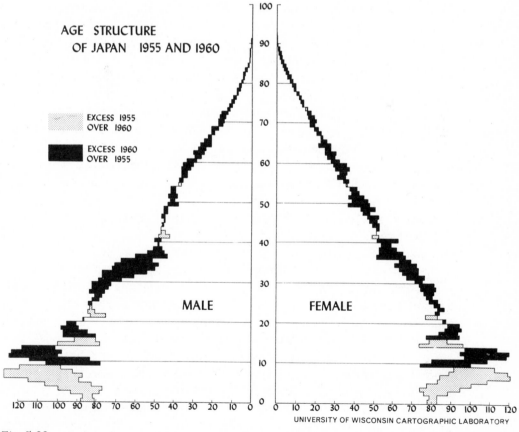

Fig. 5-22.

persed farmsteads, requires the services of a relatively large number of marketing centers.

POPULATION DENSITY AND DISTRIBUTION PATTERNS[26]

With a total area of 369,662 square kilometers, and a 1960 population of 93,407,000, the arithmetic density of population in Japan is 252.7 per square kilometer, or 655 per square mile. Only Belgium and The Netherlands in Europe, and Java-Madura in Asia, exceed this figure.

But even this exceedingly high arithmetic density is not a realistic figure with which to represent the true crowding in Japan since, owing to the prevalence of highlands, so much of the country's area is characterized by steep slopes that are without settlement, or are only thinly settled. Scarcely 16 per cent of Japan's area is classified as agricultural land (including residential lots, meadows, pastures, reservoirs, etc.), and barely 14 per cent is actually under cultivation. Thus, the population density of Japan per unit area of cultivated land, sometimes called the nutritional, or arable-land, density reaches the astounding figure of about 1800 per square kilometer or nearly 4680 per square mile. Probably in no other country are human beings crowded together so thickly on the cultivated lowlands. In part this is a consequence of the smaller percentage of land which is arable in Japan as compared to most populous countries. Furthermore, a larger percentage of Japan's population is supported by agriculture than is the case with most industrialized countries, the farm population amounting to about 38 per cent of the total population (Fig. 5-23).

A hasty observation of the population map of Japan gives the impression that density variations are chiefly a function of terrain character and of the degree of urbanization, and this no doubt is the case (see front end paper). But superimposed on this very conspicuous pattern which relates directly to terrain and urbanization, is still another pattern, of larger dimensions, connected with climatic differences, with associated variations in land productivity, and with recency of settlement. Accordingly, for the country as a whole without regard to the local variations originating in changing proportions of hill country and lowland, there is an over-all decline in generalized population density northward from about latitude 37° or 38°. So, while the national average is about 253 persons per square kilometer, that for Hokkaido is only one-fourth as great. No other prefecture approaches this relatively modest density. Tohoku, composed of the six northern prefectures of Honshu, has a density which is about three-quarters of the country average, and in the three northernmost prefectures of that subdivision the over-all density is less than half that for the nation. Still, the over-all density of Tohoku is more than twice that of more northerly and more recently settled Hokkaido. To be sure, there are a few individual prefectures in southwestern Japan, such as Shimane in northern Chugoku and Kochi in southern Shikoku, that have densities as low as the Tohoku average, or even lower, but these prefectures have meager lowland areas and are lacking in important urban development. While in that part of Japan north of about 37°, generalized population density declines with latitude, there is no similar decline in subtropical Japan. Throughout this region of Old Japan generalized population density exceeds 300 per square kilometer, with the areas of highest density bespeaking antiquity of settlement, a large proportion of lowland, or a high degree of urbanization associated with trade and industry. Accordingly, the most extensive area with a generalized density in excess of 400 per square kilometer includes the borderlands of the Inland Sea, and the region eastward into Kinki and westward into northern Kyushu (see front endpaper and Fig. 5-4).

If, in order to eliminate the effects of variable amounts of highland and lowland in the several prefectures, cultivated land is substituted for total area, the same decrease in density northward from about 37° or 38° is conspicuous. Compared with population density per unit area of cultivated land in the prefectures of southwestern Japan (which lack great cities), prefectures such as Iwate and Akita in northern Tohoku have a density 50 to 70 per cent as great, while Hokkaido's is only one-fourth to one-third. This general decline in population density northward from subtropical Japan is likewise illustrated by Table 5-13, which shows the percentages of both the area and the population of each of the great regional subdivisions of Japan which fall within various population-density categories. Thus, while in Hokkaido 77.7 per cent of the area and 32.4 per cent of the population in 1950 are included within the lowest density class, with fewer than fifty persons per square kilometer, the comparable figures for Tohoku are 41.5 per cent and 9 per cent. These in turn should be compared with Kanto's figures of 13.3 and 0.6, and Kinki's of 19.5 and 1.4.[27]

This progressive decline in population density northward from about latitude 37° to 38° reflects, for one thing, increasing remoteness from the cultural and economic heart of the country. Because Japanese culture had its origins in the subtropical southwest, the northland has always been looked upon as the provinces and the hinterland, where the environment is harsher and the ways of living cruder. As late as 1800 the number of Japanese settlers in Hokkaido may not have exceeded 40,000. Even more important in operating to reduce population density in Japan's northland is the severity of the climate which makes living conditions harsher and agricultural land less productive, especially for a people strongly bound to a subtropical type of agriculture and housing. In these northern regions with cooler summers and shorter growing season, rice-growing has been especially handicapped, so that until recently rice yields were relatively low and varied greatly from year to year, depending on the weather. Multiple and winter cropping become much more precarious in Japan's microthermal climates, and

TABLE 5-13

Percentage of the area and population of the great regional subdivisions of Japan included within certain population density categories

Subdivision	Density class (pop. per sq. km.)			
	0–25	26–50	51–75	76–100
JAPAN				
Area	20.13%	17.36%	11.85%	8.40%
Pop.	1.46	2.86	3.25	3.21
Hokkaido				
Area	53.35	24.34	10.33	3.27
Pop.	16.07	16.35	11.63	5.04
Tohoku				
Area	15.44	26.09	15.30	9.97
Pop.	1.89	7.12	6.91	6.42
Kanto				
Area	8.13	5.21	7.07	4.71
Pop.	0.25	0.36	0.79	0.70
Chubu				
Area	18.80	15.71	9.27	10.33
Pop.	1.21	2.47	2.46	3.75
Kinki				
Area	8.09	11.41	12.20	10.79
Pop.	0.36	1.05	1.94	2.35
Chugoku				
Area	2.13	15.05	18.16	15.23
Pop.	0.20	2.74	5.36	6.23
Shikoku				
Area	9.53	15.67	15.63	10.74
Pop.	0.76	2.76	4.28	4.26
Kyushu				
Area	3.79	9.15	10.19	7.23
Pop.	0.24	1.20	2.09	2.09

Source: Tatsutaro Hidaka, "Population Density of Japan," *Bulletin of the Geographical Survey Institute*, 5, Parts 1–2 (1957).

such practices are entirely absent over large parts of this northland. Similarly, a number of the country's important crops, such as sweet potatoes, tea, citrus, and mulberry, are either absent or of small importance. As a consequence largely of physical handicaps, agricultural production per unit area in Hokkaido is only one-fourth to one-third what it is in subtropical Japan, while in Tohoku the ratio is about one-half. It is significant in this respect that farms increase in average size northward from about 37° or 38° in order to compensate for this decreased productivity.

Within subtropical central and southwestern Japan population is markedly concentrated in a long, narrow, and irregular belt reaching from the Kanto, or Tokyo, Plain southwestward to northern Kyushu. Its northeastern section, in Kanto and Tokai, fronts upon the Pacific Ocean, but from about Nagoya westward, and including Kinki, it chiefly coincides with the coastal lands of the Inland Sea. In its southwestern

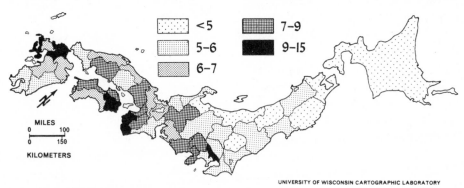

Fig. 5-23.—Rural population density per hectare of arable land, 1960.

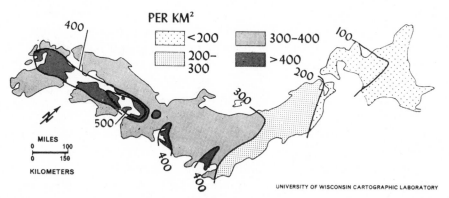

Fig. 5-24.—Rural population generalized density zones. After map by S. Inoue and others.

parts, therefore, both the Pacific and Japan Sea littorals are largely excluded. Most of this population belt was included in the region of very
early settlement, and has in the last half century come to be the most
urban-industrial part of modern Japan. Here, in a few parts, relatively
large alluvial plains face on the quiet waters of spacious bays, where
great industrial port cities have arisen. The whole region is easily accessible from the sea, for its indented coastline offers numerous havens for
ships. It contains both the ancient and the modern capital of Japan, as
well as its modern centers of industry, business, and finance. At its western end is the nation's most productive coal region, while its eastern parts
are served by the country's greatest concentration of hydroelectric development, located in central Honshu. Included within this populous
belt are all six of the country's great metropolises, all of its major ports,
and over 60 per cent of its cities of over 100,000 population. Noteworthy
nodal areas are the Kanto, Kinki, and Nagoya metropolitan regions,
whose populations of 15.8 million, 10.2 million, and 5.4 million together
represented one-third of the nation's total in 1960.

As noted earlier, it is the variations in relief and in proportions of
slopeland which play a dominant role in determining the most conspicuous local variations in population density and the most striking patterns of population distribution. In a region of such prevailingly rugged
terrain, where lowlands are usually separate and isolated, it is not unusual that a strongly clotted or cellular pattern of population distribution
should develop, with the relatively compact clusters of dense settlement almost coincident in both area and shape with sizable alluvial lowlands. It is common for very sharp, almost knife-edge boundaries to separate densely populated lowlands from very sparsely inhabited hill and
mountain land. Indeed, it is nearly possible to reconstruct the crude relief pattern of Japan from a detailed population map. In the hill lands
themselves there prevail very complex dendritic patterns of population
arrangement, with the filaments of settlement becoming finer as the valleys narrow at the higher elevations. In some places smooth upland surfaces, either volcanic ash or older alluvium, and low-level areas of weak
Tertiary rocks where dissection is well advanced are likewise moderately well settled.

In a paper entitled "Population Density by Landform Division of Japan," prepared by the Geographical Survey Institute to accompany the
map "Population Density by Landform Division, 1955" (3 sheets, scale
1:800,000), seven different types of terrain are recognized, and the population density for each type and for each subdivision of each type is
computed. Under the assumption that rural population has a closer

connection with terrain character, the calculations exclude the population (totaling 28,618,839) of all urban places with an area exceeding 0.25 square kilometers.

From Table 5-14, based upon 1955 population, it may be observed that mountains and volcanoes, which comprise over three-fifths of the total area of the country, have only 21 per cent of the total rural–small city population. Within these terrain categories population densities are lowest. Hill lands, with lower relief and wider valleys, have densities that are markedly higher than those of mountains and volcanoes. But it is the lowlands of new alluvium, and to a lesser extent the low uplands of older alluvium and volcanic ash, that support the highest densities. These two terrain types, while occupying only 24 per cent of the area, contain 65 per cent of the population, the alluvial lowlands alone, with 13 per cent of the area, accounting for 45 per cent of the population. It bears emphasizing that the latter figure would be greatly increased if all urban population were included.

From Table 5-15 it is also obvious that similar landform types support very different densities of population within the various regional subdivisions of Japan. Thus, mountains and volcanoes in Hokkaido are strikingly lower in population density than are the same terrain types in any of the other regional subdivisions, although similar very low densities are likewise characteristic of a few high and rugged mountains farther

TABLE 5-14

Population density by landform categories

Landform	Area (sq. km.)	Per cent	Number of units	Average area of one unit	Inhabitants	Per cent	Population density per sq. km. Average	Median
Mountain	203,713	55	155	1,318	12,306,237	20	61	82
Volcano	23,682	6	130	185	806,712	1	34	6
Hill land	41,586	11	157	263	7,428,751	12	179	180
Piedmont, gentle slope	4,082	1	36	113	303,752	0	74	101
Volcanic flank	9,929	3	58	171	1,124,201	2	113	81
Upland	40,403	11	331	122	11,702,205	20	290	285
Lowland	46,370	13	460	101	26,984,332	45	581	465
TOTALS AND AVERAGES	369,765	100	1,327	279	60,656,690	100	166	215

Note: Population of Japan, Oct. 1, 1955=89,275,529; excluded urban population=26,618,839.

Source: "Population Density by Landform Division in Japan," *Bulletin of the Geographical Survey Institute*, 6, Parts 2–3 (1960).

south, particularly the Central Range in Tohoku and the Japanese Alps of Tosan. Hill lands in Hokkaido also have much lower densities than those of the other regions, and the same is true of uplands, although in this category the contrast is not quite so strong. Lowlands of new alluvium in Hokkaido have about half the density of those in Tohoku, which is the lowest for any subdivision of old Japan, and only about one-fifth the density of those of Kinki in southwestern Japan, which is the highest in the nation. Not only the more rigorous climate of the northern island, but also the heavy and acid lowland soils, with extensive areas of peat bog, reduce the agricultural potentialities of the Hokkaido lowlands.

Even within the several regions of Japan, individual landform units within the same terrain category show remarkable variations in population density, depending in large part upon local physical characteristics.[28]

The strong coincidence between plains and areas of dense settlement makes for a decided peripheral or seaboard concentration of people, since most of Japan's lowlands are delta-fans. Hence it is not surprising that many elements of Japanese culture are closely bound to the sea. It is chiefly in Tohoku and Tosan that the detrital lowlands are interior basins, with the consequence that many of the population clusters are separated from tidewater by hill land barriers. Coincidence of population with alluvial lowlands is perhaps more pronounced in Japan than in most parts of the world because of the Japanese farmer's near obsession with growing irrigated rice, which requires flat inundated fields.

Like the Chinese, the Japanese tend to overcrowd the best agricultural lands and neglect the possibilities of the less fertile, or even marginal,

TABLE 5-15

Population density by landform categories within the principal regional divisions
(inhabitants per sq. km.)

Landform	Hokkaido	Tohoku	Kanto	Chubu	Kinki	Chugoku	Shikoku	Kyushu
Mountain	12	48	89	51	87	94	83	90
Volcano	9	25	123	11	...	56	...	70
Hill land	19	118	264	188	262	205	373	252
Piedmont, gentle slope	50	51	...	116	...	98	396	173
Volcanic flank	57	97	145	...	...	169	...	192
Upland	80	237	498	522	752	269	478	346
Lowland	210	391	685	640	1,071	681	785	915

Source: "Population Density by Landform Division in Japan," *Bulletin of the Geographical Survey Institute,* 6, Parts 2-3 (1960).

and more isolated upland areas.[29] This tendency to neglect the less productive upland sites is of multiple origin. In part it reflects the dominance
of irrigated rice and the difficulties associated with growing that inundated crop on uplands and slopes where irrigation is not easy to accom-
plish. In part, also, it is a consequence of Japanese farming techniques
being poorly adapted to the exploitation of hilly uplands, since they
largely exclude an intensive and rational pastoral activity and fail to
emphasize the development of fruit and nut crops. Even more, perhaps,
it is a consequence of the Japanese farmer's inability to make a living
from any but the best land, using the method of spade agriculture. It
has been estimated that it takes fifteen man-days to spade an acre of
land by hand; hence the farmer who lacks draft animals and machinery
and must depend upon human labor can cultivate only a hectare or less,
while even with an ox, horse, or hand tractor for plowing operations
the limit is still only a very few hectares. Since it takes just as long to farm
poor land as good land, and since hand labor limits the size of the area
that can be cultivated, it behooves the Japanese farmer to apply his energy chiefly to good land and avoid the poorer land as much as possible.
He can make a meager living on a hectare of fertile alluvial plain, but
he could starve on ten hectares of infertile slope land because he
is unable to handle that great an area. It is interesting to speculate about
the effect which the greatly expanding use of hand tractors may have
on the utilization of infertile upland sites, for even these simple machines
considerably increase the area that can be efficiently plowed by a farm
family. Still, both mechanical and animal power have limited useful-
ness on steep slopes.

As here used, the term "settlement" refers to the characteristic colonization or occupance unit. These units range in size and function from the simple isolated farmstead and rural hamlet to the great metropolis like Tokyo. In all cases, however, "settlement" designates an organized colony of human beings, together with their houses and other buildings and the paths and streets over which they travel.

On the basis of form and function, two principal subdivisions of settlements are recognized: (1) the dispersed type, in which the isolated, unitary, one-family residence is the distinctive nucleus, as it is on an American farmstead, for example; and (2) the clustered type, in which there is a more or less compact grouping of several or many residences, together with additional buildings serving other purposes. According to its size and the complexity of its functions, a clustered settlement may be designated a hamlet, village, town, or city. Whatever its size, the two most conspicuous features are always the buildings and the streets; yet it is human beings who are the essential and primary element of any settlement. Buildings and streets are merely the corporeal evidence of a functioning colony of people. In the United States and some other parts of the world, dispersed settlement was until recently almost synonymous with rural or agricultural population, but that is far from being true in Asia and in parts of Europe, where it is common for farmers to live in geographic villages.

142

RURAL SETTLEMENTS

Dispersed type.—In accordance with the prevailing village pattern of agricultural settlement in most of southern and eastern Asia, Japan's rural sections are dominated by clustered or grouped settlements, the isolated farmstead being characteristic only of particular regions and types of locations. The tendency toward a village type of rural settlement is strongest on the alluvial plains where rice is the prevailing crop, and it is more nearly universal in subtropical southwestern and central Japan than it is in the northeastern parts. As a general rule, the regions of grouped rural settlement are the most densely populated and maturely civilized parts of the country, where amenities are numerous, culture old, and history long. By contrast, dispersed rural settlement is more typical of isolated areas and is less likely to be continuous over extensive regions. In the main, it tends to become more widespread as both altitude and latitude increase (Fig. 6-1). This is not to say that farm villages are uncommon in the hill lands and mountains, but only that the isolated farmstead is of more frequent occurrence there than on the irrigated lowlands. Moreover, dispersed settlement is much more common in some rough lands than in others. Tanioka[1] has catalogued the distribution and occurrence of the more important areas of dispersed settlement as follows: the larger part of the northern island of Hokkaido; Kitakami Basin east of the central range, and some of the basins west of this range (for example, Yokote and Yonezawa) in Tohoku, together with the Kitakami and Abukuma hill lands in that same region; parts of the Kanto and Chubu mountains in central Japan; certain alluvial plains such as the Toyama piedmont in Hokuriku, the Oi delta-fan in Tokai, the West Izumo Plain in Sanin, and the Sanuki Plain in northeastern Shikoku; parts of the Chugoku upland in southwestern Honshu and the mountains of Shikoku; certain newly reclaimed lands along the coast of the Inland Sea and of Ariake Bay in western Kyushu. In not a few regions and areas there is a mixture of dispersed and clustered types of settlement.

It is not easy to explain the distribution pattern of dispersed settlement in Japan; different regions reflect different influences, so that it is difficult to make generalizations that have widespread application. The fact that dispersed settlement is more common in slope lands than on the lowland plains shows an association that has been noted for parts of Europe as well. It may be related to the general diffusion of resources—arable land, water, and natural sites with pleasant exposures—in rough lands. In regions of dissection and slope the scattered fragments of cul-

tivable land may be too small to support more than a few separate farmsteads, so that the inhabitants are obliged to depend on other resources
such as woodland and natural pasture that require larger holdings, which
in turn favor dispersed settlement.

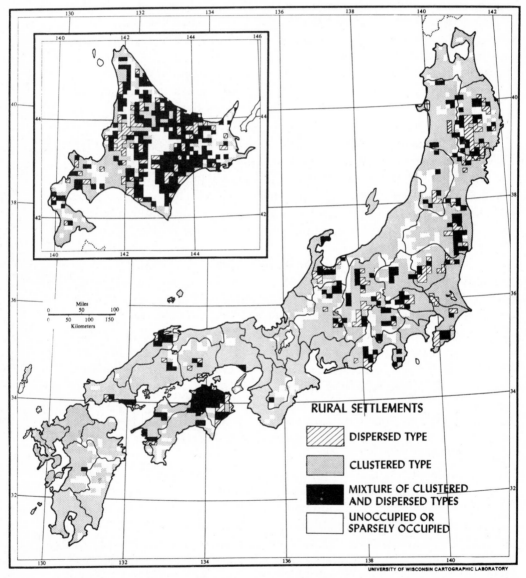

RURAL SETTLEMENTS

DISPERSED TYPE

CLUSTERED TYPE

MIXTURE OF CLUSTERED
AND DISPERSED TYPES

UNOCCUPIED OR
SPARSELY OCCUPIED

UNIVERSITY OF WISCONSIN CARTOGRAPHIC LABORATORY

Fig. 6-1.—Distribution of the types of rural settlement in Japan. After map by Takeo
Tanioka.

Recency of occupance also may be a partial answer to the dispersed settlement of certain areas, especially where this feature is combined with low quality of the agricultural land, which in turn requires larger farm units. Hokkaido is the best regional illustration of these circumstances, but they are present as well on some newly reclaimed coastal lands, and certain poor ash and diluvial uplands. On a few delta-fans such as those of Oi and Toyama, dispersion may be associated with retarded settlement due to the frequency and violence of floods, which led to the construction of isolated dry sites of low elevation for the individual farmstead.[2] In Tohoku and Hokuriku some areas of dispersed settlement have resulted from the practice followed by individual feudal lords of providing their servants with small portions of the seignorial land as private farms (*myoden*), with each of these servants subsequently building his house on a corner of his given farm lot.[3]

More common, I believe, than one would gather from the usual descriptions is a semidispersed or amorphous type of Japanese rural settlement which is intermediate between the compact village and the isolated farmstead. In these loose agglomerations a definite street pattern may be lacking, with the straggling residences, separated by small fields, being connected only by winding footpaths or cart tracks.

*Clustered or village type.**—The most elemental form of rural clustered settlement is the unincorporated hamlet or village (*buraku*). A de-

---

* Regrettably official Japanese statistics are not published by individual geographic or social settlements, but instead only by administrative subdivisions such as mura, machi, gun, and shi (in addition to prefectures), any of which may contain more than one clustered settlement. The prefectures are subdivided into over 600 gun or counties, which since 1923 have ceased to be self-governing units, though they continue to have some significance as electoral districts and as social and economic units. Population census figures are given by gun, for example, and there gun population is synonymous with non-shi, or machi-mura, population. It is predominantly rural. Since the abolition of gun as political units, the administrative subdivisions of the prefectures are mura, machi, and shi. The common English translation of mura is "village"; of machi, "town"; and of shi, "city"; but these are somewhat misleading synonyms, for as noted above, they refer not to clustered settlements of different sizes, but to administrative areas. As a result of the amalgamation of towns and villages in 1953, the number of administrative subdivisions has been greatly decreased. As of October, 1960, there were 1031 mura, 1924 machi, and 556 shi. Perhaps mura is more correctly translated "township." A mura may have within it several individual clusters of families and their dwellings which are the real rural hamlets and villages. These are called *buraku* or *aza*, but data are not published by buraku. The machi differs from the mura in that it contains at least one settlement cluster large enough to be incorporated, but in addition it may have several rural buraku as well. Each mura has a common headman, administrative office, school, and Shinto shrine. Each buraku community has its own head, and its citizens handle cooperatively such matters as funerals, festivals, road work, etc. In the Edo Period each buraku usually comprised a mura.

scription of Suye Mura, a former small rural administrative subdivision located in the Kuma River basin in southern Kumamoto Prefecture, Kyushu, which has been studied in detail, will provide kinds of information that are approximately applicable to Japan's rural compact settlements in general.[4] Suye Mura covered an area of 6.5 square miles, and in the early 1930's, when the study was made, it had a population of 1663, comprising 285 households. Within Suye Mura there were seventeen separate settlement clusters, or buraku. Each of these hamlets or small villages, averaging fifteen to twenty households, had a name of its own and was recognized as a distinct settlement unit. Each buraku had one or more headmen called *nushidori*, elected by the responsible heads of households in the hamlet, each house having but one vote. To the nushidori, who is the caretaker of buraku affairs, fall such duties as the supervision of funeral preparations, the announcing of holidays and buraku meetings, the care of hamlet property, and the supervision of road and bridge repairing. Each buraku has a shelter house, called a *do*, which houses some Buddhist deity; here the children play in the daytime, and pilgrims, beggars, and itinerant workmen find shelter. Good-natured rivalry exists between the several buraku, and there is interhamlet competition at mura parties and festivals.

Individual buraku in Suye Mura vary considerably in character. Those in the midst of bountiful rice fields on the plain are more prosperous, their households are larger, their people have common political interests and close social and blood ties. Most families have had a long history within a village; newcomers are few, and intermarriage is common. The paddy buraku are relatively compact settlements with rectangular street patterns and houses fairly close together (Fig. 6-2). Other buraku in Suye Mura are situated on upland sites, such as diluvial terraces and mountain foothills, where unirrigated crops prevail. Because of the irregular terrain, the houses of these upland buraku are somewhat scattered, and the lanes and paths that connect them are winding, so that the settlement lacks compactness. Much the larger part of the population of Suye's hamlets is engaged in farming, but at least two hamlets, both situated on the main thoroughfare, have a considerable number of shopkeepers. The shops of various kinds front directly on the highway, while the farmers' houses are usually off the main highway and irregularly situated. In the shopkeeper hamlets, where both farmers and merchants are well represented, there is much less social cohesion than in the farmer burakus, and the divergent interests of the inhabitants lead to numerous quarrels. More of the inhabitants of the shopkeeper burakus are relative newcomers to Suye Mura. Mountain

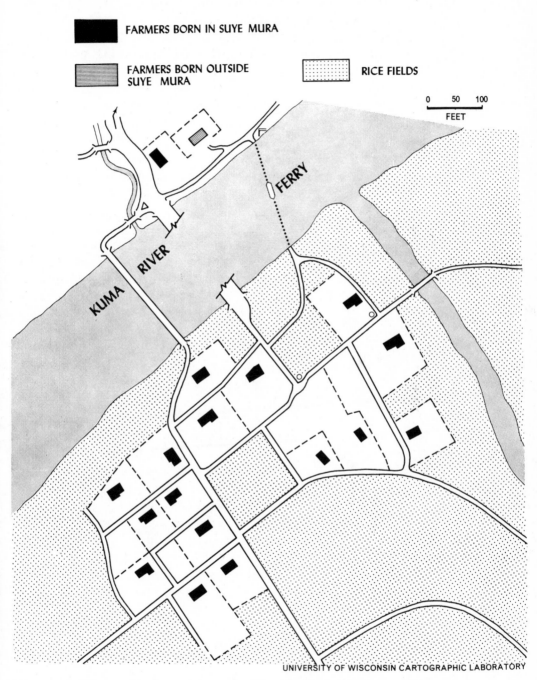

FARMERS BORN IN SUYE MURA

FARMERS BORN OUTSIDE SUYE MURA

RICE FIELDS

0   50   100
FEET

FERRY

KUMA RIVER

UNIVERSITY OF WISCONSIN CARTOGRAPHIC LABORATORY

Fig. 6-2.—Paddy-type *buraku* consisting of 19 households in Suye Mura. After a map by John Embree, in *Suye Mura: A Japanese Village* (Chicago, 1939).

burakus are more poverty-stricken than the others, and their citizens are regarded as cruder and less educated.

Although the above description applies specifically to a small portion of the Hitoyoshi Basin in the Kyushu Mountains, it is fairly representative of small agricultural settlements throughout much of Old Japan. By far the larger part of Japan's 6 million farm households, which comprise some 34.5 million individuals, live in hamlets and villages ranging from a few score up to a thousand and more inhabitants, resembling the seventeen communities in Suye Mura. In general the agricultural settlements are less compact than the larger towns, and the houses may be separated by small field plots. Some of the settlements, because they lack a

Fig. 6-3.—Residents of a *buraku* (farm village) cleaning the narrow local road passing through the settlement. Cooperative effort is characteristic of such rural places. Photograph by John Embree.

well-defined street pattern, may give the impression of semidispersion.

Living standards in the thousands of farm villages and hamlets in Japan betoken an amazing frugality, for the pressure of many men on little land has obliged the rural folk to get along on less than most of the urban population. Careful and never-ending thrift is the mode of life in most rural villages; yet the last decade has seen a distinct improvement in the living standard, as evidenced by the sharp rise in consumption levels and the adding of a number of comforts, such as bathroom facilities, and even a few luxuries, like television and foreign style clothes.[5]

This nearly universal frugality is made somewhat more endurable by the close communal life of the buraku, for the latter is a social unit to a degree unknown in American rural settlements. Many of the commodities needed are produced either in the home or by local craftsmen. Though labor is becoming increasingly specialized buraku residents may build their own houses, make their own mats, baskets, and charcoal, and clean their own rice. Villagers work and play, sorrow and rejoice, together. The result is a closely integrated and interdependent life which is attractive and warm despite the meagerness of material things.[6] This willing frugality on the part of Japan's large agricultural population has been in the past one of the country's greatest assets in its industrial competition with the Occident, for it operated to keep industrial wages low and thus permitted the Japanese manufacturer to compete on a more equal footing with nations having superior natural resources. But this advantage is diminishing as living standards rise in the rural areas and the lowered birth rate reduces the young male element in the rural labor force.

URBAN COMMUNITIES

A study of urban places in Japan is made particularly difficult by ambiguities in the definition of a city, changes in the definition, and methods of compiling census data on population. Even before the Amalgamation Act of 1953, the lack of an adequate definition made it difficult to discover the number and distribution of actual urban places. The term *shi*, usually translated city, was unsatisfactory, for it refers to an urban administrative district which is considerably more extensive and populous than the geographical city. Moreover, the minimum population of a shi was usually about 30,000, so that compact communities of 10,000 to 20,000+ people were included with the *gun*, or non-urban, population. But the confusion was multiplied when by reason of the 1953 Amalgamation Act the number of shi abruptly rose from 285 to 500 in 1957, and

556 in October, 1960; the shi area swelled from 5 or 6 per cent to 16 per cent of the national territory; while the shi population, which was 38 per cent of the total in 1950, suddenly became 56 per cent in 1955, or 64 per cent if the twenty-three wards of Tokyo-to are included.[7] Of course this is not real but only apparent urbanization, for the new shi resulting from the amalgamation of mura and machi embrace additional non-urban land and population. At the time of the 1960 census, only 65 per cent of the shi population was included in Densely Inhabited Districts (DID), or bona fide urban settlements. From what has just been said it becomes obvious that the number and population of official shi provide unsatisfactory data on which to base a geographical analysis of cities—their functions, size groupings, distribution, and changes in number and total population through time.

In an attempt to correct the defects in the urban-rural classification as a basis for presenting statistics, a new category, called Densely Inhabited Districts in the 1960 Population Census, has been created by the Bureau of Statistics. As noted in Chapter 5, a DID consists of a contiguous enumeration district with a population of about 5000 or more in 1960, in which population density is not less than 4000 per square kilometer. A DID therefore is a bona fide compact settlement unit, and while no doubt the smaller ones contain a fair number of farm families, still a majority of the population in the DID's is probably urban in employment. With this new census category, for the first time there is a fairly realistic definition of an urban place. But since it is only in the 1960 census that DID's are recognized and reported, there has been no improvement in the availability of data for analyzing the growth and distribution of bona fide urban settlements prior to 1960.

Fig. 6-4.—A main street in a Japanese town, many of whose buildings combine residential and commercial functions. Photograph by Herman R. Friis.

A total of 970 DID's were recognized by the 1960 Census, 715 of them in 94.3 per cent of all the 634 shi and ku (wards of the six great cities), and 255 in 8.3 per cent of the 2,955 machi and mura.[8] The total population of the DID's was 40,830,000, or 43.7 per cent of the nation's population of 93,419,000 in 1960. It is not far wrong then to say that about 44 per cent of Japan's population is urban, or lives in compact, densely populated communities of more than about 5000 inhabitants. The aggregate area of the DID's is only about 1 per cent of the total area of Japan, which is to be compared with 16 per cent for all shi. Some 65.1 per cent of the total shi population and 6.4 per cent of the machi and mura population reside within DID's.

Small market towns

The market town is distinguished by size and by functional and structural differences from the typical rural buraku, although the line of demarcation certainly is not sharp. Characteristically the towns are larger communities, with more than five thousand people, rather than a few score or a few hundred. In addition, non-farmers constitute a much larger proportion of the total population in the towns than in most rural buraku, for towns function primarily as the market places for the rural areas and are therefore more urban than rural in function. Farm households usually are in the minority. More commonly than buraku, towns are located on railroads or on main highways which give them direct access to bus and truck transportation. They also contain the post offices from which mail is distributed to the rural areas.

Although differences between towns are not lacking, to a foreigner they appear to have many features in common. Brightness and color are missing, and yet the rather somber hues of weathered-wood siding and gray tile roofs blend agreeably with the quiet tones of the Japanese countryside. The skyline of the closely spaced houses is regular, and in the larger and more compact towns there are no shade trees to vary the sameness of line and color. Green lawns and flower gardens are lacking also. Because of its narrower streets and the closer spacing of its dwellings, the compact town of Nippon occupies less area than one of the same population in the United States; often its residences abut against one another as they do in our commercial districts. Most small towns have no very distinct and compact commercial core and residential district. Many of the streets contain buildings serving a dual function, with a shop occupying the street side and the family living in the back or second-story rooms. Normally each small shop specializes in a single type of product—fish, eggs, produce, hardware, paper, liquor, cloth, etc. By means of sliding doors the entire front of the little shop can be thrown open to the thoroughfare to display its stock of wares to the passer-by. The narrow streets always appear crowded; pedestrians, bicycles, motorcycles, cars, animal-drawn carts, and playing youngsters all jostle one another. Sidewalks are rare; pedestrians use the street, which is usually macadamized but not smooth. Open drains or gutters carrying drainage or refuse water (not sewage) from the residences line both sides of the street.

In the mountains, because of the very obvious advantage of proximity to transportation facilities, villages and towns occupy the valley floors. These locations also contain the only patches of near-level alluvium where water is available for irrigation and the precious rice crop can be readily cultivated. If the lowest part of the valley floor is subject to serious inundation, an adjacent terrace bench or the slopes of an alluvial fan may furnish a more desirable site.

Many of the less extensive diluvial uplands, especially those that are considerably elevated and separated from the alluvial plain by abrupt slopes, have few villages, their cultivated lands being worked by farmers residing on the alluvial plain below. Where the uplands are of considerable area, as on the Kanto Plain, or are less elevated, villages may be relatively numerous. Both on the diluvial uplands and in the mountains, dispersed settlement tends to be more common than on the new alluvium.

Villages and towns are very definitely concentrated on the plains of recent alluvium. If the lowland is not too wet or subject to serious inundation, small settlements dot its surface at sufficiently frequent intervals to enable the farmer to reach his scattered plots of land without traveling excessive distances. To such settlements, surrounded as they are by inundated rice fields, the road takes on the importance of a bridge and may largely determine the linear shape of the settlement. Besides those settlements widely distributed over a plain and associated with roads in both cause and effect relationships, there are others which occupy typically strategic sites, usually elevated ones, which have the dual advantage of saving the settlement from occasional inundation and of providing superior locations for transportation lines, both rail and highway. Three types of such elevated dry sites, on or adjacent to the paddy-covered lowlands, are (1) beach ridges and dunes paralleling the coast, (2) river levees, including relics of older ones, and (3) the contact zone between wet alluvium and adjacent uplands, either hard-rock foothills or diluvial terraces. The first and third types not only provide dry points in an otherwise wet plain, but have the additional advantage of being adjacent to both rice land and dry-crop land, both of which the farmer finds it desirable to cultivate. In its proximity to the ocean the beach-ridge location has a further advantage: its settlements are frequently combination agricultural-fishing villages and often are local ports as well.

While these small cluster settlements take a variety of shapes—with site, function, and period of origin all seeming to operate as influences—two at least are sufficiently ubiquitous to warrant comment. One of the most common types as to form, especially on the long-settled plains devoted to irrigated rice, is the compact lump village, often without any regular outline or shape. Such a form seems particularly suited to paddy areas where much cooperative human effort is required and where farms are composed of scattered fields. Most lump villages are simply nodal developments at intersections of local travel routes. In some parts of the country these compact lump villages have grown up within walls or moats, which have largely disappeared, however, as time passed and the settlement grew. In areas of Jori land subdivision such villages and towns in old paddy areas may have a distinctly rectangular shape, with a similar geometric pattern of streets.[9]

Another very widespread form of village and town is that in which the linear dimension greatly exceeds the width (shoestring or *strassendorf* form).[10] Not infrequently it is a single row of houses extending for some distance along either side of a main road. Some of the longest villages of this type are located along a highway paralleling a narrow strip of coastal plain located between mountains and sea. Where this form occurs on a low alluvial plain in the midst of inundated paddy fields, or even on a well-drained diluvial upland, the road may have influenced the shape of the settlement. But many such shoestring villages, perhaps a majority of them, occupy elevated dry sites such as beach ridges and levee tops, where the linear landform itself, in conjunction with the highway along its crest, is responsible for village linearity. Most villages occupying such elevated sites are bordered on either side by strips of dry fields.

The Japanese also recognize several types of rural settlements whose form is related to the time of their origin.[11] One of these is associated with the Jori system of land subdivision, the oldest that can be traced in Japan, which was imposed on many of the alluvial plains of subtropical Japan by the Yamato Court after it attained political supremacy in the seventh century. Only Hokkaido, most of Tohoku, and southern Kyushu completely escaped Jori influence. Fields, roads, canals, and irrigation ponds were arranged in a geometric grid pattern, and compact lump villages, many of them rectangular in form and containing thirty to fifty houses, were characteristic.[12] Many of these still survive.

Shinden settlements[13] (*shin* meaning new, *den* meaning paddy fields) are relatively recent in origin, having developed during the Tokugawa Period (1603–1867). This was an era of extensive land reclamation for agriculture; and on the newly reclaimed land, part of it on upland sites and part from drained swamps and shallow lakes, fields were characteristically platted in parallel strips at right angles to a road, along which the farmers' dwellings were strung out to form a linear type of settlement. Over 1.1 million hectares (2.8 million acres) of agricultural land were reclaimed and 15,000 new shinden settlements established during the 265 years of Tokugawa rule, a goodly number of the settlements having linear form as described above.[14] This constitutes the most active reclamation period in Japan's history, and as a consequence of it Japan was able to maintain her policy of seclusion.

Tonden-hei, or militia settlements, are features of the late nineteenth century, when the interior of Hokkaido was in the process of being developed. First pioneers in this enterprise were the colonial militia, over 7300 soldier-families, grouped into thirty-nine villages, representing the vanguard of agricultural settlement. Such villages were both linear and compact in form.[15]

Cities: Origin and growth

Only during the last four centuries have cities with multiple functions been an important element of Japanese civilization. Earlier there prevailed an almost exclusively agricultural subsistence economy, and in such an environment there was little need for plurifunctional cities. This is quite in contrast to the situation in the Mediterranean lands and in

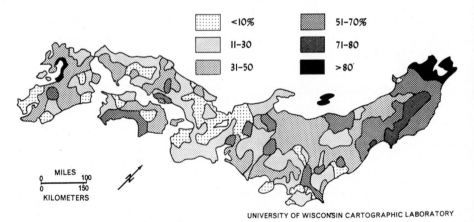

UNIVERSITY OF WISCONSIN CARTOGRAPHIC LABORATORY

Fig. 6-5.—Percentage of the total number of settlements that were *shinden,* or new, settlements during the Shogunate Period. After map by Toshio Kikuchi.

the dry lands of interior and western Asia, where such cities began far back in antiquity, even in prehistoric times. What are probably greatly exaggerated estimates indicate a population of 200,000 for Japan's first capital, Nara, in the seventh and eighth centuries; and for Kyoto, the second capital, a total of 100,000 houses and 500,000 inhabitants, in the ninth century.[16]

It was not until the mid-sixteenth century, however, when some of the more important clans began to build their castle towns on defensively and economically important sites, that cities in any number and any importance began to appear.[17] During the more than two and one-half centuries of Tokugawa rule which followed, several factors combined to favor a modest multiplication and growth of urban places. Among the most important of these were the spread of a commercial economy and the differentiation of occupational classes—craftsmen, merchants, professional soldiers—outside the predominantly farming population. Castle towns of the *daimyos,* or feudal lords, multiplied. Five great highways were constructed which radiated out to the various parts of the country from Edo (now Tokyo), the seat of Shogun power, and subhighways were built to connect the developing areas of specialized production with the great cities. Increasingly peaceful conditions reduced the needs for services of large numbers or *samurai,* or professional soldiers, who in turn tended to gravitate toward the cities. A multiplication of household industries and handicraft factories, as well as of the service and distributive occupations, resulted in increased opportunities for urban employment. A requirement of the Shogun's government that all daimyos spend alternate years of residence in Edo and that in the periods of their absence their families be maintained in Edo as hostages was a further stimulus to urban development, as traffic in men and goods along the great highways swelled in importance.

Four principal types of city origin may be noted for Tokugawa Japan: (1) castle towns which were the capitals of the daimyos' domains; (2) traffic stations, containing accommodations for travelers and goods moving along the great highways, which offered special market advantages; (3) temple and shrine towns which provided goods and services for the thousands of pilgrims and worshippers; and (4) free ports and markets, which were not controlled by any feudal lord or group of priests, but were under the direction of the merchants who inhabited them.[18] Castle and highway towns greatly outnumbered the others, but not a few combined more than one function.

The castle town (*joka-machi*) had as its central point of accretion the imposing moated and walled castle of the daimyo. This was surrounded by the residences of large numbers of samurai. Beyond the outer walls and moats of the nuclear castle, and along the principal highways leading to it, were concentrated the shops of the merchants and artisans.[19]

Such joka-machi offered at least three attractions: they had superior market facilities, they furnished a measure of protection in periods of internecine warfare, and they offered opportunities for amusement and entertainment that the country village did not. Thus, most of the first large towns in Japan had their origin as strategic political-economic centers of small semi-independent feudal states. Artisans and traders flocked to these daimyo towns, and not infrequently a center came to specialize to such an extent in a particular feature of trade or manufacture as to acquire national fame. So firmly established did these specializations become that some of them have persisted to the present time, as for example, lacquerware at Takamatsu or paper fans at Marugame.

At first the daimyo castle was a simple frame structure surrounded by turf-covered earthen embankments in the form of steps or terraces, protected and made more durable by wooden piles. No radical departures

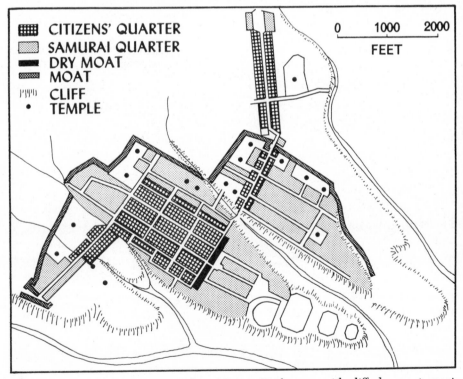

Fig. 6-6.—Plan of a castle town: Iida in Nagano Prefecture, with cliff, dry moat, moat, temple, citizens' quarter, samurai quarter, and castle. The castle is situated at the eastern extremity of the middle upland prong. After R. Isida.

Fig. 6-7.—A reconstructed Japanese feudal castle. Note the splendid dry wall. Moat in the foreground. Courtesy of Consulate General of Japan, New York.

from this simple type of structure were made until after the introduction of European firearms. The first castle having definite European earmarks was constructed under Portuguese direction in 1575. It was at this time that heavy flared walls of cut stone, usually andesite or granite, pierced by several protected gates, began to supplant earthen terraces and embankments, and encircling moats were added as a further measure of defense. The residential unit, however, still remained a frame structure, Japanese in type and appearance (Fig. 6-7). These extensive wall-

and-moat-encompassed grounds with their prominent castles, so boldly disproportionate to most other Japanese structures, were then, as now, conspicuous and attractive features in a number of cities. They have been the cores around which accretion has taken place. Today the extensive grounds of these feudal relics are commonly occupied by parks and schools. Few, if any, of the original castles remain, for repeated fires have razed the wooden structures.

Ordinarily the daimyo's castle was so located as to command an important productive area, which in Japan means an alluvial plain. Moreover, the encircling moats had to be filled with water, and usually this was possible only where the groundwater was near the surface, as it is on the plains. On the other hand, an elevated site was desirable to increase the castle's conspicuousness, facilitate its defense, and make it more impressive. Sometimes the site was an outlier or spur of hard-rock hill, but more often it was the edge of a diluvial terrace. Proximity to the seacoast was also considered of some importance as a matter of defense, and in time of war it made certain an uninterrupted supply of salt evaporated from ocean water.

According to Fujioka a total of 453 castle towns existed at one time or another in feudal Japan. Of these, 114 exist today as modern cities, 174 as towns, and 165 as villages.[20]

Towns originating as traffic stations (*shukuba-machi*) along the surfaced trunk highways, and later the subsidiary ones as well, became especially numerous during the period of the Shogunate. Some sixty-six such towns existed along the Tokaido trunk line between Edo and Kyoto-Osaka (350 miles), and seventy-four along the Nakasendo line,

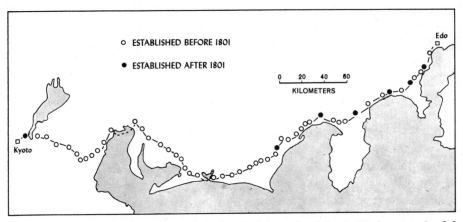

Fig. 6-8.—Location of post towns (*shukuba machi*) on the Tokaido Highway in feudal Japan. After map by Yukio Asaka.

which took an interior route between the same terminals.[21] Average spacing of the shukuba-machi along the highways was about eight to twelve kilometers. Such settlements were distinctly linear in form, thereby resembling the newly developed farm villages, or shinden settlements, of the Tokugawa period.[22] The town lots of the shukuba-machi residents were laid out on either side of the highway. Hotels and inns occupied the central section of such a settlement, and shops the marginal areas. Functions of shukuba-machi were to vend services and goods to the travelers along the highway. Horse stables, tea houses, inns, pleasure houses, porter depots, shops, open market places, and workshops were included among the more distinguishing features. Since more than 1800 of these station towns were in existence at the time of the Restoration, it is probable that a great many of today's cities and towns owe their origin, in part at least, to their position on the highways of feudal Japan.[23] By far the greater number of castle towns were also traffic stations on the main highways and hence profited from both political-military and commercial functions. Below is presented an analysis of the population and functional establishment of one shukuba-machi of the late Tokugawa Period.

Hodogaya Town on the Tokaido Highway in 1850[24]

Families	590
Population	3171
Shops and workshops	222
Inns	69
Inns with maidservants	49
Other shops	18
Shops not related to post-town function	38
Workshops and factories	52

Somewhat comparable to the daimyo castle as a lodestone attracting population were certain important shrines and temples. To these holy spots came pilgrims by the thousands, even as they now visit modern Nara by the millions. Since these pilgrims had money to spend for services and goods, inns, shops, theaters, and a variety of service establishments catering to their needs sprang up along roads in proximity to the holy places. Like castles, the shrines and temples were large and set in spacious grounds, and hence they were the prominent features of towns or cities where they existed. They did not, however, commonly occupy conspicuous elevated sites, and more often they were located on the peripheries rather than at the centers of their service towns. Only eight great Shinto shrines and eight similarly important Buddhist temples were recognized by the Tokugawa Shogunate, but there were many others of lesser importance. It seems likely, therefore, that while no great number of feudal towns or cities owed their origin and development mainly to a temple or shrine, there were many in whose development these holy places played an auxiliary role. At the time of the Restoration in 1868 several hundred shrines and temples of national fame were listed in contemporary pilgrims' guides.

Although to some extent castle towns and highway station towns also performed market functions, a number of them came to specialize in that service by reason of unusually favorable locations at the intersection of highways or at river crossings. Particularly in the modern period since Tokugawa, many castle towns have revived as regional administrative centers with important commercial functions. Highway towns, except those that developed modern commerical and industrial functions, tended to decay as railroads took over the chief transportation functions. Temple towns continue to be centers of tourism. Especially during the last half-century cities with new functions have come into being. Such are the great port cities of Yokohama and Kobe, the naval bases of Yokosuka, Kure, and Sasebo, and the new industrial cities such as Muroran, Kamaishi, and Yahata.

Already in the Tokugawa Period two of the greatest population concentrations of modern Japan—Kinki and Kanto—were importantly developed; and Edo, Osaka, and Kyoto, there located, were cities of several hundred thousand, their combined populations amounting to 5 or 6 per cent of the total commoner population of the country. So, while the maximum increase in urban population has occurred in the twentieth century, the great city was a product of feudal Japan two or three centuries earlier. Recorded population figures for urban places in feudal Japan are very fragmentary, but five cities—Nagasaki, Sendai, Kanazawa, Hiroshima, and Nagoya—were reputed to have commoner populations of 40,000 to 65,000 while Okayama and Tokushima had 20,000 to 30,000.[25]

Cities in modern Japan

At the beginning of Meiji there were only five cities in Japan with over 100,000 population. These included Tokyo, formerly Edo, the capital of the Shogunate, which had a population of 596,000 in 1873; Kyoto (239,000), the residence of the Imperial family; Osaka (272,000), the principal commercial center; and Nagoya and Kanazawa, which were typical castle towns.[26] Significantly, Yokohama and Kobe are not included. None of these was an industrial city, their functions being primarily political, commercial, or both. Besides the five large cities there were ninety-four others—mainly castle or station towns, though a few were harbor, temple, or handicraft towns—that had a population of 10,000 or more.[27]

But although cities grew rapidly in numbers and population during the latter decades of the nineteenth century and the early decades of the twentieth, Japan as late as 1920 was still predominantly a rural country, for less than one-third of the population lived in communes (administrative subdivisions) with more than 10,000 people, and only one-sixth in an urban district of 50,000 or more.[28] Still, the proportion of the total population living in administrative districts of 10,000 or more population rose from 9.8 per cent in 1878, to 15.8 per cent in 1898, and 26.1 per cent in 1920.[29] Between 1903 and 1918, 9.6 million were added to the total population, and of this increase 3.7 million, or two-fifths, were added in the urban prefectures containing the great cities of Tokyo, Yokohama, Osaka, Kyoto, Kobe, and Nagoya. These six cities which in 1888 contained only 2.4 million, or 6 per cent of the nation's population, had expanded to 6.1 million by 1918, or 11 per cent of the population.[30]

It was after about 1920, however, as the country began to industrialize at a more rapid pace, that cities and towns grew most rapidly, both in numbers and in population. The absence of data by bona fide geographical cities and towns, defined as compact and dense settlement units, necessitates recourse to data by administrative units—shi, machi, mura—which are only relatively akin to geographical settlements. As a general rule, the larger the city, the more valid are the substitute figures given by administrative subdivisions. The number of persons living in communes of 10,000 population or over, which may be accepted as the minimum figure for administrative subdivisions likely to have a bona fide compact settlement of urban size and function, was 31.9 per cent of the nation's population in 1920, 40.6 per cent in 1930, 50.4 per cent in 1940,

TABLE 6-1

Progress of urbanization in Japan (excluding Hokkaido and Okinawa)

Year	Number of cities and towns*	Urban population (in millions)	% of total population
1878	99	3.43	9.8
1898	166	6.96	15.8
1920	232	14.15	26.1
1935	243	24.03	36.3
1950	418	31.20	37.5

* Urban and town administrative districts of 10,000 or more population in 1878, 1898, and 1920, but of 20,000 or more in 1935 and 1950.

Source: Ryuziro Isida, *Geography of Japan* (Tokyo, 1961).

and 54 per cent in 1950. In the prewar period between 1920 and 1940 there was a tendency for the larger communities to grow most rapidly, with cities of over 500,000 expanding their share of the population from 8.4 per cent of the population in 1920 to 19.8 per cent two decades later. Cities of 100,000 to 500,000 expanded from 3.8 to 9.6 per cent; but those of 50,000 to 100,00 only from 3.7 to 5.2 per cent; while in those from 10,000 to 50,000 the percentage of the total population remained relatively static. The rural communities with under 10,000 population declined both absolutely and relatively.[31] Clearly, urbanization was proceeding rapidly, and the major areas of population increase were the cities of over 100,000, and more especially the three great conurbations of Kanto, Kinki, and the region around Nagoya.

Many features of city growth suffered serious dislocation during the decade of the 1940's as a result of the bombing and burning associated with World War II. The larger cities suffered the most, with places of one million or more temporarily losing more than two-thirds of their populations and those of 100,000 to a million losing about 30 per cent. As a result, cities of 100,000 or more, which accounted for 29.4 per cent of the population in 1940, were still down to 25.6 per cent in 1950, five years after the war ended. By contrast, the smaller cities of under 100,000 had grown in relative importance from 21 per cent to 28.4 per cent. But this was only a temporary dislocation, for the country is irrevocably indus-

TABLE 6-2

Number and population of communes (shi, machi, mura) by size classes

Population class	1920 No.	1920 % of pop.	1930 No.	1930 % of pop.	1940 No.	1940 % of pop.	1950 No.	1950 % of pop.	1960 No.	1960 % of pop.
Over 500,000	4	8.4	6	11.9	6	19.8	6	13.4	9	20.0
100,000–500,000	12	3.8	26	6.1	39	9.6	58	12.2	104	20.7
50,000–100,000	30	3.7	64	6.8	55	5.2	91	7.6	160	11.5
10,000–50,000	489	16.0	563	15.8	628	15.8	972	20.8	1,791	37.4
Under 10,000	11,653	68.1	11,148	59.4	10,404	49.6	9,287	46.0	1,447	10.4
TOTAL	12,188		11,807		11,132		10,414		3,511	

Note: Because of the reorganization of the administrative subdivisions that occurred in 1953, as a result of which the total number of administrative units was greatly reduced, but the number and area of shi, or official cities, was increased, the data for 1960 are not comparable with those of previous decades, except possibly for the large cities with over 100,000 population. Caution should be exercised in using the data in the table too literally, for it should be kept in mind that reference is to administrative units and not to cities and towns as compact and dense clusters of population.

Source: Japan Statistical Yearbook, 1961.

trial, and so the prewar pattern has re-established itself with the maximum growth occurring in the larger cities (except Kyoto) and especially in the three greatest urban concentrations. As noted in an earlier section, the administrative reorganization in 1953 which greatly increased the number and area of shi, or official cities, lessens the comparability of the 1950 and 1960 data. It seems safe to assume, however, that the reorganization had the least effect in the larger cities. By 1960 the metropolises of over half a million had recouped their wartime percentage losses and again accounted for about one-fifth of the nation's population, as they did in 1940. Cities in the 100,000 to 500,000 class accounted for nearly 21 per cent of the total population in 1960, more than double that of 1940, but an appreciable part of the large two-decade growth of this class was probably fictitious, resulting from areal expansion associated with the reorganization of administrative subdivisions. In the cities of under 100,000, the large 1940–1960 growth rates most certainly resulted from the multiplication of urban districts associated with the above-mentioned reorganization.

Among the cities of over a million, Tokyo had by all odds the largest absolute growth between 1950 and 1960, three times that of Osaka, its nearest rival, while its percentage growth was also the highest of the big six. Osaka, Nagoya, and Yokohama did not greatly differ in their growth rates, and all three were appreciably ahead of Kobe, whose industrial growth has suffered by reason of restricted space for factory expansion. Kyoto, the single inland metropolis and the one which has lagged in attraction of large-scale modern industry, was far below the others in both absolute and relative growth. In all but Yokohama growth

T A B L E   6-3

Growth of cities with over one million population, 1940–1960 (shi area)

City	Population (000 omitted)				Increase			
	1940	1950	1955	1960	1950–1955		1955–1960	
Tokyo	6,779	5,385	6,969	8,303	1,584	29.4%	1,333	19.1%
Osaka	3,252	2,015	2,547	3,012	532	26.4	464	18.2
Nagoya	1,328	1,083	1,337	1,592	253	23.4	255	19.1
Yokohama	986	951	1,144	1,375	192	20.2	231	20.2
Kyoto	1,090	1,106	1,219	1,285	98	8.9	66	5.4
Kobe	967	814	986	1,114	166	20.4	128	12.9
TOTAL	14,384	11,354	14,202	16,680	2,848	24.9%	2,478	17.5%

Source: Peter Schöller, "Wachstum und Wandlung japanischer Stadtregionen," *Zeitschrift der Gesellschaft für Erdkunde zu Berlin*, 93 (1962). Totals are as they appear in source.

rates have slowed perceptibly in the second of the 1950–1960 quinquenniums, a slackening which may reflect in part only a further removal from
the war period, but in part also the growth of satellite towns as a result
of the increasing congestion in these cities. The satellites take the form
of burgeoning industrial centers as well as the dormitory type of community in which a conspicuous feature is the extensive areas of planned
residential developments.[32]

Functions of urban settlements

The kinds and proportions of the functions concentrated in urban settlements vary with the size of settlements and vary also from one part of
the country to another. Thus, in the larger cities, with over 50,000 population, about one-third of the workers are in manufacturing, one-fourth
in commerce, and one-fifth in services.[33] The proportions are only slightly
different in smaller cities, with the figures somewhat reduced for manufacturing and commerce, and increased slightly for construction and
government. In towns, by contrast, only one-third to one-fifth are engaged
in manufacturing, one-fourth to one-fifth in commerce, and a higher relative proportion in government and construction.

An unusual feature of incorporated Japanese cities (*shi*) is the very
large number which are strongly agricultural in their labor force composition. Thus of the 491 official cities (1955) 306, or 60 per cent, may
be functionally classified as agricultural cities in which 25 per cent or
more of the male population is active in agriculture, forestry, and fishing.[34] Seventy-five of the so-called agricultural cities had 40 per cent
or more of the male labor engaged in agriculture. What is indicated here
is a tendency to "overbound" urban areas although, to be sure, there is
a large agrarian element located within the genuinely urbanized sections of many Japanese cities. Characteristically agricultural cities are of
modest size, 69 per cent having a population below 50,000, and they are
widespread throughout the country. Second most numerous category of
cities is that emphasizing industrial functions, with a total of 57 which are
more exclusively industrial, and 33 others which combine manufactural
with important commercial, transport, and service functions.[35] Such
urban places are strongly concentrated within the larger metropolitan
complexes. On the other hand, cities showing various combinations of
commerce, services, and transport functions are widely scattered. Of the
six cities with over a million population, Tokyo emphasizes industry-
commerce; Yokohama, transport and services; Nagoya, industry; Osaka,
industry and transport; Kobe, transport; and Kyoto, industry and transport.

Certain broad generalizations likewise can be made concerning the functions of cities in different parts of the country. For example, commerce, government, and services, in that order, are the most common functions in a majority of the cities located in the two extremities of the country, the northeast and the extreme southwest. There urban settlements are likely to be of the local nodal type, in which commerce and services obtain their earnings by absorbing income from the local primary industries. In other words, in these regions it is the primary industries which underpin the economy and pay for services and for the manufactured goods, many of the latter being produced outside these peripheral regions. By contrast, in cities of the more central parts of Japan, from Kanto to northern Kyushu, manufacturing is a more important, if not dominant, element in the economy. Here the urban settlements derive their earnings chiefly from the processing of goods, with which earnings they purchase the products of primary industries, many of them coming from overseas. Within this more central part of Japan, so-called "urban" industry (including chemicals, primary metals, fabricated metals, machinery, transportation equipment, and printing) is concentrated in cities lying along an axis connecting Tokyo with North Kyushu by way of Tokai, Kinki, and the margins of the Inland Sea. Urban places, especially the smaller ones, lying on either side of this axis more often are specialized in what has been designated as native manufacturing (including food, textiles, lumber and wood products, furniture, and stone-clay-glass products).[36]

Distribution, location, and characteristics of cities

In 1960 there was a total of 970 DID's, or densely populated urban places, with a minimum of 5000 or more inhabitants. This is to be compared with the figure of 556 shi.* In addition to the six cities numbering over a million, each containing numerous ward DID's, there are 223 other DID cities with populations exceeding 25,000 (2 in each of the 500,000, 400,000, and 300,000 classes; 12 in the 200,000; 36 in the 100,000; 63 between 50,000 and 100,000; and 106 between 25,000 and 50,000). Of these 223 DID's, 20 are in Hokkaido, and 23 in Tohoku, while 180, or about 80 per cent, are located in subtropical Old Japan (Fig. 6-9). Within Old Japan the greatest concentration of cities lies along an axis

* The discrepancy between the number of DID's and of shi results partly from the fact that upwards of 90 of the shi contain two or more urban clusters (DID), while each of the *fu*, or wards (over 80 in total), of the six great metropolises is counted as a separate DID. Moreover, numerous small DID clusters are contained within gun, and these do not have shi rank.

which reaches from Kanto in the northeast, southwestward along the Tokai coast to Kinki and thence along the shores of the Inland Sea to northern and western Kyushu. Along this axis, which is also the manufactural belt of Japan, are located all six of the cities of over one million, which together contain about 39 per cent of the total DID urban population. Of the nation's other cities with over 25,000 population, two-thirds are likewise concentrated within this belt, as are over 80 per cent of those within Old Japan.

Within the axial belt of strong urbanization, certain nodal concentrations are to be observed. Greatest of these is in the Kanto area (seven prefectures) where in addition to Tokyo (8.1 million) and Yokohama (1.1 million) there are ten other DID's with over 100,000 inhabitants and forty-two with over 25,000. The next largest concentration is in the Kinki area (six prefectures) at the eastern end of the Inland Sea where besides Osaka (3.0 million), Kobe (1.1 million), and Kyoto (1.2 million), there are six additional DID's with over 100,000 inhabitants and thirty-two with over 25,000. On the lowlands bordering Ise Bay, in Aichi, Gifu, and Mie prefectures, is a third and smaller nodal area of cities. Here in addition to the metropolis, Nagoya (1.5 million), there are fifteen other cities of over 25,000, and this is expanded to twenty-seven if the entire Tokai area, embracing also Shizuoka Prefecture, is included.

Like population in general, Japan's cities are strongly concentrated on the restricted areas of lowland, and a great majority have developed on

TABLE 6-4

Number of metropolises and other DID cities with over 25,000 population, 1960

Region	1,000,000	300,000– 600,000	200,000– 300,000	100,000– 200,000	50,000– 100,000	25,000– 50,000	Total in region
Hokkaido	–	1	1	5	3	10	20
Tohoku	–	1	–	3	10	9	23
Kanto	2	1	–	9	15	17	44
Hokuriku	–	–	2	2	2	7	13
Tosan	–	–	1	1	3	5	10
Tokai	1	–	1	2	7	17	28
Kinki	3	1	2	3	9	17	35
Chugoku	–	1	–	3	4	10	18
Shikoku	–	–	–	4	2	3	9
Kyushu	–	1	5	4	8	11	29
TOTAL	6	6	12	36	63	106	229

Source: Bureau of Statistics, Office of the Prime Minister, *1960 Population Census, Densely Inhabited District* (Tokyo, 1961).

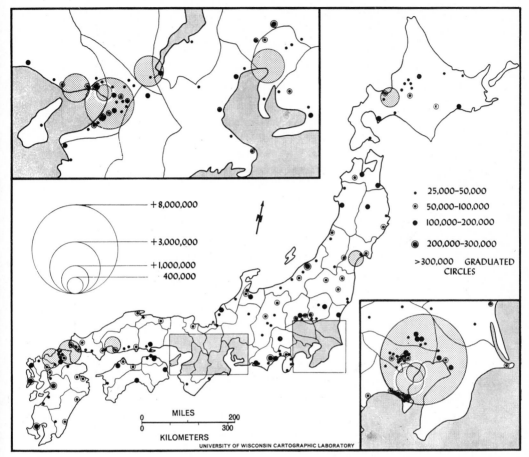

Fig. 6-9.—Distribution of cities with a DID population of 25,000 and more, 1960.

lowlands that front upon the sea, so that they can be served with relative ease by marine transportation. This importance of the sea as a factor in city growth is suggested by the fact that of the approximately three-score DID cities with over 100,000 inhabitants about 60 per cent are designated as major ports by the Japan Port and Harbor Association.

The sites of most cities are portions of flat alluvial lowlands. In contrast with conditions along the Mediterranean, Cornish, and Norwegian coasts, where picturesque towns and cities spread up adjacent steep slopes, Japanese cities are characteristically confined to flattish sites and only a few like Nagasaki, Kobe, and Hakodate have steeply sloping sections. A number of cities, like Tokyo and Yokohama, have two distinct levels of occupance: the flattish crest of a diluvial upland, as well as the lower plain of new alluvium, but the difference in elevation between the two levels is seldom great.

On the basis of size, functions, and general morphology, it may be useful to recognize two groups of cities. The first of these comprises the six great metropolises or "national cities"—Tokyo, Osaka, Kobe, Nagoya, Kyoto, and Yokohama—with national, and even international, hinterlands and a strong Western appearance, especially in the central business district and in some manufacturing areas. Contrasting with the big six are the scores of smaller indigenous cities comprising the second group, which are much more Japanese in their features and appear to be cast in much the same mold. But since the war, as a result of reconstruction following the bombing and burning, many more of these indigenous cities have taken on a distinct Western appearance in their commercial-core sections. Between the six major cities and the much larger group of indigenous cities, there is a wide population gap. In 1960 there was no DID city in the 600,000 to 1,000,000 population category and only six in the 300,000 to 600,000 category. (The new, 1963, amalgamated city of Kitakyushu probably has a DID population in excess of 800,000.) It would appear as though the local hinterlands of Japan are not able to support many cities beyond about the 300,000 class.

As a group, the local cities share the somberness of the towns described earlier in this chapter. The image is of a level slate-gray sea of tile roofs, with few green islands of trees or lawns. But recently the flatness of skyline has been changing as more multistoried steel-and-concrete commercial buildings rise up. In a number of cities two indigenous features are conspicuous because their height or mass breaks the monotony of the skyline—the daimyo castle, with its extensive wall-and-moat-encircled grounds, and the shrines and temples, the latter painted in gay colors.[37]

Most Japanese cities are unattractive in physical appearance and short on character and aesthetic integrity. In a few of the more traditional ones like Kyoto, Kanazawa, and Matsue the soft, neutral shades of tile and weathered wood are pleasing in their harmony with the quiet tones of the countryside; and in all cities a few streets, especially those lined with the indigenous small open shops, might be called picturesque. But a majority of the urban places have a cluttered and unfinished appearance: the uneven surfaces of the streets—and of the sidewalks, if any— seem forever under repair without being completed. Confusion is worse confounded by numerous poles strung with electric and telephone wires, by advertising signs, and by parked bicycles, motorcycles, and other vehicles. Individual modern buildings may not be unpleasant in appearance, but the general effect produced by the stained and crumbling stucco walls is not good. Drab shades of tan and gray predominate, so that the eye yearns for the livelier hues of blue, green, and pink that add so much beauty to certain European cities. Streets are usually narrow, making the proportion of street to total area much smaller than in Western cities. In addition, a well-ordered street pattern is often lacking, with the consequence that traffic is unbelievably congested.

A prevailingly rectangular grid of streets characterized many of the early cities and is still a fairly common feature of the old cores of numerous modern ones. Two explanations have been offered for this street arrangement. Certainly the low flattish delta sites of most of the cities made such a grid arrangement feasible. The other explanation is that the Chinese, who produced the street grid of Imperial Peking, had a widespread influence on Japanese city plans as on so many other things Japanese. Imperial Kyoto and Nara are excellent examples of this Chinese influence. Most of the rural highways, however, are not straight, nor do they follow the cardinal directions; hence, as the city expanded and encroached upon what were once agricultural areas, the street system built upon the original rural highway skeleton frequently lost the grid pattern and became confused. Thus, in not a few Japanese cities the original core has a rectangular street arrangement, whereas its later peripheral sections do not. This is the reverse of the situation in many old American and European cities.

The numerous and recurrent fires which have devastated Japanese settlements have permitted frequent and extensive modifications of the original city plans, modifications that are often made with revolutionary rapidity. Such drastic changes took place in Tokyo and Yokohama as a result of the earthquake and fire of 1923, and in many more cities following the destruction of World War II. Repeated fires and, until recently, the near universal use of wood for construction account for the almost total lack of ancient buildings.

In its numerous narrow streets with combination residences and small shops, open gutters and dearth of sidewalks, the indigenous Japanese city in many respects resembles the market town. The zoning of functions and building forms so characteristic of American cities is less prevalent in Japan. A well-defined central business district is most evident in the larger cities. In the exclusively residential sections, where most dwellings are surrounded by high wooden walls, the view from the street is uninteresting. Small one-family houses, all much alike and closely spaced, sometimes abutting against one another, are characteristic.

Because of their delta locations, not a few Japanese cities are interlaced by rivers and canals. These are arteries both of commerce and of waste disposal. Factories, warehouses, and heavy-retail establishments tend to be concentrated along their margins, largely because they offer cheap transport for bulky goods. Most of the canals are unattractive both because of the kinds of establishments bordering them and because of the refuse they carry. A corollary of the network of canals and rivers is the unusual number of bridges in Japanese cities.

The six major cities are almost alone in reflecting more than a local sphere of influence. Their distinctive landscapes are the result not so much of different Japanese elements as of a more complete westernization of their business, transportation, and manufacturing features. The stamp of foreign influence is most evident in the central business district, where large multistoried steel-and-concrete buildings, European in appearance, are numerous, even forming solid blocks in some sections. Certain factory areas likewise have a strong Western flavor. But elsewhere, the metropolis remains largely Japanese in appearance, resembling the smaller local cities.

All six of the cities with over a million population are located on or near three of the country's large alluvial plains that have developed at the heads of extensive bays on the Pacific side of the country. Each plain is a major settlement area and an important local hinterland. Only one, Kyoto, is not on tidewater. The other five compose the nation's quintet of leading ports. Three of them—Tokyo, Osaka, and Nagoya—are important rail centers. The big six are the principal foci of manufacturing within the three greatest industrial nodes of the nation, and so, with one exception, they combine port and manufactural functions. They are similarly the population centrums of the greatest metropolitan areas of the country, 15.8 million in the Tokyo-Yokohama area, 10.2 million in the Osaka-Kobe-Kyoto area, and 5.4 million in the Nagoya area.

On the basis of location and function the big six may be divided into three groups: (1) The three largest centers—Tokyo, Osaka, Nagoya, each located at the head of its respective bay, on the sea margins of an advancing delta—are the chief business, industrial, and consuming centers of their large local hinterlands. Each is a foreign-trade port of consequence, but their shallow waters do not permit them to readily accommodate vessels of deep draft. (2) Kobe and Yokohama, the genuinely deep-water ports of the great Kanto and Kinki industrial nodes, are located fifteen to twenty miles down their respective bays, where silting is less active. Nagoya has no such important down-bay, deep-water port. Kobe and Yokohama are both great manufactural cities as well. (3) Kyoto, the only one of the big six with inland location, contrasts with the others in being neither a port nor a principal center of modern industry, although the latter function is expanding. Its chief fame derives

Fig. 6-10.—Street scene in downtown Tokyo. Courtesy of Consulate General of Japan, New York.

from the fact that it was for over a millennium the nation's capital and is today a mecca for tourists attracted by its palaces, shrines, and temples. Thus, from a regional point of view, four of Japan's six metropolises are arranged in two binuclear conurbations, Tokyo-Yokohama and Osaka-Kobe, the comparable members of each duad being relatively similar in location, site, and functions.

THE JAPANESE HOUSE [38]

The prevailing type of Japanese house seems to have been designed for the never-ending summer of the tropics. This would appear to be one of the cultural features inherited from that branch of the original stock which moved northward from the tropical islands and coastlands of

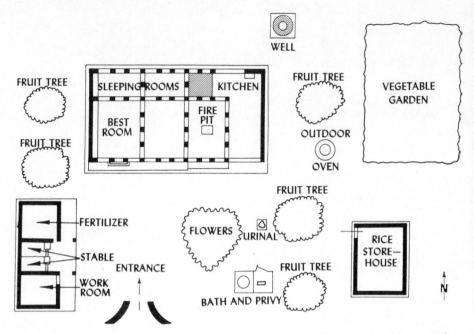

Fig. 6-11.—Plan of a representative farmyard and rural house in southern Kyushu. Redrawn from a plan by John Embree in *Suye Mura: A Japanese Village* (Chicago, 1939). No scale given.

southeastern Asia. In its present-day modified form it is perfectly adjusted to the long, hot, humid summer of subtropical Japan, which is, to be sure, the dominant season in the populous central and southwestern parts of the country. It is less well adapted, on the other hand, to the chilly, raw winters of the same sections, and still less so to northern Honshu and Hokkaido with their continental winter climates. Apparently the philosophy of the Japanese is to endure the discomforts of winter in anticipation of obtaining the fullest enjoyment of the long summer to come. Genuine fortitude is required, however, to carry on normal living throughout the winter in a drafty tropical house in which little or no provision is made for heating the rooms themselves, only the occupants. It is true that in the mountains and in northern Japan, where winter temperatures are much lower and the summer period cooler and shorter, the subtropical house has been modified to some extent, but the surprising thing is that so many of its original features have been retained.

While Japanese rural houses vary in detail, in basic elements they have considerable similarity. They are erected for service, not for display, so there is little ornamentation. Simply rectilinear in form, they are small in size, three to six rooms being most usual. One story is much more common than two, although there may be an attic. Basements are absent and ordinarily there is no continuous foundation. Wooden posts resting upon stones serve as the underpinning of the house, the floor of which usually stands a foot or two off the ground. This stilt foundation is suggestive of a tropical house. Since most Japanese farmhouses are in the immediate vicinity of flooded paddy fields, this elevation which permits free passage of air beneath aids ventilation and so prevents excessive dampness. The structure is not anchored to the ground, but is weighted down by its heavy roof of gray tile or deep thatch, a feature which tends to make the Japanese house topheavy but also enables it to withstand heavy winds. Dimensions of an average rural dwelling in subtropical Japan are about 36 feet (11 meters) long by 18 feet wide. If the measurements differ, they still retain the same general proportions in order to conform to the dimensions of the *tatami* mats (3 by 6 feet) with which the house rooms are floored.

Fig. 6-12.—House of a middle-class farm family in southwestern Japan. Photograph by John Embree.

Unlike the Chinese, who have used brick, clay, and stone, the Japanese have always built their homes of wood, despite the frequent and disastrous fires. This is attributable not only to the easy availability of timber, but seemingly also to their artistic preference for wood. They neither paint the outside of their dwellings nor varnish the inside, preferring the gray, velvety appearance that outdoor weathering gives to well-dressed timber and the rich tawny and brown tones that are obtained by hand-rubbing the interior wood. It has been said that the Japanese derive from the natural colors and textures of wood a pleasure that is akin to appreciation of a fine skin. The wooden columns which compose the skeletal framework of the house, supporting the heavy roof, and which are both finishing and construction members, are small by our standards, and complete logs of small diameter are used rather than sawed timbers. Those walls that are stationary are made of several coats of mud plaster into which short rice straw has been mixed for binder. Lathing is split bamboo, woven into a mat and bound together by straw ropes. For a finished surface over the mud plaster a coating of fine sand or clay is often used. White lime plaster is also commonly used for a top coating in some districts, notably along the borderlands of the Inland Sea. Sometimes the mud-plaster wall is overlaid with a sheathing of thin boards. Only a portion of the house walls, however, are stationary; the rest, especially on the south side and perhaps elsewhere as well, consist of light wooden sliding panels or screens (*shoji*), set in grooves at top and bottom, which on hot summer days can be slid back, thus throwing open the house to receive every slight breeze that moves. At night

Fig. 6-13.—A more pretentious rural house. Photograph by Cutler Jones Coulson.

the panels are drawn together, ostensibly as a protection against burglars, but actually for the sake of privacy. Translucent paper or panes of glass in the sliding shoji permit entrance of light. At night, light streaming through these panels gives the Japanese house the semblance of a huge lantern. Houses usually open or face toward the south, so that they get the full benefit of the south winds of the summer monsoon and of the sun in the cooler season. Often there is a narrow veranda, some three feet wide, on this side, the floor of which is an inch or two below the straw mats of the adjacent living room. Double sliding panels, one forming the outer wall of the veranda, the other the wall of the house proper, are characteristic. There is no better place to relax during sultry summer evenings than on such a veranda.

It is often said that the beauty of Japanese houses is in their roofs, whose multiple forms have such pleasing combinations of straight and curved lines. The beautiful but ponderous low-hipped roofs strike one as somewhat out of proportion to the fragile walls that support them, however. Most common roof material in the rural districts was formerly thatch, but this is being widely displaced by tile and sheet metal, while in the mountain villages, roofs of bark and shingles are also common. Thatched roofs, which are sometimes two feet thick, are steeply pitched to produce rapid runoff of the heavy rain. Bright-colored flowers growing on the ridges of the thatch roofs are not unusual. Wide eaves projecting a meter or more beyond the walls of the house protect the open rooms against the rain and the nearly vertical summer sun, while at the same time offering little obstruction to the entrance of the more oblique rays of the lower winter sun. They also add grace to the roof and give an air of stability to the house. Grain and vegetables are hung under the eaves to dry.

In an ordinary farmhouse of three to six rooms in subtropical Japan, one or two rooms are at ground level and usually have an earthen floor. One of these is an entry room which is something of a catchall for sheltering bicycles and storing footgear, rainwear, and various items of household equipment. The other is a kitchen. The elevated part of the house is divided into rooms of roughly equal size, whose dimensions must be multiples of those of the tatami mats with which they are floored. Thus, a room's size is described by giving the number of mats required to cover it. Shoes are always removed before entering the elevated rooms. Bath, urinal, and toilet are in separate compartments and, while attached to the house, are not under the main roof.

Economy and simplicity are the keynotes of the Japanese house, both of its exterior and of its interior. Inside, seen by light diffused through paper windows, the colors are muted—pale green or straw tint of the reed tatami, warm neutral tones of hand-rubbed wood. Floor mats, walls, and frames are harmoniously rectangular in outline, and the angular contours register cleanly, for there is little in the way of decorations and furnishings to detract from the simple, uncluttered pattern; there are no rugs, carpets, or curtains, and no chairs, for people kneel or sit upon flat cushions which are stored in closets when not in use. Some cooking is done on low portable charcoal braziers.

In its sequence of functions the room of the Japanese house is likewise a model of economy and efficient use of space. Whereas the Occidental house contains separate rooms for different activities, each with its own furniture, a single room of a Japanese house may serve a number of purposes. Furniture and fixtures are reduced to a minimum, beds are blanket rolls which are stored in a closet in the daytime, chairs are lacking, and meals are eaten from trays or small removable tables. Since a person carries on only one activity at a time, it appears wasteful to the Japanese to have rooms that are idle many hours of the day. Rooms are separated only by light sliding screens, which can easily be removed if a room of larger dimensions is required. It is interesting to note to what extent new Western architectural design has made use of certain features of Japanese home construction—emphasis on uncluttered surfaces and natural textures of wood and straw; open construction which maintains connection with the outdoors; employment of modular building units

Fig. 6-14.—Good quality urban residences in Kyoto.

permitting standardization of building materials; removable interior partitions, allowing for flexibility of room size and multiple functions for each room.

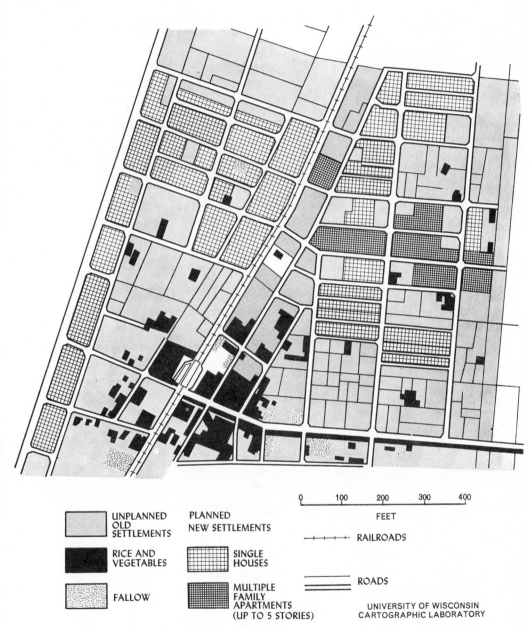

UNPLANNED OLD SETTLEMENTS

PLANNED NEW SETTLEMENTS

FEET

RICE AND VEGETABLES

SINGLE HOUSES

RAILROADS

FALLOW

MULTIPLE FAMILY APARTMENTS (UP TO 5 STORIES)

ROADS

UNIVERSITY OF WISCONSIN CARTOGRAPHIC LABORATORY

Fig. 6-15.—A new settlement area on the coastal lowland in eastern Osaka Prefecture. After P. Schöller.

Fig. 6-16.—New housing development on a diluvial upland on the outskirts of Tokyo. Courtesy of the Consulate General of Japan, New York.

Mostly local materials are used in house construction, although some materials and services must be supplied from outside the community. Paper for the sliding shoji panels, tatami for the floors, special woods for certain interior posts, bricks for the bath and stove, tile for the roof, and certain items of builder's hardware must be purchased outright. So usually must the services of a carpenter, plasterer, and roofer. Apart from the family labor, the cost of building an average rural house during the 1950's was from 400,000 to 600,000 yen ($1,117 to $1,668).[39]

The house yard of the rural dwelling plays an important part in the life of the farm family. Square or rectangular in shape, it ordinarily has dimensions of about 60 to 80 feet on a side. Sometimes it is enclosed by a mud wall, but more often a sense of enclosure is provided by the arrangement of buildings and service units, which may include a woodshed, tool and implement shed, outdoor oven, perhaps a tiny barn, together with pens for chickens, rabbits, and pigs. The stretch of bare ground in front of the house serves as a drying area for grains and other farm products, as well as a general utility yard. It may contain a few fruit trees, vegetable and flower gardens, and sometimes even a fish pond. In many parts of the country the whole farmstead is protected by windbreaks in the form of trees and bushes.

Houses in the market towns and cities do not differ fundamentally from the farmhouse, only in details. Because of the greater fire hazard thatch roofs are replaced by tile, galvanized iron, and composition products. In the house of the shopkeeper or craftsman the work or sales room is open to the street, and the residential rooms are behind it or on the second floor. Windows are commonly protected by iron bars or wooden shutters. The rich man's town house is set back from the street and is enclosed by a high frame or shrubbery wall, behind which is the garden, where small bushes, ponds, steppingstones, flower beds, and small trees are attractively arranged. This garden is an integral part of the high-class city residence, for the walls of the house fronting upon it may be slid open and it then becomes a living mural. The Japanese garden is a work of perspective art in which the spacious panoramas and vistas of nature are compressed to domestic scale.

A feature of postwar Japan associated with the great influx of population into the prospering cities is the mushrooming of large multistoried apartment structures. These may exist as single units, but more often they are multiple, while not uncommon are the extensive housing developments in the city outskirts where hundreds of symmetrically arranged apartment buildings with associated shops, banks, cinemas, etc., form populous dormitory towns.

In contrast to the unlovely appearance of so many of the hectic cities of Japan, with their confusion of Oriental and Western features, the countryside retains much of the pastoral charm which one associates with the premodern period. Here most of the man-made elements blend harmoniously with those of natural origin.

Agriculture is relatively more important in Japan than it is in most industrially advanced countries, for it employs nearly a third of the gainfully occupied, and supports a still larger proportion of the total population. And although the relative share of the national income attributable to agriculture has gradually waned as the secondary and tertiary industries have waxed greatly in importance, it still produces about 10 or 11 per cent of the nation's current income and supplies all but 15 to 20 per cent of its food requirements. But in spite of the relative importance of agriculture in the national economy, even greater in the past than at present, the agricultural segment of the population has consistently been denied its rightful share of the wealth produced, while at the same time it has been taxed exorbitantly to support the development of the other sectors of the economy. General MacArthur in his memorandum of December 9, 1945, to the Japanese Government, instructing it to legislate reform of the land system, spoke of the "economic bondage which has enslaved the Japanese farmer to centuries of feudal oppression." During the Tokugawa Period the oppressively high rice tax levied in

180

kind on the peasants of each feudal fief was the primary source of government income. The spirit of that taxation system is illustrated by sayings like "Peasants should not be allowed to die, nor yet to live," and "Peasants are like rapeseed; the more you squeeze them the more you get out of them." Sir George Sansom has written of the Tokugawa Government that it thought highly of agriculture but not of agriculturists.[1]

And while the Meiji Restoration initiated many changes, most of these were not such as to improve the lot of the oppressed peasants. To be sure, private ownership of land was established and with it the right of land alienation, but under the continuing crushing load of the land tax, which on rice land amounted to as much as 35 per cent of the value of a normal crop, an exploitive system of farm tenancy became securely fastened upon the country, the consequence of which was that the rich got richer and the poor got poorer. It was the unusual pressure of population on the land, with lack of alternative sources of employment, plus the emotional attachment felt by a farm family toward the land which it had cultivated for generations, and its disinclination to leave the security provided by the rural village, that created an environment in which such a burdensome system of taxation and tenancy could be imposed. Thus, lack of bargaining power on the tenant's part obliged him to submit to extortionate rental rates and other oppressive conditions.

Very briefly, the situation as it existed before the application of the Land Reform Law passed in 1946 was as follows: 32 per cent of the nation's farm households owned their land and tilled 54.2 per cent of the farm land, while 68 per cent were tenants or part tenants and together worked 45.8 per cent of the tilled land. About 7.5 per cent of the owners possessed 50 per cent of the land, whereas 50 per cent of the landowners owned only 9 per cent. The terms of the tenant-proprietor contract were notoriously harsh. Moreover, the lease itself rested on a vague and usually oral agreement. Rice-land rents were paid in kind and as a rule amounted to half or more of the crop. The only responsibility of the landlord was to supply the land and pay the land tax; the tenant took all the risks and was responsible for most of the expenses. Interest on farm loans was as high as 20–30 per cent.

As a consequence of the Occupation-inspired land reform legislation, all absentee landlordism was abolished and tenant-farmer land and owner-farmer land in excess of about one hectare per household became unlawful. As a result nearly 1.7 million hectares, or over 4.8 million acres, of land were purchased by the government and resold to over 4.7 million owner-farmers on easy terms. More than 27 million separate tracts were transferred, in excess of one-third of all the agricultural land of Japan. While tenancy has not been entirely abolished, its harsher features have been, and the tenant-farmer land at present is only about 13 per cent of the total cultivated area. In addition rental ceilings have been fixed by law and all leases must be in writing. These changes have greatly affected the distribution of the income of more than 2 million farm families.

To what extent the recently improved condition of the agricultural population in Japan is due to land reform is controversial. Obviously reform did not increase the area of cultivated land, so the basic handicap of a land deficiency remains unchanged. Nor did it increase the size of farms. Actually, because of an increase in the number of farm families, amounting to three-quarters of a million between 1941 and 1949, caused by wartime evacuees, displaced emigrants, and unemployed town-dwellers returning to agriculture, the average size of land holdings has been smaller since land reform than before. But doubtless some credit must be given land reform for the added productivity and increased income per unit area of land, which in turn were largely a result of improvement in land utilization and production techniques.[2] Unquestionably, there is recently an increased prosperity in the rural areas of Japan.

In many respects Japanese agriculture has changed remarkably little in the long course of its history. The main operations of planting, tilling, and harvesting as carried on today would be familiar to the pre-Tokugawa peasant, while farming is still a family enterprise as it was earlier. Still, there have been important changes, most of them occurring within the Tokugawa Period, the prime feature being the shift from cooperative to individual-family farming.[3] At the beginning of Tokugawa, farming was generally carried on cooperatively by families organized into kinship groups. By the end of this period the individual family had become the unit of production organization, and its welfare had come to be the goal of economic activity. Of special importance in instigating this change was the growth of the market, which was a powerful element in modifying the ideas and ways of men.[4] As a consequence agriculture became increasingly commercial and competitive, a feature which continues to distinguish Japanese agriculture from that of most other Asiatic countries.

With the Meiji Restoration, as Japanese agricultural products became more deeply involved in a money economy with its fluctuating markets, all feudal fief boundaries were abolished and the previous monopoly of marketing operations held by the feudal lords was dissolved. This disappearance of interferences associated with marketing their products encouraged the peasant families to improve their agricultural techniques, while the development of new and more productive varieties of rice caused a large increase in yields. Fertilizer consumption likewise expanded, better farm implements came into widespread use, the greater employment of animal power permitted deeper plowing than was possible by hand spading, while this same power could be used effectively in preparing the wet paddies for winter planting which made double cropping of more rice land possible.[5]

The distinguishing feature of contemporary Japanese agriculture is its intensity in use of land, the scarce element, and its prodigality in the use of human labor, which is, or has been until very recently, so abundant. The result is that while yields per unit area are among the highest in the world, the yields per man hour are low.

Agricultural production declined during World War II and remained at low ebb for a short time following it. Marked upturn began in about 1950, and since then striking changes have occurred that have resulted in a greatly increased production. Considering the 1950–1952 average as 100, the index of total farm production by 1961 had risen to 140 (there was a further 3 per cent increase in 1962); rice had risen to 125, making the country essentially self-sufficient in that staple food grain; pulses were up to 136; fruits, 234; poultry and eggs, 351; livestock, 241; and milk and dairy products, 460.[6] These gains are attributable in part to

TABLE 7-1

Index of agricultural production, 1961 (1950–1952 = 100)

Agriculture	143.1
Farm Crops	129.3
Rice	125.2
Wheat and barley	106.2
Pulses	135.8
Potatoes	120.4
Vegetables	130.1
Fruits	233.6
Industrial crops	154.3
Cocoons	126.3
Livestock and livestock products	312.6
Livestock	240.7
Cattle	139.2
Swine	400.1
Milk	459.8
Eggs	351.2

Source: Economic Planning Agency, Japanese Government, *Economic Survey of Japan (1961–1962)*, Tokyo, 1962.

land reform which provided for increased individual land ownership, but also to (1) government price supports and subsidies and increased aid for agricultural research, (2) large-scale land improvement especially of rice land, (3) increased use and more scientific application of fertilizers, (4) more use of chemical insecticides and weed-killers, (5) improved varieties of several crops, (6) increased mechanization, and (7) the development of earlier and stronger rice seedlings through the introduction of vinyl plastic covers for the seedbeds. This use of vinyl covers permits earlier sowing of the rice seedbeds, so that in the cooler north, harvest time is earlier and chances are improved that the maturing rice will escape cold damage in late summer. With the increase in dairying, cattle feed is coming to be supplied to a somewhat greater extent from the farm, with less dependence on the purchase of commercial feeds. More dairy cattle make for a larger supply of stable manure, which in turn offsets some of the ill effects associated with a too great reliance on commercial fertilizers.[7]

In spite of these improved conditions in Japanese agriculture, its future appears somewhat uncertain. Imbalance between agricultural and manufacturing productivity is increasing, and the growing rates of agricultural income and farm-household income are not keeping pace with the more rapidly increasing national income. Thus, the proportion of the national income provided by agriculture has been steadily diminishing since the war, from 26.1 per cent in 1950 to only 11.4 per cent in 1960 and 9.8 per cent in 1961. Moreover, the real income per farmer, although rising, has increasingly lagged behind that of the more prosperous urban employee.

In addition, there is the problem growing out of competitive foreign prices on staple food products. In the immediate postwar world of prevailing food shortages, the desirability of a policy looking toward a greater self-sufficiency in Japan's food supply was never questioned. Accordingly, it became a cardinal aim of the government to increase domestic food production, a goal which has been fostered both by subsidies and by a pricing policy designed to encourage farmers to produce more. But recently, with a fall in prices of foreign grains and a swelling tide of Japanese exports of manufactured goods, serious doubts have arisen as to whether a continuing large-scale subsidy of agriculture, looking toward greater food self-sufficiency, is economically wise. Perhaps the burgeoning export of manufactures can provide the necessary foreign exchange with which to purchase both the needed large amounts of raw materials and the more modest volume of food. In simple terms, given a certain magnitude of resources available for investment, which of two alternatives is wiser?—to invest heavily in subsidizing agriculture in the hope (a) of solving balance-of-payments problems by curtailing food imports and (b) of raising the productivity, and hence the income, of the agricultural population; or, on the other hand, to give priority to industry with the hope (a) of increasing exports of manufactures and thus solving the balance-of-payments problems and (b) of raising the productivity of the agricultural population by decreasing the number of farm families and absorbing more agricultural labor into industry. What should be the relative share of government aid going to agriculture as compared to industry? Up to now agriculture has had the top priority, but there are evidences that its relative position is declining.[8]

Significantly, the Agricultural Basic Law adopted by the National Diet in 1961 has set goals which conceivably may result in a gradual change in the geography of Japan's agriculture. Of a fundamental nature is its advocacy of an adjustment of Japan's agriculture through "selective expansion" and "structural renovation" as these are related to fundamental alterations in the food consumption pattern. In response to the expanding income of consumers, a dietary change is in progress which shows a shift from starch to protein. Thus, the planned "selective expansion" in agriculture noted above encourages an increased production of meat, eggs, milk and dairy products, and mineral and vitamin foods such as fruit and vegetables, in order to meet the needs associated with consumer demands. "Structural renovation" looks toward the development of a more efficient agriculture through (1) enlarging the farm-management scale and (2) consolidating into one unit the several scattered parcels which now comprise a farm, both of which will permit a wider and more efficient use of farm machinery, which in turn will allow a reduced, but more affluent, farm population. The plans call for a decade increase in rice production of only 10 per cent over the 1958 level, while wheat is to remain unchanged. By comparison, milk, eggs, meat, and fruit are to undergo respectively a 570 per cent, 240 per cent, 310 per

TABLE 7-2

Consumption quantum per capita in 1960 of principal foodstuffs (1951 = 100)

Rice	115.1	Eggs	218.2
Barley	37.9	Milk	441.4
Wheat	101.4	Vegetables	116.9
Total cereals	101.4	Fruit	232.1
Total meat	252.6		

Source: Economic Planning Agency, Japanese Government, *Economic Survey of Japan (1961–1962),* Tokyo, 1962.

cent, and 400 per cent expansion. If the government's plan should be accomplished, rice would decline from 50 per cent to 40 per cent of the value of agricultural production, while livestock would rise from 15 per cent to 30 per cent, although the latter would still be far below the 70 per cent characteristic of Europe and America.[9] From Table 7-1 it is clear that the largest agricultural-production increases during the decade of the 1950's were made by milk, swine, eggs, fruit, and industrial crops, which suggests that the complexion of Japanese agriculture is altering, albeit not with revolutionary speed.

In order that a considerably larger proportion of Japan's farm crops may be converted to human food by way of milk- and meat-animals, important and difficult-to-make alterations of the agricultural system will be required. Hay and forage crops and improved pastures for grazing are considered essential for successful dairy farming, and yet Japan outside of Hokkaido is largely lacking in such forms of land use. The question is posed, therefore, whether the country should import from abroad the animal products required by her more affluent population, or, on the other hand endeavor to modify her own agriculture to meet these needs. It is not unnatural that some in Japan look to Switzerland as their model, and argue that like that mountainous country, Japan should make a more efficient use of her extensive slope lands for pasturing cattle.

AGRICULTURAL LAND AND THE FARMING POPULATION

As noted in an earlier chapter, although total population in Japan has increased steadily and with moderate rapidity over the past century since the Restoration, the number of farm households has remained practically static at about 5.5 million from 1886 down to the end of World War II.* Following the war, with the repatriation of large numbers from outside Japan proper and with the decreased employment opportunities in the bombed and burned cities, the number of farm households increased to slightly over 6 million, and there it remained until 1960. A slight decrease was apparent in 1962, when the figure was 5.9 million. But on the other hand, the total number of farm-household members declined from 37.8 million in 1950, to 36.6 million in 1955, and 34.5 million in 1960. Working members engaged in farming declined 8.2 per cent between 1955 and 1960. With the above-average birth rates characteristic of farm families and a near static condition in the number

---

* The average farm family numbers about 6+, which is somewhat larger than an urban family.

of farm households, the figures indicate there has been a large exodus of population from the rural areas to find employment in the cities. It is estimated that the new population born and raised on the farms of Japan amounts to approximately 800,000 a year, of which about 360,000 (representing 180,000 new farm families, usually the first son plus his wife) remain on the farm to replace the parents, while some 440,000 leave the farms for the cities.* This represents the reservoir of new non-urban labor available to the secondary and tertiary industries. But the expense associated with raising and educating this young population to productive ages and subsequently losing it is a social and economic burden that continues to be borne by the rural areas.

Similarly, there has been no remarkable growth in the area of agricultural land or of cultivated land. An investigation completed in 1877 showed the area of cultivated land to be 4.13 million cho.[10] This figure remained the basic information concerning cultivated land for a quarter of a century. Later reports showed the cultivated area to be 5.32 million cho in 1905, 5.90 million cho in 1920, and 6.04 million cho in 1934, or an apparent increase of about 13.5 per cent in thirty years. But changes in definition and in methods of data collection make it impossible to determine whether there was actually any real and significant change in the area of cultivated land during the nearly six decades after 1877. Sample surveys by the Ministry of Agriculture and Forestry for each year since 1956, using consistent methods of data collection, indicate a near static condition in the area of cultivated land at a figure slightly in excess of 6 million hectares. This does not mean that no new crop lands have been reclaimed, for modest areas are newly brought under cultivation each year, but there have likewise been losses of cultivated fields to other uses, often urban, with the gains and losses almost canceling each other.

To be sure, the areas of agricultural land and of cultivated land are not synonymous, but they are much closer to being identical in Japan than in most countries. In 1961 cultivated lands comprised 83 per cent of the area of agricultural land, while 7–8 per cent was in meadows and pasture, nearly 5 per cent in farmsteads, and other small percentages in forests used for grazing and cutting forage, or in areas occupied by farm paths and waterways.† So, preponderantly the Japanese farm area is cultivated land. Green pastures are rare except in Hokkaido, although

---

* In the year from February 1959 to February 1960, 565,000 persons left their farm households.

† Meadows are permanent natural grasslands used mainly for cutting grass. Pastures are permanent natural grasslands used chiefly for grazing.

there is a small amount of rough grazing land. More than two-fifths of the privately owned pasture is in Hokkaido.

At the time of the Meiji Restoration in 1868 nearly 80 per cent of the employed population was engaged in agriculture and forestry. As the nation's economy has diversified, this proportion has steadily decreased —52.2 per cent in 1920, 47.9 per cent in 1930, 43.0 per cent in 1940, and only 33.4 per cent in 1960 (30–31 per cent in agriculture alone).

Because of the very large amount of land in Japan which is unfit for agricultural use, only 17.2 per cent of the country's total area was utilized for agriculture in 1960, and 14–16 per cent was classed as cultivated land.* By comparison, the proportion of cultivated land in the United States is about 25 per cent; in England, 30 per cent; and in The Netherlands, 30 per cent. As a consequence of her small area of farm land and her large agricultural population, the cultivated area per farm family is an unbelievably small figure. It currently averages under one hectare (2.45 acres), which is even less than the prewar figure, owing to a modest increase in the number of farm households since the war. Evidences of this great pressure of human beings on the meager land resource are everywhere apparent:

There are the roads always narrow and mostly at the wood's edge or the river's.

TABLE 7-3

Employment in agriculture and forestry

Year	Number of working population (000 omitted)	Proportion of total working population
1880	14,742	75.4%
1890	14,755	65.3
1900	14,405	58.2
1910	14,318	54.7
1920	14,235	52.2
1930	14,192	47.9
1940	13,950	43.0
1950	17,760	49.2
1955	16,860	41.2
1960	14,920	33.4

Sources: N. Kayo, *Basic Statistics for Agriculture in Japan* (Tokyo, 1959); and *Japan Statistical Yearbook, 1961.*

* Reckoning from the results of a sample measurement survey, the Ministry of Agriculture and Forestry gives the area of cultivated land as 6,071,000 cho (6,022,000 ha.), of which 5,755,000 cho are field and 315,700 cho are border.

There is the straw piled on brushwood bridges off the loam and the trees only growing at the god's house, never in the fields.

There are the whole plains empty of roofs, squared into flats of water, no inch for walking but the dike backs, not as much as a green weed at the foot of the telegraph poles or a corner patch gone wild.

There are the fields empty of crows after harvest: thin picking for black wings after cloth ones.

There are the men under moonlight in the mountain villages breaking the winter snowdrifts on the paddies to save days of spring.

There are the forest floors swept clean and the sweepings bundled in careful, valuable piles.

There are the houses without dogs, the farms without grass-eating cattle.

There are the millet fields at the sea's edge following the sweet water to the brackish beginning of the salt, the salt sand not the thickness of a stake beyond.

There are the rivers diked and ditched and straightened to recover a napkin's breadth of land and the hill valleys terraced til the steepest slope turns flatwise to the sun.

There are the mountains eroded to the limestone where the axes and the mattocks have grubbed roots.

All these are in the landscape. And all these—the cheese rind eaten to the brittle crust above and the careful hoarding of the crumbs below—are like Japan.

Japan is the country where the stones show human fingerprints: where the pressure of men on the earth has worn through to the iron rock.

There is nothing in Japan but the volcanoes and the volcanic wastes that men have not handled. There is no getting away from men anywhere: from the sight of men in the open houses or from the shape of their work in the made fields or from the smell of their dung in the paddy water.

In other countries a farm is meadows and a wood lot and a corner that the plow leaves: room to turn about and time to turn about in. In Japan a farm is as rigid and tight a thing as a city lot—a patch here and a triangle there and a square or so somewhere else; every road corner of land diked and leveled off even though the growing surface is less than a man's shirt; every field soaked with manure and worked and reworked as carefully and as continuously as a European farmer works a seedbed. . . .

. . . nothing thrown away, nothing let go wild, nothing wasted.[11]

It must be recognized, however, that to compare the population per unit area of cultivated land in different countries has only limited economic significance, for the productivity of land varies with such physical factors as temperature, rainfall, quality of soil, etc. The uncultivated land of some countries is practically worthless; in others it is valuable for grazing and as a source of forest products. Some countries are rich in mineral resources and others not. Japan's abundant rainfall and the relatively mild subtropical climate that prevails over about three-fifths of the total area create a climatic environment which favors multiple cropping and high yields per unit area, a factor which slightly offsets the handicap of scarcity of agricultural land. Another compensating factor is the abundance of fish in the surrounding seas. With the aid of these physical assets and by dint of the most intensive cultivation, Japan has been able to make herself largely self-sufficient with respect to foodstuffs.

Distribution of Japan's agricultural land is difficult to describe in regional terms. In Anglo-America, by contrast, a coarse pattern of distribution facilitates regional description; but in Japan the proportion of the total area that is under cultivation does not vary widely among the four main islands, there being little difference between Tohoku, in northern Honshu, and subtropical Kyushu or Shikoku. The important feature of distribution is the fact that cultivated land has a local and not a regional pattern. It exists in the form of small and fairly isolated discontinuous fragments which are strongly coincident with the numerous diminutive lowland plains of new alluvium and with adjacent areas of low diluvial upland. Such lowlands are widely distributed throughout the various islands and regions of Japan, and do not predominate in any of them. As noted in an earlier chapter, a majority of the alluvial lowlands have sea frontage, so that agricultural land is close to littorals. In the genuine hill and mountain regions agricultural land, instead of being in small compact areas, assumes a dendritic pattern coincident with the valley systems.

LAND RECLAMATION[12]

The search for new farm land and the expansion of the cultivated area in Japan has constantly been stimulated by population increase. More domestic food to support this growing population could only be provided by improved agricultural techniques and an expansion of the cultivated area.

It is believed that as of about the middle of the Heian Era (794–1185) paddy land amounted to nearly a million hectares or about one-third of its present area.[13] In feudal Japan, when the strength of a clan depended largely on the agricultural production within its boundaries, great effort was made to expand the cultivated area, especially of rice land, for rice was not only the basic food but also a substitute for money. During the Edo or Tokugawa Period, reclamation was very active, and both lowland and upland sites were reclaimed for cultivation. By the middle of the Edo Period the area of cultivated lands may have reached 3 million hectares.[14] During this period large areas of terraced paddies were developed on the slope lands bordering the Inland Sea, and numerous wet lowlands were drained and diked.

The following Meiji Period also witnessed a sizable expansion of the cultivated area, especially in Hokkaido, although Old Japan was not neglected. As noted earlier, if the figures can be relied upon there was a 22 per cent increase in the cultivated area over a period of about four decades following 1877. Thus, it becomes clear that land reclamation is not new in Japan, but has been a feature of government policy for centuries.

To meet the serious food shortage following World War II and to alleviate the mounting population pressure, new land surveys and reclamation programs were initiated by the government. A survey to determine the lands suitable for reclamation and development for agriculture produced the astonishingly large estimate of almost 3 million cho. That the estimate was unrealistic is indicated by the fact that during the decade of 1946–1955 less than 20 per cent of the lands, or 515,000 cho, were actually developed. An additional 190,000 cho were reclaimed by drainage. About 32 per cent of the land reclaimed was in Hokkaido alone, while the remaining 68 per cent was widely distributed over Old Japan. A great majority of the land reclaimed was upland, largely slope land, with infertile soils; in other words, marginal lands that previously had a cover of forest or shrub and grass. In spite of the large national investment that was made in this project, the actual accomplishments were small, indicating that reclamation of marginal lands in Japan is both difficult and expensive. It is significant also that over the ten-year period the rate of reclamation dwindled rapidly, for while it amounted to 218,000 cho in 1946 and 114,000 in 1947, it was only 27,000 cho in 1955 —and this in spite of increasing expenditures by the national government.

Since 1955 the rate of land reclamation and conversion to cultivated fields has slowed still further. In the five-year period ending in 1960 a total of only 17,200 cho were reclaimed for paddy fields and 96,750 for upland fields, a total of 114,000 cho, or less than 23,000 cho per year. Moreover it must not be forgotten that while some new lands are reclaimed for cultivation each year, other cultivated lands are lost as a result of damage from natural causes and conversion to non-agricultural uses. Thus, during the quinquennium ending in 1960, the loss of paddy area by natural and human causes was nearly three times the gain, while the loss in upland fields was about two-thirds of the gain. For paddy and upland fields combined the area lost just about equaled the area gained, so that the cultivated area remained static.

The gain in paddy area by reclamation in 1960 was highly concentrated in Hokkaido (29 per cent) and Tohoku (31 per cent), while the loss in paddy area was much more widespread, but with the largest losses in central Japan and Kinki. Gains in area of upland fields were more widely distributed than those of paddy land, but with Hokkaido accounting

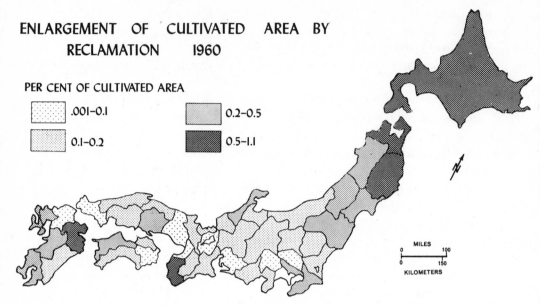

ENLARGEMENT OF CULTIVATED AREA BY
RECLAMATION   1960

PER CENT OF CULTIVATED AREA

.001–0.1   0.2–0.5

0.1–0.2   0.5–1.1

Fig. 7-1.

for 47 per cent, and northern Honshu next in order, followed by southern Kyushu. Losses in upland field area were also widespread; largest
in Hokkaido, followed by the Kanto and the Nobi areas. Losses due to
natural causes alone are concentrated in highland central Japan and
Kinki (Figs. 7-1, 7-2, 7-3).

TABLE 7-4

Agricultural land gains and losses

			1960	Total for 1955–1960
Paddy fields	Gains	Land reclamation	3,540 cho	14,220 cho
		Reclaimed from water areas	580	2,980
	Losses	Natural damage	8,120	22,998
		Non-agricultural uses	8,070	25,829
Upland fields	Gains	Land reclamation	15,300	96,095
		Reclaimed from water areas	85	655
	Losses	Natural damage	2,470	8,919
		Non-agricultural uses	15,900	52,841

Source: Norinsho Tokeihyo, 1960 [Statistical Yearbook of Ministry of Agriculture and Forestry].

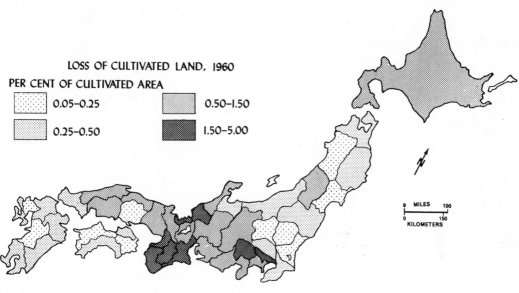

LOSS OF CULTIVATED LAND, 1960
PER CENT OF CULTIVATED AREA

0.05–0.25   0.50–1.50

0.25–0.50   1.50–5.00

Fig. 7-2.

The virtual failure of efforts to increase the cultivated area is evidence of the dearth of potentially reclaimable land that is economically usable under present conditions of agriculture.[15] The outlook for adding appreciably to the net cultivated area in the future is gloomy. Actually it is not unlikely that the land in cultivated crops may decline as a result of urban expansion and the encroachments of highways, railways, and air fields. Moreover, the land which is currently lost to agriculture is usually on lowlands and of high grade, while that being reclaimed is largely upland and of low quality. As early as twenty years ago Nasu noted, "It will be seen that the exploitation of arable land in Japan Proper has virtually reached the limit, leaving little room for further reclamation, even with highly expensive and thoughtful assistance from the government."[16] The solution to Japan's food and agricultural problems is scarcely to be sought in the expansion of the cultivated area.

Corollary to the plans for reclamation of new agricultural land are those which look toward the physical improvement of the present cultivated area. It is difficult to estimate accurately the effects of these improvements on future agricultural production but some of them appear to be reasonably important. The physical improvement of already cultivated land takes several lines: (1) extension and improvement of irrigation systems, (2) improved drainage, (3) repair of land that has been physically damaged, (4) erosion control, (5) soil dressing, and (6) consolidation of the scattered small fields.[17] The Japanese Government estimated in 1948 that physical improvements in the then cultivated

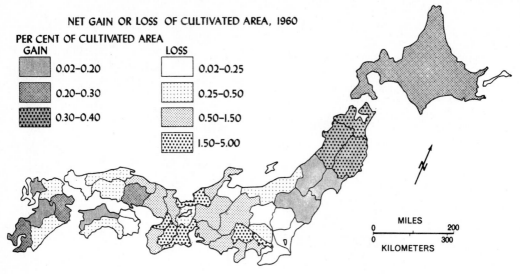

NET GAIN OR LOSS OF CULTIVATED AREA, 1960

PER CENT OF CULTIVATED AREA

GAIN

0.02–0.20

0.20–0.30

0.30–0.40

LOSS

0.02–0.25

0.25–0.50

0.50–1.50

1.50–5.00

MILES
0   200
0   300
KILOMETERS

Fig. 7-3.

land might result in a total increase in crop production of 1,322,800 tons as rice equivalents.

Paddy fields in greatest need of improvement are those that are short of irrigation water (38.7 per cent) and those that are poorly drained (23 per cent). The percentage of upland fields needing improvement is less than that of the paddy acreage and in this case the emphasis is upon such features as providing supplementary irrigation (13.8 per cent), reduction of acidity of the soil (14.1 per cent), and improvement of roads and paths (20 per cent).[18]

While programs looking toward the reclamation of new lands for cultivation, as well as the physical improvement of present cultivated lands, should be developed to the maximum consistent with economic feasibility, it is doubtful whether such programs can do more than slow down the shrinking of the cultivated area. Perhaps more is to be gained by methods looking toward increased agricultural production through (1) the introduction of new crops and the bettering of present ones by means of selection and hybridization; (2) improvement in the efficiency of fertilizer utilization; (3) better control of crop pests and diseases; (4) change in emphasis of present food crops, with a shift in acreage from low-yielding to high-yielding crops; (5) increase in livestock; and (6) further mechanization of agriculture.[19]

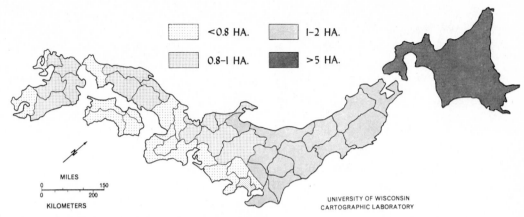

Fig. 7-4.–Average size of the farm operating unit.

FARMS AND FIELDS

Since the number of farm households in Japan at present is slightly in excess of 6 million while the area of land in farms is about 6.4 million hectares, the average size farm may be a trifle over one hectare, while the average area of cultivated land per farm probably is between 2.2 and 2.4 acres. Of some 6 million farm households about 34 per cent manage less than half a cho or hectare (1.2 acres), while over 65 per cent farm one hectare (2.45 acres) or less.

Farm size increases with latitude (Fig. 7-4). Along the Inland Sea margins of subtropical Japan farms average only about half a hectare; in northern Honshu they are commonly 1–1.5 hectares; and in Hokkaido five to six hectares is usually the average. The increasingly severe climate north of about 37° or 38° N., with shorter growing season, cooler summers, and longer duration of a snow cover makes northern Honshu and Hokkaido somewhat less productive per unit area. Thus, agricultural production per hectare or acre is lowest in Hokkaido, and next lowest in northern Tohoku where, however, it is 25–50 per cent greater than in Hokkaido. In southern Tohoku and central Japan it is more than 50 per cent greater, while in many prefectures in southwestern Japan it is double that of Hokkaido and even more. Because of the harsher climate, double cropping of fields and the growing of fall-sown cereals is more difficult in the north. The double cropping of paddies is entirely absent in Hokkaido and Aomori, and is relatively low everywhere in the northern prefectures. Where they exist at all, such subtropical crops as tea, sweet potatoes, and mulberry are of small importance in these northern regions.

By contrast with the American farm, which is ordinarily a fence-enclosed, compact, and contiguous plat of land, the tiny Japanese farm is hard to visualize since it is not a single unit, but instead is composed of several small unfenced parcels of land scattered in many directions and at varying distances from the village where the farm family lives (Fig. 7-5). The distance between the outermost farm plots may be as much as two or three kilometers. The farmstead itself usually occupies a space of about 500 to 1000 square meters and contains not only the residence but in addition one or more sheds or outhouses used for storage, for sheltering animals, or for workrooms; there is also a farmyard, a vegetable garden, and perhaps some fruit trees. Windbreaks are common.

The "open-field" system of unfenced, dispersed plots, which exists also in China and in parts of Europe, is the result of several factors. Chief of these are centuries of renting, buying, bartering, and inheriting; the farmer's desire for a diversity of crops; and the antiquated methods of irrigation. Each of the separate parcels of land is further subdivided into little fields of various sizes and shapes, called *hitsu*. These may be either flooded rice fields (*ta*), or dry upland fields (*hatake*). In Old Japan (excluding Hokkaido) the rice fields average only about one-tenth to one-eighth of an acre. The more rectangular unirrigated upland fields are somewhat larger, but 75 per cent of them include less than a quarter acre. Thus the one hectare or less of cultivated land on the average farm is subdivided into several (usually fewer than ten) non-contiguous plots; these are further subdivided into fields so that the entire farm may be composed of ten to twenty individual fields of varying sizes and shapes. On the average each farm family cultivates 8.2 hitsu of paddy and 6.6 hitsu of upland, or a total of 14.8 hitsu. The Japanese farmer is fortunate if, in addition to rice land on the wet lowlands, he has other fields satisfactory for dry crops on adjacent elevated sites—beach ridges, levees, diluvial terraces, or mountain foothills.

Partly offsetting the very obvious disadvantages of the open-field system is at least one advantage: it permits a somewhat more equitable distribution of the good and inferior lands and of those suited to only a single kind of crop. Thus, wishing to diversify his output, a farmer may choose to possess, in addition to lowland paddy land, some upland areas where he can grow dry crops and from which he can obtain fuel, grass, and fertilizer.

Considerable areas of paddy land differ from the landscape previously described in that the fields are larger, of more uniform size and more

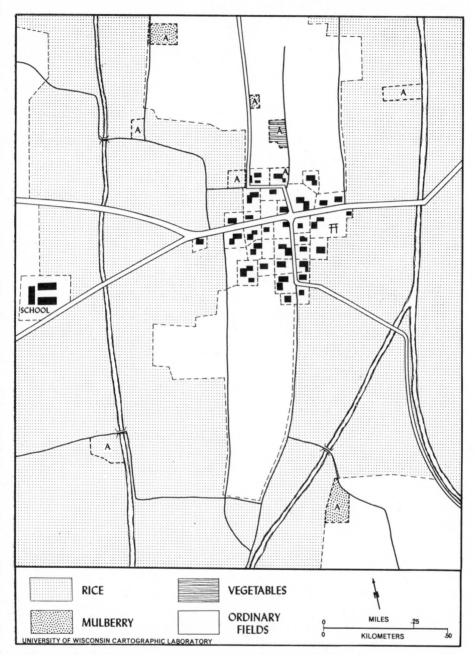

Fig. 7-5.—Layout of a representative farm. Farmer A lives in a small rural village. He has eight separate parcels of land, scattered in various directions out from the village, four of them planted to rice, and four others in dry crops. The more distant fields are about 1.4 km. from his dwelling. After S. Nasu.

rectangular shape. Where this more regular geometric field pattern prevails it usually indicates that farm lands have been "adjusted" under government supervision (Fig. 7-6). That is, a farmer is given one or two contiguous plots equivalent in area or productivity to his previously more numerous and scattered parcels. Paths, roads, and irrigation and drainage ditches are all rearranged. By this process of adjustment and consolidation a number of improvements are made: (1) The increased size of the fields and their greater rectangularity expedite farm work and the use of animals and small machinery. (2) As a result of the straightening of the field boundaries and the destroying of many useless ones, the productive power of a given area is increased 3 to 5 per cent on the average. (3) The reconstruction of the canal system allows for greater perfection in both irrigation and drainage, thus increasing the land's productivity for rice by providing greater insurance against an excess or deficiency of water. Moreover, lands which prior to adjustment were too wet for winter crops can thereafter often be sufficiently well drained for fall planting. It has been officially calculated that consolidation may increase the yield by as much as 15 per cent. Although consolidation of the scattered fields into a more compact farm unit is necessary for the rationalization of farm management, the process is made difficult because of the farmer's opposition, for he has a special attachment to each of his fields and values it higher than another's field of equal grade.

AGRICULTURAL PRACTICES

Japanese agriculture is intensive in every sense of the word—in its abundant use of human labor and fertilizer per unit area, in terms of number of crops planted in a field each year, and in the output per hectare. Probably the application of labor to land in Japan has reached the point of diminishing returns, so that augmenting productivity by increasing the input of labor is becoming uneconomical. Cultivated lands are only about 0.3 hectare per farm worker, and 0.15 hectare per member of a farm household. A single family constitutes the management unit on a farm and as a rule little outside labor is employed. Women and children do farm work as well as the men.

Slope fields including terraced fields.—Slope fields and terraced fields, which are so conspicuous a feature of Japanese rural landscapes, especially in parts of southwestern Japan, are evidence of that country's poverty in good agricultural land. On such slope fields the use of any but human labor is very difficult. Striking to a foreigner's eye are the widespread artificially terraced hillsides, where diminutive fields rise

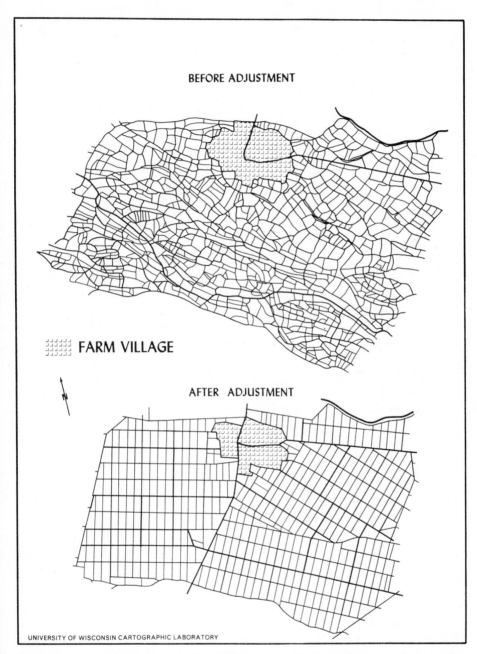

BEFORE ADJUSTMENT

FARM VILLAGE

AFTER ADJUSTMENT

UNIVERSITY OF WISCONSIN CARTOGRAPHIC LABORATORY

Fig. 7-6.—Arrangement of agricultural fields; *at top,* before adjustment; *below,* after adjustment. After map prepared by Natural Resources Section of S.C.A.P. No scale given.

one above the other for hundreds of feet in the form of enormous stairs.

Only by the most arduous labor has this terracing been accomplished, for it involves not only the carving of flattish areas out of mountain flanks and the building of retaining walls along the down-slope side of the fields, but frequently also the carrying of soil from the plains below to provide the new plot with a veneer of productive soil. The Japanese recognize two types of terraced fields: those for irrigated rice and those for dry crops, especially of the tree and bush variety. Terraced rice paddies are said to be a relic of the long period of feudalism in Japan, when the labor of farmers counted for almost nothing and the unbearable taxes levied upon the peasants forced them to expand their paddy fields up the slopes. Terracing for rice is a particularly difficult undertaking, for each tiny field must be approximately flat to permit inundation. Intricate and difficult irrigation systems are likewise required to water the slope paddies (Fig. 7-7). With the abolition of feudalism and the somewhat increased appreciation of the value of human labor, this type of terracing became less common and at present the number of slope paddies is decreasing.[20] On the other hand, terracing for unirrigated crops is increasing; such fields can be made with much less labor, since their surfaces need be far from flat, and difficult irrigation systems are unnecessary. The terraced orchards of peaches, oranges, and other fruits along the shores of the Inland Sea, and of oranges in Shizuoka Prefecture are examples of this more recent and expanding type of slope cultivation.

By no means is all cultivation on steep slopes in the form of terraced fields. Many of the unirrigated plots conform to the natural slope with little or no modification, while others are benched. Even among the terraced fields there are various degrees of slope modification. Generally speaking, bench terracing of slopes is more widespread in southwest Japan, while in central and northern Japan natural-slope fields, or slightly terraced fields, predominate. The most advanced type of terracing is the flattish bench variety with stone retaining walls, which look like contour lines as they bend with the directional change in slope. These are most conspicuous along the margins of the Inland Sea and in western Shikoku facing Hoyo Channel. Other less completely terraced fields may have a slope of 10°–15° with sod retaining banks instead of stone. Local and regional variations are great.

Terraced rice fields on slopes exceeding 15° amount to only 1.14 per cent of the total area of lithosols, but even this small percentage represents a total area of 282,442 hectares.[21] Numerous *gun* lying adjacent to densely populated plains in southwest Japan have over 30 per cent of their rice fields in the form of benches on such steep slopes. Northward from Kanto, in northeast Japan, the ratio declines and in these parts few *gun* show ratios exceeding 10 per cent.

Upland or dry fields have nearly double the paddy area on slopes exceeding 15°, and these upland fields represent 2.02 per cent of the total area of lithosols, which is nearly twice that of paddies. Ratios of the area of steep-slope upland fields to the total area of upland fields by prefectures are generally high in southwestern Japan, especially in Shikoku, where in Kochi Prefecture it reaches 60–70 per cent, but it is likewise high in the Inland Sea borderlands and in western Kii Peninsula (Fig. 7-8).[22]

Where fields are bench terraced so that their slope is slight, planting can be in any direction. But on natural-slope fields contour cultivation is carefully practiced. In southwestern Japan sweet potatoes are a particularly favored crop on slope fields because their thick growth pro-

Fig. 7-7.—Terraced paddy fields under a snow cover. Courtesy of Consulate General of Japan, New York.

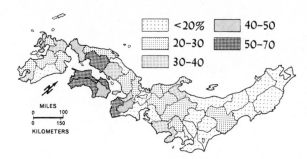

Fig. 7-8.—Proportions of steep upland fields with slope exceeding 15° (both terraced and natural slopes) to all upland fields. After map by F. Ueno.

tects the soil from slope wash. While staple food crops such as wheat, barley, rice, and sweet potatoes continue to be the most extensively grown crops on slope and terraced fields, there is an increasing tendency to plant these fields in commercial crops such as orchards and industrial crops.

Multiple cropping.—In the United States it is the usual practice to harvest only a single crop from a particular piece of land and then to allow the field to lie fallow thereafter until the next season's planting time. In Japan, on the other hand, numerous fields are obliged to support more than a single harvest during the annual cycle. In the case of the more staple crops, multiple cropping usually means one summer crop followed by one winter crop, but where quick-maturing vegetables are involved there may be even more than one summer crop. One measure of the intensity of cultivation is the ratio of the aggregate area of all crops grown during a given year to the total area of arable land. In 1960

TABLE 7-5

Areas of slopelands with greater than 15° inclination by regions

Region	Area (ha.) and percentage				
	Lithosol	Upland fields on slope land		Rice fields on slope land	
Hokkaido	4,174,880	52,325	1.25%	1,532	0.03%
Northern Japan	5,412,537	69,927	1.24	49,935	0.89
Kanto	1,475,397	39,099	2.65	12,290	0.83
Nagoya	4,286,299	88,351	2.06	47,649	1.11
Kyoto	2,787,628	51,715	1.85	52,080	1.87
Hiroshima	1,726,953	36,411	2.10	44,024	2.55
Shikoku	1,609,800	70,655	4.41	20,255	1.27
Kyushu	3,120,744	77,550	2.48	54,627	1.75
TOTAL	24,784,979	500,329	2.02%	282,442	1.14%

Source: Fukuo Ueno, "The Problems of the Utilization of Slope Land in Japan," *Proceedings of the IGU Regional Conference in Japan* (1957).

the total area of cultivated land in Japan was only 6.07 million hectares, whereas the aggregate crop area was a third larger, or 8.084 million hectares, so that the rate of arable land utilization for the whole country was 133 per cent. For 1955 the comparable figure was 159 per cent, the large variation suggesting that data on frequency of cultivation may not have a high degree of accuracy. The ratio is lowest in Hokkaido (only 99 per cent), indicating that in this northland region of relatively cold winters and short growing season, it is a nearly universal practice to harvest but one crop, and that a summer one, from a field each year (Fig. 7-9). In no other prefecture is the ratio of planted acreage to cultivated area below 100, although in Akita, Aomori, and Yamagata prefectures on the north and west sides of Tohoku it is as low as 101, 108, and 106, respectively, and even the meager winter cropping

T A B L E 7-6

Areas planted to crops, 1960 (in hectares)

Total area of arable land		6,071,000
Area of winter crops		2,266,960
Paddy fields	1,044,635	
Ordinary fields	1,222,325	
Total aggregate area of all crops		8,083,614
(Rate of utilization of arable land = 133%)		

Aggregate areas in various crops		
Rice		3,308,000
Paddy rice	3,124,000	
Upland rice	184,000	
Small grains		1,520,000
Barley	401,900	
Naked barley	436,000	
Wheat	602,400	
Oats	79,000	
Rye	870	
Other food crops		2,144,000
Miscellaneous cereals	169,577	
Sweet potatoes	329,900	
White potatoes	204,218	
Pulse	686,837	
Green vegetables	502,770	
Fruits	250,612	
Industrial crops		427,108
Tea	48,510	
Mulberry		166,163
Green manure, soiling, and forage crops		506,280
Seed-harvesting fields		11,435

Source: Japan Statistical Yearbook, 1961.

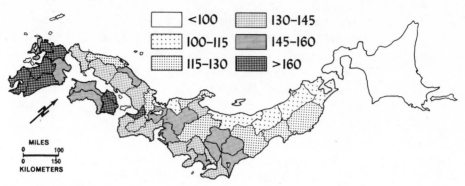

Fig. 7-9.—Annual frequency of cultivation of arable lands, or the ratio of crop area to cultivated area, 1960.

which these figures represent is scarcely cereals, but preponderantly green-manure crops. Thus, next to Hokkaido, northern Honshu—especially snowy northwestern Honshu—has the lowest amounts of multiple cropping and of winter crops. Multiple cropping tends to increase southward from Tohoku in warmer subtropical Japan, but with considerable variation among individual prefectures. Within central and southwestern Japan the highest ratios, 155–170, are reached in Shikoku and Kyushu, while the lowest are in the central highlands and along the Japan Sea side.

While the factor of temperature is fundamental in explaining the general increase in multiple cropping from north to south, it is not adequate to explain the prefectural variability within approximately the same latitudes. For example, the snowy winters with a long-continued snow cover along the Japan Sea side of the country make all winter cropping there more difficult. In that region Niigata Prefecture has an annual frequency of cultivation of only 105, Fukui 108, and Ishikawa 122, but even these low figures reflect chiefly the fall planting of green-manure and soiling crops and only very modest amounts of cereals. Also within subtropical Japan variations in the ratio of crop area to cultivated area arise as a consequence of differences in the proportions of various paddy lowlands that can be sufficiently drained to permit the planting of winter crops. The percentage of high-altitude land likewise is responsible for variations in the ratio.

There are important differences in the degree of multiple cropping, or the frequency of cultivation, between paddy fields and upland fields; about 34 per cent of the former and 57 per cent of the latter are replanted to winter crops (Figs. 7-10, 7-11). In paddy fields, where there is practically no summer fallowing, the frequency ratio is the ratio be-

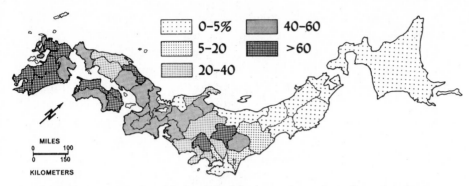

Fig. 7-10.—Percentage of paddy land double cropped, or planted to winter crops.

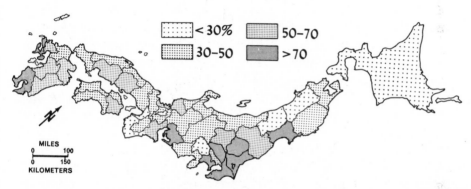

Fig. 7-11.—Percentage of ordinary or upland fields double cropped, or planted to winter crops.

tween one-crop and two-crop, or even three-crop, paddies. Rice production is of such paramount importance in Japanese agriculture that paddy fields are valued much more highly than upland fields and practically all paddy fields bear one crop of summer rice. The second crop in the irrigated rice fields may be small grains (wheat, barley, and naked barley), comprising 60 per cent of the total, or green-manure and forage crops accounting for 25 per cent, or white potatoes, rapeseed, onions, and other vegetables, most of which are planted in the fall after the rice harvest and reaped in late spring or early summer before the setting out of the rice seedlings. Even triple cropping is practiced in the warmest parts, with summer rice followed by winter wheat, barley, or naked barley, and these in turn by vegetables. The winter fallowing of two-thirds of the paddy land often represents a matter of choice on the part of individual farmers, influenced by a variety of factors. As a matter of fact, the fallowing process is becoming more widespread as rural prosperity increases, part-time farmers become more numerous, and more farm households augment their incomes by urban employment.

Whether such second crops are physically possible is determined largely by climatic and drainage conditions. Thus, in northern Japan the short growing season plus low temperatures in spring and fall, and along the Japan Sea side the deep and long-continued snow cover, discourage the planting of a second crop in the rice fields. As a consequence none of Hokkaido's paddy land is replanted in winter, and in the five northernmost prefectures of Honshu the aggregate of winter food crops grown on paddy land is only about 850 hectares. While this is raised to nearly 5500 hectares if winter green-manure crops are included, even that figure is only .014 per cent of the total rice area. The northern limit of significant winter cropping of rice fields is about 38°, or approximately the latitude of Sendai and Niigata cities, and along the snowy Japan Sea side the area of fall-sown cereals in paddies remains modest as far south as Wakasa Bay, even though winter green-manure and soiling crops are relatively important. South and west of about 38°, where climatic conditions in general permit winter cropping, the chief obstacle to fall replanting is the difficulty of sufficiently draining the summer-inundated rice fields.

Upland or unirrigated fields vary much more widely with respect to intensity of land use and kinds of crops grown, ranging from complete year-round fallow to more than two crops a year, with the variations being due in part to climatic differences. As noted previously the proportion of ordinary upland fields which are winter-cropped is considerably larger than that of paddy land. Also, significant winter planting occurs farther north in ordinary fields than in paddies; for it will be recalled that even in Hokkaido about 10 per cent of the acreage of small

TABLE 7-7

Single and double cropping of paddies

Total area of paddy fields		±3,100,000 ha.
One-crop paddies		±2,060,000 ha. (66%)
For climatic reasons	±26%	
For drainage and other causes	±40%	
Two-crop paddies		±1,040,000 ha. (34%)
Small grains	±20%	
Green manure and forage crops	± 9%	
Rapeseed, vegetables, rushes	± 5%	

Sources: Compiled from *Norinsho Tokeihyo, 1960* [*Statistical Yearbook of Ministry of Agriculture and Forestry*]; *Japan Statistical Yearbook, 1961*; and Shiroshi Nasu, *Aspects of Japanese Agriculture* (New York, 1941).

grains is fall-sown, although this amounts to only 1–2 per cent of the total cultivated area. In Tohoku winter cropping of upland fields is common, for even in the three northernmost prefectures of that region about one-third of such fields are so utilized, and in the three southernmost the figure rises to 56 per cent. Within Tohoku, the contrasts are sharper between east and west actually than between north and south, for in Akita and Yamagata prefectures only 25 per cent of the upland fields are sown to winter crops, while the comparable figure for Iwate and Miyagi prefectures in the east is 55 per cent. This would suggest that the long-continued snow cover characteristic of the Japan Sea side provides an additional handicap to fall planting.

Farther south in subtropical Japan multiple cropping of upland fields is somewhat more common, though even here the annual frequency of cultivation varies greatly depending on the kinds of crops grown. For example, where perennials such as mulberry, orchards, and tea are important, the ratio may be small even though the region has a high productivity. The kinds of crops grown on upland fields vary widely from region to region, but common as summer crops are upland rice, sweet potatoes, and miscellaneous vegetables. These are succeeded by fall-sown cereals (79 per cent of total), with smaller amounts of white potatoes, rapeseed, and green-manure and forage crops. The latter group, while comprising about a quarter of the winter-crop acreage of paddy land, amounts to only 3 to 4 per cent of that on upland fields. In the suburban market-garden districts of great cities the multiple cropping system chiefly involves the growing of a succession of vegetables. For example, in the prefectures of Osaka and Hyogo, in which the cities of Osaka and Kobe are located, the index of double cropping for upland fields is between 200 and 300.

Interculture.—Several crops from the same field are also made possible by interculture practices, a kind of simultaneous rotation in which alternate rows of different crops sown at different times are grown together in the same field. This is sometimes called catch-cropping. Commonly the intercultured crops are those having contrasting growing seasons, or light requirements. Thus, in a field bearing a crop of winter wheat or of barley planted in rows, beans of various kinds, or some other vegetable, may be sown early in spring between the rows of grain. After the winter grain has been harvested other vegetables may be planted in the vacant rows between the rows of earlier vegetables. By such a close dovetailing of several crops in time and space the land is forced to yield several harvests each year. This practice is largely confined to upland fields, and it is more common in those parts of the country where the growing season is shorter than it is in the warm southwestern districts.

Crop rotation.—Rational and methodical crop rotation systems resembling those of Western countries are precluded by the unusual emphasis Japan must place on cereal crops in order to produce a maximum of food. Instead, various irregular methods have been developed regionally and locally. Most of these involve leguminous crops, especially various kinds of beans.

Fertilizer.—Land which is forced to grow several crops a year, by such practices as have just been described, can be maintained at a high level of productivity only by frequent and abundant applications of fertilizer. This is especially true of the poor upland soils which have been subjected to the strong solvent action that is inevitable in a rainy climate, much of it subtropical. Compared with most parts of the world the application of fertilizer per unit area in Japan is large. In recent years the average outlay for fertilizer per farm family has been close to 25 per cent of the total agricultural expenditure. This intensive use of fertilizers involves a large assortment of domestic manures as well as commercial materials, both inorganic and organic. Domestic manures include animal manure, human excrement (night soil), compost made from crop residues such as straw, leaves, stems of crops, and wild vegetation cut from uncultivated land, kitchen refuse, seaweed, fish fertilizer, wood ashes, silkworm excrement, and many others. Night soil, which long has been an important form of fertilizer in Japan, especially in those areas adjacent to cities and towns, is rapidly being replaced by chemical fertilizers. It is a common practice for the farmer to dump night soil and some other kinds of manure into a shallow cement storage cistern, where, with water added, the mass is allowed to decompose. Later it is carried to the fields in large wooden buckets, and with the use of a long-handled wooden dipper it is poured over the rows of growing crops. The idea seems to be to feed the plants rather than fertilize the soil.

Gradually Japan has been shifting to a greater dependence on commercial fertilizers, and at present its consumption of these per unit area is higher than for any country with the exception of The Netherlands. Organic commercial fertilizers include soybean meal and other oil-cake meals, bone meal, and products manufactured from fish. Before World War II, organic fertilizers were imported in large quantities from other Oriental sources, especially China and Manchuria, but the relative importance of these has declined in recent years as the use of inorganic commercial fertilizers has soared.

In summary, it bears restating that, measured by any one of three standards—human labor applied per unit area, number of crops planted per year, fertilizer consumed per unit area—Japanese agriculture is unusually intensive. Thus, the per-unit-area production of rice is two to three times as great in Japan as in the countries of tropical South Asia and 40 per cent greater than in China. On the other hand, it is lower than in Italy and Spain, the reason being, according to Nasu, that Japan's national average is brought down as a consequence of an important part of its rice being grown in the northerly latitudes and in high altitudes where climatic conditions are unfavorable for this subtropical plant. Per-unit-area production of wheat and barley in Japan is 65 per cent greater than it is in the United States, is about equal to that of France, but is lower than that of Belgium and The Netherlands.

Farm labor and machines.—Until very recently, human labor has been the item in most bountiful supply on Japanese farms and hence it is the factor that did not have to be economized as did land. As a consequence of overabundant labor in the rural areas there was little incentive to develop labor-saving devices; the problem was how to squeeze the largest possible harvest out of a small area by using the maximum amount of labor.

Most farm work is still done by manual labor, but in recent years motors for pumping water, small power threshers, and even power tractors for plowing have become more common. The mechanization of threshing and hulling is now almost universal. The hand tractor for plowing is more recent and much less widely disseminated. Plowing is the most burdensome of all farm labor, and elimination of manual labor from this operation would solve the greatest physical problem of farming, even though most other farm operations, such as transplanting rice, weeding, cultivating, fertilizing, and reaping would continue to rely on manual labor. It is still a common sight to see a farmer and his wife wielding a long-bladed mattock to turn the soil; indeed, in the smaller and more irregularly shaped rice fields and on the terraced and steeply sloping fields of mountains and hills, there is no alternative to hand plowing. But on the broader plains and where fields are larger and more regular in shape, plowing is more commonly done with a horse or an ox. Still, in 1960 only 43 per cent of the farm households had one or more draft-beef cattle or a horse.

Today the real star in the mechanization of agriculture is the small hand tractor, which is a postwar innovation. As late as 1947 there were only 8000 such gasoline tractors, but by 1960 this had skyrocketed to 517,000, or one power cultivator for every eleven or twelve households. Numerous farmers rent tractor service from neighbors who own one, so that probably nearly three-quarters of a million families make use of a hand tractor. This means that in 1960 there was one tractor for about every ten hectares, or twenty-five acres, of cultivated land. The probable effects of such mechanization are still in dispute, although there is some evidence that it has resulted in increased production.[23] Perhaps the primary result, however, is that it frees the farm family from long hours of hard labor and leaves them more time to rest, to enjoy radio and television, to drink tea and talk politics.

The distribution of hand tractors is widespread, but varies greatly in numbers and density among the prefectures. They are relatively scarce in Hokkaido where the larger farms and the system of plowing with horse power has been long established. A rational pattern of distribution in Old Japan is difficult to observe. Tractors are especially numerous in Niigata, a prefecture specialized in rice growing; in Saitama in the Kanto area; and in Okayama in Sanyo. They are relatively scarce in the more backward prefectures of southern Kyushu.

Part-time farming.—A striking feature of change in rural Japan is the remarkable increase in the percentage of farm households for which farming is only a part-time occupation. Thus, the ratio of full-time to part-time farm households was 69 to 31 in 1940, about 50 to 50 in 1950, and 34 to 66 in 1960. Part-time farm households can be further subdivided into those in which farming is still the main occupation of the family, about half the total, and those in which it is a subsidiary occupation.

This marked increase in the number of part-time farm households is related to a number of factors, including the sharp uptrend in the number of people who had to be supported by agriculture immediately following the war, combined with no increase, or possibly even a decrease, in the area of cultivated land. Accordingly, the per-capita area of cultivated land shrank, leaving the farms with surplus labor which sought means of supplementing its meager farm income. In proximity to the coast, some found employment in fishing; in mountain areas, in forestry; while in urban areas job-seekers resorted to a great variety of employments. Later, as the manufacturing and tertiary industries expanded widely, the opportunities for city employment of farm labor increased.

Saito has mapped the distribution of the ratio of part-time to total farmers and sought an explanation for what appears to be its relative irrationality.[24] The distribution is complicated, partly because the reasons for part-time farming are not one but several. Doubtless there is a degree of correlation with size of farm, with the percentage of part-time farmers higher on small farms than on large ones. It is also related, however, to agricultural gross income and to working hours per farm household; so Saito concludes that both working hours per hectare (agricultural labor capacity) and productivity as well as farm size are important factors.

THE AGRICULTURAL SCENE

On a world map of agricultural types and regions Japan is usually shown as belonging to that type, so extensively developed in eastern and southern Asia, designated as "Intensive Subsistence Agriculture, Rice Dominant." It is true that Japan's is an intensive agriculture, based chiefly on cereals, with rice the dominant crop, but on the other hand Japan differs from most of eastern and southern Asia in that a much larger percentage of its agricultural production is sold on the market rather than being consumed by the producing household. Important fractions, even, of such staple crops as rice, wheat, and barley are sold in the market in the same way that cash crops like fruits, industrial crops, and vegetables are. However, despite the development of a market economy, not more than half the rice crop is placed on the market, and no more than 40 per cent of the wheat and barley. This is chiefly because government rationing of rice still persists and because subsistence farm households continue to amount to nearly 40 per cent of the total. Here again in the matter of commercial agriculture, as in so many other ways, Japan is intermediate in position between the countries around the North Atlantic Basin and most of those in eastern and southern Asia.

As noted earlier, Japan's is a food-crop agriculture with such crops occupying about 87 per cent of the cultivated acreage, while industrial and feed crops account for only 13 per cent. Among the food crops great emphasis is upon cereals, which taken together occupy about 63 per cent of the crop area. Such stress on shallow-rooted cereals reflects the necessity for obtaining a maximum of food from a limited agricultural area.

The irrigated lowlands and paddy rice

As pointed out previously, the Japanese distinguish two kinds of fields: paddy fields where irrigated rice is grown and upland, dry, or ordinary fields on which are grown other crops, very largely unirrigated. Thus, irrigated fields are almost synonymous with rice, for the amount of unirrigated or upland rice that is grown is very minor, only about 5–6 per cent of the total.

The planted acreage of various crops changes from year to year, but in 1960 rice occupied 56 per cent of the cultivated land and about 46 per cent of the crop area, while on the average it represents 50–60 per cent of the total value of all agricultural products. While Japanese agriculture scarcely qualifies as monocultural, it hints of that condition. From earliest historic times Japanese farming has revolved around paddy rice. By contrast, crops other than rice, largely grown on dry fields, have never been more than a supplement and always secondary. Moreover, forests and grasslands have been used for promoting the cultivation of paddy fields, for they have been a source of supply of green manure and of compost ingredients, of materials for fashioning agricultural implements, of fuel and building materials, and have been utilized as pastures for draft animals and as natural areas for water storage connected with irrigation and flood control. The fact that paddy fields occupy well over half of the cultivated land is all the more remarkable when one considers that paddy rice is more exacting in its climatic and terrain demands than are most upland dry crops. Rice is grown on the same fields year after year, and change in the planted area of this crop has not altered significantly over the past several decades. On paddy lands very few summer crops compete with rice.

Such a high degree of specialization in paddy rice has a historical basis. The crop was introduced from China by the Yayoi peoples some 2000 years ago, and already by the time of the Taika Reform of 645 A.D. it was firmly established throughout subtropical Japan. But the pre-eminence of rice is scarcely to be explained in terms of adjustment to a particularly favorable physical environment, although to be sure, some physical features of central and southern Japan, such as subtropical climate, abundant warm-season rainfall, and flattish alluvial lowlands that are easily irrigated, are suitable and even attractive for rice cultivation. Another factor favoring specialization in rice is that the production of calories per unit area of rice land is much higher than that for other cereals (170 per cent of that for wheat, and still larger as compared to corn), so that more people can be supported by rice than by other grain crops.* Also, it is nutritionally superior to other cereals, for it contains 7 per cent by weight of easily digestible protein. In addition, unhulled rice has excellent keeping qualities under conditions of a warm humid climate and can be stored for several years. All of these qualities recommend it as the standard food crop for Japan.

But actually it is more than a food. For several centuries it was the basis of exchange in feudal Japan. To some extent it is almost deified.

* Rice, wheat, and corn supply about the same number of calories per unit of weight, but since the yield of rice per unit area is higher than for wheat and corn, the available calories are also.

All Japanese festivals have some relations to rice and rice is the supreme offering to every god. Also it is important in the home, for in some districts a farmer's wife does not actually obtain the right to manage the house until her mother-in-law ceremoniously hands her a dry rice measure with which she will dispense the rice belonging to the household. Rice is also believed to possess magic power. The new year of the ancient Japanese calendar came immediately after rice harvest. Annual celebrations are keyed into the work-cycle of rice cultivation.[25]

The inordinate preference of the Japanese for rice as a food, reflected by its ceremonial importance, not only has led to the growing of paddy rice in locations to which it is unsuited, such as steep slopes requiring elaborate terracing, and cool northerly latitudes such as Hokkaido, but it has made the Japanese somewhat myopic concerning the value of non-rice land and the food crops that can be grown on it.

Through the breeding and propagation of early-maturing varieties of rice during the present century, even most of Japan's northern island of Hokkaido as well as the highlands of central Japan have been brought within the bounds of the rice-farming area. About 9 per cent of Japan's paddy fields are found on steep mountain slopes whose gradients exceed 15°. Close to 6 per cent are in Hokkaido, with its microthermal climate, where they occupy 20 per cent or more of the total land in crops. About 90 per cent of the farm households raise rice. The potential altitudinal limit of paddy is set by an August average temperature of 19°C. (66.2° F.), and in many highland parts of Japan it has not reached this limit. The present-day highest paddies are in the central highlands of Honshu at an elevation of about 1300 meters.[26] Strangely enough, however, some of the highest yields of rice per unit area are recently to be found in mountainous Nagano Prefecture and in northern Honshu, where summer temperatures are not subtropical (Fig. 7-12). And while Hokkaido's yields are the lowest for any prefecture, still in some years they are almost as high as those of the subtropical south. It has been argued that the harvest of rice decreases in geometric progression as the temperature of the warmest month falls in arithmetic progression below 21° C. (69.8°F.).[27]

Three important low-elevation boundaries are associated with rice culture in Japan (Fig. 7-13). The first, in extreme northern and eastern Hokkaido, marking the northern limits of its cultivation, is set by minimum diurnal temperatures in mid-May, the planting season, and in mid-October, the harvesting season.[28] Here, supplementing latitude, a cold ocean current, together with sea fog, makes for unusually cool summers. Only a small part of the country lies beyond this northern boundary. A second boundary, described previously as located at about 38° N.,

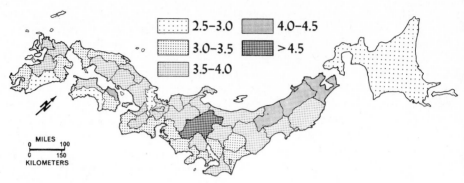

Fig. 7-12.—Yield of rice (tons per hectare) in brown-rice equivalent. Average of three normal years during the period of 1954 to 1958. After map by Takane Matsuo.

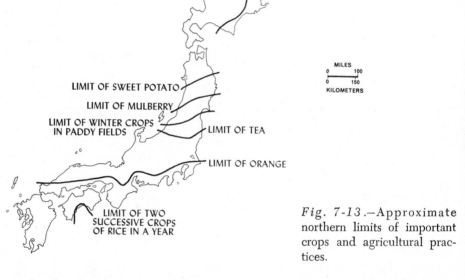

Fig. 7-13.—Approximate northern limits of important crops and agricultural practices.

marks the approximate northern limit of winter-cropped paddy fields, and hence separates the region of one-crop paddies to the north from the more southerly parts where many, but by no means all, paddies bear a second winter crop. In this latter region, because of widespread double cropping of paddies, the rice fields have a total greater production than do the single-crop paddies farther north. The third rice boundary, marking the northern limit of two successive crops of rice during a year, is of little significance, since only a very small area in Japan is sufficiently warm for long enough to grow two successive crops of rice. Only on the Kochi Plain of southern Shikoku, which is protected on the north by relatively high mountains, and in very limited areas in southernmost Kyushu is double rice-cropping practiced.

The total land area in paddy showed a slow but steady increase until about 1930, the expansion amounting to 23 per cent in the 45-year period 1885 to 1930, or about one-half of 1 per cent per year. But if this national change is examined regionally, some important areal shifts are obvious. In general, the southwestern districts registered either decreases or only slight increases, while the northern districts experienced distinct gains, so that the rice center of gravity gradually moved eastward and northward. Between 1892 and 1951, the acreage of paddy in Hokkaido increased nearly 60 times; in Aomori, Iwate, and Miyagi prefectures in Tohoku, more than 20 per cent; and in Yamagata in the same district, 10–20 per cent.[29]

The area in paddy is about the same now as it was in the 1930's, and there is little likelihood of any important increase in the foreseeable future. But further improvements in rice species, greater use of artificially warmed rice-seedling nurseries, and betterment of peatbog soils will likely result in larger and more dependable yields of rice, even on an area that, for reasons discussed earlier, is fairly static or even declining. The two most extensive types of flattish or mildly sloping land seemingly suitable for rice, but still unreclaimed, are some diluvial uplands and volcanic-ash slopes. But the latter type usually lacks sufficient water resources for paddy, and while the elevated diluvium may be somewhat better supplied, even such sites frequently require long and expensive irrigation canals or costly pumping. In both, soils are notably infertile. Nevertheless, diluvial uplands account for about 20 per cent of the flattish land in subtropical Japan, and it is this diluvium which appears to offer the greatest hope for expanding the paddy area.[30]

A major factor in the distribution of paddy land is the availability of water for irrigation or, more correctly, inundation. A variety of water sources are employed—rivers, lakes, artificial ponds collecting runoff, underground water, etc. Rivers, together with minor amounts from lakes and marshes, provide irrigation water for 74 per cent of the paddy area; ponds and reservoirs are responsible for 18 per cent; and underground water, pumped and from springs, about 8 per cent. Pond irrigation is especially important in the Kinki and Inland Sea districts where rainfall is less abundant, and stream volume small. Ponds are numerous both on plains and in the nearby foothills and diluvial lands where the damming of ravines and small valleys is relatively easy. Quality of paddy-land irrigation and drainage varies from locality to locality and from region to region. Water-logged rice fields, for example, are nu-

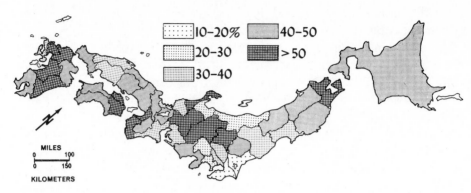

Fig. 7-14.—Drainage conditions of paddy fields. Percentage of all paddy land that can be adequately drained for fall planting. After map by F. Ueno.

merous in Kanto and Hokuriku, and to a somewhat less extent in Tokai (Fig. 7-14). Such conditions are much less common in Tosan, Kinki, Shikoku, Kyushu, and Chugoku.

As might be expected, the distribution of rice shows a high degree of coincidence with the plains of new alluvium—deltas, floodplains, and fans (Fig. 7-15). On the flattish delta-fans multitudes of diminutive fields, increasingly rectangular in outline, and each enclosed by tiny dikes a foot or more wide and of equal elevation, form a landscape which is a faceted mosaic in pattern, but rather uniform in coloration (Fig. 7-16). At certain locations on some of the plains one can gaze for a mile or more over continuous expanses of paddy fields whose monotony is virtually unbroken, but this is the exception rather than the rule. More frequently the expanse is interrupted by villages, sometimes tree- or hedge-enclosed, rising like islands above the inundated paddies, and by the elevated footpaths and roads which traverse the plains, usually in straight-line courses. Here and there intersecting irrigation and drainage canals and ponds add a note of variation, and in some places scattered dry fields, artificially elevated one or two feet above the paddies, stand out conspicuously above the lower level of rice fields. On the sandy and elevated beach ridges which commonly border the sea margins of the delta-plains, as well as on the river levees, paddies give way to dry fields. Not infrequently there are riverine belts, where laterally shifting streams have somewhat roughened the surface of the land and have deposited coarser materials; and these also are generally non-rice areas.

A well-defined grid pattern of paddies, a remnant of the Jori system of land allotment, which was imposed upon the agricultural lowlands of Japan by the Yamato rulers during the Taika Reform in the seventh century, is very conspicuous in some parts of southwestern Japan. In such regions the grid pattern is evident not only in the arrangement of paddy fields, but also in the road and canal patterns and in the somewhat regularly spaced compact villages (Fig 7-17). The basic unit of Jori was one *ri* square, or about 640 meters on a side. This was divided into thirty-six square units, each of which was one cho, or about one hectare. Such was probably one family's holding. Thus the square *ri* must have provided for a village of about thirty to forty families. The Jori system originally was imposed over most of Old Japan, for evidences of it are found from northern Honshu to southern Kyushu, although, as the distance from Nara, the seat of the Yamato Court, increased, the system was less intensively applied (Fig. 7-18). Evidences of it are especially widespread in the paddy lands of Kinki, the Inland Sea borderlands, and northern Kyushu.[31]

The march of the seasons produces a succession of contrasting paddy landscapes. Early spring sees the preparation and sowing of the rice seedbeds, for the method of direct seeding of paddies is seldom used. In May or June the rice seedlings are transplanted, and set out by hand in the flooded paddy fields. Transplanting has a number of advantages over direct seeding: (1) In late spring and early summer many paddy fields are still growing winter crops. (2) May and June are especially busy months for farmers and the schedule of agricultural operations is tight, so that the work schedule would be badly disarranged if the paddies had to be prepared as well as planted at this time. (3) Weeding is facilitated by having the rice in rows. (4) Through the growing of seedlings the weaker plants can be eliminated before transplanting occurs. After transplanting, a flooded alluvial lowland becomes a much subdivided water surface, pricked by the rows of tiny rice seedlings.

By midsummer the scene has changed; lush green is the prevailing color over large expanses of the plain, although the boundaries of the individual fields are still obvious, being set off from one another by the interruptions at the small dikes which enclose each field. As green changes to yellow in autumn and the ripened grain is harvested, the fields swarm with human beings engaged in cutting and threshing the precious crop (Figs. 7-19, 7-20). Autumn is a season of great anxiety, for September and October are typhoon months, and the heavy rain and wind that accompany the storms may do great damage to the ripening rice heavy with seed. Most of the rice is harvested in late October and November.

As noted earlier in this chapter, only about one-third of Japan's paddy area is replanted after the rice harvest to a second, or winter, crop. Of the over 2 million hectares which remain fallow in winter, probably about two-fifths are in Hokkaido and northern and western Honshu, where

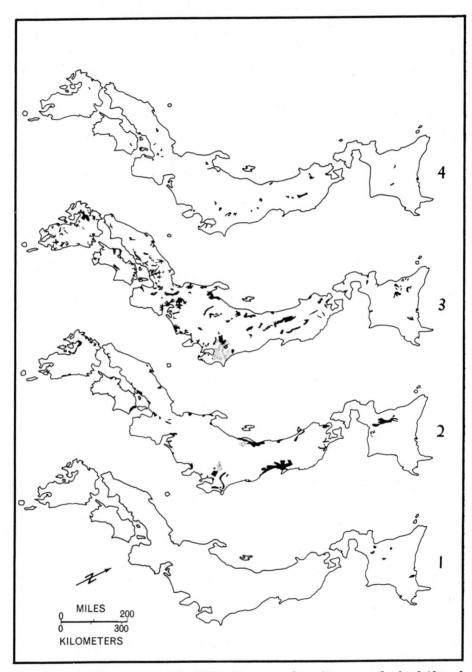

MILES
0   200
0   300
KILOMETERS

Fig. 7-15.–Distribution of paddy fields: (1) on peat bog, (2) on marshy land (dotted areas represent dispersed small patches), (3) on well-drained alluvial plains, (4) on slightly elevated (under 10 meters) alluvial plains, (5) on diluvial uplands, (6) on steep

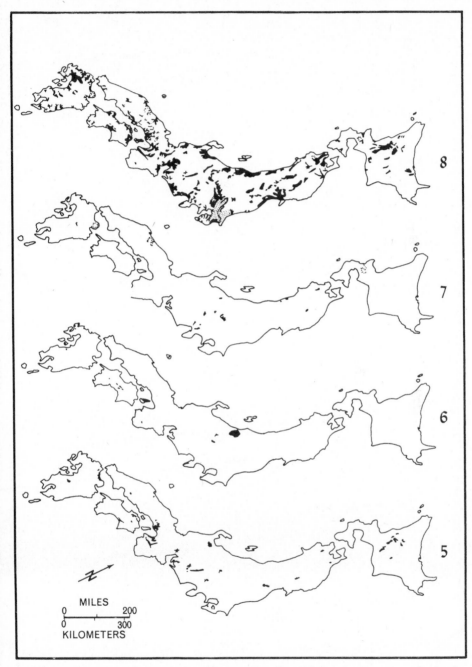

mountain slopes, (7) on flanks of volcanoes, (8) all paddy land. After maps by Yoshi-katsu Ogasawara.

Fig. 7-16.—Inundated paddy fields in late spring at the time of rice planting.

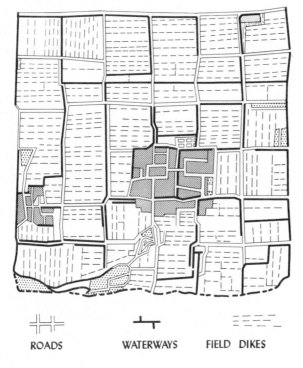

ROADS   WATERWAYS   FIELD DIKES

Fig. 7-17.—A Jori village and the rectangular arrangement of its roads, canals, and fields. After map in *Regional Geography of Japan*, No. 3, *Kinki Guidebook*. No scale given.

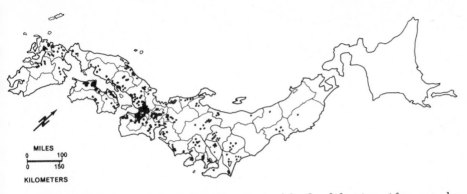

Fig. 7-18.—Distribution of the Jori grid pattern of land subdivision. After map by M. Fukaya and others.

Fig. 7-19. Harvesting rice.

for climatic reasons the replanting of paddy fields is impossible, or at least unprofitable. The other three-fifths of the winter-fallow rice fields are located in subtropical Japan, where difficulties both natural (chiefly poor drainage) and economic restrict winter planting.

If the fields are to be fall-planted their surfaces are usually worked up into a series of narrow parallel ridges and troughs, the former twelve

Fig. 7-20.—The rice fields after harvest.

to eighteen inches wide on top and perhaps eight to fourteen inches high, and on the flat ridges grain is sown very thickly in single or double rows. The practice of "ridging" for winter crops is quite essential, for the poor natural drainage plus the cool-season precipitation tends to make many of the planted paddies fairly wet (Fig. 7-21). It is not uncommon to see water standing several inches deep in the troughs between the rows of winter grain or rapeseed. The planted sections of some delta-plains are never more colorful than in late spring, when the dark green fields of winter grain in rows are intermingled with plats of brilliant yellow rapeseed and bright pink *genge* (a green-manure crop). Other advantages of ridging the paddy fields before planting them in winter crops in rows are that (1) the fields can be more easily weeded by hoeing; (2) the furrows or troughs are kept clear so that other crops can be intercultured between the ridges in the spring; (3) the cultivator can walk between the rows of winter grain and apply fertilizer at the base of the plant; and (4) sidewise tillering and spreading of the grain are prevented. Usually the winter grain is sown in clumps along the top of the ridge rather than spaced evenly and continuously. The Japanese believe this helps to prevent lodging and that the grain is less likely to be beaten down by heavy wind and rain.

Near cities the paddy lands are likely to get little or no rest between the major summer crop of rice and a winter crop of cereals. For example, immediately after the rice harvest a field may be ridged, and a quick-maturing crop, such as the giant radish (*daikon*), may be planted on the ridges, and eggplant along the bottoms of the furrows. To give them a head start these vegetables may have been sown earlier in seedbeds and the young plants subsequently transplanted to the paddy fields. Usually the daikon is mature before the eggplant interferes. After the daikon is harvested winter grain is planted on the ridge tops. The eggplant is pulled when it begins to interfere with the growing grain. In the spring a quick-maturing crop, such as dwarf taro, cucumbers, or seedling melons may be planted in the troughs between the rows of young grain. To be sure, the vegetables are somewhat shaded by the grain, but they struggle upward for light and finally, after the grain harvest, have the field to themselves until the time comes to relevel the field and transplant rice seedlings on it.

Although it is not within the scope of this book to describe the regional variations in rice culture in Japan,[32] it is worth while to emphasize that these variations are both great and areally complex in reference to varieties grown; injury from insect pests and diseases; seeding, transplanting, and harvesting dates; types of rice nurseries; transplanting

Fig. 7-21.—Ridged paddy fields with rows of winter cereals and rapeseed. The scene is in spring.

methods; cultivating and weeding operations; and fertilizing practices.

The rice cropping season in Japan extends from March to late November. Normally there is as much as a month's difference between Kyushu and Hokkaido in times of transplanting, harvesting, and other operations, but this differential is becoming less as short-season rice varieties capable of high yields have been developed and artificially warmed vinyl-covered seedbeds have become more widespread. Such protected nurseries permit the transplanting period to be advanced as much as fifteen to twenty days. These two developments, plus the breeding of rice varieties which are resistant to summer cold, quite naturally are of greatest importance to rice growing in the cooler and shorter-growing-season regions of Japan's higher latitudes and higher altitudes, where as a consequence yields have been increased as much as 10–20 per cent. But the effects are being felt, although in lesser degree, in the subtropical parts as well, where earlier rice harvests make it possible to avert damage associated with typhoons, which are most common in late August and September and October. As a general rule, rice varieties change from late-maturing to early-maturing kinds with increase in latitude and altitude. But on the other hand, since different maturing periods enable the farmer to better distribute farm labor requirements numerous rice varieties are grown even in the same locality. Moreover, the distribution of rice varieties varies not only with temperature, but also in response to soil fertility, resistance to insect pests and diseases, irrigation and drainage efficiency, and whether or not winter crops precede rice on a particular field.

Extensive transfers of harvested rice occur within Japan. Two powerful and complementary forces—increasing regional specialization and expanding urban markets—are responsible for the interprefectural movement of large quantities of the staple. Regions of greatest surplus are Tohoku, Hokuriku, and northern and eastern Kanto, which together form a large and contiguous area. No other large area so qualifies, but there are individual prefectures elsewhere which also have important surpluses, such as Kumamoto, in Kyushu, and Shiga, in Kinki. In general, however, the greater part of central and southwestern Japan forms a region of rice deficiency, as did Hokkaido until very recently. Major direction of rice flow, therefore, is southward from Tohoku, Hokuriku, and northern Kanto, toward central and southwestern Japan, and more especially toward the great urban centers of Kanto and Kinki.[33]

Production of rice in any particular year varies with weather conditions prevailing in different parts of the country, so that annual fluctuations are considerable. Thus, the yield in 1955 was nearly 36 per cent greater than in 1954, while the harvest in 1956 was about 12 per cent less than in the preceding year. Still, over the past decade, a distinct upward trend in output is observable. Thus, the average output of the five-year period 1950–1954 was about 9 million tons, while in the following five-year period, 1955–1959, it was about 11.6 million tons. This upward trend in rice production may be due partly to a slight increase in the planted area, but much more to the increased yield per hectare, which was 3.12 tons in the earlier quinquennium and 3.76 tons in the latter one. Increase in rice yield per hectare and also in total production has been greatest in cooler Tohoku and in the Hokuriku district. Before the war northern Honshu was low in rice productivity per unit area compared with southwestern Japan, but since the war Tohoku has been above the national average, while highland Nagano Prefecture is the highest in the country. In part, the more rapid productivity increase in backward cooler northern Japan and in retarded Hokuriku stems from land reform, which in these regions of particularly oppressive prewar tenancy had the effect of greatly stimulating the new farm owners to improve farm management, especially fertilizer application. It is likewise due to new and improved rice varieties which are able to withstand cold and mature faster, to stronger rice seedlings developed in artificially warmed seedbeds, and to land improvement in general.

Upland fields and unirrigated crops

Upland, or dry, fields (*hatake*) depend almost exclusively upon natural rainfall for their water supply, although recently there has been an extension of supplementary irrigation to a very small minority of non-rice crops on upland fields. The term "upland field" is not entirely satisfactory as a descriptive synonym for dry or unirrigated crops, or for dry fields, for while a great majority of the unirrigated crops are grown in fields that are on upland sites, such as hill slopes, diluvial terraces, and volcanic slopes, still, much unirrigated winter grain is grown as a second crop in paddy fields, while on many lowland irrigated plains there is a minority of fields not in rice but in so-called dry crops. Perhaps the term "ordinary field," used in statistical volumes to describe a non-irrigated, non-rice field, is to be preferred, since it also suggests the lower value which the Japanese associate with non-paddy land.

In 1960 the area in ordinary or upland fields was some 2.7 million hectares as compared with 3.4 million in paddy fields, the ratio being about 44 to 56. But the area in upland fields (chiefly ordinary fields planted to annuals such as cereals, vegetables, etc., but also orchards, grass fields, and fields used in shifting cultivation) is by no means identical with the sum of the aggregate area of unirrigated crops. This discrepancy arises as a consequence of the replanting of over one million hectares, or about one-third of the paddy land, in unirrigated winter crops, and the practice of multiple cropping many of the ordinary upland fields. Thus, the aggregate area of dry crops may be close to 5 million hectares, even though the area of upland fields is only 50–60 per cent of that figure. So while the total area of paddy fields is greater than that of ordinary fields, the aggregate area of harvested dry crops greatly exceeds the area of paddy rice.

The unirrigated small grains—wheat, barley, naked barley,* and oats —the combined acreage of which is nearly half that of paddy rice, are grown under three contrasting conditions involving locational and planting differences: (1) on both upland and lowland sites in Hokkaido, chiefly (90 per cent) as spring-sown crops; (2) in paddy fields of Old Japan, as fall-sown crops following the late-summer or autumn rice harvest (46 per cent); (3) on upland fields in Old Japan, as fall-sown crops following sweet potatoes, legumes, vegetables, and other summer crops (54 per cent). In extreme northern Honshu the small grains are both spring-sown and fall-sown. Oats, chiefly a feed crop and much less important than the other three, is almost exclusively confined to Hokkaido (92 per cent), although small amounts are grown in southern Kyushu and northeastern Honshu.

The barleys and wheat are primarily food crops. Of the two barleys, the common variety is the more widely grown, although 60 per cent of its acreage is concentrated in Kanto and eastern Tohoku, so that it has something of a northeastern orientation (Fig. 7-23). By contrast, naked barley is much less important in Tohoku and Kanto and conversely reaches its maximum development in southwestern Japan, more especially Tokai, the Inland Sea borderlands, and Kyushu (Fig. 7-24). The reasons given for this regional separation of the two barleys are the common barley's somewhat better adaptation to colder climates and the fact that naked barley was a later development in Japan and was adopted to a higher degree in the more progressive southwest where a greater over-all diversification of crops is typical. Although both barleys are grown on paddy fields as well as upland fields, more of the common barley occupies upland or ordinary fields, while a larger share of the naked barley is produced on paddy lands. Wheat, the most important of the un-

---

* Naked barley is true barley. It differs from ordinary barley chiefly in that the chaffy scales that enclose the kernel are not fused to the seed, so that the latter is readily freed from the scales. In ordinary barley a special hulling operation is necessary to free the seed.

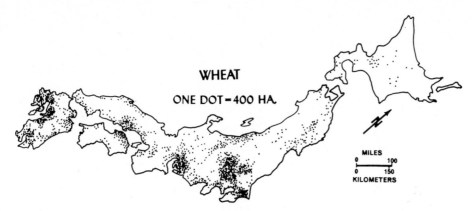

Fig. 7-22.

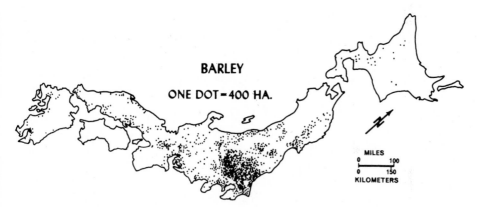

Fig. 7-23.

Fig. 7-24.

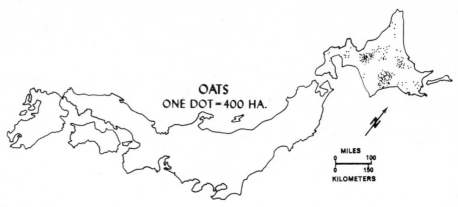

OATS
ONE DOT = 400 HA.

MILES
0   100

0   150
KILOMETERS

Fig. 7-25.

irrigated small grains, with an acreage two-thirds that of the two barleys combined, is, like common barley, widely distributed (Fig. 7-22). Fifty-five to sixty per cent is grown on upland fields. All of the small grains are unimportant along the Japan Sea side of northern and central Honshu where the heavy and long-continued snow cover makes all winter cropping, except for green manures, difficult. There has been a decline of small-grain acreage amounting to about 11 per cent between 1955 and 1960. This probably reflects the low price of the domestic product and the fact that it can be imported more cheaply from abroad, although the increasing shortage of labor on the farms may likewise play a part. Other non-irrigated cereals, all of minor importance, are upland rice, maize, several kinds of millets, and buckwheat.

Special cash crops and industrial crops

Japan's agriculture is a combination commercial-subsistence type, for although it operates within a market economy, subsistence farm households, defined as those which consume more than 80 per cent of what they produce, comprise nearly 40 per cent of the total. Not more than half of the rice, 40 per cent of the wheat and barley, and 38 per cent of the potatoes are placed on the market. Thus, many of the staple food crops are both cash and subsistence in character. But there is another group, including vegetables, fruit, tea, mulberry, tobacco, and miscellaneous industrial crops, more than 80 per cent of which is sold on the market, and it is these specialized cash crops that are here under discussion. The combined acreage of such special crops represents about 16 per cent of the total planted crop area.

With the exception of Hokkaido, there are no specialized cash crop regions on a significant scale anywhere in Japan, and except for a lim-

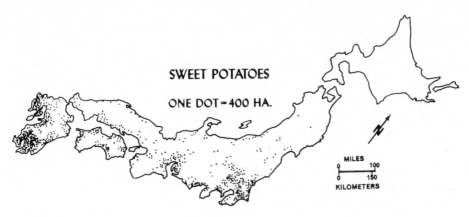

SWEET POTATOES

ONE DOT = 400 HA.

MILES
0   100
0   150
KILOMETERS

Fig. 7-26.

ited number of orchardists, there are no specialized cash-crop farmers. In contrast to the situation in Western countries, areas planted to cash crops in Japan do not form a continuous block of land, but are widely scattered in tiny patches located in the midst of staple crops. In this respect Hokkaido with its larger farms is something of an exception. In addition, only a few cash crops are concentrated on lands particularly suited to their cultivation; instead they are relegated to lands which are least suitable to the cultivation of staple food crops.[34] Thus, they commonly occupy such sites as the middle and upper parts of steep alluvial fans where soils are coarse, coastal dunes and beach ridges where soil is sandy, diluvial uplands, volcanic slopes, and the steep slopes of hills and mountains. These in general are considered inferior agricultural lands in a country so rice-oriented as Japan.

Ogasawara points out that cash crops in Japan have the following characteristics: (1) They require an unusual amount of human labor —e.g., tobacco, tea, and mulberry. (2) They require relatively advanced husbandry techniques and large amounts of fertilizer—e.g., horticulture. (3) They must not interfere with the growing of staple food crops. So, cash-crop farming is secondary in Japan and is a consequence of a developing market economy in which low farm incomes resulting from limited farm areas are partially offset by diversification in the direction of supplementary cash crops.

The distribution of cash crops reflects the influence of multiple factors, including the characteristics mentioned just above, the prevailing local agricultural systems, historical influences, and degree of access to great urban concentrations. For instance, cash-crop farming is conspicuously meager along the Japan Sea side of Honshu as a result of that region's high degree of specialization in paddy rice, its small development of winter crops because of a deep and long-continued snow cover, and the absence of great metropolitan areas. Similarly, the northern part of Tohoku is meagerly developed in cash crops, the chief exception being apple orchards in Aomori Prefecture, for the more severe climate of this northerly region excludes many of Japan's subtropical market crops such as tea, citrus, and mulberry. In this same region, because of the lower productivity of the land—since winter crops in paddies are not possible and even the frequency of cultivation of ordinary fields is reduced—most farmers are unable to cultivate any land above and beyond what is needed to produce staple food crops.

Within subtropical Japan the distribution of specialized market crops is very uneven (Fig. 7-27). They are relatively unimportant in central and northern Chugoku, in backward southern Shikoku, and in central and eastern Kyushu. Of more than average importance are the Fukushima Basin (mulberry and tobacco), the Kanto area (variety), the basins of Nagano Prefecture (mulberry and fruit), Shizuoka (tea and fruit), Nagoya area (variety), the margins of the Inland Sea (fruit and flowers), and the Kumamato Plain in western Kyushu (varied). Generally, most cash-crop farming does not coincide with important paddy areas but is in dry-crop locations, and hence more often is coincident with upland fields on hill slopes and diluvial uplands. Flowers and vegetables are highly specialized crops in close proximity to large cities, but otherwise they are widely scattered. In closest proximity to the metropolitan markets are the truck gardens, particularly those specializing in more perishable products. Beyond these are zones specializing in flowers, less perishable garden products, and orchards. In the specialized market garden areas there is practiced an intensive kind of multiple cropping by which four or five harvests of vegetables are obtained from a field during the course of a year.[35]

Japan's fruit acreage has more than doubled within the decade of the 1950's and since 1955 has exceeded that of mulberry. Leader in acreage is the citrus crop, predominantly mandarin or tangerine oranges, followed by apples and persimmons. Citrus is confined to subtropical Japan west and south of Tokyo, excluding largely Hokuriku-Sanin facing the Sea of Japan and the highland areas (Fig. 7-28). Main centers of concentration, all of them in close proximity to the coast, are (1) Shizuoka-Aichi area in Tokai and adjacent parts of western Kanagawa Prefecture in southwestern Kanto, which has about 30 per cent of the national acreage; (2) Kinki area, especially Wakayama and Osaka prefectures at the eastern end of the Inland Sea, with nearly 15 per cent; (3) Inland Sea

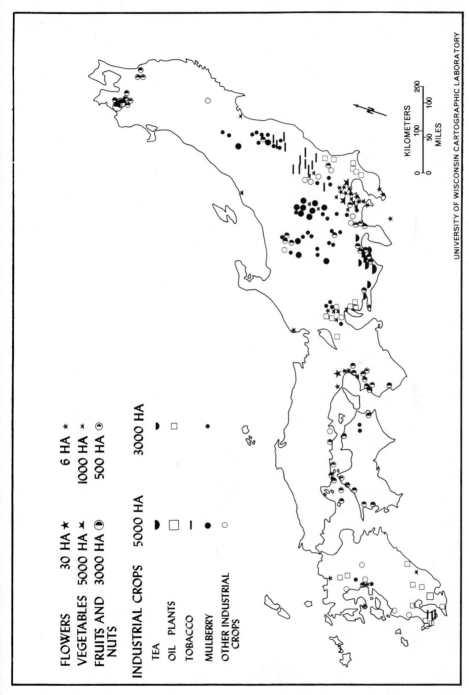

Fig. 7-27.—Distribution of the principal cash crops. After map by Yoshikatsu Ogasawara.

borderlands of Shikoku and Chugoku, with a high degree of concentration in the Ehime-Hiroshima district, accounting for nearly 27 per cent of the acreage; and (4) scattered areas in northern and western Kyushu whose combined acreage is about 15 per cent of the total.[36] A large percentage of the citrus orchards are located on hill slopes, some of them exceeding 30° in inclination and therefore requiring terracing.

By contrast with citrus, apple orchards are concentrated in cool northeastern and highland central Japan (Fig. 7-29). Sites also are usually less steep, the commonest ones being diluvial uplands, the upper and middle parts of alluvial fans, and the riverine belts of alluvial plains. Sixty to seventy per cent of the apple acreage is in northernmost Honshu in Aomori Prefecture, and nearly 60 per cent in the Tsugaru or Iwaki Basin of that subdivision. Other more modest concentrations are in the Nagano Basin located in the highlands of central Japan, and on and around the margins of the Ishikari Plain in western Hokkaido.

Tea is a subtropical crop, with little grown north of about 37°, and although it is widely raised throughout subtropical Japan, 58 per cent of the acreage is located in the Tokai–Southwest Kanto region of central Japan, with 43 per cent in Shizuoka Prefecture alone (Fig. 7-30).[37] Like mulberry, it is a crop of the foothills and diluvial terraces.

Mulberry, whose leaves are used exclusively for feeding silkworms, at present occupies only one-quarter of the acreage that it did in the early 1930's before the serious decline in raw silk (Fig. 7-31). It is widely grown in most parts of the country except in cold Hokkaido and northernmost Honshu, but 60 to 70 per cent of the acreage is concentrated in a few regions such as western Kanto, the upland basins of mountainous Tosan and the basins of southern Tohoku (Fig. 7-32). Since mulberry does not require irrigation and is soil tolerant, it is usually relegated to the less desirable upland sites such as the mountain foothills, diluvial uplands, and the upper coarse-textured parts of alluvial fans. Other common, though less extensive, dry sites are the sandy beach ridges and the levee belts of alluvial plains. Mulberry is unimportant on the Japan Sea side of the country in western Tohoku and in Hokuriku.

Other industrial-cash crops, mostly of minor or local importance, are rapeseed and sesame, tobacco, sugar beets, peppermint, Igusa mat reed, and Japanese-paper bush. Some of these will be referred to in a later section of this book dealing with the geography of particular regions and localities.

Animal industries

Animal industries in Japan deserve brief mention chiefly because of their conspicuous underdevelopment, and this in spite of the preponder-

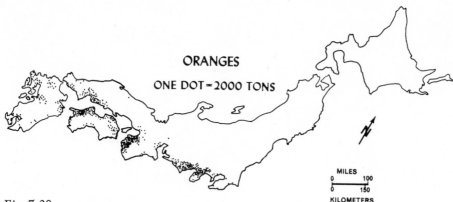

Fig. 7-28.

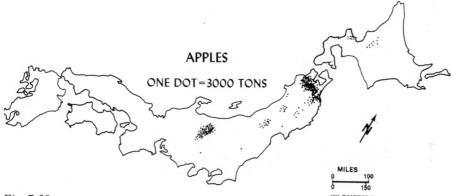

Fig. 7-29.

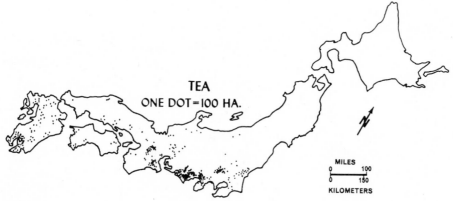

Fig. 7-30.

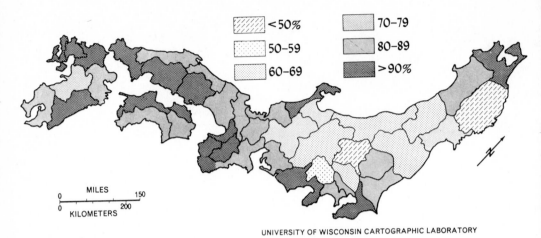

Fig. 7-31.—Percentage decrease in area of mulberry between 1929 and 1960. After map by Richard Hough.

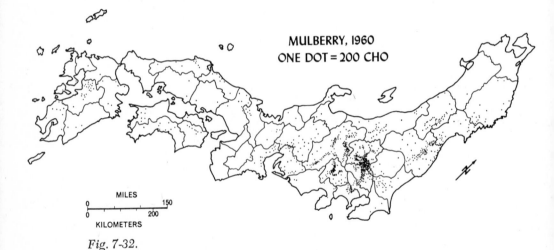

Fig. 7-32.

ance of rough slope land that is unfit for normal cultivation. Herds and flocks are generally lacking, and what cattle there are, are characteristically kept at the farmstead, where their highly valued manure can be salvaged for use on cultivated fields. Only about half the total number of farm households possess a farm animal, and 80 per cent of these have only one animal. The livestock figures for Japan in 1960 were as follows: draft and beef cattle, 2,340,000; dairy cattle, 824,000; horses, 673,000; sheep, 788,000; and swine, 1,918,000.[38]

This underdeveloped state of the animal industries is characteristic not only of Japan but of all of eastern and southern Asia, no important

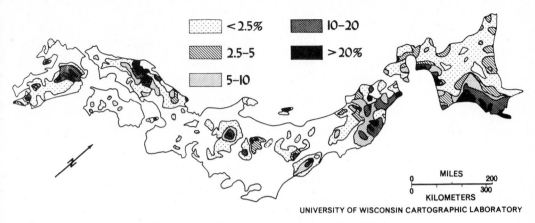

UNIVERSITY OF WISCONSIN CARTOGRAPHIC LABORATORY

Fig. 7-33.—Ratio of grazed land to total area. After map by H. Yasuta.

part of which has developed a mixed crop-and-livestock type of farming. Here an intensive cereal agriculture became the mainstay of the economy far back in the past and has persisted even down to the present. Indeed, it is this type of agriculture, emphasizing as it does a reliance on the vegetable rather than the animal kingdom, and thereby producing a maximum of food from a given area, which has permitted the growth of such large and dense populations in these Asiatic regions. And it is the large food requirements of these same dense populations which now make so difficult a shift to a less productive mixed type of agriculture involving animals. It is relatively impossible to raise good beef and dairy cattle without resorting to feeds other than wild pasture, most of which in Japan is of poor quality, while on the other hand Japan has felt it unwise to reduce her meager food-crop area in order to grow feed crops. Only in Hokkaido are fodder and other feeds produced as special crops. Elsewhere there is heavy reliance on straw, wild grasses cut from slopes or the margins of fields, and purchased feeds, expenditures for the latter amounting to 30–40 per cent of the income from the sale of the milk. At present it is cheaper to import butter and cheese from abroad than to produce them at home.

Only about 6–7 per cent of the Japanese farm households keep a dairy cow, although this figure rises to nearly 25 per cent in Hokkaido. During the decade of the 1950's there was a nearly 400 per cent increase in the number of dairy cattle, and there was a similar increase in swine, but there are still (1963) only 1,145,370 of the former and 3,296,000 of the latter animals.

Nearly 22 per cent of Japan's dairy cattle are in Hokkaido, in the easternmost parts of which there is a relatively specialized dairy agriculture. In Old Japan some degree of dairy cattle concentration is to be found in parts of eastern Tohoku, Kanto and vicinity, Tokai, highland Nagano Prefecture, the Inland Sea borderlands, and western Kyushu. Cattle are particularly scarce along the Japan Sea side of the country and on the Pacific side of southwestern Japan in Kyushu, Shikoku, and Kii Peninsula. Beef-draft cattle are chiefly concentrated in the Kanto area and in those parts of the country west and south of about Nagoya. They are very few in Hokkaido. By contrast, horses are the principal work animals and are most numerous in northeastern Japan, but are also numerous in western and southern Kyushu.[39]

Milk production in 1960 amounted to only 1,887,000 tons, but this is double what it was five years previously, and four times what it was in 1952. About 21 per cent of the total originated in Hokkaido where, because of distance from markets, most of the milk is processed into dairy products. For the country as a whole about half of all milk produced is for fluid use, and the proportion is much larger in Japan Proper. Close to 45 per cent of the milk originates in parts adjacent to metropolitan areas, 20 per cent in Kanto alone, and 11 per cent in Kinki.

AGRICULTURAL REGIONS OF JAPAN

A division of Japan into a system of agricultural types and regions based upon main crops and crop combinations is less meaningful than it is for the United States.[40] This arises from certain peculiarities of Japan's agriculture. One such is that agricultural land occupies only about 17 per cent of the total area, and this land is coincident with numerous non-contiguous small plains and basins separated by more extensive areas of non-agricultural hill and mountain land. Moreover, Japan's agriculture in almost all parts, from the far south even into central Hokkaido, emphasizes the growing of paddy rice, and it is rare for lowland areas to depart from this specialization. Except in northern and eastern Hokkaido the non-rice areas are chiefly elevated dry sites in the form of scattered fragments lying in fairly close proximity to paddy-dominated lowlands. But while the general emphasis upon rice creates a kind of over-all agricultural uniformity, still, soil, water resources, topography, and socio-economic conditions vary from plain to plain and basin to basin, and even within the different parts of a single plain, so that the pattern of land use is a complicated mosaic representing the intermingling of almost innumerable small units, and the development of a broad system of land-use types and regions is difficult.

What is probably the most serviceable regionalization of Japan's agricultural land use has been developed by Y. Ogasawara, of the Geographical Survey Institute in Tokyo (Fig. 7-34).[41] He attempts to create a system of land-use regions employing the following indicators: (1) kinds of crops, (2) intensity of land use, (3) degree of land use in relation to surface features, (4) distribution and use of grasslands, and (5) distribution and use of pastures. On the basis of these elements, Ogasawara recognizes two principal subdivisions—Old Japan and Hokkaido.

In Hokkaido grassland occupies relatively large areas, even on arable lands where, in rotation, it is intermingled with other cultivated crops. Pasture is likewise widespread. Farms and fields are relatively large, rectangular in shape, and the farm is a contiguous block of land. Associated crop and livestock farming is best developed in Hokkaido, as is commercial agriculture.

In Old Japan, by contrast, grassland occupies only a small area and is limited to non-arable sites. Pasture is likewise limited in extent; and livestock, other than draft animals, are relatively few. Cultivated area per

T A B L E 7-8

Agricultural land use regions of Japan (after Ogasawara)

A. Old Japan (Honshu, Shikoku, Kyushu)
  I. Central Zone
    1. Core Area
      1a Kinki-Setouchi District
      1b Northern Kyushu District
      1c Chukyo District
      1d Western Kanto District
      1e Tokai District
      1f Tosan District
      1g Hokuriku District
    2. Peripheral Area
      2a Hida District
      2b Sanin District
      2c Kii District
      2d Southern Shikoku District
      2e Middle Kyushu District
  II. Frontier Zone
    1. Eastern Kanto District
    2. Dewa District
    3. Mutsu District
    4. Southern Kyushu District
B. Hokkaido
  1. Western Hokkaido
  2. Eastern Hokkaido
    2a Eastern Hokkaido Proper
    2b Pioneering Region

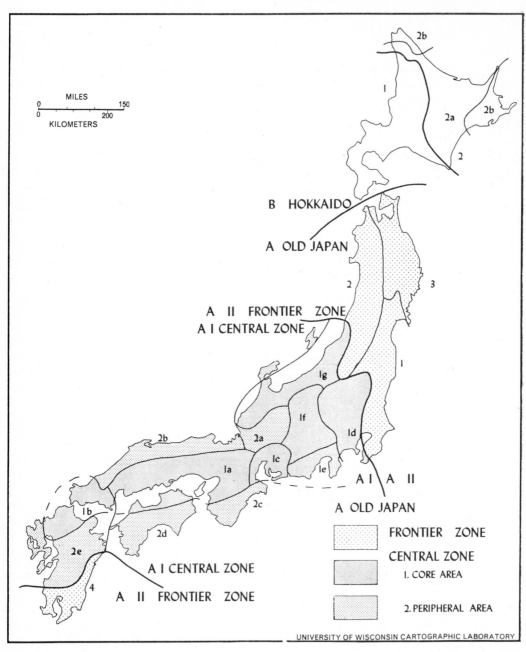

Fig. 7-34.—Agricultural land-use regions. See Table 7-8 for names of individual districts. After map by Yoshikatsu Ogasawara.

farm household and areas of individual fields are only one fourth what they are in the northern island; farming is more intensive, crops more varied in character, and agricultural management more complicated.

Each of the two principal agricultural regions just noted is further divided into subregions. Thus, Hokkaido is divided into a western half where grain farming dominates, paddy rice is the most important crop, farms are smaller, and grassland and pasture are less important. Eastern Hokkaido, by contrast, is more a region of associated crop and livestock farming, with larger farms and greater emphasis on grassland and pasture.

Old Japan (Honshu, Shikoku, Kyushu) is much more complicated in its subdivisions. It is first divided into a Central Zone and a Frontier Zone. The latter is composed of two far separated parts, a much larger northern one which includes Tohoku and eastern Kanto, and a smaller

TABLE 7-9

Land use of Hokkaido

Characteristics	Western Hokkaido	Eastern Hokkaido	
		Eastern Hokkaido proper	Pioneering region
Farming system	Grain farming, rice dominant	Associated crop and livestock farming	Livestock ranching, immature crop farming, subsistence crops dominant
Crops	Multiple cash crops, paddy rice, fruit, white potatoes, beans, wheat, pyrethrum	Specialized cash crops, wheat, white potatoes, beans, peppermint, pyrethrum, sugar beets	White potatoes, oats, buckwheat, sugar beets
% of farmers cultivating more than 10 ha.	<10%	30–50%	10-15%
Presence of grasslands, pastures, and rows of trees bordering fields	Some	Abundant	Dominant
Period of agricultural occupance	Since 1875	Since 1900	Since 1915
Pioneering possibilities	Almost absent, some peat-bog and volcanic-flank areas already reclaimed	Present in some places, where transportation facilities are inadequate	Present over most of area

Source: Y. Ogasawara, "Land Use of Japan," *Bulletin of the Geographical Survey Institute, 2,* Part 1 (1950).

southern one in southernmost Kyushu. The Central Zone, therefore, includes most of central and southwestern Honshu, Shikoku, and central and northern Kyushu. It may be differentiated from the Frontier Zone largely in terms of the degree of land reclamation. Thus, in the Central Zone paddy fields are to be found on almost all types of land, including such inferior ones as steep alluvial fans, most natural levees, some volcanic and mountain slopes, and diluvial uplands. In the Frontier Zone these same inferior types of land are much less utilized for rice, while upland fields as well occupy a far smaller proportion of these less desirable lands than they do in the Central Zone.

TABLE 7-10

Degree of land reclamation, Old Japan

Distribution	Central Zone	Frontier Zone
Cultivated areas on less favorable land		
Paddy fields		
Alluvial fans	Almost all	Parts of some
Reclaimable water areas and marsh	Almost all	A few
Diluvial uplands	Some	A few
Volcanic flanks	Some	A few
Natural levees	Most	A few
Mountain slopes	Some	A few
Upland fields		
Undulating hills	Some parts	Limited areas
Volcanic flanks	Some parts	A few
Patches of flat land in mountains	Almost all	Some
Mountain slopes	Some parts	A few parts
Non-cultivated land		
Grassland	Tops of lofty mountains, volcanic flanks	Non-arable lands, some places on diluvial uplands and undulating hills
Pasture	Tops of lofty mountains, elevated peneplain remnants higher than 800 meters, volcanic flanks	Non-arable lands, some places on undulating hills, some diluvial uplands
Forest	Abrupt mountain slopes	Mountain slopes, almost entire surface of undulating hills, parts of some diluvial uplands

Source: Y. Ogasawara, "Land Use of Japan," *Bulletin of the Geographical Survey Institute, 2,* Part 1 (1950)

The Frontier Zone was a pioneer area prior to 1600, for agricultural occupance in an important way did not occur until the seventeenth century. Even now, land utilization is less intensive and less diversified than in the other parts of Old Japan, and there is a less complete use of the marginal lands, while double-cropped paddy fields are far less common. Agriculture is much more of a subsistence type than elsewhere.

Four separate districts of the Frontier Zone are recognized. Southern Kyushu District in feudal times was one of the most isolated from the cultural center of Japan. Here land reclamation is only modestly developed and agriculture is of the self-sufficient type. Rice yields are some of the lowest for the entire country but, on the other hand, 85 per cent of the paddies are double cropped. Dewa District, in western Tohoku, is highly specialized in rice, so that there is a large surplus for export to other regions; but only a very small proportion of the paddies bear a winter crop, because of adverse climatic and drainage conditions. Land reclamation, especially of upland sites, is less complete than in areas farther south, since agricultural colonization largely occurred after 1600. Mutsu District, in northeastern Tohoku, is still regarded as a pioneering region, in which considerable areas of peat bog and diluvial upland remain to be reclaimed. Here the cool summers have retarded rice cultivation, so that the ratio of paddy area to cultivated area is as low as 40 per cent. Only 3 per cent of the paddy fields are double cropped and even a large majority of the upland fields yield but a single harvest. Millets and oats are relatively important, and the raising of horses has been something of a specialization. This was formerly one of the more important regions of fire-cleared fields and primitive migratory agriculture, but this has largely disappeared, while dairying has increased in importance. Eastern Kanto District presents something of an anomaly, for in spite of its proximity to the great Tokyo-Yokohama market, it has remained relatively backward in its agricultural development. Subsistence farming is still strong; considerable areas of marshy land and diluvial upland remain unreclaimed; only 6 per cent of the paddy fields are double cropped; and rice yields are relatively low. An increasing number of cash crops such as vegetables, fruit, and tobacco have begun to develop within the region, but emphasis on commercial agriculture is much less than would be expected considering the proximity to Tokyo.

Ogasawara's Central Zone has two principal subdivisions, the Core Area and the Peripheral Area. The latter chiefly comprises the hilly and mountainous lands of southern Kii and southern Shikoku facing the Pacific Ocean, rugged central and western Kyushu, and the northern side of Chugoku, or Sanin, facing the Japan Sea. It may be thought of as the hinterland of the Core Area, where, although agricultural occupance is very ancient, there is evidence of the handicaps associated with separation from the main culture and economic centers—a separation due to difficult accessibility and inadequate communication. Land use is less intensive, especially so in the more inaccessible areas, and subsistence crops such as millets, buckwheat, and corn play a more important role than they do in the Core Area. Land reclamation also is less complete than in the Core Area, but, on the other hand, the use of mountain slopes is here better developed than anywhere else in Japan. Ninety per cent of the country's fire-cleared fields and 90 per cent of its most imposing terraced fields are found within this Peripheral Area. Mountain grasslands are widely used for pasturing draft-beef cattle and as a source of compost materials. For centuries the dense forests mantling the steeper slopes have been carefully managed so that they yield large amounts of charcoal and firewood, as well as 80 per cent of the country's high quality lumber.

The Core Area is the agricultural heartland of Japan. Here land reclamation has reached its highest level, agricultural population is densest, the intensity of land use is at a maximum, and commercial agriculture is more prevalent than elsewhere.

On the basis of intensity of land use and certain other distinguishing features the Core Area is divided into seven districts and the Peripheral Area into five. The characteristics of these districts are concisely summarized in Table 7-13, so that only brief supplementary comments are necessary at this point. Within the Core Area, the Kinki-Setouchi District represents the heart. For a millennium and a half before 1868 it had been the political-economic-cultural center of Japan, so that land reclamation has been pushed to the limit. Paddy fields occupy 80 per cent of the cultivated area, while a similar percentage of the paddies are double cropped. Agriculture is highly diversified, and cash crops are relatively important. The Northern Kyushu District resembles Kinki-Setouchi in most respects, but it differs chiefly in that its agriculture is less diversified and there is less emphasis on tree crops. The Chukyo District likewise has a rather limited development of tree crops, and compared with the two previous districts its yield of rice per unit area is lower, as is the ratio of double-cropped paddy fields, while farm machinery and draft animals are utilized to a lesser degree.

Western Kanto shows important departures from the three districts previously noted in that upland fields, chiefly on diluvium, greatly predominate, so that the ratio of paddy fields to the whole cultivated area is only 30 per cent. Moreover, only 40 per cent of the paddy fields are

T A B L E 7-11

Intensity of land use, Old Japan

Characteristics	Central Zone		Frontier Zone
	Core Area	Peripheral Area	
		Farms	
Crops	Cash crops dominant, combined cash and subsistence crops important	Cash and subsistence crops dominant, cash crops important, subsistence crops only in mountains	Cash and subsistence crops dominant, subsistence crops important, cash crops in limited areas
% double-cropped paddies	60%	50%	5%
Green manure winter crops on double-cropped paddies	Only in districts of heavy snowfall	On 20% of double-cropped paddies	On 90% of double-cropped paddies
Yields of rice per acre	10.8–12.0 koku*	9.5–10.8 koku	9.0–10.5 koku
Intensity of farm use	Intensive everywhere	Intensive in some areas	Intensive in a few areas
Irrigation facilities	Intricate networks of irrigation ditches and ponds, thousands of motor pumps, thousands of dikes and drainage lines	Networks of irrigation ditches, hundreds of irrigation ponds, hundreds of motor pumps, hundreds of miles of dikes and drainage lines	Irrigation ditches not very well developed, a few irrigation ponds, a few motor pumps, dikes and drainage lines not well developed
		Grassland and pasture	
Livestock industries	Dairying in the suburbs	Raising meat and draft animals	Raising military animals (until recently)
Grazing	Seasonal movements of draft animals in almost every district	Seasonal movements of draft animals in some districts	No seasonal movements of animals
		Forest	
Kind of trees	Pine and Japanese cypress	Cedar, Japanese cypress, and pine	Deciduous trees dominant
Management	Reserve forest dominant	Private forest dominant	National and public forests dominant
Afforestation	70%	60%	30%
Products	Good timber	Large quantity good timber, small quantity fuel	Large quantity fuel, small quantity timber

* 1 koku = 1.8 kiloliters, or 0.236 cubic yards.

Source: Y. Ogasawara, "Land Use of Japan," *Bulletin of the Geographical Survey Institute, 2,* Part 1 (1950).

double cropped. Proximity of large urban markets has stimulated the development of cash crops, and consequently agriculture is predominantly commercial. In the Tokai District agricultural land use is somewhat less intensive than in most other districts of the Core Area. Land reclamation has not been pushed as far, rice fields occupy only 46 per cent of the cultivated area, and the percentage of double-cropped paddy fields (30 per cent) is the smallest within the Core Area. A distinctive feature of Tokai is its emphasis on cash crops, including tea, fruit, and commercial vegetables. Indeed, Shizuoka Prefecture is the nation's largest producer of citrus fruit and of tea. Mountainous Tosan District's agricultural land is chiefly confined to a series of small, elevated, fault basins and adjacent volcanic slopes. In spite of physical handicaps, however, land use is highly developed and the yield of rice per unit area is the highest in the country. Still, only 30 per cent of the cultivated area is in paddy rice, although considering the cool summers of the upland basins, an amazingly high percentage (60 per cent) of the paddy fields bear a winter crop. Tosan leads all other districts in mulberry acreage and is nationally famous for such fruit crops as apples and grapes. Snowy Hokuriku District facing the Japan Sea is highly specialized in paddy rice, with a large surplus for domestic export. And despite its climatic handicap, about 40 per cent of the summer paddy fields are replanted to winter crops, but chiefly green-manure crops rather than cereals.

TABLE 7-12

History of agricultural settlement and agricultural population, Old Japan

Features of agricultural history	Central Zone		Frontier Zone
	Core Area	Peripheral Area	
Principal periods of settlement	From ancient time to 1614 A.D.	From ancient time to 1720 A.D.	I. 1615–1720 II. Since 1868
Expansion possibilities of cultivable land	Almost absent since 1720	Almost absent since 1868	Present in some parts of Kagoshima, Iwate, and Aomori
Density of population	Almost saturated by 1614	Almost saturated by 1720	Less densely populated than Central Zone
Trends in agricultural population	Declining	Generally declining	Variously static or on increase

Source: Y. Ogasawara, "Land Use of Japan," *Bulletin of the Geographical Survey Institute, 2,* Part 1 (1950).

Characteristics	Districts in Core Area			
	Kinki-Setouchi	Northern Kyushu	Chukyo	Western Kanto
Type of agriculture	Commercial	Commercial	Commercial	Predominantly commercial
Ratio paddy area to all cultivated area	80%	70%	80%	30%
Ratio double-cropped paddy area to all paddy area	80%	75%	60%	40%
Av. rice yield per acre	12.2 koku	12.5 koku	10.5 koku	11.0 koku
Characteristic winter crops of double-cropped paddies	Vegetables Wheat Pyrethrum Poppy Rapeseed Tatami reed	Vegetables Wheat	Wheat Barley	Wheat White potatoes
Over-all characteristic crops	Vegetables Fruit Paddy rice Tea Wheat	Vegetables Paddy rice	Vegetables Paddy rice Wheat Sweet potatoes	Vegetables Mulberry Wheat Sweet potatoes Tea Fruit **Barley** Upland rice
Auxiliary activities	Dairying in suburbs, movement of draft animals between mountain pastures and plains (farm work), fish culture in moats and canals, seaside salt beds	Seaside salt beds	Dairying in suburbs, poultry farms, seaside salt beds	Dairying in suburbs

(Continued on following pages)

TABLE 7-13—*Continued*

Characteristics	Tokai	Districts in Core Area	
		Tosan	Hokuriku
Type of agriculture	Predominantly commercial	Commercial	Predominantly commercial
Ratio paddy area to all cultivated area	46%	35%	80%
Ratio double-cropped paddy area to all paddy area	30%	60%	40%
Av. rice yield per acre	10.5 koku	13.0 koku	11.6 koku
Characteristic winter crops of double-cropped paddies	Barley Wheat	Wheat Barley	Green manure crops
Over-all characteristic crops	Tea Fruit Vegetables Sweet potatoes Barley Upland rice	Mulberry Fruit Vegetables Wheat Barley	Paddy rice Fruit Vegetables
Auxiliary activities	....	Dairying in mountains, movement of draft animals between mountain pastures and plains (farm work), fish culture in paddy fields	Movement of draft animals between mountain pastures and plains (farm work)

TABLE 7-13—*Continued*

Characteristics	Districts in Peripheral Area				
	Hida	Sanin	Kii	Southern Shikoku	Middle Kyushu
Type of agriculture	Subsistence	Commercial and subsistence	Commercial and subsistence	Subsistence, and commercial	Commercial and subsistence
Ratio paddy area to all cultivated area	30%	65%	40%	50%	55%
Ratio double-cropped paddy area to all paddy area	5%	40%	50%	75%	80%
Av. rice yield per acre	9.0 koku	11.0 koku	9.5 koku	8.9 koku	11.5 koku
Characteristic winter crops of double-cropped paddies	Green manure crops	Green manure crops Barley	Green manure crops Barley	Green manure crops Barley Paddy rice	Reed Wheat Barley Green manure crops
Over-all characteristic crops	Buckwheat Barley Millet German millet	Paddy rice Fruit Mulberry	Sweet potatoes Fruit	Sweet potatoes Barley Tobacco Corn German millet Mitsumata	Fruit Paddy rice Sweet potatoes Upland rice Tobacco Corn German millet Barley
Auxiliary activities	Lending draft animals to Hokuriku District, timber (Japanese cypress)	Raising draft and meat animals, employment of draft animals of mountain villages on plains, lending draft animals to Kinki-Setouchi District	Lumber (cedar)	Lending draft animals to Kinki-Setouchi District, employment of draft animals of mountain villages on plains, lumber	Raising draft animals, timber

TABLE 7-13—*Continued*

Characteristics	Districts in Frontier Zone			
	Eastern Kanto	Dewa	Mutsu	Southern Kyushu
Type of agriculture	Subsistence and commercial	Subsistence and commercial	Subsistence	Self-sufficiency
Ratio paddy area to all cultivated area	55%	80%	40%	40%
Ratio double-cropped paddy area to all paddy area	6%	5%	3%	85%
Av. rice yield per acre	10.5 koku	11.5 koku	9.5 koku	8.8 koku
Characteristic winter crops of double-cropped paddies	Barley Green manure crops	Green manure crops	Green manure crops	Green manure crops Barley Wheat
Over-all characteristic crops	Tobacco Vegetables Sweet potatoes Barley Paddy rice Upland rice Wheat Corn Fruit German millet	Paddy rice Fruit Vegetables Mulberry Soybean Barnyard millet Tobacco	Barnyard millet White potatoes Oats Millet German millet Fruit	Sweet potatoes Paddy rice German millet Upland rice
Auxiliary activities	....	....	Dairying, production of fuel	Raising animals for military use, production of fuel and lumber

Source: Y. Ogasawara, "Land Use of Japan," *Bulletin of the Geographical Survey Institute, 2,* Part 1 (1950) (some data revised).

Within the Peripheral Area, mountainous Hida District coincides with the most extensive meagerly inhabited area in Old Japan. Subsistence agriculture prevails, only 30 per cent of the cultivated area is in paddy fields, and a scant 5 per cent of the paddies are double cropped. Upland slope fields predominate, while the forest industries are of great importance. Sanin District, facing the Sea of Japan and containing Izumo, one of the oldest culture centers, has nevertheless remained rather backward, in part because of its isolation by highlands from the Core Area. Intensity of land use is below that of Hokuriku, even though snowfall is less in Sanin, and there is less emphasis on paddy rice. Highlands are specialized in the rearing of cattle for draft and beef purposes. The three remaining districts (Kii, Southern Shikoku Middle Kyushu) are all parts of the lofty, rugged mountains of the Outer Zone of Southwest Japan, so that forest industries are relatively of great importance and agricultural land is limited in extent. Subsistence and commercial agriculture are both represented, but the intensity and nature of the land use varies greatly from one locality to another, depending on accessibility to adequate transportation facilities.

FISHERIES

Japan leads all other countries in its annual fish catch, with about twice that of the United States, which holds second rank. In most of the important countries of the world fishing is a minor element of the national economy; in Japan it rates much higher. As one of the primary industries, fishing contributes considerably more to the national income than either mining or forestry; its value is about one-sixth that of agriculture. In addition, Japan exports aquatic products valued at about $175 million, while sea foods provide 90 per cent of the protein requirements of Japan's rural population, and 80 per cent of those of its urban people.

This extraordinary importance of fisheries and the intensity of exploitation of sea resources reflect, in part, the limitations of land resources. Quite naturally, the insular character of the country, with population concentrated on littoral plains, is a condition which likewise facilitates the exploitation of marine resources. But the most positive element, certainly, is the fact that the seas surrounding Japan abound in numbers and varieties of fish, this being one of the earth's three richest fishing grounds. Here a cool current from the north mixes with a warm current from the south, providing a variety of temperature and plankton environments suitable for diverse species of marine life.

Several important recent developments have occurred in Japan's fish-

Fig. 7-35.—Part of Japan's fishing fleet in a congested harbor of central Japan. Courtesy of the Consulate General of Japan, New York.

ing industry. One such is the remarkable increase in the volume of the annual catch, which in recent years has been more than a million tons greater than that in the best prewar years. Even more striking is the relative increase in the take by pelagic fisheries, reaching a figure six to eight times what it was in the 1930's. This reflects the fact that Japan is currently obliged to carry on more of her fishing activity at a greater distance from the homeland, which in turn requires larger boats, more expensive equipment, and is altogether a large-scale operation. Relatively, coastal fishing has declined, for while in the prewar period it provided 77 per cent of the total catch, this had waned to 42 per cent by 1960. But even now the yield from domestic waters, coastal plus off-shore,

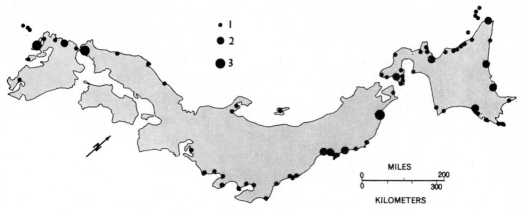

Fig. 7-36.—Principal fishing ports and landing bases: (1) 10–50 thousand metric tons of fish, (2) 50–100 thousand metric tons, (3) more than 100 thousand metric tons. After map by Ryuziro Ishida and others.

amounts to about 81 per cent of the total. Coastal fisheries, which employ 88 per cent of those engaged in fishing, are characterized chiefly by small-scale operations, with nearly 200,000 fishermen's households using small boats of less than three tons, many of which lack motors.

The main pelagic fisheries are (1) salmon and crab fishing in the Okhotsk Sea, (2) tuna and skipjack fishing in the Pacific and Indian Oceans, (3) tuna fishing in the Atlantic Ocean, (4) trawling in the South China Sea and East China Sea, (5) whaling in the Antarctic and North Pacific, and (6) pearling in the Arafura Sea north of Australia. Most important of the Japanese bases for pelagic fisheries are Hakodate, in Hokkaido, and Shimonoseki, in extreme western Honshu.

TABLE 7-14

Fish catch and aquicultural production in 1960 (tons)

Marine fisheries			5,817,939
Pelagic		677,021	
North Pacific	545,042		
South China Sea	5,647		
Tuna long-line fishing	113,362		
Pearling in Arafura Sea	390		
Domestic		5,140,918	
Inland water fisheries			74,063
Aquiculture			300,769
TOTAL			6,192,771*

* Total fishery catch in 1961 was 6,710,000 tons.
Source: Japan Statistical Yearbook, 1960.

Several scores of important fishing ports dot the coasts of Japan and there are hundreds of smaller ones. Nearly half of the important ones are in Hokkaido and there over one-quarter of the nation's total annual catch is landed. Other concentrations of the fishing industry are in eastern Tohoku; eastern and southern Kanto; Shizuoka Prefecture, in Tokai; Yamaguchi Prefecture, in westernmost Honshu, containing the great fishing port of Shimonoseki; and northern and western Kyushu. Unimportant are western Tohoku, Hokuriku, and the Inland Sea borderlands.

8 · MANUFACTURING
AND TRANSPORTATION

Japan was a late-comer into the field of capitalistic industry, for prior to the Meiji Restoration (1868), which approximately marks the beginning of her modernization, manufacturing was of the workshop handicraft type. But once she had been obliged to open her doors to Western influence, the acceleration in her industrial development was rapid, especially after 1900, and it was little short of phenomenal after 1930, with the result that she has attained the unique distinction of being the only country with a non-Western culture that has achieved a high degree of industrialization. Reforms introduced by the new Meiji Government, involving a unification of the monetary and currency systems, the institution of land taxes to be paid in cash and not in kind, the establishment of a national bank organization, and the rapid development of a railway system after 1890, all were preparatory for the subsequent industrial expansion, for which the Sino-Japanese War (1894–1895) was a more specific stimulant. Not to be minimized, either, are the consequences of the fact that universal compulsory education was early established by the Meiji Government, so that by 1900, it is believed, 95 per cent of the Japanese population was literate. Japan's "economic take-off" had to be executed in the face of a lack of accumulated capital and trained manpower and a population that already was large and dense. For this reason during the early preparatory period the government acted to supply the necessary capital and to foster the introduction of Western science and technology, including the training of high-level manpower.

Repeating the example set by England in its industrial revolution, early emphasis was placed upon cotton spinning and weaving, with the result that textiles—cotton and silk in particular—became the first modern industry of Japan. Thus, a modern cotton industry was already well established by 1890, while the first successful government iron and steel plant, at Yahata in northern Kyushu, was not installed until 1900.

It is not the purpose here to write a history of Japanese manufacturing.[1] Suffice it to say that as late as the early 1930's Japan, although possessed of some modern industry, was not by Western standards a highly industrialized country. Agriculture still supported half the nation's population, whereas manufacturing's share was less than 20 per cent and that of commerce about 17 per cent. Light industries and consumers' goods greatly predominated, with textiles and clothing employing about 40 per cent of the wage earners in manufacturing, while textile products represented two-thirds of the total value of exports. Heavy metal and machine industries, by comparison, employed only 13 per cent of the factory workers.

But manufactural composition changed rapidly during the decade of the thirties as Japan prepared for the contingency of war. Light industries and consumers' goods declined relatively as metals, machinery, and chemicals soared in importance, while at the same time the country's technical efficiency in factory industry greatly improved. Without these structural changes in her manufacturing, Japan's early successes in the Pacific war would have been impossible.

The bombing and burning associated with World War II had devastating effects on Japan's manufacturing. By the end of the war 44 per cent of the country's factories had been destroyed, and the national wealth had been depleted by 41.5 per cent. Production in April 1945, four months preceding the end of the war, dropped to one-fourth that of the peak years 1935–1936.[2] Postwar recovery in manufacturing was slow at first, and it was not until 1951 that the rate of production equaled the 1934–1936 average. But between 1950 and 1955 factory output more than doubled, and it more than tripled between 1955 and 1962 (Table 8-1). As in the immediate prewar years, so during the postwar recovery period, it has been the secondary and tertiary industries that expanded more rapidly than the primary. Thus, between 1953 and 1962 employment in agriculture and forestry declined from 37.4 per cent to 25.8 per cent, while non-agricultural employment increased from 62.6 per cent to 74.2 per cent. In 1961 the proportion of the national income from manufacturing (30.7 per cent) was higher than that from any other production sector, and manufacturing was also the most rapidly growing sector of the economy (Table 8-2). Also, within manufacturing, it is the heavy and chemical industries that have experienced the greatest growth, so that in 1959 they represented 62.4 per cent (66.0 in 1961) of the value added by manufacturing, compared with 50 per cent in 1951 and 49.4 per cent in 1955.[3] Emphatically Japan's present manufactural structure is strongly weighted on the side of heavy industries.

It is notable that for centuries population growth paralleled closely economic expansion within the country. Only during the last decade have the two elements shown strong divergence in rates of increase, as population growth has slowed down, while simultaneously the national economy, and more especially the manufacturing sector, has soared to unprecedented heights. The economic growth rate, from the postwar recovery down to the present, is the most outstanding in the world, even surpassing that of West Germany. Between fiscal years 1957 and 1961 gross national product at constant prices increased 58 per cent, an

TABLE 8-1

Production indexes in manufacturing (1955=100)

1934–36	62.5	1955	100.0
1946	16.1	1956	123.5
1947	20.1	1957	146.4
1948	26.9	1958	147.0
1949	35.7	1959	185.2
1950	44.7	1960	236.3
1952	68.0	1961	290.1
1954	92.4	1962	333.1

Source: Foreign Capital Research Society, *Japanese Industry* (1961); *ibid.* (1962).

TABLE 8-2

Japan's national income by industrial divisions and production sectors in fiscal 1961 (in billions of yen)

Primary industry		2,025.3	14.3%
Agriculture	1,396.0		
Forestry	313.1		
Fisheries	316.2		
Secondary industry		5,485.4	38.7
Mining	226.9		
Construction	906.8		
Manufacturing	4,351.7		
Tertiary industry		6,671.7	47.0
Wholesale and retail	2,219.6		
Finance, insurance, and real estate	1,055.3		
Transport, communication, and public utilities	1,434.8		
Services, government and others	1,962.0		
Total domestic national income		14,182.4	
Net income from abroad		−64.7	
Total national income		14,117.7	

Source: Japan Statistical Yearbook, 1962.

TABLE 8-3

Relative importance of the principal manufacturing groups in Japan, based on value added

Group	1935	1951 (fiscal)	1955 (fiscal)	1960 (fiscal)	1961 (fiscal)
Food	12.0%	7.8%	11.4%	9.2%	6.6%
Textile	19.1	19.9	14.8	11.1	8.6
Paper and pulp	3.4	5.8	4.3	4.0	3.7
Chemical	10.0	15.1	13.8	12.8	12.0
Petroleum	0.1	1.8	1.6	1.4	1.3
Ceramic	4.5	4.8	5.5	4.6	4.7
Metals	16.0	15.4	19.3	16.3	17.8
Machinery	26.0	17.7	18.7	28.7	34.8
Other	8.8	11.7	10.6	11.9	10.5
Combined heavy and chemical industries	52.2%	50.0%	49.4%	59.2%	66.0%
Other industries	47.8	50.0	50.6	40.8	34.0

Source: Foreign Capital Research Society, *Japanese Industry* (1962); *ibid.* (1963).

average annual rate of 12.1 per cent. As a consequence Japan has entered upon a new era of heightened prosperity and rapidly improving standards of living which, because these benefits are far from equally distributed throughout the country, are resulting in remarkable internal shiftings of the population, especially of the young adults.[4] One of the most important recent developments in the commodity markets of the free world has been the emergence of Japan as one of the greatest consumers of industrial raw materials, with what appears to be almost an insatiable appetite for ores and fuels to stoke her burgeoning heavy and chemical industries.

CHARACTERISTICS OF JAPAN'S MANUFACTURING

Japan's manufacturing,[5] like her agriculture, has certain distinctive characteristics, some of which set it apart both from manufacturing in Western countries and from that of the rest of Asia. For one thing, industrial raw materials are extremely scarce and, in many instances, even lacking. Although one of the world's greatest cotton textile nations, both in production and exports, Japan grows almost no cotton. Likewise one of the world's leaders in the production of steel, Japan imports 90 per cent of her iron ore and nearly half of her coking coal. A similar paucity of raw materials prevails for most other manufactures. Hence, Japan's basic economy is associated with the processing of preponderantly foreign raw materials, using Japanese labor, technical skill, and

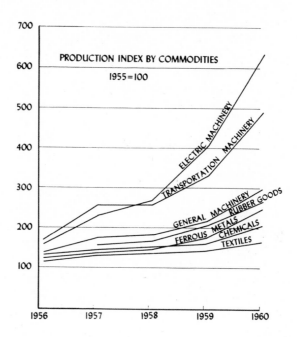

Fig. 8-1.

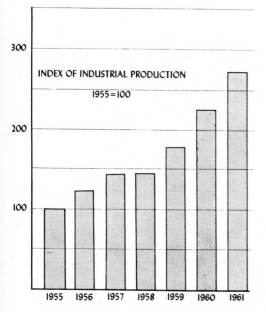

Fig. 8-2.

power resources, and subsequently selling a large share of the finished product in foreign markets.

A second feature is that small enterprises greatly predominate in Japan. Of the country's manufactural establishments, 73 per cent had nine workers or under in 1960, while only 0.58 per cent employed three hundred or more. Moreover, small plants with nine or fewer workers account for nearly 17 per cent of the country's manufactural employees, and those with under one hundred, about 55 per cent. But on the other hand, the less than 2 per cent of the factories that employ over one hundred workers account for about 68 per cent of the total value added by manufacturing in plants with four or more employees. As Table 8-4 shows, the trend is toward larger establishments, a not unexpected development in view of the increasing emphasis on heavy industry. That small plants are reasonably efficient is suggested by the success Japan has had in its manufactural growth. They represent an effective use of cheap local labor as well as a saving in capital goods, which are scarce in Japan. Wage levels in the smaller establishments are only about half what they are in the larger factories, partly because a considerable proportion of their labor is recruited from the ranks of youths in their midteens who have just completed their compulsory education. In many instances the products of these small plants may not be of the highest quality, but, at any rate, they appear to satisfy the needs of the domestic market.

While many of the small establishments are engaged in producing traditional wares for home consumption, large numbers, nevertheless, do process goods for export, such workshops often being closely integrated with large establishments for which they may produce a single part or carry out a single process, within a broad and complicated production

TABLE 8-4

Proportion of Japan's manufacturing establishments and their workers by the scale of management, 1957 and 1960

	Establishments		Workers	
Number of workers	1957	1960	1957	1960
1–9	75.7%	72.9%	20.2%	16.7%
10–29	17.4	18.5	20.5	18.6
30–99	5.4	6.5	19.4	20.1
100–199	1.1	1.1	13.5	9.6
200 and over	0.4	0.9	26.4	34.9

Source: Bureau of Statistics, *Establishment Survey Reports* (Japan, 1960).

scheme. Not uncommonly such small-parts plants are spatially well re-moved from the large central factory and are so situated as to be able to make use of cheap local labor. Thus, there exists a dual structure in Japanese industry, the two components being (1) the large corporation-owned factories, chiefly located in the larger urban centers, with gen-uine attention paid to locational factors, and (2) the much more nu-merous plants of small and medium size, usually privately owned and operated, which are widely scattered, existing in the cities as well as in towns and villages. The latter accept orders from the large enterprises for the processing of small parts, or for the partial processing of what will be a finished product. This dual structure is especially strong among the machine industries.

Still another distinctive feature of the manufactural economy is the generally low prices of Japanese goods, a characteristic made possible by the low wages paid to labor. The low wage scale in turn reflects the great reservoir of surplus labor which has been available in the overpopu-lated rural areas, as well as the generally low standard of living. In recent years, with the very rapid expansion of manufacturing and the shrink-ing surplus of labor in the rural areas, labor costs have risen, and with them the prices of manufactured goods, so that the price differentials between similar products processed in the West and in Japan have been somewhat reduced.

COMPOSITION AND DISTRIBUTION OF JAPAN'S MANUFACTURING

Industry groups and their distribution[6]

That Japan's manufacturing has a highly diversified structure can be observed from Table 8-5. Of nearly a score of industry groups listed, seven (foods, textiles, chemicals, primary metals, both ferrous and non-ferrous, general machinery, electrical machinery, and transportation equipment) are of such importance that each represents over 5 per cent, and all together about 70 per cent, of the total national value added by manufacturing.

The final column in Table 8-5, showing the proportion of manufactur-ing belonging to each of the four main types—rural, light, heavy, and machine industries—provides further evidence of the prevailing manu-factural diversification, although a strong and growing emphasis on the heavy and machine types is indicated by the fact that they make up 64 per cent of the total, according to Doi's method of inclusion. This situation testifies to the technologically advanced nature of Japan's present industrial structure which is markedly in contrast to the condition in the early 1930's, when light industries, especially textiles, strongly predominated.

Type *R* (rural industries, 15.3 per cent of value added by manufacturing) includes the food, lumber–wood, and pulp–paper industry groups. In them the raw materials produced locally by agriculture, forestry, and fishing are processed, mostly by traditional methods, in small or medium-sized factories located with regard to the supply of raw materials, al-

TABLE 8-5

Composition of Japan's manufacturing in terms of value added, 1960
(Only establishments with four or more employees are included.)

Major industry group	Letter symbol	Industry type (after Doi)	% of national total	
Food and kindred products	Fd	R (rural)	8.6	
Lumber and wood products	Lw	R	2.9	
Pulp, paper, and paper products	Pp	R	3.8	
			—	
				15.3%
Chemical and related industries	Ch	H (heavy)	11.9	
Petroleum and coal products	Pt	H	1.4	
Stone, clay, and glass products	Sc	H	4.8	
Iron and steel	Ir	H	8.8	
Non-ferrous metals	Nf	H	3.5	
			—	
				30.4
Fabricated metal products	Fm	M (machinery)	4.5	
Machinery (except electrical)	Mc	M	9.6	
Elec. machinery and equip.	Em	M	9.1	
Transportation equipment	Tr	M	9.1	
Precision instruments	In	M	1.5	
			—	
				33.8
Textile mill products	Tx	L (light)	9.6	
Apparel and other finished fabric products	Ap	L	1.0	
Furniture and fixtures	Fr	L	1.1	
Printing and publishing	Pr	L	4.0	
Rubber products	Rb	L	1.7	
Leather and leather products	Lt	L	0.4	
Miscellaneous	X	L	2.7	
			—	
				20.5

Source: Ministry of International Trade and Industry, *Census of Manufactures, 1960: Report by Cities, Towns, and Villages* (1962).

though large modern plants dominate paper-making and some branches of the food industry.

Type *L* (light industries, 20.5 per cent) includes chiefly textiles but also a number of other consumer goods such as apparel, furniture, rubber, leather, and printing, whose locational orientation is toward labor and markets rather than toward raw materials. Although they evolved from traditional handicraft industries, most of them have been greatly modernized.

Type *H* (heavy industries, 30.4 per cent) here is made to include primary metals (ferrous and non-ferrous), chemicals, petroleum and coal products, and clay-glass-stone products, all of them, with the exception of indigenous pottery, modern factory industries which were transplanted from the West following the Meiji Restoration. Most heavy industry, but not all, is dominated by modern large and medium-sized factories, some of which are locationally oriented with respect to domestic raw materials and fuels, but with the attraction of markets becoming increasingly stronger. The necessity for importing large amounts of the raw materials used causes tidewater location to be particularly favored.

Type *M* (machine industries, 33.8 per cent), embracing not only general and electrical machinery but transportation equipment and precision instruments as well, is also a product of modern Japan. Labor-market orientation is strongly developed, so that there is an unusual concentration in a few great metropolitan districts. Although large modern factories are well represented, small plants also are very numerous.

In his original paper on industrial structure, Kikukazu Doi, using a mode of combination analysis modified and simplified from a method employed by John C. Weaver[7] in a study of Midwestern crop combinations, presented a table[8] which set forth the hierarchy of industry-type and industry-group combinations for each prefecture based upon value added by manufacturing in factories with four or more employees in 1953. A similar modernized table (Table 8-6), here presented, has been prepared by Doi using 1960 data, but differing from the original one in that it includes only factories with thirty or more employees. On the other hand, the figures for total value-added are for establishments with four or more workers. Table 8-6 provides information making it possible to analyze in a coarse way by prefectural units the distribution both of manufactural intensity and of industrial structure as indicated by the industry-type and industry-group combinations.

Figure 8-3 represents a very generalized distribution of manufacturing intensity based upon the total value-added data in Tables 8-6 and

TABLE 8-6

Industry-type and industry-group combinations, by prefecture, 1960

Prefecture	Industry combinations	Other industries of significance	Total value added (100 million yen)
1 Hokkaido	R(Fd* Pp* Lw*) H(Ir Ch)	. . . .	1,236
2 Aomori	R(Fd Lw) H(Ch Ir Sc)	. . . .	122
3 Iwate	R (Fd Lw*) H(Ir Sc Ch) Em	. . . .	212
4 Miyagi	R(Fd Pp) L(Pr Rb) M(Fm Em)	. . . .	247
5 Akita	R(Lw* Pp Fd) H(Ch Pt Nf) Em	. . . .	213
6 Yamagata	L(Tx) H(Ch Ir Nf Sc) M(Mc Em) R(Fd)	. . . .	206
7 Fukushima	R(Fd) H(Ch Sc Nf) M(Em) Tx	. . . .	392
8 Ibaraki	M(Em) H(Nf Sc)		468
9 Tochigi	L(Tx) M(Em Tr Mc) H(Nf* Sc) Fd		434
10 Gumma	L(Tx) M(Em Tr Mc) H(Ch) Fd	. . . .	502
11 Saitama	L(Tx X*) M(Tr Mc Em Fm) H(Ir Sc* Ch) Fd	Ap	1,104
12 Chiba	R(Fd*) H(Ir) M(Em Mc)	. . . .	632
13 Tokyo	L(Pr*) M(Em* Mc* Tr* Fm* In*) H(Ch* Ir*) Fd*	Tx Ap Fr Rb Lt X Sc Nf Lw Pp	8,356
14 Kanagawa	M(Em* Tr* Mc*) H(Ir* Ch*) Fd*	Fr Rb X Pt Nf Fm Lw Sc	4,546
15 Niigata	M(Mc Fm) H(Ch* Ir) Tx	. . . .	855
16 Toyama	L(Tx) M(Mc) H(Ch* Ir) Pp*	. . . .	628
17 Ishikawa	L(Tx) M(Mc)	. . . .	412
18 Fukui	L(Tx*) H(Ch)	. . . .	282
19 Yamanashi	R(Fd Lw Pp) L(Tx) M(Mc Em) Sc	. . . .	115
20 Nagano	L(Tx) M(Em Mc In*) H(Nf Ir) Fd	Lw	612
21 Gifu	L(Tx*) M(Tr) H(Sc* Nf* Ch)	Lw	729
22 Shizuoka	L(Tx*) H(Ch* Nf*) M(Tr* Mc) R(Pp* Fd*)	Lw Fr X	2,006
23 Aichi	L(Tx*) M(Tr* Mc* Em) H(Ch* Sc*) Fd	X Nf Fm Lw Pp Ap Fr	4,550
24 Miye	L(Tx*) M(Em Mc) H(Ch Sc*)	Rb	891
25 Shiga	L(Tx*) M(Em Mc) H(Ch)	. . . .	388
26 Kyoto	L(Tx*) M(Em Tr Mc) H(Ch Sc*) Fd	Ap	1,215
27 Osaka	L(Tx* Pr*) M(Em* Mc* Tr* Fm*) Fd	Lw Pp Ap Fr Rb Lt X Sc Nf	6,646
28 Hyogo	M(Mc* Em* Tr*) H(Ir* Ch*) Fd* Rb* Tx*	Lw Pp Ap Lt X Sc Nf Fm	3,562
29 Nara	L(Tx X) Sc Lw Fd Mc	. . . .	159
30 Wakayama	L(Tx) H(Pt Ir Ch)	Lw	342
31 Tottori	R(Pp Fd Lw) M(Em Mc) Ir	. . . .	90
32 Shimane	R(Pp) H(Ir) M(Mc) Tx	. . . .	151
33 Okayama	L(Tx Ap*) M(Tr) H(Ch* Sc*)	. . . .	160
34 Hiroshima	M(Tr* Mc*) H(Ch Ir) Fd	Lw Ap Fr Rb	1,243
35 Yamaguchi	H(Ch* Pt* Ir)	Sc	1,158
36 Tokushima	R(Pp Fd) L(Tx) H(Ch)	. . . .	185

Note: Asterisks indicate national significance as defined by the areal distribution analysis. Industry combinations are for factories with 30 or more employees. Figures for value-added are for factories with 4 or more employees.

Sources: Ministry of International Trade and Industry, *Kogyo Tokei Sokuho* [*Preliminary Report of the Annual Survey of Manufacturing for 1960*], Tokyo, 1961; Ministry of International Trade and Industry, *Census of Manufactures, 1960: Report by Cities, Towns, and Villages* (1962).

TABLE 8-6—*Continued*

Prefecture	Industry combinations	Other industries of significance	Total value added (100 million yen)
37 Kagawa	L(Tx) H(Nf Ch) M(Mc) R(Fd Pp)	....	219
38 Ehime	L(Tx) M(Mc) H(Ch* Nf*)	Pp	652
39 Kochi	L(Tx Pr) H(Sc Ir Ch) M(Mc) R(Pp Fd)	....	98
40 Fukuoka	H(Ir* Ch* Sc*) Rb* Fd* Mc	Pp Lw	2,203
41 Saga	R(Fd Pp) H(Sc Ch) M(Mc Em)	....	114
42 Nagasaki	M(Tr)	....	343
43 Kumamoto	R(Pp* Fd) H(Ch)	....	268
44 Oita	R(Fd Pp) H(Sc Nf Ch)	....	227
45 Miyazaki	R(–) H(Ch)	....	272
46 Kagoshima	R(Fd Lw Pp) L(Pr Tx)	....	138

8-7. Here the prefectures are grouped into three general industrial areas—metropolitan, central, and peripheral—and each of these in turn is further subdivided. What becomes evident is that manufacturing intensity in Japan is very unequally distributed. Thus, about 86 per cent is concentrated in what are here called the central and metropolitan areas, located within central and southwestern Japan and accounting for only 35–40 per cent of the national area. The two metropolitan subdivisions of Kanto and Kinki together account for about 56 per cent of the country's value added by manufacturing; and Kanto alone, for nearly one-

TABLE 8-7

*Distribution of the total value added by manufacturing, 1960**

Area	Value added (100 million yen)		% of total	
Metropolitan		28,009		55.8
Kanto (7 prefectures)†	16,039		32.0	
Kinki (4 prefectures)†	11,970		23.8	
Central		15,138		30.1
Tokai	7,447		14.8	
Central Setouchi and N. Kyushu	6,235		12.4	
Tosan	1,456		2.9	
Peripheral		7,023		14.1
Hokkaido	1,236		2.5	
Tohoku	1,382		2.8	
Hokuriku-Sanin	2,418		4.8	
Outer Zone of Southwest Japan	1,987		4.0	

* Includes establishments with 4 or more employees.
† See Fig. 8-3 for prefectures included.
Source: Ministry of International Trade and Industry, *Census of Manufactures, 1960: Report by Cities, Towns, and Villages* (1962).

third. Actually the intensity of concentration is greater than here shown, for while in Figure 8-3, Kanto comprises seven prefectures and Kinki five, within each of these two metropolitan areas manufacturing is overwhelmingly concentrated in the two prefectures where the great cities of Tokyo-Yokohama and Osaka-Kobe are located, so that the four

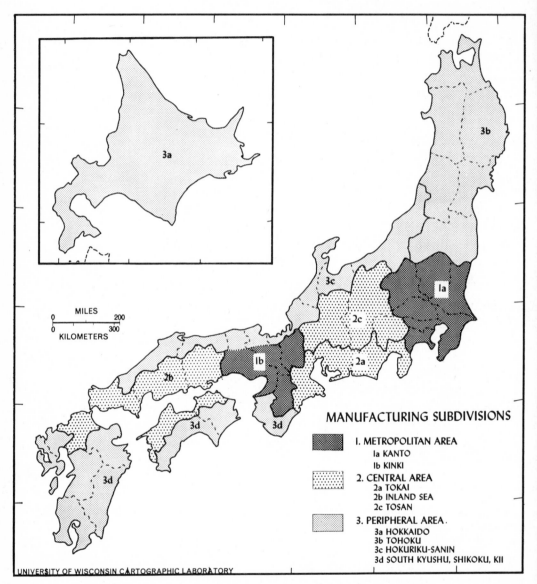

MANUFACTURING SUBDIVISIONS

1. METROPOLITAN AREA
   1a KANTO
   1b KINKI
2. CENTRAL AREA
   2a TOKAI
   2b INLAND SEA
   2c TOSAN
3. PERIPHERAL AREA
   3a HOKKAIDO
   3b TOHOKU
   3c HOKURIKU-SANIN
   3d SOUTH KYUSHU, SHIKOKU, KII

Fig. 8-3.—Manufacturing subdivisions by prefectures.

prefectures of Tokyo-Kanagawa and Osaka-Hyogo include 46 per cent of Japan's manufacturing.

Within the more extensive central area, which accounts for about 30 per cent of Japan's industrial output, manufacturing is still a highly developed sector of the economy, but much less intensively so than in the metropolitan area. Three subdivisions are recognized: Tokai (2a), located between Kanto and Kinki; the Inland Sea basin (2b), located west of Kinki; and Tosan (2c), the weak member of the triad, situated in the mountains of central Honshu. In its intensity of manufactural development Tosan really belongs in the peripheral area, although its location can scarcely be called peripheral. Within the central area, to a greater extent than in the metropolitan area, industry is concentrated in dispersed small centers, there being only two genuinely large centers, one in and around Nagoya city in western Tokai, and the other, Kitakyushu city, in the westernmost part of the Inland Sea subdivision.

Within the so-called peripheral area which comprises 60–65 per cent of the country's area, manufacturing is modestly or even meagerly developed, for from it is derived only 14 per cent of the total value added by manufacturing. All parts of the area are relatively isolated from the country's heartland either by distance or by terrain barriers. Four general subdivisions are recognized: Hokkaido (3a) and Tohoku (3b), occupying the northern outlands; Hokuriku-Sanin (3c), on the Japan Sea margins; and southern Kyushu-Shikoku-Kii (3d), in the southwestern mountain lands. Least industrialized are the two northern subdivisions, for while in area they are distinctly larger than the other two, their percentage contributions to total national value added by manufacturing are clearly smaller (Table 8-7). Least retarded of the four subdivisions is the Hokuriku-Sanin area, more especially the Hokuriku or northern part, whose output in proportion to area is four to five times that of Hokkaido or Tohoku, and more than double that of the southwestern area (3d).

Turning now to the question of whether there exist any recognizable gross patterns of distribution as it applies to manufactural composition, from Table 8-6 and Figure 8-4 it may be observed that of the sixteen prefectures in which rural industries (R) stand foremost in importance, all but two are located in what has been designated the peripheral manufactural area, while of the twenty-two prefectures included within the peripheral areas, fourteen show rural industries as having highest rank and only six show a predominance of light consumer-goods industries (L), with three of these in the more advanced

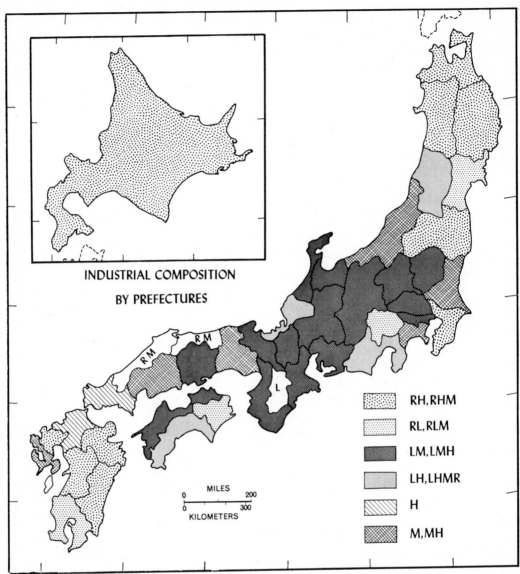

INDUSTRIAL COMPOSITION
BY PREFECTURES

RH, RHM	
RL, RLM	
LM, LMH	
LH, LHMR	
H	
M, MH	

MILES
0   200
0   300
KILOMETERS

Fig. 8-4.—For explanation of letter symbols see Table 8-5.

Hokuriku area and only one in the manufacturally more retarded northern areas of Hokkaido and Tohoku. In the peripheral area, no prefecture shows a first specialization in heavy industry; and only two, in machinery. Of the fourteen prefectures in the peripheral areas where rural industries hold first rank, nine have heavy industry as the second most important group; two, machinery; and six, light industry.

Light industries (L) predominate in twenty-one prefectures, but only one of these is in northern Japan and none in Kyushu, so that there is a high degree of concentration in central and southwestern Honshu and in Shikoku. Eight of the twenty-one are located in the metropolitan area, seven in the central area, and six in the peripheral area, all but one of the latter six in subtropical Honshu and Shikoku. Clearly, the light consumer-goods industries, although widely distributed, avoid the northern and southern extremities of the country and instead concentrate closer to the nation's heartland and not too far removed from the great metropolises and their markets. Of the twenty-one prefectures where light industries hold first rank, machinery is next in importance in fifteen (LM), while only six show the combination LH, although twelve are LMH. There is no instance of the combination LR.

In only eight prefectures are the machine (M) or heavy (H) industries dominant, and these prefectures are widely distributed, although significantly none are located in the northern parts of the country. Of the eight, five are represented by the combination MH, two of them in Kanto and one in Kinki. Only two prefectures—Fukuoka, in northern Kyushu, and adjacent Yamaguchi, in extreme southwestern Honshu—have a first-degree specialization in heavy industry (H), chiefly primary metals and chemicals. Nagasaki Prefecture alone deserves the single symbol M, a reflection of the great shipbuilding yards that dominate the industrial structure of that somewhat isolated area.

In summary, the metropolitan area emphasizes light industries other than textiles (printing and publishing, rubber and leather and rubber products, furniture), primary metals, fabricated metals, machinery, and transportation equipment, while it is slightly below average in textiles and more so in rural industries and stone-glass-clay products. The central area gives somewhat stronger emphasis to textiles and a moderate one to rural industries, but is below average in primary and fabricated metals and machinery of various kinds. The peripheral area, chiefly the northern and southern extremities of the country, is strong relatively in the rural industries, stone-clay-glass products, and chemicals, but is weak in textiles, fabricated metals, and machinery.

The manufactural belt and its nodal areas

Within Japan manufacturing is strongly concentrated in a thin belt, 1000–1100 kilometers (600–700 miles) long, which extends from the Kanto region at its northeastern extremity, southwestward through Tokai, Kinki, and the borderlands of the Inland Sea to northern Kyu-

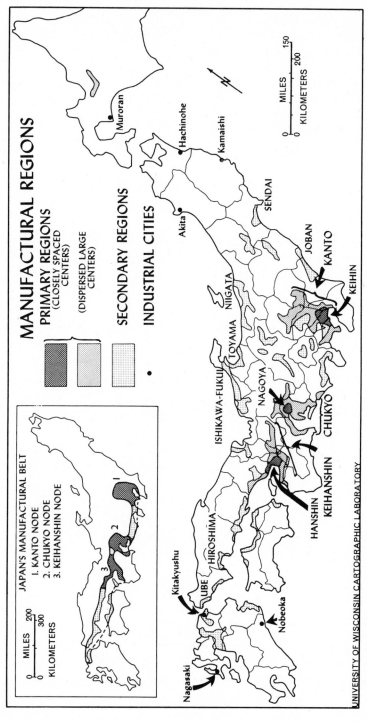

MANUFACTURAL REGIONS

PRIMARY REGIONS
(CLOSELY SPACED CENTERS)

(DISPERSED LARGE CENTERS)

SECONDARY REGIONS

INDUSTRIAL CITIES

JAPAN'S MANUFACTURAL BELT

1. KANTO NODE
2. CHUKYO NODE
3. KEIHANSHIN NODE

MILES
KILOMETERS

Nagasaki

Kitakyushu

UBE

HIROSHIMA

Nobeoka

HANSHIN

KEIHANSHIN

CHUKYO

NAGOYA

ISHIKAWA-FUKUI

TOYAMA

NIIGATA

Akita

SENDAI

JOBAN

KANTO

KEIHIN

Kamaishi

Hachinohe

Muroran

MILES
KILOMETERS

UNIVERSITY OF WISCONSIN CARTOGRAPHIC LABORATORY

Fig. 8-5.—Modified from Thompson and Miyazaki.

shu (Fig. 8-5).[9] As might be expected, this manufactural belt likewise represents a concentration of total population and of the nation's cities. Within it are employed nearly 80 per cent of the total number of workers in manufacturing establishments with four or more employees, and there also originates about 85 per cent of the total national value added by manufacturing. The belt is always narrow, in most parts only a few kilometers wide paralleling the coast. In a very few places it expands into nodes and in such areas it may attain a width of 35–50 kilometers or more. Although it is drawn as a continuous belt in Figure 8-5, awareness of the nature of Japan's terrain will make it obvious that such an uninterrupted form is unlikely to prevail, for the manufacturing plants are concentrated on small, discontinuous littoral plains which are separated by spurs of relatively barren hill land. Even on the individual plains most of the factories are clustered in relatively small, unequally spaced, discontinuous centers, sometimes very close together, but in other places separated by gaps of 15–25 kilometers. The northeastern part of the belt faces the Pacific Ocean, while the southwestern half largely fronts on the Inland Sea.

Specific advantages of this region for attracting industry may not be so obvious. Certainly no single factor largely accounts for the localization. In part the causes are historical in nature, for during a millennium and more preceding the Meiji Restoration in 1868, this coastal strip on the Pacific and Inland Sea side of Japan, lying between Kanto at the northeast extremity and North Kyushu at the southwest, was very much the focus of political power and of the nation's economic activities. It included the old imperial capitals of Kyoto and Nara, as well as Edo (now Tokyo), the capital of the Tokugawa Shoguns. Numerous important castle towns and traffic centers grew up along the famous Tokugawa Tokaido Highway, connecting Edo with Imperial Kyoto, and along the Sanyodo, which followed the northern shore of the Inland Sea from Kyoto to northern Kyushu. It is not surprising, therefore, in the light of history, that when Japan was finally opened to Western influence, industry and commerce should be attracted first to this populous southwestern coastal strip on the Pacific side where numerous important cities, including many ports, were already in existence. As a consequence, the old established towns and cities waxed mightily in size and functions, while new specialized ones such as Yokohama, Kobe, and smaller ones in northern Kyushu, sprang to life with special modern functions. Also, some of the first railway lines in Japan were built in this same region because of its already-established importance as an area of concentrated population and numerous cities. From Tokyo to Kobe ran the Tokaido Line, which was continued to Shimonoseki as the Sanyo Line. At present this Tokaido-Sanyo route is by all odds the longest double-tracked section of rail line in Japan, has the best service and fastest trains, and carries the heaviest freight and passenger loads.

Perhaps the most important single advantage of the Manufactural Belt, common to almost all parts, is its easy approachability by sea. No other similarly extensive part of Japan is so readily accessible, either for foreign or domestic sea trade. The whole belt is coastal in location, while the three great bays (Tokyo, Ise, Osaka) along its northeastern part, with their large bay-head plains, offer protected waters, large populous local hinterlands, and extensive flat areas for factory construction. In its southwestern half the quiet waters of the nearly land-enclosed Inland Sea provide a protected water corridor which is highly attractive to shipping. Within the Manufactural Belt are located all of the more important foreign-trade ports of Japan, and there are far more numerous smaller ones, engaged chiefly in domestic trade, which supplement the services of the larger ones.

By contrast, the power resources located within, or close to, the Manufactural Belt are only modest. At its extreme southern end—in northern Kyushu and adjacent southwestern Honshu—are Chikuho and several lesser coal fields, whose combined output makes it the greatest coal-producing area of the country. By way of the Inland Sea, especially, this North Kyushu coal is distributed to the whole southwestern part of the Manufacturing Belt. The less important Joban field a short distance northeast of Tokyo is a source of supply for a portion of Kanto's coal requirements, while the whole northeastern part of the belt, especially Kanto and Tokai, is favorably located with respect to the country's greatest focus of hydroelectric development, situated in the highlands of central Japan.

As noted previously, there are three principal nodes of industrial concentration within the attenuated Manufactural Belt, all of them significantly located at the heads and along the adjacent margins of three great bays—the Kanto node on Tokyo Bay, the Chukyo node bordering Ise Bay, and the Keihanshin node adjacent to Osaka Bay. At the heads of these bays, especially, are relatively extensive and populous depositional plains which have become the nation's great hubs of transport lines. All six of Japan's cities of over a million, as well as her five most important foreign-trade ports, are to be found within the three great nodal areas of industrial concentration.

Kanto, Keihanshin, and Chukyo together accounted for over two-thirds of the national total of value added by manufactures of all kinds in 1960

—74 per cent of textiles, 68 per cent of metals, nearly 74 per cent of machinery, and 55 per cent of chemicals (Table 8-8). Kanto held first rank, with 31 per cent of the country's total, followed by Keihanshin with 24.5 per cent and Chukyo with 12.3 per cent of the total.

Restricted parts of each of the three nodes are composed of closely spaced centers, large and small, including such giants as Tokyo-Yokohama in Kanto, Nagoya in Chukyo, and Osaka-Kobe-Kyoto in Keihanshin, while other sections are made up of dispersed centers of various sizes. Each of the triad has a highly complex and diversified industrial structure (see Tables 8-6, 8-8, and 8-9). Thus, food, metals, machinery, chemicals, and textiles are importantly represented in each of the three. Yet there are significant differences. All three strongly emphasize ma-

TABLE 8-8

Proportions of total national value added by manufacturing contributed by each of the manufactural nodes in the more important industrial sectors, 1960

Node	Metals	Machinery	Chemicals	Textiles	All industries
Kanto	32.4%	40.8%	23.3%	11.7%	31.0%
Keihanshin	29.8	20.2	21.4	31.8	24.5
Chukyo	6.1	12.5	10.5	30.5	12.3
All three nodes	68.3	73.5	55.2	74.0	67.8

Note: Kanto node includes six prefectures: Tochigi, Gumma, Saitama, Chiba, Tokyo, and Kanagawa. Keihanshin node includes six prefectures: Shiga, Kyoto, Osaka, Hyogo, Nara, and Wakayama. Chukyo node includes parts of three prefectures: Gifu, Aichi, and Mie.
Source: Ministry of International Trade and Industry, *Census of Manufactures, 1960: Report by Cities, Towns, and Villages* (1962).

TABLE 8-9

Composition of manufacturing in the three manufacturing nodes based upon value added, 1960

Industry group	Kanto	Keihanshin	Chukyo
Food	8.0%	6.8%	6.1%
Textiles	3.6	12.5	24.0
Printing and publishing	6.5	3.8	1.9
Chemicals	9.0	9.9	12.2
Ceramics	3.8	4.1	8.9
Metals	17.3	20.2	8.3
Machinery	38.7	29.6	30.2
Others	13.1	13.1	8.4

Source: Ministry of International Trade and Industry, *Census of Manufactures, 1960: Report by Cities, Towns, and Villages* (1962).

chinery, but Kanto more so than the other two. Kanto is likewise relatively more important in printing-publishing, but is markedly below the other two in textiles. Chukyo stands out from the others in its much greater emphasis on textiles and ceramic products, and somewhat greater emphasis on chemicals. On the other hand it is well below the others in printing-publishing and metals.

Not uncommonly the Kitakyushu area in Fukuoka Prefecture in northern Kyushu is designated as a fourth manufactural node. But even if all of Fukuoka Prefecture, which embraces a number of isolated industrial centers in addition to Kitakyushu city, is included the area still contributes only 4.4 per cent of the total national value added by manufacturing, while the new city of Kitakyushu, which is the main industrial focus, accounts for less than 3 per cent. Clearly, the northern Kyushu area is scarcely a manufactural node in the sense that the other three are. No doubt one reason why Kitakyushu city is sometimes designated as one of the four manufactural nodes is that it is the fifth or sixth city in Japan in industrial rank, but in this case there are no satellite centers whose industries act to supplement importantly those of the local primary city as is true in the three nodes described previously.

Outside of the three industrial nodes, which together account for nearly 68 per cent of the national value added by manufacturing, an additional 17 per cent is contributed by the other parts of the Manufactural Belt—roughly 4 per cent by Shizuoka Prefecture, or Sunen, situated between Kanto and Chukyo; and 12–13 per cent by the Inland Sea borderlands in Chugoku, Shikoku, and northern Kyushu.

Outside of the Manufactural Belt are dispersed smaller concentrations of industry, both areas and individual cities, whose locations and names are provided in Figure 8-5. Consideration will be given to some of these individually in the regional section of this book. In addition, of course, there are the widely scattered industries of the rural areas. Altogether these areas and industries outside the Manufactural Belt probably account for some 15 per cent of the total national value added by manufacturing.

Recent changes in the distribution of manufacturing

It has been noted earlier in this chapter that manufacturing in Japan experienced a great upsurge during the decade of the fifties, with manufacturing employment in 1960 being 153.6 per cent of what it was in 1955, value of shipments 231 per cent, and value-added 236 per cent. While this burgeoning of manufacturing has affected nearly all parts of the country, still there were important differences in the regional rates

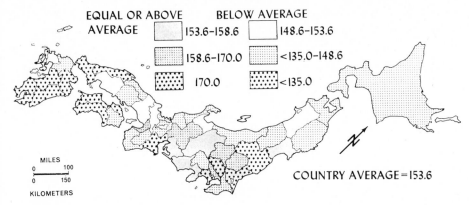

EQUAL OR ABOVE AVERAGE
BELOW AVERAGE

153.6–158.6
148.6–153.6

158.6–170.0
<135.0–148.6

170.0
<135.0

MILES
0 — 100
0 — 150
KILOMETERS

COUNTRY AVERAGE=153.6

Fig. 8-6.—Growth of manufacturing by prefectures, 1955 to 1960, based upon number of employees. In per cent.

of growth, so that the map of manufacturing intensity is gradually changing. Figure 8-6 shows the change in manufactural employment in factories with four or more workers during the five-year period 1955–1960. While only one prefecture (Oita) showed an absolute loss, thirty-three of them, or 72 per cent, registered gains that were below the national average (153.6 per cent), indicating that the strong gains were occurring in somewhat restricted areas. Six of the below-average prefectures were only slightly below (5 per cent); fifteen were well below, and twelve strongly below. The distribution of these thirty-three slow prefectures and their degree of lag as represented in Figure 8-6 reveals that it is the northern (Hokkaido, Tohoku), Japan Sea side (Hokuriku, Sanin), and southwestern (most of Kinki, Shikoku, Chugoku, and Kyushu) parts of the country which are falling behind in rate of manufactural growth, with the strongest lags registered in the far southwest in Kyushu, Shikoku, and western Chugoku.

Of the thirteen prefectures showing a rate of manufactural-employment increase above the national average, eleven are located in central Japan, with Miyagi in eastern Tohoku and Osaka in the southwest being the only exceptions. Seven of the thirteen showing an accelerated increase lie within the Kanto District, while six are within the Kanto Manufactural Node. Moreover, all of the Kanto prefectures show gains that are not merely in excess of the national average but are notably strong. Comparing the three manufactural nodes, Keihanshin's rate of growth (154.6) was only very slightly above the national average, and Chukyo's (151.9) slightly below, but Kanto's (170.3), by contrast, was well above. What appears to be happening is that the center of gravity of Japan's manufacturing is shifting gradually northeastward, with the southwestern parts lagging behind the central parts in rate of growth, and Kanto outdistancing all other districts in its relative acceleration.

If manufactural growth is viewed in absolute as well as relative terms a similar pattern emerges, for in the same five-year period, 1955–1960, about 74 per cent of the whole national increase in manufactural employment was in the three great nodes, 37 per cent in Kanto, 23 per cent in Keihanshin, and 14 per cent in Chukyo. Thus considering both relative and absolute growths, it becomes clear that the inequalities in the distribution of manufacturing intensity have been increasing, for industry has been piling up more rapidly in the three already-great metropolitan centers than it has elsewhere.

This excessive rate of growth—especially in the Kanto marine area, but also in the Osaka-Kobe, and to a less degree the Nagoya, areas—has produced something of a crisis in these older and larger industrial centers, which is reflected in their shortages of industrial water, deficiencies of tidewater land for new heavy-industry plants, harbor overcrowding and inadequate port facilities, and traffic confusion and congestion within the metropolitan areas. Traffic paralysis within Tokyo has resulted in a 30 per cent decline in the turnover rate of freight trucks in the two-year period 1958–1960, while the average speed per hour has decreased from 20 to 12 kilometers.[10] Congestion on commuter electric trains during the morning rush hours is unbelievable, with the Chuo Line in Tokyo packing in three times the allowable number of persons.

Such abnormal industrial growth in the great manufactural nodes, and particularly in their metropolitan parts, has led to a zealous search for new lands on which to erect the factories of the giant heavy-industry combines, such as steel and petrochemicals, which require tidewater location. Of the 3888 hectares (9600 acres) approved as reclaimable for industrial-land use during the seven years 1954–1960, 76.5 per cent was located in the four most highly developed industrial districts—32.3 per cent in the Kanto marine district, 35.1 per cent in the Kinki district, 3.6 per cent in Tokai, and 5.5 per cent in northern Kyushu.[11] Much of the above is land to be reclaimed from the sea by dredging and filling.

The increasing handicaps to expanding industry, and especially heavy industry, experienced within the present highly developed metropolitan areas have led to comprehensive national development planning, particularly as it applies to manufacturing. A report submitted to the Prime Minister by the National Land Planning Conference, July 5, 1961, the draft of which was prepared by the Economic Planning Agency, urges planning with a view to checking the overconcentration of population and industries in a few metropolitan areas, especially Keihin and Hanshin, and to spreading them more broadly over the country, giving particular attention to the less well developed sections.[12] As the planners conceived it, the country was to be divided into a number of economic blocks or districts, and recommendations were made relative to the desirable rate of future manufactural growth in each, with 1958 as the base year and 1970 as the target year (see Table 8-10 and Fig. 8-7). Particularly to be noted are the contrasting percentages, actual 1958 and recommended 1970, for the several regions and also the contrasting projected rates of manufacturing growth for the individual regions. Thus, Kanto and Kinki, and especially their congested coastal sections, are planned for only a relatively slow growth which will importantly reduce their proportions of manufactural production in 1970 compared to what they were in 1958. By contrast, considerably increased growth rates are

T A B L E 8-10

Manufactural production, based on value of shipments, for various districts of Japan: actual for 1958 and recommended for 1970

Districts	1958	1970	Rate of increase
NATION	100.0%	100.0%	4.3%
Hokkaido	2.9	3.0	4.4
Tohoku	3.3	4.2	5.3
Northeast	(0.9)	(0.9)	(4.1)
Southeast	(1.4)	(2.3)	(7.3)
West	(1.0)	(1.0)	(3.8)
Kanto	37.1	35.1	4.0
Inland	(5.5)	(9.1)	(7.0)
Coastal	(28.9)	(23.0)	(3.4)
Shin Etsu	(2.7)	(3.0)	(4.8)
Tokai-Hokuriku	14.1	16.9	5.1
Tokai	(12.2)	(14.6)	(5.1)
Hokuriku	(1.9)	(2.3)	(5.3)
Kinki	25.9	20.2	3.3
North	(1.2)	(1.6)	(5.5)
Inland	(2.6)	(3.3)	(5.4)
Coastal	(22.1)	(15.3)	(2.9)
Chugoku-Shikoku	9.2	12.5	5.8
Sanin	(0.5)	(0.5)	(4.3)
Sanyo	(6.2)	(8.7)	(6.0)
Shikoku	(2.5)	(3.3)	(5.6)
Kyushu	7.5	8.1	4.6
North	(6.0)	(6.2)	(4.4)
South	(1.5)	(1.9)	(5.5)

Source: Chiiki Kaihatsu Kankei Shiryo Shu [Materials Concerned with Regional Development], Nippon Kogyo Ritchi Senta (Japan Institute of Location Economy), 1962.

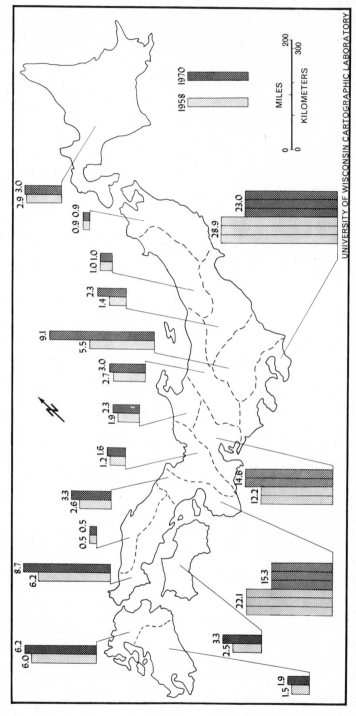

Fig. 8-7.—Planned goals of regional changes in manufacturing between 1958 and 1970 based on value added. After map prepared by Nippon Kogyo Ritchie Center.

planned for a large number of districts, including southeast Tohoku, inland Kanto, north and inland Kinki, Tokai-Hokuriku, Sanyo, Shikoku, and south Kyushu, so that their 1970 relative output of manufactures will be importantly increased over that of 1958.

It is still uncertain how much of the comprehensive plan for industrial dispersion will be accomplished, but already there is concrete evidence of its workings in the increasing number of new industrial centers coming into being and the still more numerous already-existing small ones which are undergoing expansion. A few examples of burgeoning new tidewater foci of heavy industry and port and harbor development are Tomakomai in southwestern Hokkaido, Chiba-Goi on the northeast shore of Tokyo Bay, Yokkaichi on the west side of Ise Bay, Mizushima in Okayama Prefecture and Iwakuni in Yamaguchi Prefecture facing the Inland Sea, and Oita-Tsurusaki on Beppu Bay in northeastern Kyushu.[13] The most noteworthy change in industrial composition expected and planned for by 1970 is a sharp rise in the relative importance of the machinery sector and a further waning of textiles.

IMPORTANT INDIVIDUAL INDUSTRY GROUPS

Location and site factors as they apply to Japan's manufacturing may be considered both for manufacturing in general and for specific industry groups, also. Thus, manufacturing in general is strongly concentrated in close proximity to the coast, not merely as a necessary consequence of Japan's elongated insular shape, but even more because workshops and factories are attracted to the populous depositional plains which are themselves predominantly littoral in location. Quite naturally, the prevailingly rugged interior has acted to discourage factory industry there. More emphatically, it is the modern industries which were transplanted from the West that are to an unusual degree focused on the important port cities, for such are most heavily dependent upon foreign sources for their raw materials (including iron ore, coking coal, petroleum, salt, cotton, and wool) and also upon overseas markets for their processed wares. Accordingly, the main port cities have served to attract those industries that are raw-materials oriented, those whose products enter importantly into foreign trade, and also those catering to an important domestic market, since the larger ports, at least, have populous hinterlands. These same hinterlands provide large reservoirs of industrial labor. Already in an earlier section emphasis was placed upon the port functions of Yokohama, Tokyo, Kobe, Osaka, and Nagoya as a factor leading to the unusual concentrations of industry in Kanto, Keihanshin, and Chukyo.

Extending inland from the ports, the network of rail lines makes connection with the immediate and more distant hinterlands and induces secondary industrial concentrations in the form of lines and clusters reaching inland from the coastal cities. Moreover, Japan's principal coal fields are located close to tidewater, at the northern and southern extremities of the islands, facts which favor the use of sea routes for transporting coal to industrial centers. For the indigenous or traditional industries coastal location and proximity to ports has less attraction, and consequently they are more ubiquitous in location. Among the modern industries it is particularly the heavy groups, especially steel and petrochemicals, that seek out the seaside sites where they can be served by direct marine transportation, many times from their own company piers.[14] The attractiveness of tidewater location, especially within the great metropolitan manufactural areas adjacent to the important port cities, has created a zonation of industrial classes owing to land-use competition, with the coastal zone used principally by the metallurgical and heavy engineering industries and large petrochemical combines—chiefly by their large plants engaged in primary processes and in some finishing stages such as shipbuilding. Significantly, 70–80 per cent of all industrial workers in the coastal areas of Tokyo, Osaka, and Nagoya are employed in metal and engineering industries.[15] In the next zone back from the coast, manufacturing is much more varied in character and the plants are generally smaller.[16] While Japan's manufacturing has always been concentrated not far from the coast, it is recently, as heavy and chemical industries have shot ahead in growth, that the impetus toward tidewater sites has become so strong.

Hasegawa has studied the distribution of manufacturing in Japan by means of statistical analysis in which he calculates the "coefficient of localization" and "the location factor" for each major industry group, using employment data. His coefficient of localization for a given industry measures the degree to which it is geographically concentrated. Table 8-11 shows the coefficient of localization of manufacturing by major groups as computed by Hasegawa. If the geographic distribution of a given industry is completely uniform the coefficient is zero, and the more remarkable the geographic concentration becomes, the nearer the coefficient approaches unity. From the table it may be observed that the least concentrated, or most uniformly distributed, industries are general machinery, paper and paper products, food, apparel, and textiles. A majority of these belong to the indigenous and residentiary industries and so tend to follow population distribution. Industry groups with the highest degree of geographic concentration are rubber, petroleum products, and precision instruments. In between, but still relatively high, are chem-

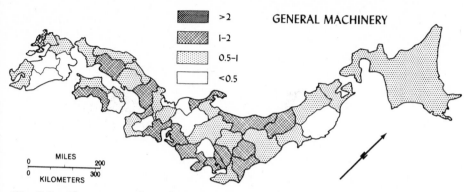

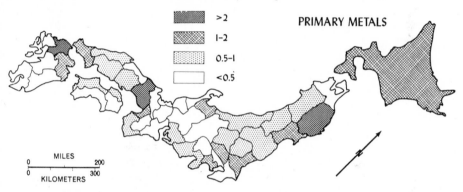

Fig. 8-8.—Above, general machinery, location factor as defined by Hasegawa. *Below,* primary metals, location factor as defined by Hasegawa.

icals, primary and fabricated metals, electrical machinery, and transportation equipment.

From Hasegawa's Table 3 showing the location factor by industries, maps may be produced showing the distribution of the location factor by prefectures for each industry group, based upon number of employees. Two such maps (Figs. 8-8a, 8-8b), one for an industry with a low coefficient of localization—machinery—and the other for primary metals, with a relatively high coefficient, are here reproduced.

Space does not permit an analysis of each of the major industry groups, so that in the discussion to follow attention will be directed to only the four most important—textiles, chemicals, metals, and machinery—which together represent close to 68 per cent of the value added by manufacturing in 1960.

Textiles

Textiles were the first modern Western industry to develop in Japan after the Meiji Restoration and by the early 1930's represented nearly one-quarter of the industrial employment. In the face of the remarkable recent expansion of heavy manufactures such as metals, machines, and chemicals, textiles have declined relatively, so as to represent, in 1960, only 11.1 per cent of total national value added by manufacturing and 15.3 per cent of the manufactural employment. Still, textiles, although only slightly exceeding machinery, continue to represent the single largest group of exports (30 per cent in 1960, and 25 per cent in 1962).

The industry lost about 70 per cent of its equipment as a consequence of World War II, but has recovered to the point where it now has such a large surplus of equipment, much of it highly modern, that production has had to be officially curtailed. One of the noteworthy and recent developments in Japanese textiles is the disproportionate growth in the chemical-fiber industry. In prewar years the ratio of chemical textiles to natural textiles was about 20 to 80, while it is now nearly 50–50. This reflects the fact that Japan must rely on foreign sources for natural fibers other than silk, while the raw materials for chemical fibers are for the most part domestically available. Consequently the shift toward the latter is a factor in improving the balance of international payments.

Considering all textile mill products, by far the two greatest centers are Keihanshin and Chukyo, each accounting for over 30 per cent

TABLE 8-11

Coefficient of localization of manufacturing by major groups, 1951

Major group	Coefficient
Food and kindred products	0.207
Textile mill products	0.281
Apparel and other fabricated textile products	0.268
Lumber and wood products	0.327
Furniture and fixtures	0.301
Paper and other products	0.165
Printing and publishing	0.311
Chemicals	0.301
Petroleum and coal products	0.445
Rubber products	0.517
Leather and leather products	0.343
Stone, clay, and glass products	0.275
Primary metals	0.308
Fabricated metals	0.338
Machinery (general)	0.139
Electrical machinery and equipment	0.315
Transport equipment	0.310
Scientific and precision instruments	0.423

Source: Norio Hasegawa, "Distribution of Manufacturing in Japan: A Macroscopic Analysis of Localization of Manufacturing," *The Science Reports of the Tohoku University,* 7th Series (Geography), No. 8 (1959).

of the total national value added by textiles, and together representing 62–63 per cent (Fig. 8-9). Next in importance, but well below the other two, is the Kanto node with 11–12 per cent of the value-added. Elsewhere textiles are relatively important in the Niigata, Toyama, and Ishikawa-Fukui industrial areas facing the Japan Sea; Shizuoka Prefecture in Tokai; along the margins of the Inland Sea in southern Chugoku and northern Shikoku; the southernmost prefectures of Tohoku; and the highland basins of Tosan.

Cotton textiles, the leading industry of the early 1930's, has lost weight in Japan's postwar economy, for the production level has been restored to only about 80 per cent of its prewar record years. This failure to expand commensurately with the swelling postwar economy reflects, among other things, the growing competition of synthetic textiles, the contraction of overseas markets resulting from the rise of new cotton-producing nations, and the world-wide tendency to impose import restrictions. In addition, it has become increasingly more difficult to attract cheap female labor to the cotton industry because of better employment compensation in other sectors. Still, Japan stands first in the export of cotton goods, accounting for about 30 per cent of the total world export in 1960.

Characteristically the cotton spinning operation is housed in large modern factories, but weaving is done both in conjunction with the large spinning establishments and also in much more numerous small-scale plants, many of cottage-industry size. The latter are widely scattered geographically, and are common to large cities and to rural villages.

TABLE 8-12

Textile industry

Textile	Yarn production, 1962 (1000 tons)	Fabric production, 1962 (millions of sq. m.)	Textile goods	Exports, 1962 (millions of dollars)
Cotton	484	3083	Cotton fabrics	$ 341
Woolen	148	333	Woolen fabrics	50
Silk	20	181	Raw silk	54
Hard and bast fiber	84	135	Spun rayon fabrics	95
Rayon filament	137	660	Rayon fabrics	58
Spun rayon	185	841	Apparel	204
Synthetic	135	645	Synthetic fiber	67
TOTAL	1,193	5,878	TOTAL	$1,257

Source: Foreign Capital Research Society, *Japanese Industry* (1963).

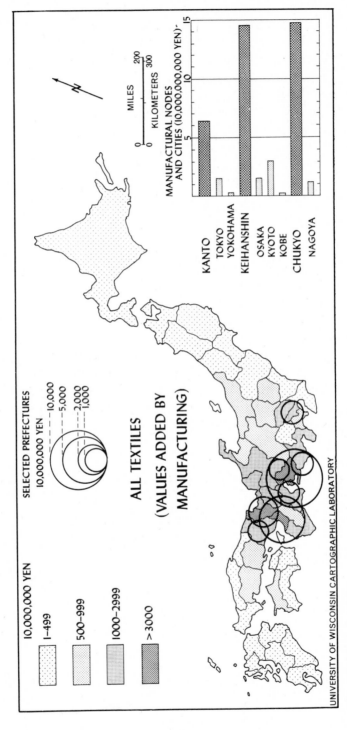

UNIVERSITY OF WISCONSIN CARTOGRAPHIC LABORATORY

Fig. 8-9.—Data for 1960.

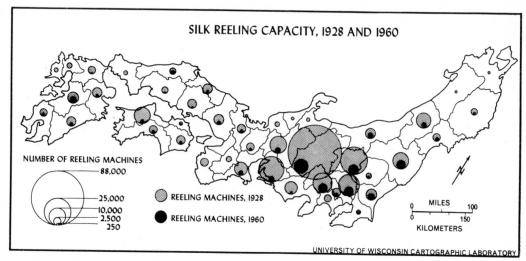

Fig. 8-10.—After map by Richard Hough.

The greatest cotton processing centers are Chukyo and Keihanshin. Lesser but still important concentrations are the Kanto area, Shizuoka, Toyama-Niigata prefectures along the Japan Sea, the Inland Sea borderlands, and Fukuoka and Kumamoto prefectures in northern and western Kyushu.

Silk reeling, like all other aspects of the silk industry, has shrunk markedly over the last several decades. During the 1920's silk—chiefly raw silk, but including some silk fabrics—accounted for 30–40 per cent of Japanese overseas exports. But beginning in about 1930, the industry lost ground due to the introduction of low-priced chemical fibers at home and the remarkable development of synthetic fibers abroad. World War II imposed additional handicaps as foreign markets were lost, while at the same time mulberry fields had to be converted to food crops. Immediately after the war, production and export dropped to about one-third of the prewar level. Since the war, raw silk production has only slowly revived, so that in recent years it is still somewhat less than half what it was in the 1930's. But while production of raw silk has not changed greatly between 1956 and 1960, and exports of raw silk have increased only modestly, the export of silk fabrics has doubled.

Filatures, engaged in reeling raw silk, are highly concentrated in the mountainous prefectures (Nagano, Gifu) of Tosan in central Honshu, Gumma and Saitama in western Kanto, and Shiga and Osaka prefectures in the Kinki region (Fig. 8-10). Silk and silk-rayon fabrics are most strongly centered in western Kanto, and Ishikawa-Fukui-Toyama prefectures in Hokuriku. Lesser, but important centers are the basins of

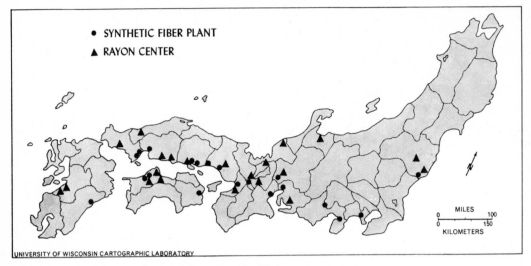

● SYNTHETIC FIBER PLANT

▲ RAYON CENTER

MILES
0   100
0   150
KILOMETERS

UNIVERSITY OF WISCONSIN CARTOGRAPHIC LABORATORY

Fig. 8-11.—After map by Alan D. Walton.

southern Tohoku (Fukushima, Yamagata), Niigata Prefecture on the Japan Sea side, and Kyoto and Shiga prefectures in Kinki.

In contrast to other natural-fiber industries the Japanese woolen industry has shown a progressive growth, regaining its prewar level by 1955, and continuing to expand since then. This is partly due to the underdeveloped nature of the woolen industry prior to the war, but also to the increased home market as a result of a change in dress from kimonos to Western-style clothes. As of about 1961, woolen textiles employed some 20 per cent of the workers engaged in all textiles. It is the worsted yarn and fabrics, rather than their woolen counterparts, that have expanded so remarkably. Manufacture of woolen yarn and cloth is strongly concentrated in two major centers, Chukyo a strong first, and Keihanshin a poor second. Kanto is a much less noteworthy center.

In output of chemical fibers Japan ranks second in the world, next to the United States, while this same group accounts for 40 per cent of Japan's textile production and nearly one-third of its total textile exports. Spectacular is a fitting term to describe the boom in synthetic fibers, whose output increased seven to eight times between 1955 and 1960. Large modern plants are characteristic. Four main production centers of the new synthetic fibers are the Inland Sea borderlands, the Keihanshin and Chukyo industrial nodes, and the eastern part of the general manufactural belt in easternmost Shizuoka Prefecture and adjacent Kanagawa Prefecture in Kanto (Fig. 8-11).[17]

Iron and steel industry

Japan in 1961 produced 28.3 million tons of crude steel (31.5 million in 1963), which gave it fourth position among the world's steel-producing countries (very likely third place in 1963). The pre-1945 high point was 7.65 million tons in 1943, attained under wartime stimulation, and this figure was not equalled after the war until 1953. In the decade 1951 to 1961 Japan's output of crude steel increased 435 per cent, while during the three years terminating with 1961 the annual increments were 32 per cent, 29 per cent, and 25 per cent. What is more, all this has been accomplished in the face of such handicaps as having to import 85–90 per cent of its iron ore and 40–50 per cent of its coking coal. Partially offsetting this raw-material handicap is the locational advantage of having all of its large integrated steel mills situated on the coast, so that transportation costs are minimized. At each large integrated plant, ships of up to 10,000-ton capacity can dock directly at the berths.[18] Thus, Japan appears to be in a fairly strong competitive position when compared with other steel-producing countries. As a result of modernization of equipment and technological developments, she has been able to reduce appreciably the consumption of raw materials used in the production of a ton of blast-furnace pig iron, 12 per cent in the case of iron ore, and 32 per cent in the case of coke, over the 1950–1960 decade. At present the consumption of coke per ton of pig is the lowest for any of the large iron-producing countries.[19] That Japan ranks with the Western nations in degree of emphasis on heavy industry is indicated by her ratio of steel output to total manufactural output, which is higher than the comparable figures for the United States and Britain and is only exceeded by West Germany. The iron and steel industry accounts for the highest ratio, namely 20 per cent of the total fixed assets of Japan's manufacturing industries.

A noteworthy feature of Japan's steel industry is the relatively small ratio of pig iron to crude steel, about 58 per cent in 1960. This reflects the dearth of domestic raw materials for pig and the ready availability of foreign scrap iron, especially from the United States. But with the diminution of available scrap, Japan is increasingly placing more stress upon integrated pig iron and steel-making operations, with a relatively greater emphasis on pig iron, so that by 1970, when the steel industry anticipates an output of 48 million tons of steel, pig iron production is planned for 33.5 million tons, or a ratio of pig to steel of 70 per cent or more.[20]

The slightly over one million tons of iron ore produced in Japan each year appears to be about the optimum production for the immediate future based upon present proved reserves.[21] Iron sand, a second domestic source of ferrous materials, together with iron ore, has been discussed in an earlier chapter on mineral resources. Other than these two natural ores, there is also pyrite cinder, a by-product from sulfuric acid plants, which has become the single most important domestic source of ferrous materials. Close to 1.5 million tons were produced in 1960, almost all of which was used for sintering.[22] Thus in 1960 the Japanese iron and steel industry consumed 14.52 million tons of iron ore, only about one million tons of which were domestically produced; 1.57 million tons of domestic iron sand; 1.45 million tons of domestic pyrite cinder; and 2.3 million tons of other ferrous materials, most of it scrap iron.

As noted earlier, in the chapter on minerals, only about half of the 12 million tons of coking coal consumed is of domestic origin (three-fifths from northern Kyushu; the remainder from Hokkaido), and it is of a semicoking type which must be mixed with heavy coking coal imported chiefly from the United States and Australia.[23]

Japan's pig iron production is strongly concentrated in the central and southwestern parts of the country, within what has been previously called the Manufactural Belt, where 81.4 per cent of the capacity was located in 1960 (Fig. 8-12). Five great regional centers accounted for about 94 per cent of the capacity in 1960—Kitakyushu (33.4 per cent), Hanshin in Kinki (24.2 per cent), Keihin in Kanto (22.4 per cent), Kamaishi in eastern Tohoku (5.1 per cent), and Muroran in Hokkaido (9.1 per cent). It is the latter two which lie outside the Manufactural Belt. Not quite 6 per cent of the nation's pig iron, almost all of it from electric furnaces, is produced in scattered small centers (Table 8-13).

Even more of the crude steel (88.5 per cent) and steel products (90.6 per cent) capacities are concentrated in the Manufactural Belt of central and southwestern Japan than is true of pig iron (81.4 per cent), a situation which reflects the market attraction of this region. All five of the large pig iron centers are also steel producers, but their ratios of pig to steel vary. Thus, the two northern centers in Hokkaido and Tohoku, which are well removed from the principal market areas, represent 14.2 per cent of the nation's pig iron capacity, but only 8.5 per cent of the crude steel (Kamaishi, 2.7, and Muroran, 5.8) and 6.4 per cent of the steel products, indicating that considerable proportions of their pig iron and crude steel are shipped south to the market areas to be converted into ships, machines, tools, etc. The same is also true of the Kitakyushu center, likewise outside the main market areas, for although it boasts the

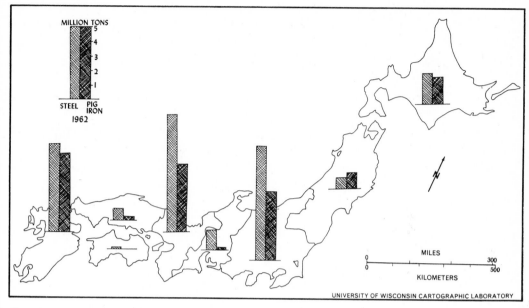

Fig. 8-12.—Pig iron and steel production by principal regions.

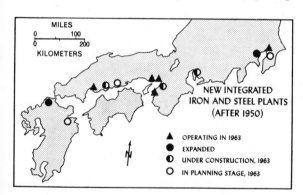

Fig. 8-13.—After map by Sadao Yamaguchi.

largest pig iron capacity (33.4 per cent), its crude steel is only 24.6 per cent, and its steel products 16.8 per cent. By contrast, the combined capacity of populous Hanshin and Keihin is only 46 per cent for pig iron, but 57.6 per cent for crude steel and 62.2 per cent for steel products. These two great centers are about on a par with respect to pig iron and steel capacities, but with Hanshin somewhat larger in 1960.

About 9 per cent of the crude steel capacity and 15 per cent of the steel products capacity—significantly more than for pig iron—lie outside the five great regional centers noted above. Most of this outlying steel capacity is located within the Manufactural Belt and in two general areas, the Inland Sea borderlands and Chukyo. Both of these outlying areas greatly exceed Kamaishi in steel products capacity and, together, represent 6.3 per cent of the nation's crude steel potential and 12 per cent of that of steel products. About 3 per cent of the crude steel and steel products capacities are in widely scattered small centers located outside both the Manufacturing Belt and the five main regional pig iron centers (see Table 8-13 for details).

Two of the country's great iron-steel centers, Hanshin and Keihin, may be thought of as market-oriented in their locations, while in contrast

TABLE 8-13

Locational changes in iron and steel production in Japan, 1926–1960 (in %)

| | The five great centers | | | | | | Intermediate districts (semi-market areas) | | |
| | Raw material area | | | Market area | | | | | |
Year	Kita-kyushu	E. Tohoku (Kamai-shi)	Hokkaido (Muro-ran)	Han-shin	Keihin	Total	Seto-uchi	Chu-kyo	Total
	Pig iron								
1926	79.4	8.0	12.2	0.4	....	100.0	...	...	...
1932	72.5	12.0	7.7	0.7	6.4	99.4	...	...	...
1951	37.3	9.3	18.1	18.3	17.0	100.0	...	...	...
1954	33.4	9.0	14.7	17.6	19.6	94.3	0.3	0.7	1.0
1957	31.0	7.3	12.2	21.4	19.9	91.8	0.8	0.7	1.5
1960	33.4	5.1	9.1	24.2	22.4	94.2	0.6	0.8	1.4
	Crude steel								
1926	65.4	3.3	2.7	16.4	11.8	99.6	...	0.4	0.4
1932	56.2	3.2	1.5	21.6	17.0	99.5	0.1	0.4	0.5
1951	29.8	6.8	11.3	29.0	21.2	98.1	1.6	0.3	1.9
1954	21.2	8.5	8.7	31.5	24.8	89.7	3.4	2.7	6.1
1957	25.5	3.6	6.5	30.8	23.1	90.5	3.3	3.1	6.4
1960	24.6	2.7	5.8	31.4	26.2	90.7	2.8	3.5	6.3
	Steel products								
1926	58.2	3.8	4.6	15.6	16.0	98.2	1.4	0.4	1.8
1932	46.0	3.2	2.4	23.0	18.0	92.6	1.8	0.3	2.1
1951	20.0	1.7	4.9	30.0	35.6	92.2	4.2	2.0	6.2
1954	19.5	3.8	1.2	36.0	24.8	85.4	7.5	2.6	10.1
1957	21.2	2.2	5.5	36.2	25.4	90.5	3.9	3.0	6.9
1960	16.8	1.1	5.3	32.0	30.2	85.4	7.4	4.6	12.0

(Continued on facing page)

TABLE 8-13—*Continued*

				Scattered minor areas					
Year	Sanin	Hokuriku	South Shikoku	Kinki (except Hanshin)	Tosan	North Kanto	Tohoku (except Kamaishi)	Total	Miscellaneous
				Pig iron					
1926	...	...	...	...	...	...	...	...	...
1932	...	...	...	...	...	...	...	...	0.6
1951	...	...	...	...	...	...	...	...	...
1954	0.6	3.4	...	...	...	0.4	...	4.4	0.3
1957	0.1	2.3	0.1	...	0.3	0.6	3.1	6.5	0.2
1960	...	0.9	...	0.1	0.3	0.6	2.2	4.1	0.3
				Crude steel					
1926	...	...	...	...	...	...	...	...	...
1932	...	...	...	...	...	...	...	...	...
1951	...	...	...	...	...	...	...	...	...
1954	0.3	0.7	0.1	0.3	...	1.9	0.8	4.1	0.1
1957	0.5	1.3	0.1	...	...	1.3	0.9	4.1	...
1960	0.3	1.1	0.1	0.1	...	0.7	0.7	3.0	...
				Steel products					
1926	...	...	...	...	...	...	...	...	...
1932	...	...	...	...	...	...	...	...	5.3
1951	...	...	...	...	...	...	...	...	1.6
1954	0.2	1.1	...	...	...	1.1	...	2.4	2.1
1957	0.3	1.6	...	...	...	0.6	...	2.5	0.1
1960	0.4	1.2	...	...	...	0.1	0.6	2.3	0.3

Source: Compiled by Sadao Yamaguchi. Data for 1926 and 1932 are for actual production as reported in *History of the Japanese Iron and Steel Industry, 4.* Data for later years are for capacities as reported in *Reports of the Japanese Iron and Steel Industry* (issued at three-year intervals).

the other three, Kitakyushu, Kamaishi, and Muroran are classed as raw-material oriented. Hanshin has no local coal or iron ore, although coal is readily obtained by cheap barge transport from northern Kyushu via the Inland Sea route. Keihin likewise has no local iron ore or coking coal, but non-coking coal needs are partially supplied from the Joban field not far to the northeast of Tokyo. Of the three centers locationally oriented toward raw materials, Muroran lies in close proximity to Ishikari coking coal and also to local, but inadequate, supplies of low-grade iron ore. Kamaishi is adjacent to the country's best domestic iron ore reserve but is removed from coal, while Kyushu is situated close to the

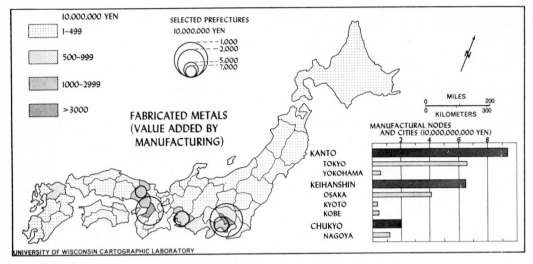

Fig. 8-14.—Data for 1960.

Chikuho coal field and the largest supply of coking coal, but must import all of its iron ore and high-grade coking coal. Pig iron is probably produced more cheaply in northern Kyushu than at any of the other centers.

Over the past thirty to thirty-five years the three centers oriented toward raw materials have waned in relative importance as the market-oriented centers of Hanshin and Keihin have waxed. Thus, in 1926, the former three centers produced almost all of the pig iron made in Japan proper and about 71 per cent of the steel. North Kyushu alone produced 80–85 per cent of the pig iron and nearly two-thirds of the crude steel. By 1960 the comparable figures for the three raw-material centers were 47.6 per cent of the pig capacity and only 33.1 per cent of the crude steel. By contrast, Hanshin and Keihin, which produced almost no pig iron in 1926, together had 40.6 per cent of the capacity in 1960, and crude steel had risen from 28.2 per cent to 57.6 per cent.[24] Clearly there has come to be an inceasing dominance of the influence of the large markets in the location of primary iron and steel.

Machinery

The machinery industry is Japan's fastest growing manufacturing group. With 1955 as 100 per cent, the index of production for all machinery was 442.4 in 1960 and 605.9 in 1961. In value added by manufacturing, machinery represented 18.7 per cent of the national total in 1955, but 29 per cent in 1960 and 34.8 per cent in 1961, the highest by far of any of the industrial groups. Of the several subdivisions of the

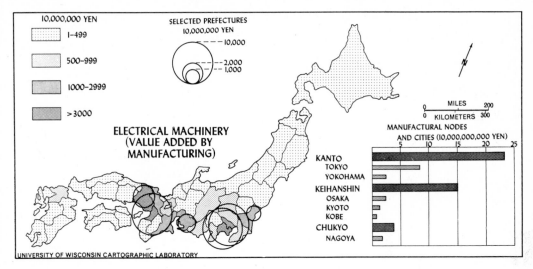

Fig. 8-15.—Data for 1960.

machinery industry, the greatest expansion has occurred in electrical machinery and in transportation machinery, more especially in motor vehicles.

Distribution of the machinery industry is widespread, but to an unusual degree it is coincident with the Manufactural Belt. Three of the conspicuous centers outside the belt are Toyama-Ishikawa and Niigata on the Japan Sea side of the country, Ibaraki Prefecture in northeastern Kanto, and Nagasaki in the extreme southwest. Within the Manufactural Belt there is an unusual concentration in the Kanto (41 per cent) and Kinki (20 per cent) areas, with Chukyo (12–13 per cent) running a strong third (Table 8-8). The margins of the Inland Sea, especially Hiroshima-Okayama prefectures in Chugoku, are important secondary areas, as is Fukuoka Prefecture in Kyushu.

On the whole the various kinds of machinery industries are well represented throughout Japan, but still there are certain regions that are more specialized in one type than another. Thus, the Kanto area is notably specialized in electrical machinery and contributes over half of the total national value-added in that industry. It is also the greatest center for the manufacture of precision instruments, contributing 68 per cent of the nation's value-added in that machinery sector. Shipbuilding, in which Japan leads the world, with steel vessels being one of her greatest exports, is concentrated in numerous centers, of which Nagasaki, Kure, Harima, and the general Kanto and Kinki areas are of unusual importance. Motorcar manufacturing is highly focused in the three greatest industrial nodes.[25]

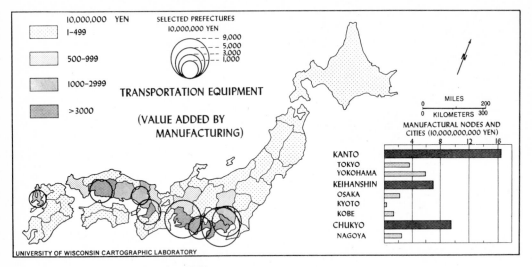

Fig. 8-16.—Data for 1960.

The chemical industry

Although postwar growth in chemicals has been rapid, during the period 1955–1961 its rate has not been as spectacular as that for machines and metals. Counting 1955 as 100, the production index of the chemical industry in 1961 based on value-added was 229.5 (259 in 1962).

During recent years Japan's chemical industry has been undergoing a structural change as it gradually shifted emphasis from inorganic to organic products. Thus, inorganic fertilizers, which were given top priority after the war, are gradually diminishing in relative importance as petrochemicals, synthetic resins, and synthetic organic chemicals are increasingly emphasized. As a result, in 1960 the above three branches represented 55 per cent of the equipment investment of the whole chemical industry.

Although Japan's domestic production of crude petroleum is minuscule, in 1961 she ranked third in oil refining, after the United States and Russia. Ninety to ninety-five per cent of the refining capacity is located on the Pacific southwestern side of the country and hence is well removed from the domestic oil fields (Fig. 8-17). Petrochemicals, a newcomer among Japan's chemicals, having been launched as late as 1957, is one of the most expansive of the country's growth industries, its output in 1961, four years after its inception, amounting to 8 per cent of the value of all chemicals. Two huge refinery-petrochemical complexes had been developed at Yokkaichi in Chukyo and Kawasaki in Kanto by the beginning of 1960, and there were twenty-four petro-

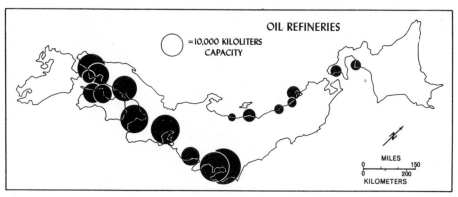

Fig. 8-17.

chemical plants of various sizes in operation at the end of 1961, all of them new and relatively large-scale. All have tidewater location, chiefly occupying land reclaimed from the sea by suction dredging and filling.

Distribution of chemicals is very widespread, but with the Kanto (23.8 per cent) and Kinki (20.4 per cent) centers well ahead of all others (Fig. 8-18). Less important concentrations are in Chukyo (10.5 per cent), Niigata-Toyama on the Japan Sea side, Okayama-Hiroshima-Yamaguchi on the northern margins of the Inland Sea, Ehime in northern Shikoku, Fukuoka in northern Kyushu, and Nobeoka city in Miyazaki Prefecture in eastern Kyushu.

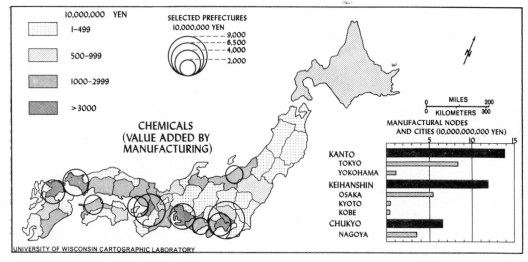

Fig. 8-18.—Data for 1960.

TRANSPORTATION AND TRADE

In its several types of transportation systems Japan exhibits remarkable contrasts. Thus, while it has a large and very modern merchant marine, about half of its extensive coastwise traffic is carried in small craft depending on combined sails and motors for power. And while its rail service is modern and efficient, highways have been neglected and motorcar traffic, especially that of trucks, is poorly developed except in and around cities.

Land transportation is handicapped by the hilly and mountainous nature of the interior, so that the principal routes are coastal in location. Still, of the total volume of freight carried in 1960, land transportation represented 55 per cent of the ton-kilometers, of which the national railways accounted for about 70 per cent and motor vehicles some 30 per cent. Private rail lines, chiefly around large cities, transported less than 1 per cent. National lines carried 51.6 per cent of the passenger traffic, private lines 25 per cent, buses 18.3 per cent, and motorcars 4.8 per cent.[26]

Railways and their traffic

Railroads in Japan are of two classes, the national railways with an aggregate length of 20,400 kilometers (12,670 miles) and privately owned lines with a length of 7,473 kilometers (4,640 miles). Privately owned railways are chiefly commuter lines in the vicinity of large cities, plus other lines connecting cities with important tourist centers. Japanese railroads are narrow gauge, 3 feet 6 inches, a width which was originally adopted for economy of construction in a mountainous country where bridges and tunnels are numerous. As a consequence, train speeds are relatively slow, the rolling stock is small in size and light in weight, and

TABLE 8-14

Domestic freight traffic, 1960

System	Tons (million)	Kilometer-age (100 million T.-km.)
National railways	195.2	536
Private railways	42.7	9
Trucks	1,156.3	208
Coastwise ships	54.8	466
Barges	53.5	149

Source: Ministry of Transportation (Japan).

TABLE 8-15

Domestic passenger traffic, 1960

System	Persons (million)
National railways	5,124
Private railways	7,275
Buses	6,291
Passenger cars	1,610
Aircraft	1,258
Boats	101

Source: Ministry of Transportation (Japan).

the steep grades and sharp curves limit individual trains to a few cars, while the proportion of engines to cars is large.

Nearly 3000 kilometers, or 15 per cent of the Japanese national railway's route-kilometers, are double-tracked, and almost all of this is on the Pacific side of the country, the Tokaido and Sanyo lines which connect Tokyo with North Kyushu accounting for 1200 kilometers of the total (Fig. 8-19). Shikoku is the only one of the four principal islands that has no double track, but the lengths in Kyushu and Hokkaido are small. About 20 per cent of the total rail mileage operates under an automatic block system control, three-fourths of this being on the double-track lines. While steam locomotives still predominate in number, none have been built since 1948, so that they will be entirely replaced by electric and diesel traction in the near future. Although by 1961 only 15 per cent of the total route-kilometers had been electrified, still 56 per cent of the total passenger-kilometers and 35 per cent of the total ton-kilometers are dependent on electric traction. Chief of the electrified lines are the Tokaido in its entirety, the Sanyo as far west as Hiroshima, and two northern lines out of Tokyo, one to Sendai and the other to Niigata. Diesel engines multiplied rapidly after 1957 and in 1961 represent 6 per cent of the total number of locomotives, which makes them 35 per cent as numerous as electric locomotives. Diesel traction is very widespread, operating on both double- and single-track lines.

Considering the national railways only, Japan has 5.5 kilometers of

TABLE 8-16

Train kilometers by powered rolling stock per day, 1961

Train	1000 train km.	Per cent
Passenger		
Steam locomotive	236	26
Electric locomotive	104	10
Diesel locomotive	35	3
Electric car	293	31
Diesel car	289	30
TOTAL	957	100
Freight		
Steam locomotive	277	68
Electric locomotive	125	30
Diesel locomotive	8	2
TOTAL	410	100

Source: Japanese National Railways, *Japanese National Railways: A General Description* (1962).

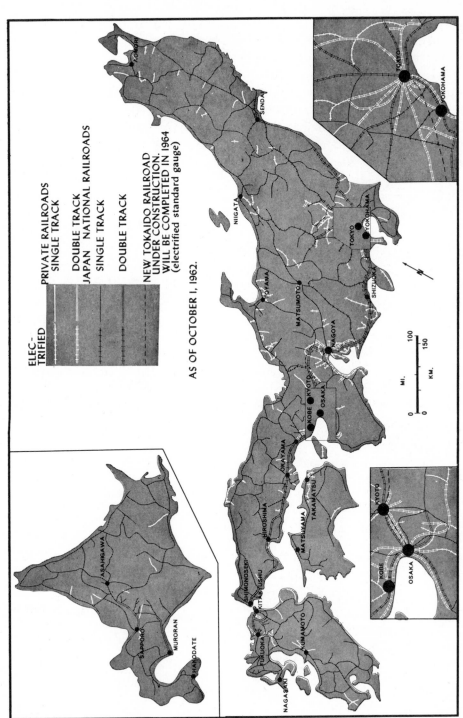

Fig. 8-19.—Modified from maps compiled by Yasuo Masai and others.

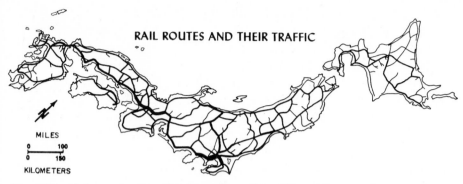

RAIL ROUTES AND THEIR TRAFFIC

MILES
0 ___ 100
0 ___ 150
KILOMETERS

*Fig. 8-20.—*Volume of rail traffic shown by width of line. After map by A. Isida.

main track per 100 square kilometers of area, a figure which, although high for a mountainous country, is well below that of the countries of Western Europe and is more on a par with that of Italy. Because of the large population, length of track per capita is relatively small. Japan has the largest railway passenger-kilometer volume anywhere in the world. Measured in terms of either area or population, the Japanese rail net is not the equal of Western Europe's, but it is far superior to that of any other Oriental country.

Trunk rail lines in general follow the coasts, thus avoiding mountain grades as much as possible and at the same time serving the lowland littorals where population is concentrated. In Honshu and Kyushu the rail pattern is crudely trellis in character, with the trunk lines oriented with the long axis of each island, and the less-used transverse lines connecting at right angles.

The Tokaido Line, linking the great metropolitan industrial nodes of Kanto, Chukyo, and Keihanshin, is by all odds the principal transportation artery of the nation. About 40 million people, or 43 per cent of the total population, live in areas served by this line, and the recent largest increases in population and industry are concentrated in these same areas. Probably 70 per cent of the nation's industrial output originates in areas tributary to the Tokaido. And, although the length of the Tokaido Line is only 590 kilometers, or 2.4 per cent of Japan National Railway's total route-kilometers, on the other hand, the volume of freight and passenger traffic carried is 23 per cent and 24 per cent, respectively, of the total (Fig. 8-20). Hence, the number of trains operated daily on this line is unusually large—eighty or more passenger trains and fifty to sixty freight trains, one way.[27] The projected superhighway between Tokyo and Kobe may divert as much as 10 per cent of the passenger traffic and 5 per cent of the freight load from the Tokaido Line, but con-

sidering the rate of increase of traffic on the latter, this anticipated diversion will not ease the congestion. The limit in transport capacity has
already been reached in some sections of the line, and this same condition will be true in other sections in the near future.

In order to relieve the traffic deadlock on this vital trunk line, construction of a new Tokaido rail line, roughly paralleling the old one,
was begun in 1959, and completion is expected by about 1964. Upon
its completion, the transport capacity of the Tokaido will be more than
trebled. Superexpress passenger trains will cover the 515 kilometers (320
miles) between Tokyo and Osaka in three hours, limited expresses in
four hours, and freight trains, operating only at night, in five and one-
half hours. The new Tokaido is standard gauge, which requires the
boring of a total of forty miles of new tunnels.

A second large traffic flow is on the Sanyo Line following the northern
shore of the Inland Sea and joining populous and industrial Kinki with
northern Kyushu. Others are (1) from the Kanto or Keihin area northward through eastern Tohoku and (2) over transverse lines joining
the Hokuriku area along the Japan Sea side with Kanto, Chukyo, and
Hanshin on the Pacific side.[28]

In contrast to the situation in the United States, rail passenger revenue
in Japan (including mail and baggage) exceeds that from freight, the
ratio being 56 to 41. Coal accounted for 23 per cent of the rail freight
tonnage carried in 1961, followed by timber and lumber, 7 per cent;
cement, 6 per cent; limestone, 5 per cent; gravel and sand, chemical fertilizer, metallic ores, pig iron and steel, rice, and oil.

In order to meet the increasing volume of freight and passengers, the
National Railways have formulated a five-year plan which among other
things calls for tripling the existing transportation capacity through the
double-tracking of other principal trunk lines, an increase of rolling
stock, the electrification of passenger trains, adoption of more diesel
locomotives, more frequent operations of high-speed trains, increased
number of freight cars, and a more integrated transportation system
linking train and motorcar services.[29]

Highways and their traffic

The Taika, or Great Reform, of the middle seventh century included
a provision for a regular development of roads, barriers, ferries, and
post horses. This plan for an integrated post-road system was the product
of a centralized government, whose chief goal was to facilitate the movement of the men in its service. During the two and a half centuries of the
Tokugawa or Edo regime (1602–1867), the post-road system was laid

out and developed. The shoguns in what is now Tokyo recognized that if they were to control the domains of the great feudal lords, or daimyos, and maintain peace throughout the country, good communications must be kept up between the capital of the realm and its outlying parts. Like the Roman roads, therefore, those of early Japan at first served chiefly political rather than commercial ends.

The trunk lines of the Tokugawa road system were five great roads: the *Tokaido,* Edo to Kyoto, along the coast, 310 miles; the *Nakasendo,* Edo to Kyoto, through the interior, 324 miles; the *Nikko-kaido,* Edo to Nikko, 89 miles; the *Ushu-kaido,* Edo to Aomori, 465 miles; and the *Koshu-kaido,* Edo to Shimosuwa, 132 miles.[30] It was these trunk highways of feudal Nippon, located along carefully selected natural routes, that set the pattern for Japan's modern transportation system. Many of them were chosen at a later date as the routes of rail lines, some of which retained even the names of their post-road ancestors. A number of the motor roads of a still more recent period also coincide in general with the routes of the ancient Tokugawa post roads.

Compared with her railway system and her merchant marine, Japan's highway system and its motor traffic have lagged well behind. Chief obstacle to an expanded motor transport is the inadequate highway system. As of 1960, there were about 25,000 kilometers of national highways, only 22 per cent of which have a width of 7.5 meters or more and are adequate for two-lane motor traffic. Only one-third of the highways have a hard top of cement or asphalt, but nearly all are surfaced and considered passable for motor vehicles (Fig. 8-21). Of the nearly 122,000 kilometers of provincial highways, less than 8 per cent are surfaced with cement or asphalt, but 90 per cent are considered passable for motor vehicles. However, roads are being improved at a quickening rate as highway budgets have soared. Thus the ratio of road budget to national budget was only 0.77 per cent in 1950, but 2.61 per cent in 1955, and 7.73 per cent in 1962.

In a 1959 survey of the traveling speed of motor vehicles in Japan, it was discovered that the mean speed on the best highways was only 50–60 kilometers per hour (30–35 miles), which is far below that on modern highways in the United States. Such low speeds are attributable to (1) the impediment of numerous slow-speed vehicles such as handcarts, ox-carts, and bicycles, as well as pedestrians, using the roads; (2) very numerous grade crossings, so that even National Highway 1 is intersected by prefectural roads on an average of every 0.6 kilometers, or 0.37 miles; (3) houses and factories abutting the roads on either side; and (4) structural deficiencies of the highways themselves, such as nar-

PRESENT STATUS OF ROADS

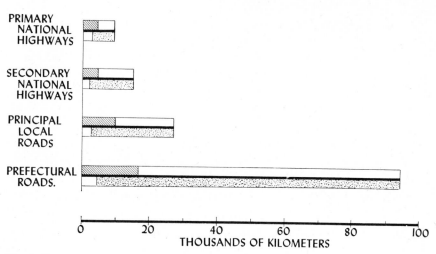

Fig. 8-21.

row widths, steep grades, short sight-distances, and small radii of curves.

The urgent demand, following the 1959 survey, for the construction of highways which would eliminate such deficiencies and permit high-speed performance of motor vehicles, prompted planning of a new sys-

TABLE 8-17

Roads of Japan, 1961

Classification	Length (km.)	Amount by width (km.)		Amount by surface (km.)		
		Over 5.5 m.	Under 5.5 m.	Gravel	Cement	Asphalt
National	24,937.33	11,992.1	12,945.2	16,796.3	3,669.4	4,471.6
Primary	(9,901.6)	(6,224.5)	(3,677.1)	(4,978.5)	(2,422.4)	(2,500.7)
Secondary	(15,035.7)	(5,767.6)	(9,268.1)	(11,817.8)	(1,247.0)	(1,970.9)
Prefectural and local	122,017.95	17,051.1	104,966.8	112,711.8	2,289.5	7,016.7
Important	(27,473.3)	(7,724.8)	(19,748.5)	(23,528.3)	(1,093.1)	(2,851.9)
Others	(94,544.6)	(9,326.3)	(85,218.3)	(89,183.5)	(1,196.4)	(4,164.8)
TOTAL	146,955.3	29,043.2	117,912.1	129,508.1	5,958.9	11,488.3

Source: Bureau of Roads, Ministry of Construction, *Roads in Japan, 1963.*

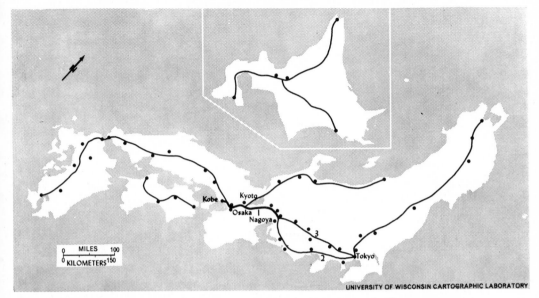

Fig. 8-22.—Routes of national highways, planned and in process of construction. Routes 1, 2, and 3 are at present under construction. After map in *Roads in Japan* (Bureau of Roads, Ministry of Construction, 1963).

tem of national expressways (Fig. 8-22) with a total length of about 4000 kilometers. Construction of the Nagoya-Kobe Expressway (119 km.) is now in progress and the target date of its opening to traffic is 1964. Similar four-lane expressways connecting Tokyo with Nagoya (357 km.) and Tokyo with Fuji-Yoshida city are presently under construction, with plans for completion in 1968 and 1967, respectively.[31]

Most of the hard-surfaced highways are concentrated on the Pacific side of central and southwestern Japan. Only 400–500 kilometers of hard-top road are to be found in all of Hokkaido, and some 1200 kilometers in Tohoku (Fig. 8-23). No continuous hard-top highway connects Tokyo with Aomori in northernmost Honshu, but only discontinuous lengths between much longer stretches where loose gravel is the surfacing material. Hokuriku, Sanin, Shikoku, and southern Kyushu are likewise meagerly served by hard-surface roads. Much the largest concentrations of hard-top are in the metropolitan Kanto, Kinki, and Tokai areas. Thus, according to the recent road maps, it is possible to drive on National Highway 1, following the Tokaido route between Tokyo and Osaka, and on National Highway 2, from Osaka to Shimonoseki, and constantly be on either cement or black-top road, almost all of it over 5.5 meters in width. Highway 1 is the major trucking route in Japan. But even on such

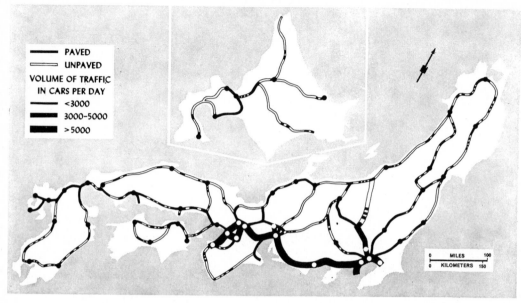

Fig. 8-23.—Main highways: their surfaces and their volumes of traffic. After map in *Nippon no Chiri,* Vol. 8.

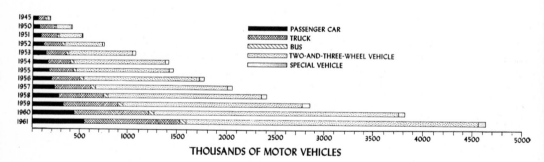

Fig. 8-24.—Growth in number of motor vehicles in Japan.

main highways as 1 and 2, cities and towns are so numerous that speeds are relatively slow.

In spite of Japan's inadequate highway system, motor vehicles have multiplied at an astounding rate in recent years, from 388,000 in 1950 to over 4,280,000 in 1961 (Fig. 8-24). However, nearly two-thirds of these are two- and three-wheel vehicles, followed in numbers by trucks and passenger cars in about the ratio of 2:1. Buses and trucks have become powerful rivals of the railroads in both number of passengers and tons of freight hauled. But because hauls are much longer on railroads their passenger-kilometers and ton-kilometers still are much greater.

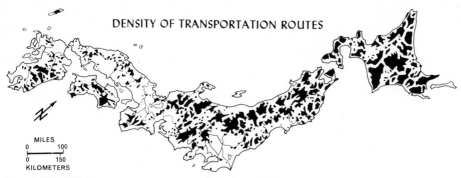

DENSITY OF TRANSPORTATION ROUTES

MILES
0   100

0   150
KILOMETERS

Fig. 8-25.—White areas are not more than four kilometers from a transportation route (rail, train, or bus). After map by Y. Ogasawara.

For example, although in 1960 trucks moved six times the freight tonnage of railroads, in terms of ton-kilometers they moved less than 40 per cent as much. But although short-haul is still characteristic of trucks, the range is being extended, for the commercial-carrier truck has increased its average haul from 26.5 kilometers in 1957 to 32 kilometers in 1960.[32]

Railway and bus lines form a network over the entire country, such that on the flattish lowlands no place is farther than 10 kilometers (6 miles) from a rail or bus stop (Fig. 8-25). The network is densest in the Kanto and Tokai areas of central Japan, and in Sanyo, northern Shikoku, and northern and western Kyushu within southwest Japan. A coarse network is characteristic of Hokkaido, Tohoku, Tosan, and the Outer Zone of Southwest Japan (Kii Peninsula, southern Shikoku, and southeastern Kyushu).

Marine and air transportation

Japan made her debut in international air services about ten years later than other countries, so that she still lags by comparison. Sixteen foreign air lines are operating in and out of Japan, and in 1960 these carried about 83 per cent of the international air-passenger traffic. Japan Air Lines at present has several flights a week to San Francisco, Los Angeles, and Seattle, and also to Southeast Asia. Domestic air service links all the major cities of the country from Hokkaido to Kyushu, with a plane leaving for Osaka from Tokyo every hour.

In marine, as contrasted with air transportation, Japan has by no means lagged. Prewar Japan ranked third in merchant marine tonnage, after the United Kingdom and the United States. Reduced by three-quarters due to wartime losses, recovery was slow until about 1955, but

TABLE 8-18

Tonnage of Japanese merchant fleet
(steel vessels of over 100 gross tons)

Vessels	Number		Tonnage (in 1000 G.T.)	
	1955	1961	1955	1961
Passenger carriers	167	202	160	176
Tankers	240	704	737	1,758
Freighters	877	1,763	2,828	5,045
TOTAL	1,284	2,669	3,725	6,979

Source: Foreign Capital Research Society, *Japanese Industry* (1962).

by 1961 the merchant fleet was larger in number of vessels and in total tonnage (6.98 million) than before the war. As a consequence, Japan's share in the world's total merchant marine tonnage has risen to 5.8 per cent, placing her in fifth place, after the United States, the United Kingdom, Liberia, and Norway.

Not only has there been a tonnage expansion, but important changes in age, kind, and quality of vessels have also taken place. Thus, Japan's present merchant fleet is predominantly of new ships, 70 per cent being less than ten years old. At the same time, the capacity and speed of the vessels have been increased, while a smaller proportion are passenger-cargo ships and a larger share of the tonnage is in tankers and freighters.

But in spite of the recent expansion in quantity and quality of Japan's merchant fleet, and a rise in its absolute earnings, the percentage of Japan's import trade carried in Japanese vessels has declined, so that in 1961 there was an excess freight payment of $456 million above what the merchant fleet earned.[33] In prewar days, the freight earnings by Japan marine transportation were more than twice the payments to foreign shipping, so that a further expansion of marine transportation is highly desirable in order to improve the balance of payments.

In 1960 the total tonnage of goods moving in coastwise trade in Japan was more than double that carried by her merchant fleet engaged in overseas trade. Despite the tremendous increase in truck and rail transportation over the decade of the 1950's, domestic shipping lines, which carried 39 per cent of the domestic freight in 1955, increased their share to 44 per cent in the period 1960–1962.[34] As indicated in an earlier connection, the coastal seaway is an important highway for insular Japan. Regular shipping service is maintained between the principal coastal cities, and this line-service is supplemented by the tramp services of several thousand small steamers and sailer-steamers. One of the most important elements of the coastwise service is the carrying of heavy bulky cargo, such as coal, ores, logs, and lumber, from Hokkaido and northern Kyushu to the great industrial centers of Kanto, Chukyo, and Keihanshin. Coal and coke represent about 43 per cent of the total coastwise cargo, followed in importance by sand-gravel-stone, pig iron, non-ferrous ores, cement, logs, and lumber. Nearly a thousand ports are served by local coastwise traffic, for close to 60 per cent of Japan's mining and manufacturing is in littoral districts, while about 30 per cent of her population and 45 per cent of the country's labor force is similarly located.[35] The Inland Sea and its borderlands are a particularly important focus of coastwise traffic, reflecting the protected nature of that nearly land-enclosed body of salt water, and the fact that its margins are so intensively developed, with the great Keihanshin industrial node at its eastern extremity, the Kitakyushu heavy industrial area and the Kyushu coal fields at its western extremity, and all of it part of the nation's Manufactural Belt.

Although foreign trade is important to all major industrial countries, this is especially true in the case of Japan, with her modest resource base and a population of over 95 million to support. Actually, Japan's foreign trade amounts to one-fifth of her gross national production. It is a truism that Japan must trade to survive, for she is constantly faced with a huge deficiency in domestically produced raw materials which must be greatly supplemented from outside. To pay for the imports, chiefly of industrial raw materials but also of some foodstuffs and fabricated wares, she is obliged to manufacture and subsequently to sell an important part of the processed goods overseas. If this flow pattern should be halted, a catastrophe would result.

In 1960 about 70 per cent of Japan's imports were either raw materials (58.5 per cent) or semimanufactures requiring further processing (11.9 per cent). Nearly 14 per cent was foodstuffs, so that foods and materials for processing together accounted for about 84 per cent of the total imports. Manufactured goods amounted to only 15.5 per cent. By contrast, Japan's exports are heavily weighted on the side of finished manufactures (70 per cent), while foodstuffs and raw materials together constitute only 9–10 per cent. For every year between 1945 and 1960 the value of Japan's imports exceeded her exports so that she had an adverse balance. This adverse balance was much reduced in 1958 and 1959 and became a slight favorable balance in 1960, but the situation reversed again and there was a large unfavorable balance once more in 1961 and 1962 and especially in 1963 when imports increased 20 per cent over the preceding year, compared with 11 per cent for exports. Particularly notable was the 47 per cent increase in imported foodstuffs.

In terms of individual items of imports, petroleum and raw cotton greatly exceeded all others, followed in order by wool, iron ore, wheat, lumber and allied products, coal, crude rubber, and sugar. Exports show a composition which is quite in contrast to that of imports, with iron and steel first, followed by cotton fabrics, vessels, clothing, rayon fabrics, aquatic products, and toys.

Japan's overseas trade is most strongly developed with North America (chiefly the United States), which in 1961 accounted for 40 per cent of imports and 30 per cent of exports, followed by Asia (27 per cent of im-

T A B L E 8-19

Foreign trade by kind of goods, 1960 (in million yen)

Goods	Exports	Imports
Foodstuffs	94,331 (6.5%)	224,679 (13.9%)
Crude 24,910		159,869
Manufactured 69,421		64,810
Raw materials	42,273 (2.9%)	946,443 (58.5%)
Semimanufactured goods	288,594 (19.8%)	192,964 (11.9%)
Finished goods	1,028,388 (70.4%)	249,669 (15.5%)
Miscellaneous, including foreign products		
re-exported	6,046	3,052
Total	1,459,633	1,616,807

Source: Ministry of International Trade and Industry.

T A B L E 8-20

Important commodities in Japan's foreign trade, 1960

Exports	Value (million yen)	Imports	Value (million yen)
Iron and steel products	139,698	Petroleum	167,411
Cotton fabrics	126,507	Raw cotton	155,305
Vessels	103,726	Wool	95,487
Clothing	79,212	Iron ore	76,944
Rayon and glass-fiber fabrics	77,713	Wheat	63,666
Aquatic products	63,136	Lumber products	61,291
Toys	32,405	Coal	50,848
		Crude rubber	45,305
		Sugar	40,408

Source: Ministry of International Trade and Industry.

ports, 39 per cent of exports) and Europe (12 per cent of imports, 13 per cent of exports).

Japan has about two thousand trade ports and, in addition, two thousand fishing ports. Sixty-eight are open to foreign trade, and those which play particularly important roles in this respect are called specially-designated major ports. These include the three Kanto ports of Tokyo, Kawasaki, and Yokohama; Shimizu in Shizuoka Prefecture; Nagoya and Yokkaichi in Chukyo; Osaka and Kobe in Kinki; and Shimonoseki, Moji, Kokura, and Dokai in Kammon in North Kyushu.

A majority of the Japanese foreign trade ports are not particularly

TABLE 8-21

Japan's regions of trade, 1961 (value in millions of dollars)

Region	Exports	Imports	Balance
Asia	1,236	792	+444
Middle East	193	254	− 61
Europe	588	728	−140
North America	1,244	2,192	−948
U.S.A.	(1,125)	(1,978)	−853
Latin America	276	382	−106
Oceania	137	448	−311
Africa	318	128	+190
TOTAL	3,992	4,924	−932

Source: Information Office, Consulate General of Japan, *Japan Report,* 8 (New York, 1962).

TABLE 8-22

Value and tonnage of Japan's trade by principal ports, 1961

Port	Exports		Imports	
	Value (million yen)	Tonnage (1000 T.)	Value (million yen)	Tonnage (1000 T.)
Kobe	546,323	3,622	380,452	8,088
Yokohama	328,228	3,061	398,710	14,825
Osaka	140,822	880	179,256	7,766
Nagoya	116,926	1,776	177,101	5,118
Tokyo	114,111	66	153,694	5,451
Yokkaichi	6,781	132	111,853	5,106
Dokai	26,869	507	90,097	11,317
Moji-Kokura	28,618	1,326	41,230	3,467
Shimizu	34,740	419	25,819	2,886

Source: Foreign Trade Survey of Japan, 1961.

well equipped, while most of their harbors are relatively shallow and are able to accommodate vessels of only 10,000 tons or less. In nearly all of them loading and unloading depends chiefly upon barges, although postwar port improvements are permitting an increasing use of piers.

The two deep-water ports of Kobe in Kinki and Yokohama in Kanto handle close to 60 per cent of the country's export trade (by value) and almost 40 per cent of its import trade. Kobe is the more important of the two, for although the value of imports is nearly the same in both, Kobe's exports are two-thirds again as great as Yokohama's. Each of these ports serves a great industrial area, and their cargoes are varied in character. Kobe's pre-eminence as an export port reflects the fact that it serves as an outlet, not only for the great Kinki industrial area, but for Chukyo and the Inland Sea industrial areas as well. Both cities are located on relatively deep water, down-bay from the delta-plains at the silting bay-heads. In importance ranking well below the deep-water ports of Kobe and Yokohama, but roughly comparable among themselves in value of trade, are Osaka, Nagoya, and Tokyo, each with about 7–9 per cent of the nation's foreign commerce. Each of the three serves one of the country's great manufactural nodes, but all are handicapped by bay-head location on an advancing delta where water is relatively shallow. They have greater difficulty, therefore, accommodating deep-draft vessels than do Yokohama and Kobe. Together, the two Kinki ports of Kobe and Osaka handle 46 per cent of the nation's exports and over 28 per cent of its imports. While nearly 83 per cent of the country's exports pass through the above five ports, they handle only 64 per cent of its imports.

PART III

REGIONAL SUBDIVISIONS
OF JAPAN

9 · HOKKAIDO

Parts I and II, where the organization is topical, focus attention upon Japan as a whole and upon the country-wide patterns of distribution, rather than upon regional differences in detail. By contrast, in Part III the organization is regional and sectional. The land of Nippon is a mosaic of regions and areas of varying sizes, which show different degrees of contrast and similarity in one or more of the important geographic elements. In some areas the dominant and unifying feature is terrain; in others, climate; in still others, it may be location, a type of land use, or a population-settlement feature.

Three principal divisions of the country are recognized: (1) Hokkaido, (2) Tohoku (Owu), which is the northern part of Honshu, and (3) central and southwestern Japan. These in turn are broken down into a number of subdivisions of various ranks.

The scheme of regional subdivisions here presented is only one of several equally satisfactory ones which might be employed in describing Japan sectionally.[1] It is not based on any rigid methodology, but is simply a convenient device for organizing regional description. The boundaries of the subdivisions are not to be regarded as precise, for the emphasis is upon the intermingled and interdependent features that together determine the geographic character of the areas, rather than upon their exact boundaries. Since the country is so fragmented geomorphically, terrain character becomes one of the most convenient criteria for delimiting subdivisions.

As Japan is the exception within eastern and southern Asia, so Hokkaido represents the exceptional within Japan. Consequently, in any scheme of geographic subdivision for the country, a major boundary may be drawn at Tsugaru Strait, separating the northern island from the other three main islands which together comprise what is sometimes called Old Japan. Omitting for the moment certain physical dissimilarities, many of the cultural contrasts with Old Japan seem to stem from the more severe climate of this northern island, together with the fact that it is a new land, most of whose settlement and development have taken place within the last one hundred years. Distinguishing features are its sparser population, dispersed rural settlement, larger and more mechanized farms dependent upon animal power, greater commercialization of its agriculture, and a manufacturing structure which slights those light industries specializing in consumers' goods such as clothing and household utensils. Per capita income is slightly above the national average, only those of metropolitan and industrialized southern Kanto and Kinki being higher. Agricultural income per farm is the highest for any of the large subdivisions of the country.

TABLE 9-1

Some comparisons of the main regional subdivisions of Japan

Region	Per capita income in 1957 (1000 yen)	Index (Japan = 100)	% of population supported		Agricultural income per farm	Industrial production income per capita (Japan=100)	Population change during 1955–60
			Primary industry	Secondary industry			
S. Kyushu	67.1	72	58	18	72	31	− 2.5%
Tohoku	72.0	78	53	20	127	35	− 0.3
Sanin	73.8	80	52	17	76	32	− 3.6
N. Kanto	74.0	80	53	17	114	49	− 1.7
Tosan	76.9	83	53	16	83	43	− 2.3
Shikoku	79.3	85	48	20	84	56	− 2.5
Sanyo	82.2	89	42	21	83	99	+ 0.1
Hokuriku	87.1	94	41	21	107	81	+ 0.7
N. Kyushu	89.5	96	33	31	82	90	+ 1.9
Tokai	90.3	97	33	21	88	144	+ 6.3
Hokkaido	93.5	101	40	27	172	53	+ 6.5
Kinki	113.4	122	23	31	84	179	+10.4
S. Kanto	123.1	133	18	29	111	157	+15.8

Source: Ryuziro Isida, *Geography of Japan* (Tokyo, 1961).

THE RESOURCE BASE

Terrain

Next to Honshu, Hokkaido is the largest of the Japanese islands (78,508 sq. km., or 31,200 sq. mi., which equals 21.2 per cent of the nation's area). Subdued mountains and hills predominate, extensive areas of which are volcanic in origin. Natural terraces, both coastal and interior, are here the most extensively developed in Japan, a majority of the coastal terraces being wave-cut surfaces thinly veneered with sand and gravel and terminating in sea cliffs. Lowlands of new alluvium are typically wet and poorly drained and parts are characterized by peaty soils.

Two contrasting landform divisions make up the island; a large compact central and eastern part, to which is appended on the west hilly Oshima Peninsula, formed by a northward extension of the central and western ranges of northern Honshu (Fig. 9-1). The larger division, or Hokkaido proper, roughly rhombic in outline, has had its general shape and terrain character determined as a result of the intersection of two mountain arcs, the Karafuto Arc forming the north-south Yezo Mountain axis extending from Cape Soya on the north to Cape Erimo on the south, and the east-west Chishima Arc and volcanic chain, forming the island's great eastward projection. Juncture area of the two arcs in central Hokkaido is represented by the highest elevations in the island, where the Daisetsu volcanic group surmounts the Kitami Mountains.

The east-west Chishima Mountain chain is composed largely of several great volcanic groups, the easternmost of which forms the broad peninsula of that part of the island. Some of the individual volcanoes are still active.

The north-south Yezo Mountain system forming the backbone of Hokkaido proper, a southward continuation of the highlands of Sakhalin Island, is composed of two main ranges, a western and an eastern, enclosing between them a meridional chain of detrital basins. In the north these mountains are subdued in character. Farther south the relief is greater, with peaks reaching altitudes of 1500–2500 meters above sea level. The two mountain chains are composed of several individual mountain masses separated from each other by antecedent valleys and low saddles. In the Tertiary rocks of the western range, more especially in its southerly parts, is one of Japan's two principal coal fields, the Ishikari. These Hokkaido mountains likewise are one of three major sources of hydroelectric power in Japan.

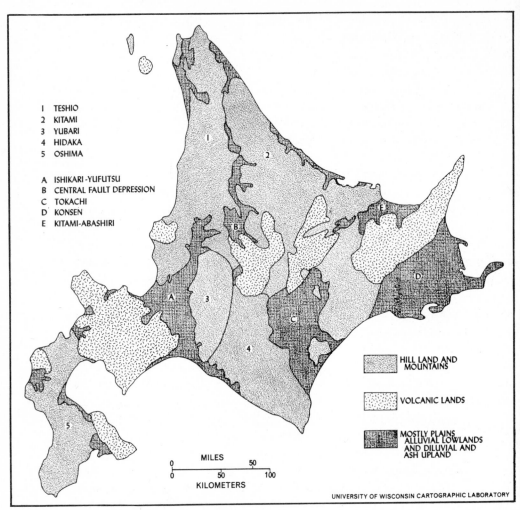

1 TESHIO
2 KITAMI
3 YUBARI
4 HIDAKA
5 OSHIMA

A ISHIKARI-YUFUTSU
B CENTRAL FAULT DEPRESSION
C TOKACHI
D KONSEN
E KITAMI-ABASHIRI

HILL LAND AND MOUNTAINS

VOLCANIC LANDS

MOSTLY PLAINS ALLUVIAL LOWLANDS AND DILUVIAL AND ASH UPLAND

MILES
0   50
0   50   100
KILOMETERS

UNIVERSITY OF WISCONSIN CARTOGRAPHIC LABORATORY

Fig. 9-1.—Terrain subdivisions of Hokkaido.

Plains are not extensively developed in Hokkaido proper. The meridional basins lying between the eastern and western ranges of the north-south Yezo system have been mentioned already. Here gravelly lowlands, parts wet and peaty, with bordering terraces, are characteristic. The small lowlands bordering the Okhotsk Sea are similar in character, as is the interior Kitami Basin, also, in the northeast. Tokachi Plain, located in the island's southeast quadrant, is composed chiefly of youthfully dissected fans of older alluvium, with smaller amounts of wet floodplain. Konsen Plain in the extreme southeast, the most extensive on the island, is for the most part a mildly sloping ash upland, which grades into flattish marine terraces along the coast. Serving to attach hilly, fishtail Oshima Peninsula to Hokkaido proper is the relatively large Ishikari-Yufutsu Plain. Extensive low-lying areas of Ishikari are characterized by low-grade peat soils. But in spite of handicaps associated with drainage and soils, the Ishikari Plain, chiefly composed of new alluvium and therefore unlike Tokachi and Konsen, which are largely upland plains with highly infertile ash soils, has become the population and economic centrum of the island.

Climate, vegetation, and soils

The most noteworthy general features of Hokkaido's relatively severe continental climate have already been described in Chapter 2. Further details on the climatic characteristics of different sections of the islands are presented in the sectional analyses to follow. It may be useful at this point to recall that the nearest climatic counterparts of Hokkaido in Anglo-America are New England and the Maritime Provinces of Canada. Thus, following a revised scheme of the Köppen system of climatic classification, all of lowland Hokkaido has humid continental, cool-summer (*Dfb*) characteristics, with the average coldest-month temperature below 32°, and the average warmest-month temperature below 71° or 72°. Actually, in maritime eastern and northeastern Hokkaido where the cool Okhotsk current prevails, producing summer fog and low summer temperatures, the climate almost qualifies as *Dfc*, for there are barely four months with average temperatures over 50° (Fig. 9-2). Coastal drift ice plagues the east and northeast coast in spring and occasionally packs the harbor of Abashiri. Only in local areas that are sheltered by terrain from the sea fog, do cropped fields extend down to the coast. Normally the sea fogs do not penetrate far inland, so that much of the Tokachi and Konsen upland plains inland from the coast are warm enough in summer to permit cultivation. The warmest summers are experienced in the west and southwest, while the east has cooler summers, colder winters, and usually less snow. Snow lies deep in the west, especially in those parts facing the Sea of Japan.[2] (See Table 2-2, Chapter 2, for regional climatic data on Hokkaido.)

Climatic differences between east and west are emphasized by associated contrasts in native vegetation and soils. In the western and southwestern part a broadleaf deciduous forest predominates, which consists chiefly of beech, oak, ash, and elm interspersed with maple, linden, alder, and willow. Conifers are by no means lacking, and most authors describe the forest as a mixed one. A thick undergrowth of bamboo grass hinders the natural propagation of this forest. In central, northern, and

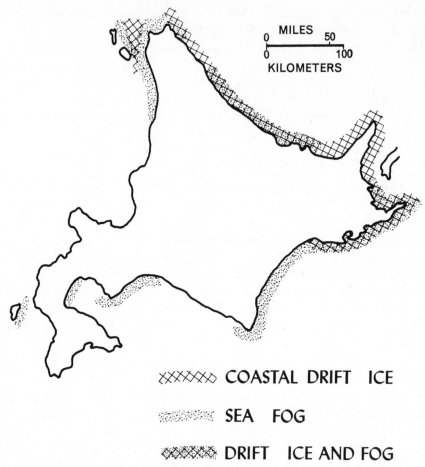

✕✕✕✕✕✕ **COASTAL DRIFT ICE**

░░░░░░ **SEA FOG**

✕✕✕✕✕✕ **DRIFT ICE AND FOG**

Fig. 9-2.—After map in Regional Geography of Japan, *No. 1,* Hokkaido Guidebook.

eastern Hokkaido the broadleaf trees decline in importance, and needle trees, principally fir and spruce, increase. The upper limit of forest growth is at 1000–1300 meters elevation; and forests cover an estimated 70 per cent or more of the north island. Large areas of moor and swamp vegetation are characteristic of the alluvial lowlands, the drier moor areas having a cover of grasses and sedges, while the wetter ones are sphagnum in character.

Hokkaido's soils are mostly young to immature, and they bear the effects of having developed under a cover of coniferous and mixed forest (Fig. 9-3). Two-thirds of the island has moderate to steep slopes, so that shallow, stony lithosols occupy nearly 54 per cent of the total area. These are of little agricultural significance. Since Hokkaido lies on the southern margin of the belt of "podzol with much bog," it is to be ex-

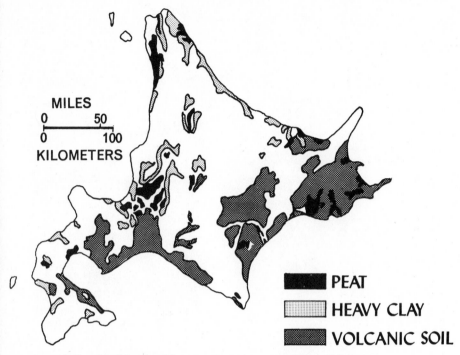

MILES

0   50

0   100

KILOMETERS

■ PEAT

▨ HEAVY CLAY

▨ VOLCANIC SOIL

Fig. 9-3.—Soils of Hokkaido. After map in *Nippon no Chiri*, Vol. 1.

pected that podzolic soils will be important. Omitting new alluvium, three main agricultural soils, all of them of low quality, characterize Hokkaido. First are the gray-brown podzolic soils which are dominant to the north and west of a line drawn from Abashiri in the northeast to Sapporo in the southwest. Here young volcanic ash is largely absent. These acidic heavy clays occur on rolling uplands and on high dissected terraces or benches in river valleys.[3] While they occupy nearly 14 per cent of the island's area, because they are difficult to till they are not importantly used at present. A second main type, the infertile black Ando soils (13 per cent of the area), derived from acidic volcanic ash, are concentrated in southern and eastern Hokkaido where they occupy the level to moderately sloping surfaces of ash aprons and dissected diluvial fans and terraces. They are most extensively developed on the Konsen and Tokachi upland plains in the southeast and the Yufutsu upland plain in the southwest. A third type is the peat or bog soils (2.2 per cent of the area) which occupy the downstream floodplains of rivers and the swampy floors of interior basins. The most extensive development of bog soil is to be found on the Ishikari lowland in the west, Teshio in the northern peninsula, and Konsen in the east.

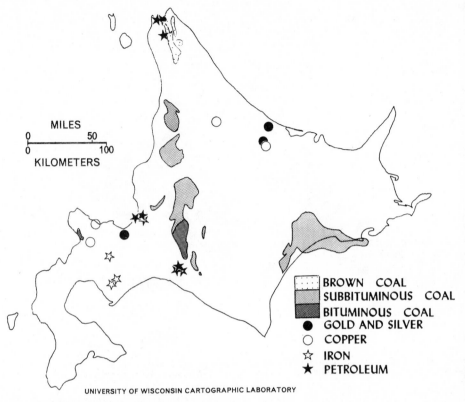

Fig. 9-4.–After map in *Regional Geography of Japan*, No. 1, *Hokkaido Guidebook.*

In addition there are the soils of new alluvium (15–16 per cent of the area) coincident with floodplains and deltas in the stream valleys. They vary greatly in texture, drainage conditions, and fertility, but doubtless they are the most productive of the island's prevailingly infertile soils.

In Hokkaido, where agriculture is both recent and relatively extensive in character, the soils utilized for crop production are much more in their natural state than they are in Old Japan.

POPULATION AND SETTLEMENTS

Colonization

Up to about the twelfth century, Hokkaido was occupied by primitive Ainu tribes dependent largely upon hunting and fishing, whose numbers probably never exceeded 30,000.[4] Although penetration of the southwestern peninsula by Japanese first occurred late in the twelfth century, consolidation of these coastal settlements to constitute a fief did not oc-

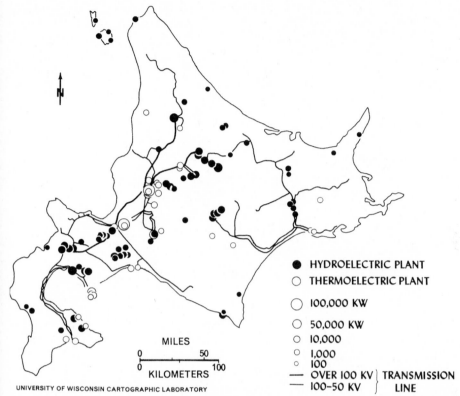

HYDROELECTRIC PLANT

○ THERMOELECTRIC PLANT

○ 100,000 KW

○ 50,000 KW

○ 10,000

○ 1,000

○ 100

— OVER 100 KV ⎤ TRANSMISSION
— 100–50 KV ⎦ LINE

MILES

0   50

0   100
KILOMETERS

UNIVERSITY OF WISCONSIN CARTOGRAPHIC LABORATORY

Fig. 9-5.—Electric generating plants and transmission lines. After map in *Regional Geography of Japan*, No. 1, *Hokkaido Guidebook.*

cur until after the sixteenth century. From these early littoral settlements three marine products—herring, salmon, and commercial seaweed—moved southward in trade to Old Japan, composing the single economic tie between the two regions. Accordingly, the Hokkaido fief was the only one in the whole country whose economy was not based on agriculture. Gradually Japanese fisheries were extended beyond the Oshima Peninsula to other parts of the island's coast, but for a time these latter settlements were only seasonal in character, and the permanent residences of the Japanese during the Tokugawa Period remained in the Peninsula.

Serious colonization of Hokkaido did not occur until after the Meiji Restoration, at which time it had both political and economic motives. Twice during the preceding Tokugawa Period, because of apprehension over Russian intentions, the Shogunate had taken over direct control in this northern island, and early in Meiji an Imperial Rescript declared the strategic importance of Hokkaido and asserted the necessity for its rapid colonization and development. Government action was prompt,

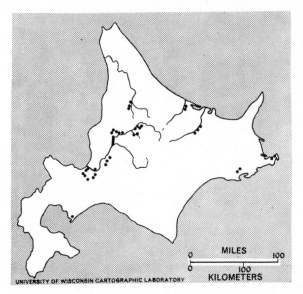

Fig. 9-6.—Location of militia settlements in Hokkaido between 1876 and 1899. After map in *Regional Geography of Japan*, No. 1, *Hokkaido Guidebook*.

for early in the 1870's the Hokkaido Land Development Board was established in Sapporo, and almost immediately there began a planned settlement of former samurai and their families on the land. These militia settlements, combining the twofold objectives of defense and economic development, were continued throughout the remainder of the nineteenth century and resulted in the establishment of thirty-nine separate communities, most of them numbering approximately two hundred households.[5] A great majority of these were concentrated in the Ishikari-Asahigawa region of central-western Hokkaido, but with upwards of ten in the more remote eastern parts (Fig. 9-6). Ground plans of the military settlements varied, but a common one had the houses strung out in two rows, one on either side of a highway, so that a shoestring village was the result. In the beginning the chief motive of the Meiji Government in settling Hokkaido was to check the thrust of Russian expansion southward, but in addition there was the wish to exploit the resources of this northern island for the benefit of the rest of Japan.[6]

But military colonization was far from being the only type, for the large subsidization program initiated by the national government—for land reclamation, exploitation of mineral resources, and the processing of products from primary industries—attracted free immigrants, also, in much larger numbers, to this frontier area. These first subsidized settlers were a sorry lot who represented the dregs and rejected of Old Japan.[7] Purely agricultural settlement proceeded very slowly, so that by 1881 there were only 11,742 hectares under cultivation, and between 1869 and 1881 the number of new immigrants amounted on the average to

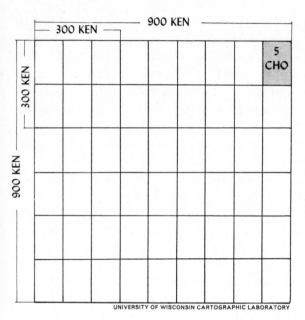

Fig. 9-7.—Standard pattern of land subdivision in Hokkaido based upon land-development scheme adopted in 1889. Normally 5 cho, or a similar number of hectares, became the standard amount of land per farm family. From *Regional Geography of Japan*, No. 1, *Hokkaido Guidebook.*

only a few thousand each year.[8] In the quarter century ending in 1900, when military colonization was abandoned, nearly 40,000 soldier-colonists and 557,000 civilian immigrants had entered Hokkaido, and the total population had reached nearly a million. The significance of the numerically less important military colonies lies in the fact that they were the frontier nuclei that served to attract the civilian settlers. Still, at the turn of the century, fishing remained the island's chief economy, and 70 per cent of the population was engaged in some phase of that extractive industry and was, therefore, strongly concentrated at the coastline.

After 1885 when Hokkaido was granted a central government, direct aid to prospective settlers was discontinued, and colonization benefits took the exclusive form of subsidies designed to improve and develop the island's economy. A land survey was made to determine what areas were best fitted for cultivation. Subsequently the survey land was divided into a rectangular grid, patterned after the range-and-township system of the United States, although in Hokkaido neither the roads nor the boundary lines necessarily follow the cardinal directions, but instead are adjusted to the local terrain. The basic unit of the survey was a rectangular plot whose dimensions were 100 by 150 ken (1 ken = 5.965 feet, or 1.818 meters) and whose area was 5 cho (about 5 hectares, or 12.25 acres). This unit of area was judged to be the amount of land required to support a family, and it therefore tended to become the common farm unit, corresponding somewhat to the quarter section in the

Upper Mississippi Valley.[9] On these rectangular 12-acre farms it seemed most convenient to build the farmstead facing the road and about midway between the property boundaries, which gave rise to the dispersed type of settlement so characteristic of rural Hokkaido today. It is this relatively coarse rectangular pattern of roads, fields, and dispersed farmsteads that makes rural Hokkaido look so different from Old Japan.

The northward movement of Japanese to Hokkaido involved substantial, but not unusual, numbers, with the annual number of new arrivals averaging nearly 70,000 over the five-year period 1905 to 1909, and about 60,000 in the years between 1921 and 1930. Some of these came for temporary employment and so represented only a seasonal migration. Others were unable to secure employment or became discouraged with the frontier environment and left, so that all of these elements composed a return flux which considerably reduced the annual net gain by migration. Between 1921 and 1930 the annual net gain as a result of in- and out-migration varied between 44,000 and 14,000. To be sure, this net loss to Old Japan was not of great significance in solving its problem of increasing population pressure, for the annual emigration to Hokkaido has always been only a small fraction of the country's annual increase. But the fact that Hokkaido was settled largely by migrants from backward rural areas in northern Honshu gave more weight to the effects of the out-migration upon these areas. In 1920, of the 916,000 persons born in other prefectures but enumerated in Hokkaido, nine-tenths came originally from the Tohoku region of northern Honshu or from the Hokuriku region facing the Japan Sea.[10]

Population

In 1960 Hokkaido had a population of 5,039,000, which gave it an average density of 64.2 persons per square kilometer. This is about one quarter of the average density of the country and is the lowest for any prefecture in Japan. Density of rural (*gun*) population per hectare of cultivated land was likewise well below that of any other prefecture, reflecting its more extensive type of agriculture. Hokkaido is the only non-metropolitan, non-industrial prefecture that in the period 1955–1960 increased in population more rapidly than the national average, a fact largely attributable to the relatively low death rate which kept the rate of natural increase well above the national average. In fact, it was next to the highest for any of the prefectures. In previous quinquenniums a part of the population increase in Hokkaido could be attributed to net in-migration. Thus, in the period 1950–1955 the net in-migration

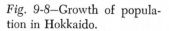

Fig. 9-8—Growth of population in Hokkaido.

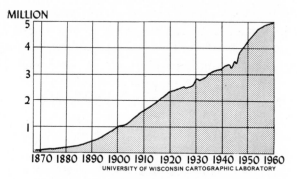

UNIVERSITY OF WISCONSIN CARTOGRAPHIC LABORATORY

amounted to 48,745, all but 3,233 being males and a great majority of them young males. But in the following five-year period there was a net out-migration amounting to 49,191, the bulk of which was young males. If one analyzes the migration pattern for 1955–1960 by smaller political subdivisions, it becomes clear that the net in-migration was chiefly to cities, more particularly Sapporo, Kushiro, Muroran, and Asahigawa, while the net out-migration was largely from rural areas, but also from Otaru and Hakodate cities.[11]

Turning to other characteristics of population, Hokkaido in 1960 had a larger percentage (48 per cent) of its working population engaged in the primary industries than did the country as a whole (42.4 per cent). But what is more striking is the much lower percentage so employed in Hokkaido compared with those prefectures in northern Tohoku just to the south (Aomori, 63 per cent; Iwate, 65 per cent; Akita, 64 per cent). Hokkaido's percentage in secondary industries (16 per cent) is considerably below the national average of 22.1 per cent, but it is still markedly higher than that of northern Tohoku, indicating that manufacturing is better developed in the northern island than in northern Honshu. Sex ratios show a preponderance of males (102) compared with a deficiency for the country as a whole (96.5). Here again Hokkaido contrasts with Tohoku where the male deficiency is above the national average. This contrast in sex ratios between the two northern areas probably reflects what has been until very recently a selective male in-migration in Hokkaido and a selective male out-migration in Tohoku. The percentage of Hokkaido's population living in cities (official *shi*) in 1960 was 51 per cent, which was distinctly below the country average of 63.5 per cent. But if the degree of urbanization is measured more realistically in terms of people living in Densely Inhabited Districts, Hokkaido's 42.1 per cent almost equals the country average of 43.7 per cent. This is strikingly higher than the figures for the prefectures in northern Honshu. In fact, only seven prefectures in Japan had a higher percentage of DID population than Hokkaido and all of these were in the three great industrial-metropolitan areas, plus industrial Fukuoka in northern Kyushu. Measured in terms of bona fide urban settlements of over 5000 population, Hokkaido appears to be one of the more urbanized prefectures.

Eighty to ninety per cent of Hokkaido's population is concentrated in the western part of the island, especially on or near the Ishikari Plain, the most accessible, most fertile, and one of the largest of the lowlands. The Ishikari is the major center not only of rural population but of urban population as well, for here is located the island's metropolis of Sapporo, a city of 455,000, while on its western margin is Otaru, a port city of 162,000. Another principal concentration of population in western Hokkaido is the strongly linear one coincident with the interior meridional basins just to the east and north of Ishikari; and there are several others on Oshima Peninsula, most noteworthy of which are those in the vicinity of the port cities of Hakodate and Muroran.

Eastern Hokkaido east of the Yezo meridional range contains only a modest 10–12 per cent of the population, a circumstance attributable not only to greater remoteness from Old Japan, but even more to the nature of the terrain and climate. The plains are chiefly of the upland variety, plagued with unfertile volcanic ash soils, while the summer climate is cooler and less suited for rice culture. Here the chief centers of population are the Tokachi Plain, with the city of Obihiro as its metropolis, and lesser centers in the vicinity of Kushiro, on the southeast coast, and Kitami and Abashiri in the northeastern part. Significantly, the Konsen upland plain, chiefly developed on an ash upland, is not yet a consequential center of population. A fragmented, patchy population pattern is obvious in Hokkaido as it is in Old Japan, and for the same reasons, although it is not so emphatic because of the lower over-all population densities of Hokkaido's plains.

Settlements and houses

Since the first colonists to arrive in Hokkaido engaged in fishing, the oldest settlements are littoral-oriented. Character of the fishing settlements varies with the nature of the coast and its immediate hinterland: where the coastal lowland is not too restricted in width, compact villages with well-defined street patterns are common; but where the area of level land is restricted by the sea in front and a wave-cut cliff behind, the houses are usually grouped along the main highway paralleling the coast to form a very narrow elongated settlement.

In general, fishing villages do not have a tidy and prosperous appearance. Frame houses with small windows and roofs of heavy wood shingles are characteristic. Cluttering the shore on the sea side is an array of drying sheds, small boats, fishing nets, trays of drying fish, boiling kettles, and racks of kelp. The odor of fish is inescapable.

Earliest agricultural settlement developed on the small lowlands of the Oshima Peninsula, where because reclamation preceded the rectangular land survey characteristic of most of the island, the rural landscape is one of small irregular-shaped fields and compact farm villages resembling those of Old Japan. The agricultural village is not uncommon in other parts of Hokkaido as well, although dispersed single farmsteads are more usual. Many of the earliest settlements, including the militia settlements, were of the village type, since they provided the greatest security for the newcomers. It was easier for a band of colonists, often originating in the same village, to face the hazards and problems of frontier life as a group, rather than as separate families. Elongated highway villages as well as the more compact type were common.

But in those areas where the rectangular land survey prevails and farms are relatively large, compact, and rectangular, the isolated farmstead predominates. Density of these farmsteads varies with the prevailing size of the farm unit, which in turn depends upon the quality of the land and the intensity of agricultural land use. For example, they are more widely spaced in the Tokachi and Konsen areas of relatively extensive farming than they are in most of the Ishikari lowland or the Kamikawa (Asahigawa) Basin. Each farmstead has close ties with a small market town (*shigai-chi*), established at the time of the land subdivision. These are spaced more or less equidistant and contain retail and repair shops, plants for storing and processing farm products, as well as schools, a village office, and post office.

Partly as a consequence of the larger landholdings, requiring as they do more animal labor, the Hokkaido farmstead differs in appearance from the village farmstead in Old Japan, resembling more nearly that of an American farm in its greater area and its larger and more numerous barns and sheds. In this region animals must be protected against severe winter weather, and space must be provided both for housing them and for storing bulky animal feed over several months. There is commonly a well-defined barnyard, sometimes fence-enclosed, which contains in addition to a barn of modest proportions, multiple sheds, wagons, farm machinery, and manure piles. The sleds one commonly sees in Hokkaido barnyards are a reminder of the deep and long-continued snow cover.

Farmhouses are of various types, reflecting the different influences which have impinged upon the settlement process in this semifrontier land. When the Japanese Government turned its attention seriously to

Fig. 9-9.—Farmstead on a dairy farm on the Ishikari Plain near Sapporo.

the peopling of this northland region, American agricultural experts were called in to advise on such matters as housing, land division, crops, and animals. Their influence is apparent in the clapboard residences, glass windows, sheet-iron stoves, and occasional split-rail fences. Russian, German, and Danish influences are also observable. Moreover, the settlers in Hokkaido, who came from various parts of Old Japan, have done what most colonists do—tried to duplicate their old homes as far as possible. In general, house construction is sturdier and offers better protection against cold weather than elsewhere in Nippon—for instance, heating by stoves is nearly universal—but it is surprising that even greater modifications of the subtropical house have not been made, considering the severity of Hokkaido's winter weather. One wonders at the tenacity with which the Japanese have clung to their traditional mode of living in this colder land. Tile and thatch are sometimes used as roofing materials, although far less commonly than wood shingles and galvanized iron. Solid immovable walls made of wood siding and containing glass windows usually replace the mud and wattle walls and the sliding *shoji* so common in subtropical Japan; so that externally, at least, the Hokkaido farmhouse and farmstead bear more resemblance to their American counterparts than to those of subtropical Japan. Houses and other farm buildings are small, unpainted, and none too substantial.

As pointed out earlier, 51 per cent of Hokkaido's 1960 population lived in official cities, or *shi*, while 42.1 per cent lived in compact densely populated districts (DID) with about 5000 inhabitants or over. Nearly all of the urban places are of recent origin; only three were in existence before the Meiji Restoration, and these were all located in the Oshima Peninsula. According to the 1960 census there were only twenty-seven official *shi*, but on the other hand there were fifty-six DID. One DID, Sapporo, had a population of 455,000; another, Hakodate, was in the 200,000 class, while four (Asahigawa, Otaru, Muroran, Kushiro) were between 100,000 and 200,000, and two were between 50,000 and 100,000. Only one DID of over 100,000 was situated in eastern Hokkaido. Of the eight DID of over 50,000, half have coastal locations and are ports of some consequence.

Coastal cities and towns for the most part had their origins as fishing centers in Tokugawa times. Some of these still continue as small centers combining fishing and marketing functions, though a few, like Nemuro and Wakkanai, have declined in recent decades. The port cities that have significantly grown and prospered are those that have taken on industrial functions in addition to their maritime ones. Muroran, Kushiro, Otaru, and Hakodate are of this group, the last two being the most important commercial ports of the island.

Many of the inland urban places developed as centers of agricultural colonization. Because in most cases their sites were determined and patterns planned in conjunction with the rectangular land subdivision which preceded actual colonization, their streets have a grid pattern and are wide and straight. Most of the smaller centers are essentially local market towns. It is those that have added other functions—manufacturing, commercial rail centers, mining—that have grown to city size. Sapporo, Asahigawa, Kitami, and Obihiro are examples.

AGRICULTURE AND FISHING

About 26 per cent of the Hokkaido households—more than for any other single economy—are supported wholly or in part by agriculture. Yet the value of agricultural production is exceeded by that of manufacturing. This latter statement may not do full justice to agriculture, however, for the food-processing industries which depend upon agricultural raw materials account for more than 30 per cent of the value of manufactures. In a broad sense, therefore, agriculture is still the main pillar of the Hokkaido economy.

Agricultural land

About 15 per cent of Hokkaido's area is classed as agricultural land while only 10 per cent is cultivated. These figures are less than their counterparts for the country as a whole, only slightly so for agricultural land but more strikingly so for cultivated land, since greater proportional areas are devoted to grazing land, pastures, and meadows in Hokkaido. Thus, in 1960, 23 per cent of the country's natural meadow and 90 per cent of its privately owned natural pasture were located in Hokkaido.* It is probably true that a larger part of this northern island, because of physical handicaps, is unsuited for the cultivation of crops than is the case with Old Japan. Still, in all of the national government's plans for land reclamation, Hokkaido's large share of the total land area judged to be reclaimable was out of proportion to its size. For example, in the 1947 ten-year program, aiming for a reclamation of 1,537,135 hectares (3,798,261 acres), 35 per cent was in Hokkaido. Much of this is located in eastern Hokkaido where the cool, short summers present one serious handicap, which is aggravated by the prevalence of low-grade ash soils. The remaining lands are largely low-grade peat-bog soil, comprising about 203,000 hectares (500,000 acres). Over 30 per cent of this is in the Ishikari Plain, 22 per cent in the Kushiro area, 15 per cent in the Teshio River valley, and 15 per cent in the Kitami District.[12]

To what extent Hokkaido's area of cultivated land can be expanded through reclamation procedures, and how many new farm households can be settled on such new lands, is a controversial question. As it bears on this topic, the history of land reclamation and of new agricultural settlement since World War II may be significant. During the decade 1945–1955 the plans of the national and Hokkaido governments called for a reclamation of 312,724 cho (1 cho = 2.45 acres) of new land to be placed under cultivation, but the actual accomplishment amounted to only 148,264 cho, or 47 per cent of the plan.[13] The above figures do not, however, represent the new area actually reclaimed and settled, for they fail to include the losses due to abandonment of farms which subsequently reverted to a wild state. It is noteworthy, also, that the annual amounts reclaimed showed a marked decline between the beginning and the end of the decade, for the most active reclamation occurred shortly after the close of the war, when the general food shortage and an increase in the agricultural population acted as stimuli. More recently, as the cities have increasingly drained away the surplus labor on farms, reclamation has greatly slowed up or even become static.

The area of uncultivated land in Hokkaido purchased by the national government after the war for reclamation purposes amounted to 770,250

---

* Meadow and pasture are defined as natural grasslands privately operated. Meadow is used mainly for mowing; pasture, mainly for grazing. National and publicly owned lands are excluded.

hectares, of which 451,035 had been sold to settlers by 1960, but only 166,900 had been brought under cultivation. And while nearly 44,300 farm households attempted settlement on these newly reclaimed lands, only about 30,600, or 69 per cent, had persisted until 1960.[14]

The large percentage of settlement failures and farm abandonments reflects in part the lack of discrimination in selecting the farm households who were to be permitted to occupy the new lands; many of these were not qualified to undertake frontier farming operations. In recent years as the qualifications of applicants have been scrutinized with more care, the survival rate has risen sharply, but at the same time the total number of new settlers is small.

The farm economy of the new settlers operating recently reclaimed land is generally in an unstable state. Almost all new settlement is on submarginal land which is handicapped by a rigorous climate and inferior soils (peat, new volcanic ash, or heavy acid clays) and which often has the further disadvantage of sloping sites. Characteristically, the new settlers lack capital sufficient to improve the poor soils, buy livestock and machinery, or to tide them over a seasonal crop failure. To aid the situation, pilot farms are now under development in the Nemuro

TABLE 9-2

Farm households settled on newly reclaimed lands in Hokkaido, 1945–1960

Year	New households settling	Households remaining	Settling ratio
1945	3,980	1,502	37.7%
1946	7,875	4,904	62.3
1947	6,268	3,808	60.8
1948	4,417	2,228	50.4
1949	4,671	3,330	71.3
1950	3,185	2,213	69.5
1951	1,784	1,394	78.1
1952	1,240	1,889	88.3
1953	2,038	1,861	91.4
1954	2,014	1,898	94.2
1955	1,895	1,817	95.9
1956	1,788	1,625	88.0
1957	689	654	95.0
1958	873	836	91.0
1959	681	681	100.0
TOTAL AND AVERAGE	44,295	30,640	69.0%

Source: Hokkaido Japan-American Rural Cooperation Society, *Agriculture of Hokkaido* (Sapporo, 1962).

and Kushiro districts, financed by loans from the World Bank and directed by the Agriculture Development Cooperative.

Distribution of the new land reclaimed during the decade 1945–1955 is widespread throughout Hokkaido, but about half is in the frontier eastern sections, where it is very largely of the upland type. Peat bog reclamation was minor. Within the frontier east, nearly half of the new land was in the extreme east in Konsen (Nemuro and Kushiro districts); about one quarter, in Tokachi south of the volcanic range; and nearly one quarter, in Abashiri, north of the volcanic range. In western Hokkaido a disproportionate share of the newly reclaimed lands was in Kamikawa district, which includes the series of five meridional intermontane basins, and in Sorachi and Ishikari districts which include the extensive Ishikari Plain.

Over 40,000 hectares, or about 100,000 acres, of peat bog have been reclaimed, most of it in the western parts of Hokkaido, and with emphasis on the Ishikari Plain. Two kinds of peat land are recognized, *niedermoor* and *hochmoor,* the former representing a more advanced stage in the cycle of bog development, in which the humus is better decomposed. For this reason most reclamation has been of niedermoor and not hochmoor. The government opposes any further attempt at hochmoor reclamation and does not plan to open any large new projects looking toward the reclamation of niedermoor, although it may continue with some projects already started.

Altogether, the converting of peat bog into agricultural land is a discouraging, costly operation involving the use of large-scale and expensive equipment such as pumping machinery, bulldozers, diesel engines, and large trucks, as well as the laying of drainpipes. The expenditure of $14 million to bring some 75,000 acres of new peat land into cultivation suggests that it is a dubious economic enterprise.[15] It seems especially doubtful whether reclamation of peat lands in the cooler east and north is worth the expense and effort required.

One school of thought maintains that it would be better planning for most of the upland reclaimable lands to be devoted to livestock farming rather than to an extension of the crop area. But this, too, will encounter serious obstacles in the form of a shortage of feed concentrates, poor quality of the natural pasture, and the limited home market for processed dairy products.[16] Altogether, the outlook for an expansion of either crop or livestock agriculture in Hokkaido, based upon the reclamation of profitable new lands, is gloomy. All that remains for possible future reclamation is low-grade land located in the less productive climates, so that the handicaps faced by the pioneering farm family are genuinely discouraging.

Crops and livestock

In a number of respects Hokkaido's agriculture contrasts with that of Old Japan. The crops grown differ in their various proportions, with the emphasis on paddy rice greatly reduced and that on forage crops and pasture strikingly increased. Numerous crops important in subtropical Japan—sweet potatoes, tea, mulberry, citrus fruits—are absent. In addition, the agriculture is more commercially oriented. All paddy land and nearly all upland fields produce only one crop each year, except in market garden areas, so that the ratio of crop acreage to cultivated area is slightly below 100. The combined acreage of fall-sown cereals, mostly winter wheat, is only 13,000–14,000 hectares, or 1.4 per cent of the crop area. Crop yields per unit area are distinctly lower than in subtropical Japan, but on the other hand, because of the larger farms, the agricultural income per household is 150–175 per cent of that in Old Japan. Farms are not only several times larger than those of Old Japan, but they are also in one unbroken piece rather than in scattered fields. Farms average slightly over 5 hectares per farm household, while the area of crop land is nearly 3.5 hectares. The comparable figures for Old Japan are 1.04 and 0.86 hectares. While some farmers do live in agricultural villages, the isolated dispersed farmstead is most common. As described earlier, the Hokkaido farmsteads are in considerable contrast to those farther south, being more substantial in appearance, with sturdier houses, more and larger outbuildings, barns for storing hay, larger barnyards, and often a silo. There is also a greater importance of animals, chiefly dairy cattle, in the farm economy. Draft animals, required by the more extensive acreage, are much more numerous than elsewhere, and they are almost exclusively horses, in contrast to the preponderance of oxen in the subtropical parts of the country. More of the agricultural land is on upland dry sites, such as ash and diluvial uplands, terraces, and fans, while the poorly drained valleys and coastal lowlands in many parts are not reclaimed. There are also greater contrasts in the characteristics of agriculture from region to region than is true in Old Japan.

Two somewhat unlike groups of crops are to be found in Hokkaido, one resembling those of Western Europe and the other those of the subtropical Orient. In the former are included oats, wheat, barley, rye, white potatoes, sugar beets, flax, hay, and pasture—all of them adapted to Hokkaido's cooler climate. They also reflect the early influence of American and European agricultural advisers. Among the Asiatic crops rice is foremost, followed by soybeans and other beans. These reflect the cultural background and experience of the Japanese people. As a result of the combination of these two classes of crops, absolute crop failures are rare. Although both groups grow in the same areas, it is a valid generalization that the European type of crop becomes more predominant in the eastern regions of cooler summers and diminished areas of good alluvium, as rice rapidly declines in importance.

In recent years paddy rice has occupied about 20–23 per cent of the land under cultivation, which is to be compared with the figure of 54 per cent for the country as a whole. Still, the remarkable thing, considering the latitude and severe climate of Hokkaido, is that rice should remain the single most important crop, even though its relative position is greatly reduced. A land unsuitable for rice was far from attractive to Japanese settlers and the stabilization of agricultural colonization of Hokkaido had to await the development of new varieties and improved methods of cultivation which would assure a reasonably successful rice culture.

As late as 1872 the rice acreage in Hokkaido was not over 200 hectares, and its cultivation was confined to the southern part of Oshima Peninsula. A quarter century later the acreage was still less than 5000 hec-

TABLE 9-3

Cultivated land in Hokkaido by districts, 1959

District	Total area	Paddy field	Upland field
Ishikari	64,943 ha.	34.4%	65.6%
Sorachi	127,884	57.3	42.7
Kamikawa	144,365	40.4	59.6
Shiribeshi	44,683	22.1	77.9
Hiyama	18,652	27.9	72.1
Oshima	29,469	22.8	77.2
Iburi	34,995	23.1	76.9
Hidaka	25,015	21.3	78.7
Tokachi	209,683	2.0	98.0
Kushiro	34,608	0.9	99.1
Nemuro	30,154	....	100.0
Abashiri	149,984	6.4	93.6
Soya	16,393	....	100.0
Rumoi	22,602	26.9	73.1
ALL HOKKAIDO	953,400 ha.	22.0%	78.0%

Source: Hokkaido Japan-American Rural Cooperation Society, *Agriculture of Hokkaido* (Sapporo, 1962).

TABLE 9-4

Land use in Hokkaido, 1960

Major divisions			% of arable land in various crops	
Cultivated land		826,456 cho	Rice	22.6
Upland fields	77%		Small grains (oats, wheat, bar-	
Paddy fields	23%		ley, rye, millets, etc.)	15.3
Grasslands		193,849 cho	Pulses	22.0
Permanent	20%		Forage crops	16.7
Grazing land	50%		White potatoes	8.8
Areas cut for grass	30%		Industrial crops	7.5
			Vegetables and fruit	3.7

Area in various crops			Distribution of paddy fields	
Paddy rice		187,263 cho		
Wheat		14,071	Ishikari	11,899 cho
Winter	9,089			
Spring	4,982		Sorachi	46,956
Naked barley		8,141		
Winter	1,775		Kamikawa	38,987
Spring	6,366			
Barley		10,078	Shiribeshi	7,435
Winter	800			
Spring	9,278		Hiyama	4,140
Oats		71,067		
Winter	1,429		Oshima	5,024
Spring	69,638			
Rye		371	Iburi	6,124
Maize (for grain)		14,441		
Millets		12,512	Hidaka	4,571
Buckwheat		10,309		
White potatoes		72,834	Tokachi	2,979
Soybeans		63,700		
Azuki (rea) beans		42,517	Kushiro	10
Beans		63,470		
Peas		6,707	Nemuro	0
Vegetables		26,545		
Fruit		4,152	Abashiri	5,475
Apples	3,035			
Hay		92,056	Soya	0
Forage crops		137,991		
Dent corn	33,376		Rumoi	4,432
Peppermint		4,726		
Pyrethrum		856		
Sugar beets		37,339		
Flax		10,044		
Rapeseed		8,497		

Source: Hokkaido General Affairs Department, *1960-Nen Sekai Noringyo Census, Noka Chosa Kekka Hokokusho* [*1960 World Agricultural Census, Report of Survey of Farm Households*], Sapporo, **1961.**

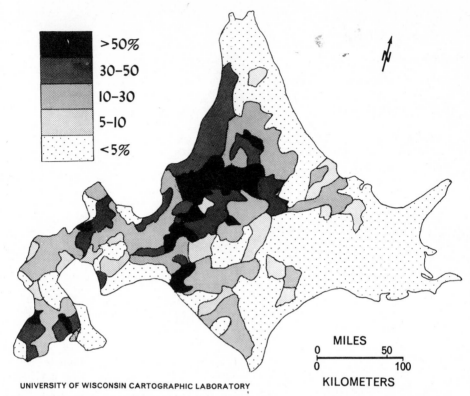

> 50%
30–50
10–30
5–10
< 5%

MILES
0   50
0   100
KILOMETERS

Fig. 9-10.—Ratio of paddy area to total cultivated area in 1957. After map by Hideo Fukui.

tares, largely in the southwestern peninsula.[17] But between 1898 and 1919 expansion was rapid, and more than 70,000 hectares were added in twenty-two years. It was during this period that the rice regions of western Hokkaido, in the Ishikari Plain and the Kamikawa Basin, were firmly established. Between 1920 and 1932 expansion of the rice acreage was most rapid, and 130,000 hectares of paddy land were added in twelve years as rice cultivation advanced northward and eastward into the regions of more severe climates. After a period of nearly static acreage between 1932 and 1940, there was a sharp decline during the war, followed by a steady increase since 1947. Still, in 1960 the paddy area was only about 86 per cent what it was in 1934, the year of maximum acreage. Yet another indicator of the increased importance of rice in Hokkaido is the change in the ratio of rice-field acreage to the total cultivated area. Only 4 per cent in 1896, it was 11 per cent in 1922, and reached a maximum of 24 per cent in the early 1930's. Since 1937 it has usually remained between 20 per cent and 23 per cent.[18]

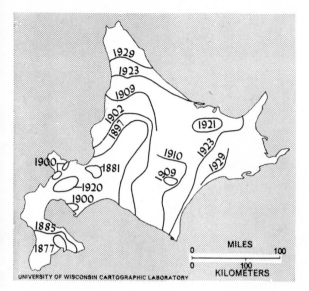

Fig. 9-11.—March of the rice frontier. After map in *Regional Geography of Japan,* No. 1, *Hokkaido Guidebook.*

But although the area planted to rice declined after 1934 and has not quite regained its earlier level, the productivity of rice, or yield per unit area, has consistently increased—from 2.0–2.85 tons per hectare in the years of good crops prior to 1945, up to 3.82 tons per hectare in 1958. Every year after 1957 the total yield of rice has been above the previous maximum yield (1938), and Hokkaido has changed from an importer to an exporter of rice.[19]

If one analyzes the maps showing change in the distribution of rice acreage between 1932 and 1957 (1957 acreage only 81 per cent of 1932), it becomes clear that while much the larger part of the island shows some degree of decline, it is in the cool-summer parts with poorer soils, including the island's northern and eastern extremities, that the acreage reduction has been greatest, exceeding 60 per cent. By contrast, it is the western parts with their warmer summers and better alluvial soils that declined least, and in some parts of the west there has been a slight expansion of the rice acreage in recent years.

In large measure, the increased rice acreage in Hokkaido after the turn of the century was related to the breeding of new cold-resistant varieties, which at the same time were capable of making good yields and were also relatively disease resistant. However, the cold-resistant strains developed before World War II were not also consistently high yielders. Since the war a number of superior new varieties have been developed, and these show an increasing tendency toward regional concentrations. Thus, in the Oshima Peninsula late-maturing and high-yielding va-

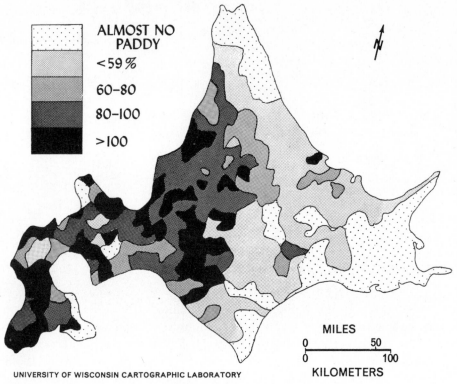

ALMOST NO
PADDY

<59%

60–80

80–100

>100

MILES

0   50

0   100

KILOMETERS

UNIVERSITY OF WISCONSIN CARTOGRAPHIC LABORATORY

Fig. 9-12.—Change in area of paddy fields between 1932 and 1957 expressed as a ratio of 1957 acreage to 1932 acreage. After map by Hideo Fukui.

rieties still prevail, while in the Abashiri region of cool northeastern Hokkaido, Norin No. 20 and Norin No. 34, both early-maturing and relatively cold- and disease-resistant varieties, together comprise nearly 53 per cent of the paddy area.[20]

Not only new varieties, but also new methods of cultivation which are better adapted to the unfavorable climatic conditions, were influential in the post-1920 expansion of the rice area in Hokkaido. Direct sowing of seeds in the paddies was tried, thereby avoiding the traditional, labor-costly procedure of transplanting from seedbeds, but in recent years transplanting has been revived because of the benefits associated with artificially warmed seedbeds, so that now, once more, most rice is transplanted. It is possible, though, that increasing shortages of farm labor will force a return to direct sowing. Unlike the situation in Old Japan, a great majority of the rice seedbeds are dry in Hokkaido, so that the seedlings do not develop in an inundated environment. The establishment of *Doko Kumiai,* or cooperative rice-irrigation societies, was still another factor stimulating the expansion of the paddy area. Outside the old rice lands in southern Oshima Peninsula, most of the island's rice-growing areas are served by this system.

Currently the rice acreage is strongly concentrated in the western and southwestern parts of the island, 57 per cent within the three *shicho* of Sorachi, Kamikawa, and Ishikari, which include the specialized rice growing areas of the Ishikari Plain and the more southerly of the interior meridional basins, especially Kamikawa. In cool eastern and northern Hokkaido paddy lands are scarce, and in three *shicho,* Nemuro, Kushiro, and Soya, they are practically absent. The only fairly extensive area in eastern Hokkaido where the ratio of paddy land to total cultivated land is 10 per cent or greater is in the Kitami-Abashiri region facing the Okhotsk Sea.

But although it is the successful raising of rice that stimulated colonization and gave stability to Hokkaido's agriculture, this subtropical crop still suffers from the adverse climatic conditions. To be sure, yields have improved, but they still are the lowest for the country, and wide fluctuation in harvests is characteristic. A small decrease in summer temperatures or a slight shortening of the growing season causes a sharp decline in output. Yields in poor years may be only two-thirds of what they are in years of favorable weather.

Small grains other than rice are widely distributed, but over half their acreage is in Tokachi, Abashiri, and Kamikawa. Wheat is especially concentrated in Abashiri; the barleys, in Tokachi; while oats are more ubiquitous. The latter are almost exclusively spring sown, and the barleys are largely so, but nearly 65 per cent of the wheat is fall sown. Millet is grown largely in the east, chiefly Tokachi. White potatoes are widely distributed. Beans are especially important in Tokachi, and sugar beets in Abashiri and Tokachi. Industrial crops as a group are concentrated in Abashiri-Tokachi. While forage crops are widely dispersed, 54 per cent of the total acreage is in the four eastern provinces of Tokachi, Nemuro, Kushiro, and Abashiri, where the dairy industry is best developed. As noted in an earlier section, a number of subtropical crops so important in the agriculture of Old Japan, including sweet potatoes, tea, citrus fruits, and mulberry, are absent in Hokkaido, and their places have been taken in a measure by white potatoes, beans, apples, and a variety of industrial crops.

Livestock-raising as an adjunct of farming is more emphasized in Hokkaido than elsewhere in Japan, and this is reflected in the greater abundance of hay and forage crops and of pasture. Thus, one-third of the country's total acreage of land devoted to forage crops is to be found

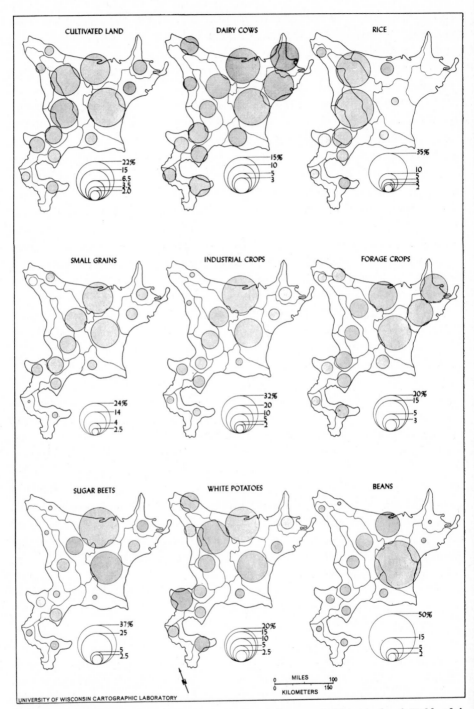

Fig. 9-13.—Distributions of various agricultural features within each of Hokkaido's fourteen districts, according to the percentage of the total for Hokkaido.

in Hokkaido, as is 90 per cent of its pastured fields, nearly a quarter of the privately owned hay meadow, and 91 per cent of its farmer-owned grazing land.

Beef-draft cattle are completely unimportant, but by contrast this northern island possesses nearly 23 per cent of the country's dairy animals. It likewise has about 19 per cent of the nation's sheep but, on the other hand, only 5 per cent of the hogs. Horses being the exclusive draft animal, it is not surprising that Hokkaido should have close to 40 per cent of the country's horse population. But while Hokkaido accounts for nearly 23 per cent of the dairy cattle, only 27 per cent of the farm households possess one or more such animals. By far the greatest share of the dairy cattle are in the possession of farm households which own only one or two animals, and herds of ten or more dairy cattle are uncommon. Both dairy cattle and hogs have more than doubled in numbers during the period 1956–1960, while sheep have declined.

Hokkaido's approximately 200,000 dairy cattle are distributed widely over the island, but with 50–55 per cent located in the four easternmost districts of Tokachi, Abashiri, Nemuro, and Kushiro, where dairying is a more important element of the farm economy, and especially in frontier Konsen (Nemuro, Kushiro). But this same eastern part of the island that contains over half of the dairy cattle has less than a quarter of the island's farm households, so that the number of milk cows per farm household is more than twice what it is in most other areas.

Milk production in this northern island amounts to about 21 per cent of the nation's total, but by contrast its output of processed milk products, such as condensed and powdered milk, butter, and cheese, much of which is marketed in Old Japan, is 80 per cent of the total. Compared with the rest of the country, a smaller share of the milk in Hokkaido is consumed in fluid form. In 1960 the milk not directly consumed was used in processing 22,600 tons of condensed milk, 16,600 tons of powdered milk, 7,300 tons of butter, and 3,600 tons of cheese. Between 60 and 70 per cent of the nation's cheese and butter is processed in Hok-

TABLE 9-5

Numbers of farm animals in Hokkaido, 1961

Dairy cattle	201,490
Draft-beef cattle	3,290
Horses	247,100
Sheep	152,930
Goats	11,970
Swine	122,240
Chickens	2,641,000

Source: Norinsho Tokeihyo, 1960 [Statistical Yearbook of Ministry of Agriculture and Forestry].

kaido, more than two-fifths of its powdered milk, and somewhat over 30 per cent of its condensed milk. Processing plants are widely distributed.

Regional agriculture

Although in most land-use classifications Hokkaido is set apart from Old Japan and is made to constitute a single agricultural division, actually the island shows a considerable amount of regional diversity in crop associations, intensity of land use, methods of cultivation, and in the marketing of produce.

One simple type of differentiation can be made between the littoral areas and the interior plains and basins. In the former, cool summers, in parts accompanied by fog and poor drainage, tend to discourage agriculture. Lonely isolated fishing villages and animal pastures are here characteristic. The Pacific and Okhotsk coastal belts, which are particularly foggy and cool, are almost lacking in crop agriculture. On the Japan Sea side cultivated fields do in places reach down to the coast but it is marginal agriculture. By contrast, it is the warmer summers and better soils and drainage of the interior that make that general region superior in its agricultural development.

As noted previously, in Chapter 7, Hokkaido may be separated into two main agricultural subdivisions—west and east.[21] In the western part agriculture developed earlier and consequently shows evidences of greater maturity. The west has a diversified agriculture, but with emphasis on grain farming, rice being the dominant crop. Ninety per cent of the paddy fields of Hokkaido are in this region. Farms are smaller than in the east, with under 5 per cent of the farmers cultivating more than ten hectares or twenty-five acres. Farm machinery is less used here and livestock industries are only modestly, or even meagerly, developed except for dairying in the general vicinity of Sapporo. Hence, grasslands and pastures are not so important in the farm economy, and where they do occur they occupy the less desirable sites such as volcanic uplands and hill slopes. By Hokkaido standards, agricultural land use is highly developed, and even some peat soils and ash uplands are under cultivation. Several kinds of crops are raised on each farm, and double cropping is even practiced in a few places.

In eastern Hokkaido, by contrast, where important agricultural settlement did not begin until after 1900, there is greater immaturity and simplicity in land use than is true in the west, with some lingering pioneer characteristics. Greater emphasis upon livestock permits the region to be designated as a region of associated crop-and-livestock farming. Farms are larger than in the west, and grassland and pastures occupy large areas on both the desirable and less desirable lands.

Ogasawara further subdivides the eastern Hokkaido agricultural region into two subregions—eastern Hokkaido proper and the Pioneering Region of the extreme east, chiefly the Konsen Upland. Here a significant indicator is the proportion of fields that annually remain unsown, this being negligible in eastern Hokkaido proper, but amounting to 50 per cent of the meager paddy area and 10 per cent of the upland fields in the Pioneer Region. In the latter the cool and foggy summers make crop harvests unreliable, so that it is a district of immature land use where livestock ranching, with associated pastures and grasslands, is a specialized development. It is the only region of livestock ranching in Japan. Crop farming is only modestly developed, and subsistence crops such as potatoes, oats, and buckwheat are dominant. Considerable areas of what is classed as potential agricultural land still await reclamation, but the low quality of such lands assures a slow rate of settlement.

Fishing

It was marine products that first attracted Japanese settlers to Hokkaido, and until the beginning of this century they continued to be the most important production sector in the island. Even now fishing supports about 5 per cent of the population. The annual take, amounting to about 22 per cent of the nation's total tonnage of marine fisheries, includes a great variety of fish and other aquatic animals, but with emphasis on squid, cod, and mackerel. Three somewhat distinctive fishing regions are recognized. (1) Along the Japan Sea coast fishing is of greatest importance, and there a close spacing of fishing villages exists. Until recently the principal catch of this region was herring, but at present it is more diversified. (2) The Pacific coast has fewer fishing villages and the catch is smaller than along the Japan Sea coast, while seaweed is of greater importance here than elsewhere. (3) Fishing is of least importance along the Sea of Okhotsk, and as a consequence fishing villages there are smaller and fewer. Scallops and crabs are important elements of the catch.

In addition to the coastal fisheries Hokkaido has a number of important bases for deep-sea fishing in the North Pacific. Most outstanding of these is Hakodate, but also included are Otaru, Muroran, Kushiro, Rumoi, Nemuro, and Wakkanai.

MANUFACTURING

Only about 9.7 per cent of Hokkaido's labor force is employed in manufacturing: the contrast between this figure and the country-wide one of 17 per cent points to a lag in industrial development which results in part from a late start but also from the semicolonial status accorded Hokkaido, under which it was viewed primarily as a source of food and of raw and semiprocessed materials for Old Japan. The island lacks many of the indigenous and traditional manufactures which are so ubiquitous elsewhere, while ordinary consumer goods necessary for daily living, including textiles, clothing, and home utensils, are not widely processed. As a consequence Hokkaido remains heavily dependent upon Old Japan and in a measure operates as a kind of colonial outland.[22]

Hokkaido's manufactures have a high raw-materials base, much of the processing being carried out in essentially rural communities. Large-scale manufacturing that is market oriented does not find here a sufficiently large body of consumers to attract development, so that private investment has been drawn chiefly to the extractive and primary economies, or to a kind of manufacturing closely associated with these activities, such as fish canning, lumbering, pulp and paper, dairy products, brewing, sugar refining, and the like. Profits from many Hokkaido industries continue to flow toward Old Japan, the major source of capital. Based upon

TABLE 9-6

Composition of manufacturing in Hokkaido, 1960, in terms of value added (unit: 10,000 yen)

Industry group	Value added	Per cent
Foods	3,102,153	25.1
Textiles	262,647	2.1
Clothing	48,053	
Lumber and wood products	1,643,849	13.3
Furniture	217,690	
Pulp and paper	2,082,666	16.9
Publishing and printing	608,919	4.9
Chemicals	725,010	5.9
Petroleum and coal products	198,014	
Rubber products	119,348	
Leather and leather products	6,766	
Ceramic, stone, and clay products	491,559	4.0
Iron and steel	1,744,256	14.1
Non-ferrous metals	193,666	
Fabricated metals	337,803	2.7
Machinery, except electrical	336,557	2.7
Electrical machinery	14,574	
Transportation equipment	111,746	
Precision machines	5,017	
Others	81,129	
TOTAL	12,361,453	

Source: Ministry of International Trade and Industry, *Census of Manufactures, 1960: Report by Cities, Towns, and Villages* (1962).

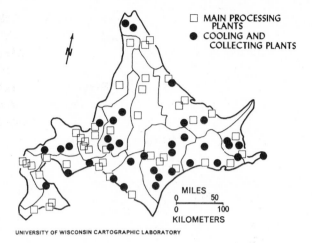

Fig. 9-14.—Dairy plants in Hokkaido. After map in *Nippon no Chiri*, Vol. 1.

☐ MAIN PROCESSING PLANTS
● COOLING AND COLLECTING PLANTS

MILES
0   50
0   100
KILOMETERS

UNIVERSITY OF WISCONSIN CARTOGRAPHIC LABORATORY

value-added, the ranking manufactures are (1) food and kindred industries, (2) pulp and paper, (3) iron and steel, and (4) lumber and wood products.

Industries with an agricultural base are numerous. Some, such as sugar refining from beets, spinning and weaving flax, brewing, and flour milling, originated as a consequence of systematic land development and colonization by the Meiji Government and began as government-subsidized operations. A few plants, including those refining sugar and brewing beer are relatively large, but the flax is processed in small plants, and the same is true of dairy products in Konsen, starch in the potato-growing region of Tempoku in the far north, and peppermint in the Kitami region of the northeast. Small plants also produce such universally used products as *shoyu* (soy sauce,) *miso* (bean paste), and *sake* (rice wine).

Industries based on forest resources are of two kinds. Small-scale plants, which are widely scattered, do the sawing. Plants for further processing of logs into veneer wood, shingles, furniture, sporting goods, pencil sticks, chopsticks, etc., are generally located in permanent settlements and are of small size except in a few special instances. Paper and pulp manufacturing, on the other hand, is done in large and modern establishments. The Tomakomai plant produces over 80 per cent of the newsprint, and the Asahigawa and Ebetsu plants more than 70 per cent of the craft paper, of Japan—statistics which show the preponderant position of Hokkaido in the paper industry.[23]

Industries closely tied to mineral raw materials are iron and steel, centered at Muroran, and those processing cement and clay products. The

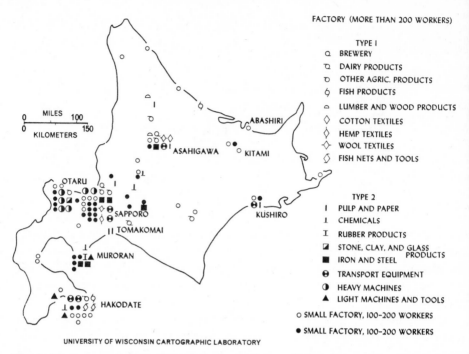

FACTORY (MORE THAN 200 WORKERS)

TYPE I
- Ɑ BREWERY
- ꓷ DAIRY PRODUCTS
- ꭂ OTHER AGRIC. PRODUCTS
- ⬧ FISH PRODUCTS
- ⌒ LUMBER AND WOOD PRODUCTS
- ◇ COTTON TEXTILES
- ◈ HEMP TEXTILES
- ✦ WOOL TEXTILES
- ◌ FISH NETS AND TOOLS

TYPE 2
- I PULP AND PAPER
- ⊥ CHEMICALS
- I RUBBER PRODUCTS
- ◪ STONE, CLAY, AND GLASS PRODUCTS
- ■ IRON AND STEEL
- ❸ TRANSPORT EQUIPMENT
- ◖ HEAVY MACHINES
- ▲ LIGHT MACHINES AND TOOLS

O SMALL FACTORY, 100–200 WORKERS

● SMALL FACTORY, 100–200 WORKERS

UNIVERSITY OF WISCONSIN CARTOGRAPHIC LABORATORY

Fig. 9-15.—Important manufacturing plants. Of small factories, open circles represent Type 1 and closed circles represent Type 2. After map in *Regional Geography of Japan*, No. 1, *Hokkaido Guidebook.*

iron and steel plants depend largely on Ishikari coal and local flux, but must import the greater share of their ore as well as their better coking coal.

There are no manufacturing regions or belts in Hokkaido as there are in Old Japan. The large modern manufacturing plants are widely scattered and often isolated; and in numerous instances the plant is the reason for the existence of the adjacent town. Tomakomai, Muroran, Ebetsu, Yufutsu, and Toyonuma are illustrations of centers which exist almost solely because of a single large factory.[24] Also common is another pattern in which a city represents a cluster of diversified industries, a situation best illustrated by Sapporo, Hakodate, Otaru, and Asahigawa. Perhaps the nearest approach to an industrial district is that in western Ishikari where the larger Sapporo and Otaru centers are combined with smaller nearby Ebetsu, Nopporo, Tobetsu, and Kotoni.[25] This district or center has the greatest diversification of industry anywhere in Hokkaido. As might be expected, most of the important manufacturing centers are in the western part of the island. In the east, only Kushiro stands out, and then chiefly by contrast with the non-industrial character of the general region.

GEOGRAPHICAL SUBDIVISIONS

Since the population of Hokkaido is very much centered in the plains regions, only these will receive serious individual attention.

The Ishikari-Yufutsu lowland

This asymmetrical Y-shaped depression (area, 4289 sq. km.), which separates Hokkaido into two very unequal eastern and southwestern parts, is the northern counterpart of the Kitakami-Abukuma tectonic lowland which in northern Honshu marks the boundary between the Inner and the Outer Zones. It is the heart of Hokkaido, for it comprises by far the most extensive alluvial lowland on the island and as such has become the focus of agriculture and its attendant industries and settlements. On it is located Sapporo, the provincial capital and the island's metropolis.

With respect to terrain, the Ishikari-Yufutsu Lowland is composed of four parts, discussed in more detail below. They are (1) the extensive Ishikari Lowland proper facing the Japan Sea and forming the northern and western parts (1886 sq. km.), (2) the diluvial terraces fringing Ishikari on most parts of its land margins (883 sq. km.), (3) the Chitose Ash Upland which occupies most of the southern arm (1213 sq. km.),

TABLE 9-7

Composition of manufacturing in the main industrial cities of Hokkaido, 1960, in terms of value added (unit: 10,000,000 yen)

Industry group	Muroran	Sapporo	Toma-komai	Asahiga-wa	Otaru	Hakodate	Kushiro
Foods	34	434	18	229	255	261	131
Textiles	1	60	...	37	13	97	3
Lumber and wood products	13	82	145	128	65	38	25
Pulp and paper	1	62	1,031	335	66	10	359
Chemicals	28	24	5	15	7	37	48
Iron and steel	1,603	30	...	2	16	14	2
Fabricated metals	18	33	6	19	147	40	4
Machinery	4	93	3	24	35	88	20
Transportation equipment	28	31	...	2	7	3	7
TOTAL	1,940	1,455	1,234	916	761	680	662

Source: Ministry of International Trade and Industry, *Census of Manufactures, 1960: Report by Cities, Towns, and Villages* (1962).

and (4) the small Yufutsu Lowland fronting on the Pacific Ocean (307 sq. km.).

1. The Ishikari Lowland, the product of rapid aggradation by adjacent volcanoes and by the river Ishikari and its tributaries, is monotonously flat, parts of it being poorly drained peat bog. Rivers wander sluggishly over it in very shallow channels, forming broad riverine belts with somewhat uneven surfaces characterized by remnants of levees, meander scars, and occasional ox-bow lakes. Significantly the name "Ishikari" is an Ainu word meaning to meander or to wander. Some areas of what were originally much more extensive peat bogs are still unreclaimed for agriculture and support a cover of tall wild grasses and reeds, although larger areas of the none-too-fertile peat soils have been converted into cropland. Unquestionably the drainage handicaps of the Ishikari Lowland present the most serious obstacle to its more complete occupance, for in spring, with the melting of the heavy snow cover, the plain is a quagmire. Where it fronts upon the Japan Sea the coast is bordered by a belt of parallel barrier beaches with dunes, its smooth contour providing no natural harbor sites for ports. A recent uplift of the Ishikari amounting to several meters is suggested by the fact that the present Ishikari River is entrenched below the level of the peat bogs. The highest and driest parts of the plain are the beach ridges, the riverine belts composed of natural levees, and a belt of confluent fans along the base of the hills at the southwestern margin of the plain. The most poorly drained alluvium and the peat bogs are located on the interfluves away from the stream channels.

2. Diluvial terraces form an almost continuous belt of low upland benches around the northern and eastern margins of Ishikari. Most of this slightly higher land has extensive areas of smooth to rolling upland surface. Volcanic ash is a large constituent of the diluvial soils.

3. The southern arm of the Ishikari-Yufutsu Lowland is mostly occupied by a low volcanic-ash upland known as Chitose. The upland surface slopes gently downward from Shikotsu Volcano, to the west, which presumably supplied the fresh volcanic ejecta of which Chitose is composed. For the most part the surface is smooth or undulating, but to the east and south, and especially in proximity to rivers, it is more rolling. The upland averages only about twenty-five meters in height, but still it forms a divide between Japan Sea and Pacific drainage, and is something of a climatic divide as well. An inferior soil derived from fresh, coarse volcanic materials, together with a meager ground-water supply, has acted to discourage agricultural settlement, so that the rural population density is only about a quarter that of the Ishikari alluvial low-

land. Large areas of the volcanic upland surface remain out of cultivation and are covered by cutover woodland and wild grasses, so that it has a wilderness appearance. Such a situation prevails especially on the higher central portion surrounding Chitose city and the southern slopes facing the Pacific Ocean. Here there are only occasional farmsteads with patches of cultivated land adjacent, and the area is definitely frontier-like in appearance. A part of this higher, smoother upland surface in the vicinity of Chitose city was originally used by the Japanese, and later by the Allied Occupation, for an airfield, drill grounds, and other military installations. The Sapporo airfield continues to be located there as is also an American air base. Northwestward from Chitose city toward the Ishikari Lowland and Sapporo, the elevation decreases, soils are not quite so coarse, and cultivated land becomes more extensive. Here spurs of ash upland with steep margins are intermingled with alluvial-floored valleys, the latter largely in paddy fields, while the upland surfaces are planted in a variety of dry crops, including small grains, potatoes, maize, cabbages, turnips, and hay, and even paddy fields are not absent. Much of this ash upland would be classed as a marginal agricultural area, and while reclamation and new settlement were active immediately after the war, at present the situation is static.

4. The fourth terrain unit is the Yufutsu Lowland bordering the Pacific Ocean. It has a rather desolate appearance in most parts; poor drainage, cool summers, and the presence of sea fog act to discourage agricultural utilization, so that the rural population is not more than one-sixth as dense as it is on the Ishikari Lowland.[26] Beach ridges border the coast, and back marshes covered with coarse grasses occupy considerable areas of the lowland intermediate between the ocean and the terraced sea cliffs on the land side. Forlorn-appearing fishing villages dot the coastline, and there has been some agricultural colonization of drier parts of the plain, although extensive areas appear to be used for pasturing horses and cattle. Focus of settlement on this relatively barren and foggy coast is the city of Tomakomai (44,500 population), which is essentially a product of Japan's largest pulp and paper mill specializing in newsprint. Tomakomai harbor is being actively improved and new water-front industrial sites are in process of being reclaimed from the sea, with the intention of making the port one of country's new local industrial centers.

Ishikari's accessibility to Old Japan, its relatively warm summers, and its comparatively large areas of moderately fertile new alluvium suitable for paddy development are important reasons why it was one of the first parts outside Oshima Peninsula to be settled and why at present it contains the single largest population cluster in Hokkaido, including one-quarter to one-third of the island's people. After the first railroad was built in 1880, running from Otaru to Sapporo and soon extending into other parts of the Ishikari Plain, there was a large influx of population from Old Japan into Hokkaido, which tended to concentrate on this newly accessible lowland.

All of Ishikari Plain is now occupied except certain peat bog areas and portions of the bordering diluvial uplands. Rectangular land-subdivision and road systems, with dispersed isolated farmsteads, are characteristic of most parts. Such isolated farmsteads tend to locate along the roads, while farm villages and market towns are commonly situated at the intersections of main highways. Surrounding the homesteads and paralleling the highways, rows of trees, often poplars, break the force of the strong winter winds as they sweep across the flat terrain, piling up the snow. Highways, broader than in Old Japan, are elevated and are paralleled by drainage ditches. Large horses in single harness drawing heavy two-wheeled carts equipped with wide, high wheels are conspicuous on the highways. Teams of horses are less common. The environs of Sapporo are modestly specialized in dairying, evidences of which are to be observed in the Holstein cattle around the farmsteads, in the hay barns and silos, and in the goodly number of milk-processing plants.

Over the lowland as a whole, the areas of paddy rice and of unirrigated upland crops probably are fairly equal. Oats are the most important dry crop, but fields of soybeans, potatoes, buckwheat, peas, and wheat are also numerous. Ishikari is one of the two most important rice-growing regions of Hokkaido, with the crop occupying as much as 50–75 per cent of the total cultivated area in some places, though this is the exception. In the parts devoted exclusively to rice the monotonous rectangular pattern of paddy fields constitutes a quite different and a far less diversified landscape as compared with the dry-crop sections. It is difficult to perceive any very rational pattern of distribution of the more exclusive paddy and dry-crop areas. One might expect rice to greatly predominate on the low areas, such as the reclaimed peat bogs, but this is not consistently the case. Actually, there exists a very complicated arrangement of paddy and upland fields, the two being well intermingled in some parts and strongly exclusive in others. Availability of irrigation water appears to be one important factor influencing the distribution of paddy lands. Riverine belts, of variable width, with their sandier soils and slightly higher and more undulating surfaces, form conspicuous zones of non-paddy land. In the immediate environs of Sapporo there is a concentration of market gardens.

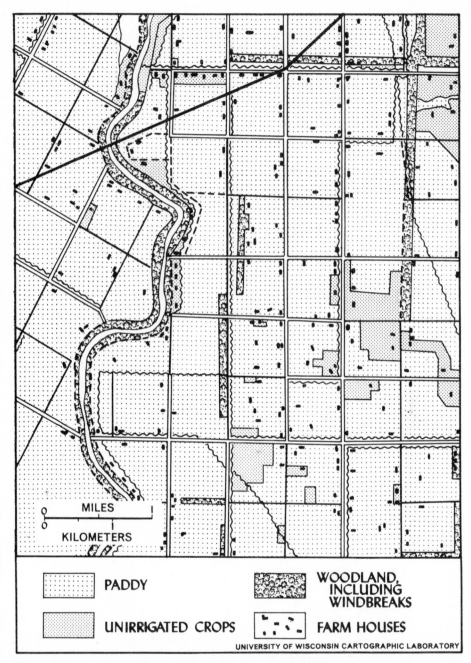

MILES

KILOMETERS

PADDY

UNIRRIGATED CROPS

WOODLAND, INCLUDING WINDBREAKS

FARM HOUSES

UNIVERSITY OF WISCONSIN CARTOGRAPHIC LABORATORY

Fig. 9-16.—A section of the Ishikari Plain where paddy fields prevail. Rural settlement is chiefly of the dispersed type, and windbreaks are common. After map by Yoshikatsu Ogasawara.

On the peripheral diluvial terraces, which have elevations of 20–40 meters above the alluvial plain, agricultural land use is not quite so intensive, but still rural population density is about 80 per cent of that on new alluvium. The rolling upland surface of the diluvium, the poor ash soils, and the incised streams cause these locations to be somewhat less satisfactory for paddy rice, yet that crop is by no means absent.[27] Woodlands are conspicuous.

Apple orchards, which comprise about 75 per cent of the total orchard acreage of Hokkaido, are highly concentrated in the Ishikari region, occupying dry sites on diluvial uplands and river terraces. Three specialized apple-growing areas are observable—in the northern arm of the Ishikari Plain, near Sapporo city, and in the Yoichi area just west of Otaru.[28]

Sapporo, the metropolis and capital city, with a DID population of 455,000, is situated on a sloping alluvial-fan dry site, which provides a marked contrast with the wet peat bog area of the lowland just to the north. Significantly, the name "Sapporo" is from the Ainu language and means extensive dry land. This, the largest urban place north of Tokyo, is a planned city on a flattish site, unhampered by surface configuration. Avenues 160 feet wide intersect at right angles with streets 100–120 feet wide, forming a grid pattern which was laid out on paper before any buildings were constructed. Downtown Sapporo, with its numerous multistoried, Western-type buildings, has a distinctly metropolitan appearance (Fig. 9-17). The city is plurifunctional in character, for besides performing industrial and commercial functions it is an administrative and university center. Unlike the other urban places, it serves the entire island. Its industries, chiefly concentrated in the eastern part, include a large brewery, dairy plants, agricultural implement and machine factories, sawmills, woodworking plants, repair shops for locomotives, and a linen-weaving establishment.

In the Tertiary hills along the eastern margins of the Ishikari Lowland, and therefore actually a part of the Yubari Mountain land, is the Ishikari coal field, ranking first both in reserves and production among the coal fields of the country (Fig. 9-18). Its northern part is called Sorachi and its southern part, Yubari. In dimensions the whole Ishikari field extends about 80–100 kilometers north-south by 16–24 kilometers east-west. Coal seams, totaling 150 or more if thin ones are included, vary greatly in thickness, from less than one foot to as much as sixty feet. Geological structures within the coal field are complicated, and the coal-bearing strata have been so disturbed by foldings and faultings that the dips of the coal seams are steep, usually from 15° to 50°. Major handicaps include the thin and steeply-inclined character of the coal

Fig. 9-17.—Sapporo, the largest city in Japan north of Tokyo, has a distinctly metropolitan appearance in parts of its downtown central business district.

seams, the generally mediocre quality of the coal, and the absence of high-grade coking coal. At a number of points along the north-south rail line paralleling the eastern margin of the Ishikari Lowland, branch lines take off from the main line and work back along the river valleys to the coal mines in the adjacent hills. Coal is responsible for a number of mining towns, among the most important being Yubari, Ikushumbetsu, Bibai, and Sunagawa. Ishikari coal serves chiefly that part of Japan north of Nagoya on the Pacific side and of Niigata on the Sea of Japan, and it is essential to the industrial development of both Hokkaido and northern Honshu. Its chief point of exit is the port of Muroran on the south coast.

The central fault depression and its five basins

This elongated meridional depression, bounded on the east and west by the scarps of the Kitami and Teshio highlands, is composed of five detritus-filled basins (from north to south: Tombetsu, Nayoro, Shibetsu, Kamikawa, and Furano). These are separated by much more constricted segments of valley in which there is almost no level land. Diluvial upland is widespread. A railroad follows the meridional depression throughout its entire length. Interior location and basin configuration combine

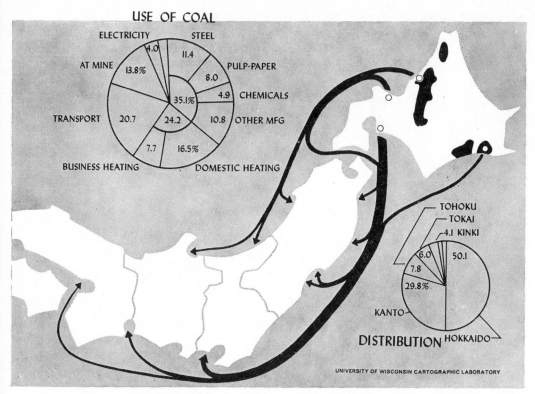

USE OF COAL

ELECTRICITY   STEEL

4.0   11.4

AT MINE   13.8%   PULP-PAPER

8.0

35.1%   4.9   CHEMICALS

TRANSPORT   20.7   24.2   10.8   OTHER MFG

7.7   16.5%

BUSINESS HEATING   DOMESTIC HEATING

TOHOKU
TOKAI
4.1 KINKI
6.0   50.1
7.8
29.8%

KANTO

DISTRIBUTION   HOKKAIDO

UNIVERSITY OF WISCONSIN CARTOGRAPHIC LABORATORY

Fig. 9-18.—Uses of Hokkaido coal, and regions where it is consumed. After map in *Nippon no Chiri,* Vol. 1.   No scale given.

to give this subregion one of the most severe continental climates anywhere in Japan. At Asahigawa the average January temperature is 14.5° F.; a low of −42° has been recorded; and the mean minimum temperature is 3°. In all Hokkaido only Obihiro, located east of the principal meridional mountain range, seems to have a colder climate and even there it is only slightly more severe. Bright sunshine is rare in winter; at Asahigawa the average number of clear days in December is only 0.1 and in January 0.5, whereas the cloudy days number 23.4 and 20.7, respectively. Average dates of the first and last frosts are October 2 and May 27, so that the growing season is only 118 days. Again, only Obihiro has a shorter growing season. Snow falls on 131 days of the year. Midsummer temperatures are pleasant, averaging 67° and 68.5° at Asahigawa in the months of July and August. The mean maximum temperature in August is only 80°. In these basins, surrounded as they are by hills and mountains, there is a good deal of cumulus cloud development in summer, so that genuinely clear days are rare (0.9 in July and August) and cloudy days frequent (17.8 in July and 15.8 in August). Annual precipitation amounts to only 109 millimeters, which is relatively low for Japan. The months of January through June constitute the driest period, and late summer the period of maximum precipitation, September being the wettest month. Fall is distinctly wetter than the spring, a condition not ideal for cereal agriculture.

The *Tombetsu,* or *Bifuka, Basin* (124 sq. km.), farthest north, is unlike the others in that it has no sizable area of flattish floodplain. Instead, the narrow basin appears to be nearly filled with diluvial terraces and fans. Agricultural population is small, a number of the isolated farmsteads look new, and some of the fields still contain stumps and girdled trees, evidence that this is a semifrontier region. Villages seem to be primarily logging and lumbering centers, for sawmills and huge piles of logs are conspicuous features of the landscape. Density of rural population in Tombetsu is the lowest for any of the five basins.

The *Nayoro Basin* (379 sq. km.), next, to the south, is 70–75 kilometers long and has a distinct alluvial floor 3–7 kilometers wide, in some places twice this width if the diluvial terraces are included. Parts of the basin floor are so wet and poorly drained as to be unused, although in recent years new settlers have reclaimed large areas and planted them to paddy rice. On drier portions of the basin floor, and on the rolling diluvial uplands adjacent to it, larger fields and dry crops prevail. The rectangular system of highways and of land subdivision and the isolated farmsteads remind one of Ishikari, but settlement here is more recent, and forest industries continue to support a considerable number of the people. While the density of rural population is double that of Tombetsu just to the north, it is, on the other hand, well below those of the two southernmost basins. Paddy land is more extensive than in Tombetsu. Nayoro city, the metropolis, located at a railroad junction point, is a lumber town with a pulp and paper mill.

Shibetsu, or *Kembuchi, Basin,* the smallest of the five (109 sq. km.), has also the smallest population cluster and, after Tombetsu, the lowest rural population density.

Next to the south is the *Kamikawa,* or *Asahigawa, Basin* (555 sq. km.), less elongated and more octopus-shaped, where several valleys converge radially upon a central core or basin. It is distinguished from the other basins of the Central Fault Depression in being the largest, the most populous, and the most completely utilized. Development of the Kamikawa Basin was begun as early as 1890 by military colonizers, and since the railroad thrusting northward from Otaru and Sapporo reached this region in 1898, an accelerated agricultural settlement began relatively early. At first, dry fields prevailed, but after a short pe-

Fig. 9-19.—Dispersed farmsteads in an area of paddy fields on the Kamikawa Plain not far from Asahigawa. Courtesy of the Hokkaido Press.

riod of successful experimentation with rice growing, there was a rapid change-over to irrigated rice, so that by the first decade of the twentieth century much of the arable land had been converted to paddy. The success with rice growing in the Kamikawa Basin hastened its introduction into other parts, which in turn activated and stabilized the colonization of the entire island.[29] As a part of the rich rice-producing lands of central-western Hokkaido, Kamikawa is one of the island's most densely peopled and most completely utilized sections, even exceeding Ishikari. The basin floor is satisfactorily drained except in the northwest corner, so that there is little wasteland, and large parts are covered with paddy fields laid out in rectangular pattern (Fig. 9-19). This is one of the most exclusive rice areas in Hokkaido. Dry fields prevail both on the diluvial uplands which fringe the basin floor and in the riverine belts. A rectangular road pattern and dispersed farm dwellings, the latter spaced rather uniformly in some parts and in others aligned along the main highways, are characteristic features.

Asahigawa city (155,000 population) resembles Sapporo in its rectangular system of broad streets, its importance as a rail center, and its variety of manufactures. Similarly, it is more than a local center of trade and industry, for it serves most of northern Hokkaido. Among its processing plants are those making *sake,* soybean paste, soy sauce, pulp and paper, tanning machinery, rubber products, cotton and wool textiles, and lumber and wood products.

Furano, most southerly of the five interior basins, is small in area (185 sq. km.) but next to Asahigawa it has the highest density of rural population. Its northern part was largely peat bog, some of which has recently been reclaimed for rice paddy. Several thousand acres of peat bog have been improved by bringing in clay soil to mix with the peat, and fields thus treated show a marked increase in the yield of rice. On the low diluvial uplands and gentle foothills the fields, devoted chiefly to upland crops, are older than those on the basin floor. The southern half of the basin is mostly composed of slightly dissected confluent fans forming an alluvial piedmont zone where dry fields predominate. Here the rectangular pattern of roads and fields so common in Hokkaido prevails.

The Tokachi Plain

The Tokachi Plain is not essentially a lowland of new alluvium like Ishikari, but instead consists largely of unconsolidated fan and coastal plain deposits of diluvial age resting unconformably upon an irregular surface of Tertiary rocks. In some places the Tertiary basement rocks are exposed along the valley sides, and in others they rise above the diluvial upland surface as islands or spurs of higher and rougher land. The fans have pushed out chiefly from the highlands on the west and north. Tokachi, therefore, is mostly an upland plain, and in that respect resembles the Kanto Plain of central Honshu. Much of the diluvial surface is described as undulating, although the eastern parts, beyond the main river, are more uneven and, in places, hilly. Rivers have incised valleys into the diluvial upland, developing wide floodplains with river terraces. Dimensions of the plain are about eighty kilometers in a northwest-southeast direction and fifty kilometers in width, while the total area is 3827 square kilometers, of which 77 per cent is classed as upland surface.

Over most of the plain the diluvial upland is composed of extensive terrace benches whose surfaces represent several different levels, indicative of successive stages of uplift. At least four main surface levels may be recognized, the lowest being the floodplain of newer alluvium and the three others being upland diluvial surfaces, the highest reaching elevations of 500–600 meters. Thus, the entire plain is composed of relatively smooth flattish or sloping surfaces at a succession of levels.[30] From certain vantage points one can view the conspicuous and even skylines of the sloping higher terraces, which extend for miles unbroken. The margins of the several terrace levels are occasionally precipitous, though more frequently the transition from one level to another is less abrupt. Commonly the steeper transition zones are sharply defined by their sinuous belts of woodland. The immediate floodplains of the rivers, especially along the lower courses of the main stream, include a great deal of swamp and peat bog. Along the coast, except at the mouths of rivers, the front of the terrace ends abruptly almost at the water's edge.

Mantling the flattish diluvial upland surfaces of Tokachi is a layer of fresh volcanic ejecta which is highly infertile. Not only is the Ando ash soil low grade, but it is also fine textured, so that it blows badly, with the result that soils and seeds may be blown away by the spring winds. Tall windbreaks, chiefly of larch trees, bordering the fields are a characteristic feature of the landscape.

Along the Tokachi littoral the summer months are cool and foggy, as they are in the Nemuro-Kushiro region, although on the average fog penetrates inland to a depth of only about ten kilometers. Because of these climatic handicaps agriculture is excluded for a distance of about five kilometers inland from the coast. Here the landscape is a desolate one, but occasional fenced fields, with horses and cattle, suggest that there is some livestock grazing in this foggy coastal belt. At Obihiro in the interior of Tokachi, on the other hand, the climate is of a distinctly more continental type, early summer temperatures being 5° to 8° higher and winters 6° to 10° colder than along the immediate coast. Obihiro has a January mean temperature of only 13° and an August mean of 67°, or an annual range of 55°. The frost-free season is only 121 days. Tokachi vies with the basins of the Central Fault Depression for having the most severe climate in Hokkaido. Deciduous and mixed forest originally covered the region, this being the only part of eastern Hokkaido where such a woodland predominated. While the ash soils admittedly are infertile, they are less coarse, and altogether somewhat better, than those of the cool, foggy Kushiro-Nemuro region farther east, which is in the boreal forest zone, where needle trees are prominent. Moreover, the accumulated summer temperatures are a little higher than farther east.

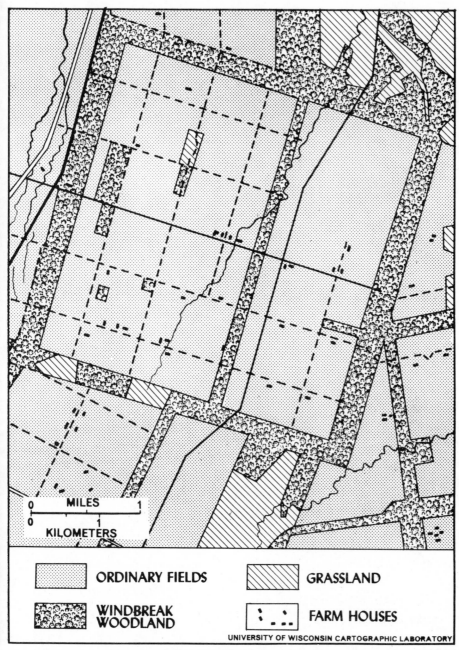

ORDINARY FIELDS   GRASSLAND

WINDBREAK
WOODLAND   FARM HOUSES

MILES

KILOMETERS

UNIVERSITY OF WISCONSIN CARTOGRAPHIC LABORATORY

Fig. 9-20.—A section of the Tokachi upland plain where dry crops or ordinary fields prevail. Note the dispersed rural settlement and the prevalence of windbreaks. After map by Yoshikatsu Ogasawara.

Fig. 9-21.—Extensive dry fields with protective windbreaks on the Tokachi upland plain. Note the higher terrace levels in the background. Courtesy of Hokkaido Prefectural Government.

Fig. 9-22.—An isolated farmstead on the Tokachi upland plain. Note the protective windbreak on three sides. Courtesy of the Hokkaido Prefectural Government.

In completeness of occupance and density of rural population, Tokachi appears to be intermediate between the Ishikari region to the west and frontier Konsen to the east, with about twice the density of Konsen but only one-fourth to one-third that of Ishikari. Not until 1910, when the railroad from Asahigawa reached Obihiro, did active settlement of the region begin. Present concentration of agricultural population is on the lower diluvial, and the alluvial, terraces, and on the newer floodplains where the latter are not too poorly drained. Dispersed settlement and rectangular systems of land subdivision and highways prevail. But, depending on the time of settlement as well as local drainage features, not all parts of Tokachi have their rectangular grids oriented with the cardinal directions, a feature which provides some regional diversity.

About 12,000 hectares are estimated to be potentially reclaimable for agriculture, but at present reclamation activity is relatively stagnant. Immediately after World War II, however, colonization of new lands in Tokachi was very active, with over 19,000 cho being reclaimed and 4,758 new farm households settled between 1945 and 1955. Subsequently 1,808 households abandoned their new enterprises, leaving a net gain of 2,950.[31] Land still to be reclaimed is located chiefly on the higher and more remote diluvial uplands and hill slopes.

Considering the upland character of the Tokachi Plain and its severe climate, it is not surprising that dry-crop agriculture should predominate. Actually only 2 per cent of the total cultivated area is in paddy. Even on the drier alluvial lowlands of the warmer inland portion of the plain, dry crops still prevail. In reality there is much less rice land on Tokachi now than there was some decades back, the acreage in 1957 being but 50 per cent of what it was in 1934.[32] Only in the frontier areas of extreme northern and eastern Hokkaido is irrigated rice of less importance than here. Tokachi is an area of true mixed farming, combining dairying; food and feed crops, such as oats, maize, potatoes, and hay; and cash crops, including soybeans, kidney beans, flax, and sugar beets. Farms average ten to twenty hectares, the size depending somewhat on the distance from a railroad station. This is two to four times the average size for all Hokkaido, indicating a relatively extensive type of farming. Each farmer, on the average, has two or three work horses and two to five milk cows.[33]

Within Tokachi important regional differences in the cropping systems are not observable. On the other hand, there are noteworthy contrasts in the degree of occupance, for it is the higher and more remote diluvial upland levels that are least used. Such sites are made less desirable by reason of their slightly cooler summers and higher wind velocities, their poverty of underground water, the lower fertility of the ash soils, and their greater distance from roads, rail lines, and markets.

Since there is no port along the smooth and abrupt Tokachi coast, Kushiro, farther to the east, serves as the port for this area as well as for Konsen. The Tokachi coast is the least developed of any part of the Hokkaido littoral, for not only does cultivation avoid the cool, foggy coastal belt, but even fishing villages are few. Obihiro (61,000 population), the metropolis of Tokachi, is primarily a market center for the local agricultural region and a processor of its raw products. Most important of the city's factories is a beet-sugar plant, but there are also smaller ones engaged in processing flax fiber and dairy products, canning beef, making alcohol, and sawing lumber.

The Konsen Plain

The Konsen Plain resembles Tokachi in that it too is chiefly an upland plain, although its origin is somewhat different. For perhaps a score of kilometers inland from the coast the upland plain is a broad marine abrasion platform covered with diluvial sediments, much of it reworked volcanic ash, several tens of meters thick, and these in turn veneered with deposits of recent volcanic ash. Farther inland the undulating to rolling upland surface represents the flanks of a volcano. Dissection is somewhat greater, and the surface more irregular in the plain's western parts. Much of the upland surface is less than seventy-five meters in elevation. Shallow, steep-sided valleys have been incised into the upland, the valley forms as well as the interfluves indicating an early stage in the erosion cycle. Characteristically the valley floors are excessively swampy, a result probably of recent slight subsidence.

Adverse climatic and soil conditions have greatly retarded the development of the region, so that large areas are still cutover timberland and moors. The low-grade soils, of the Ando type derived from recent volcanic ash, blow badly and stubbornly resist improvement. Tall windbreaks are common along the field boundaries. Bottom-land soils are imperfectly drained and peat bog is prevalent.

The entire eastern and southeastern coast is washed by the cool Oyashio Current from Bering Sea, the southward surge of cold waters being most marked in summer. As a result the region has a deficiency of both sunshine and heat in summer, and dense coastal fogs are common in June and July. In both Kushiro and Nemuro an average eighty-six days of the year are foggy, including about half of the days of June, July, and August. Inland portions suffer less. The average temperature for May at both Nemuro and Kushiro is only 43° or 44°; for June, 49° to 51°; and

Fig. 9-23.—A frontier farm in eastern Hokkaido, in an area recently reclaimed from the forest. Note the stumps still remaining in the fields.

for July, 57° to 59°. Only in August does the average temperature rise above 60° (Nemuro, 62°; Kushiro, 64°). The mean maximum temperature for May, at both stations, is only 51° and for June 57° or 58°. The period between frosts is about 130 days. Because of the low spring temperatures, cereals must be sown here at least a month later than elsewhere in Hokkaido, while the cool summers retard plant growth.

This is the most extensive area of plains frontier land in Hokkaido. Agricultural settlement has been very recent, most of it occurring during the past two or three decades. At present new agricultural colonization is almost at a standstill, the boom in land reclamation having occurred immediately following World War II. During the decade 1945–1955, 30,000 cho of agricultural land were reclaimed and settled in Konsen, the maximum for any natural region in Hokkaido. Over 6000 new farm households originally undertook to occupy these new lands, but only about 4500 of them became permanent.[34] It is significant, however, that in the Kushiro section of western Konsen, over the sixteen-year period of 1945–1961, 41 per cent of the new and durable farm households settled during the first two years and 81 per cent during the first eight, while only 19 per cent arrived during the latter eight years. Future reclamation and settlement are likely to be very slow indeed.[35] Rural population density is the lowest for any plains area in the island, about 40 per square kilometer, and in spite of the poor drainage it is denser on the floodplains than on the highly infertile ash upland. Actually it is lower in Konsen in general than in some of the hill and mountain lands.

Agriculture here is more extensive in character than elsewhere in Japan, with farms averaging more than thirty hectares, divided between grassland, pasture in rotation, and cultivated fields. The cool foggy summers are a great handicap to cereal cultivation, and all agriculture declines toward the coast where these climatic handicaps operate at a maximum. Main emphasis is on dairying, with the average farmer having five to ten head of milk cows, and work horses in addition. Feed crops, to support the livestock, dominate the cropping system, which specializes in hay, legumes, oats, barley, and potatoes, as well as grassland. Silos are numerous, about half of them being of the trench type. Milk collecting and processing plants are in fair number. Compared with Tokachi, there is much more emphasis on feed crops and dairying and less upon small grains, legumes, and industrial crops, including sugar beets. Konsen lies outside the rice zone so that the familiar sight of paddy fields is here absent. On the wet floodplains, the best agricultural sites are the natural levees and terrace benches, but much of their surface has a vegetation cover of swamp grass studded with deciduous trees. The swamp grass is cut and preserved chiefly for animal bedding.

A major contrast prevails between the eastern (Nemuro) and western (Kushiro) parts of Konsen. In the latter, where the interfluves are more dissected, most of the cultivated land is confined to the valleys at some distance inland from the coast where the bottom lands are better drained, temperatures higher, and fog less frequent. Here the pattern of cultivated land is strikingly linear. But in the Nemuro section farther east, cultivated land is concentrated on the undulating to rolling upland surfaces, and linear pattern is absent.

Kushiro port (133,000 population), the metropolis of eastern Hokkaido, occupies one of the few natural harbors along this desolate coast, and the busy city is somewhat out of harmony with its wilderness surroundings. Kushiro's importance has been enhanced by the proximity of a small local coal field and by the expansion of agricultural settlement within its hinterland, for it serves not only Konsen, but Tokachi as well. Locally produced coal and lumber find their outlet through Kushiro, and the city has some fame as a fishing base as well. Of the labor force, 12 per cent are engaged in fishing, 20 per cent in mining, and 12 per cent in manufacturing.[36] Local coal provides the necessary power for a modest industrial structure based upon locally produced raw materials —a structure which includes sawmills, a paper mill, plants processing marine and agricultural products, a fertilizer factory, and a shipyard.

Fig. 9-24.—Dairy cattle pasturing on an extensive wave-cut terrace upland in eastern Hokkaido.

Nemuro city (23,000 population), at the eastern extremity of Hokkaido, was until about 1930 a fairly prosperous, although squalid-appearing city whose economy was based almost wholly upon the fishing industry. Its decline began when the locally accessible fishing waters became gradually less productive, and since the war this decline has been accelerated by the loss of more distant fishing waters to the north due to Russian occupation of Sakhalin and the Kurile Islands.[37] It has no other important economic base than fishing, for its local hinterland is small and unproductive.

The Kitami-Abashiri plains along the Okhotsk Sea

This northeastern region of Hokkaido fronting on the bleak Okhotsk Sea is not a compact unified area as are Tokachi and Konsen. Any unity the region may have lies chiefly in the fact that this new settlement area, composed of several contrasting terrain units, stands out as the single extensive and relatively populous region facing the Okhotsk Sea. Physically it belongs both to the Kitami Mountains and to the eastern volcanic lands. On the basis of terrain characteristics, three areas of occupied land may be distinguished: (1) the Shari ash upland to the east, (2) the wet alluvial lowlands of the Abashiri and other rivers, as well as additional lowlands surrounding the lagoon-lakes lying back of beach ridges and dunes close to the coast, and (3) the nearby inland Kitami Basin. The Shari volcanic upland, which continues almost down to the coast east of Abashiri city, is a recently reclaimed agricultural area which looks relatively prosperous in contrast to the more desolate Konsen upland to the south and which has a population density twice as great as Konsen's. Dry crops prevail except in some of the valleys.

On the wide floodplain of the Abashiri River and those of its tributaries, as well as on the adjacent lowlands surrounding the coastal lagoons, paddy lands are fairly extensive. This Abashiri region took part in the remarkable expansion of rice acreage which occurred prior to 1934, but has declined markedly since that time, so that the recent acreage is less than half what it was a quarter century earlier. At present paddy rice represents only about 6–7 per cent of Abashiri's total crop area. The remarkable thing is that rice is grown at all, considering the cool summers, for July at Abashiri has an average temperature of only 62° and August 66°. On the river terraces dry fields predominate.

Kitami Basin (474 sq. km.) is the agricultural heart of the Abashiri district. The basin floor consists of coarse alluvium, some of it poorly drained, together with river terraces and diluvial uplands covered with volcanic ash. Rice is extensively grown on the wet alluvium where irrigation is easy, but on the whole the drier sites are more completely utilized than is the wet floodplain. Large areas of the wet basin floor are still unreclaimed.

Agriculture in this northeastern area bordering the Okhotsk Sea is of a mixed crop-livestock variety which resembles greatly that of Tokachi. Rice is of greater importance in Kitami-Abashiri, where 6.4 per cent of the cultivated land is in paddy compared with only 2 per cent in Tokachi, and industrial crops, especially sugar beets, also receive greater emphasis. Feed crops such as hay, legumes, oats, other small grains, and corn are extensively grown, for a significant part of the farm income is derived from the sale of milk. Strong emphasis is placed on leguminous crops, used for a variety of purposes, and especially upon beans of various kinds. No other section equals Abashiri in acreage of sugar beets, and the same is true of white potatoes. Farms average about five to seven hectares in size, which is only half to two-thirds that for Tokachi, and one-fourth to one-third that for Konsen.

Kitami city (35,000 population), in Kitami Basin, which now surpasses the port city of Abashiri in importance and size, is chiefly a marketing and processing center for a local, inland agricultural region. Abashiri city (26,000 population) is a fishing base of importance, the largest along the entire coast of the Okhotsk Sea. Not surprisingly, its industries are specialized in marine and forest products. As a market town it serves a hinterland in which recent land reclamation and settlement have been important.

The hill and mountain lands of Hokkaido

The highland areas of steep slopes are much more sparsely occupied in Hokkaido than are their counterparts in Old Japan, the ratio of population density being about one to ten. Still, the rough lands are the principal source of lumber and wood products, as well as of hydroelectric power. To be sure, there are small lowlands scattered along the hill land coasts and in the interior, where the settlement pattern thickens somewhat, but generally speaking agricultural occupance is meager. Small logging settlements are widespread, while a few are supported by the exploitation of mineral resources. Shikotsu-Toya National Park in southwestern Hokkaido and Akan National Park in the east, both in mountainous volcanic areas with numerous hot springs, are tourist attractions of some consequence.

Most importantly occupied of the highland areas is the Oshima Peninsula forming the southwestern extremity of Hokkaido, the region of earliest settlement and the route by which settlers from the south reached the main part of the island, as well as a focus of important exploitation of marine resources. With regard to terrain, two parts of Oshima may be recognized: a western part, composed of weak Tertiary rocks where dissection is great and relief moderate, and an eastern volcanic part, comprising a small southeastern and a much larger northeastern segment, the latter dominated by ash and lava cones. Associated with the igneous activity are a number of small deposits of copper, gold, silver, and tin. Also related to the vulcanism are several actively exploited sulfur deposits and three or four bog deposits of limonite iron ore, all of them in the northern volcanic subdivision. The total amount of iron ore mined is only about 200,000 tons, much of which goes to the blast furnaces located at Muroran. The beach deposits bordering Uchiura Bay, which separates the two volcanic areas, are an important source of iron-sand ore.

On the seaward margins of Oshima Peninsula are located three of Hokkaido's six cities with over 100,000 population, all of them ports. Largest of these is Hakodate (236,000 population), most important fishing base on the island, the northern terminus of the Honshu-Hokkaido ferry line, as well as the southern terminus of the Hokkaido railway system, and hence the main gateway to this northland region. Its importance as a commercial port is minor, ranking well below Muroran or Otaru.

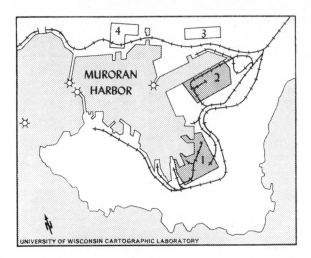

Fig. 9-25.–Types and locations of the main industrial plants at Muroran: (1, 2) iron and steel, (3) cement, (4) oil refinery. After map in *Nippon no Chiri*, Vol. 1. Scale not given.

As one of the more significant local manufacturing centers, its factories produce a variety of items including cement, small vessels, agricultural and fishery foods, fishing equipment, chemical fertilizer, pulp, tannin, and cotton thread.

Otaru (162,000 population), located on the Japan Sea coast of Oshima Peninsula at its easternmost end, profits by proximity to the populous Ishikari region. Since the smooth seaward margins of the Ishikari Plain are without natural harbors, Otaru, located on a small indentation in the adjacent hill lands, has developed as the port for Hokkaido's principal lowland. Its main exports include Ishikari coal, agricultural products, fertilizer, lumber, and fish meal. As a part of the general Sapporo-Otaru industrial area, it contains factories making light machinery, rubber products, and a variety of food items.

Muroran (132,000 population) has the distinction of being the most important iron and steel center north of the Tokyo area. The plants located there, both on the water front, produce about 9 per cent of the nation's pig iron and nearly 6 per cent of its steel, which gives this Hokkaido center fourth rank among the nation's five principal areas of concentration. Local coal from Ishikari and local iron ore and iron-sand ore from nearby Oshima Peninsula provide some of the necessary raw materials, but much ore and high grade coking coal must be imported. In addition to iron and steel, there is an oil refinery and a cement factory (Fig. 9-25). The city is located on the concave side of a recurved spit and its land-tied island, the whole being shaped like a fish hook and enclosing a protected harbor. Bulk cargo is characteristic of the port's trade, imports consisting largely of oil, iron ore, coal, and salt, while exports are pig iron, steel, lumber, and coal.

Tohoku, or Owu,* with 18 per cent of the nation's area and 10 per cent of its population, is that part of Honshu north of about latitude 37°. Six prefectures are included: Aomori, Iwate, Akita, Miyagi, Yamagata, and Fukushima. Sometimes referred to as Northeastern Japan, Tohoku occupies an intermediate position between Hokkaido, on the other side of Tsugaru Strait, and subtropical Japan, on the south—a position reflected in the intermediate and transitional character of climate and many socioeconomic features in the region. In fact, it is this transitional character that gives Tohoku its chief claim to geographic distinctiveness. Still, being a part of Old Japan, admittedly it resembles the south in more ways than it does Hokkaido. Because of its harsh climate, living conditions are less attractive for the Japanese, whose culture evolved in milder subtropical environments. Consequently, Tohoku is looked upon by the peoples of southwestern Japan as a relatively backward area where the amenities of life are scarcer, urban living less well developed, and the rustic virtues are prominent. Here the per capita income is next to the lowest for any of the large subdivisions of the country; primary industries are paramount, industrial production is low, and the proportion of urban population is below the national average. (See Table 9-1, in Chapter 9.) Agriculture alone employs nearly 56 per cent of the labor force, while by contrast manufacturing occupies only slightly more than 7 per

* The regional name of Owu originated from the names of two feudal areas, O-shu in the east and Wu-shu (Dewa) in the west.

cent, the above figures being respectively the highest and the lowest for any of the principal regions of Japan.

Colonization of this northeastern region was later than for any other part of Old Japan. Its active exploitation by the Yamato Court did not begin until the seventh century, after which the frontier was pushed steadily northward so that it had covered the entire region by the ninth century. Like the rest of Old Japan, Tohoku for many centuries was the domain of various feudal lords, but even after the Meiji Restoration the feudal traditions of the earlier period continued to persist here in greater strength than elsewhere in Japan, the effect of which was to retard social and economic improvement, so that Tohoku moved less rapidly than most other regions into the modern industrial period. As might be expected, a large majority of the present cities of Tohoku originated as feudal castle towns. It was in this region, also, that agricultural landlordism was most strongly developed, with all of its attendant ills, and it was here that land reform as a consequence has had some of its most marked and beneficial effects.

During the Meiji Period comparatively more emphasis was placed upon the development of Hokkaido because of its military significance, so that Tohoku was neglected. And in the earlier decades of the twentieth century the exploitation of new colonies acquired as a consequence the Sino-Japanese and Russo-Japanese wars drained off the energy that otherwise might have gone to developing backward northern Honshu.

Like the country in general, Tohoku is a hilly and mountainous region, in which areas of level lowland are relatively scarce. More so than in any other part of Japan the terrain features have a striking meridional trend. Three roughly parallel north-south systems of highland are separated from one another by longitudinal structural depressions, parts of which widen into conspicuous basin-like areas (Fig. 10-1). Each of the three highland systems has some features of an elongated upwarped arch, with margins that are flexure- or fault-scarps.

The easternmost of the three longitudinal highland areas is not continuous but is composed of two spindle-shaped masses—the Kitakami Highland to the north and the Abukuma to the south—separated by Sendai Bay and Lowland. Backbone of Tohoku is the more elevated Central Highlands, which extend almost without interruption throughout northern Honshu. However, a number of fairly low saddles, where traces of river piracy exist, make communication between the lowlands on either side of the Central Range not too difficult. Several volcanic clusters rise above the general upland level to form the highest elevations. The Western Highlands vary greatly in elevation and relief, in

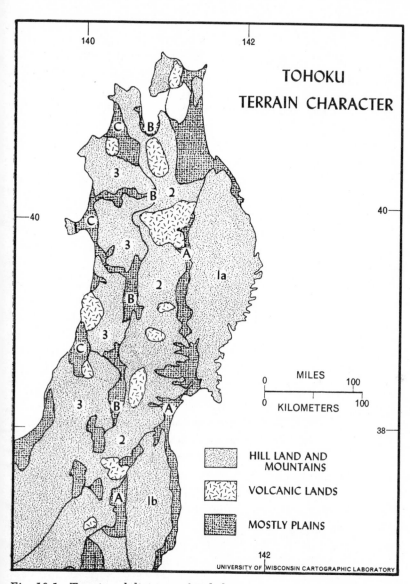

Fig. 10-1.—Terrain subdivisions of Tohoku:

Highlands	*Plains* *(alluvial lowlands and* *diluvial and ash uplands)*

1. Eastern Highlands
   a. Kitakami
   b. Abukuma
2. Central Range
3. Western Highlands
	A. Kitakami–Abukuma Depression
	B. Median Groove
	C. Western Basins

some parts giving the appearance of low hill land, in others rising to elevations of more than 1000 meters. Several rivers originating in the higher Central Range cross the Western Highlands, their transverse valleys providing natural routes of communication between the west-coast lowlands and the interior basins. A few huge youthful volcanic cones along the coastal margins of the Western Highlands stand out by contrast with the prevailingly non-volcanic terrain.

Separating the Eastern from the Central Highlands is the Kitakami-Abukuma meridional depression, composed of several distinct interior basins. In the far northeast, and again in mid-Tohoku in the gap between the Kitakami and Abukuma Highlands, this depressed zone reaches the Pacific Ocean, and there two extensive plains, Sambongi in the north and Sendai in the midsection, have developed. Elsewhere along the east coast of Tohoku plains are very limited in extent and the highlands descend with considerable abruptness to the sea. Basins within the Kitakami-Abukuma depressed zone usually exhibit restricted areas of floodplain, but more extensive areas of alluvial fans and diluvial upland. A broad alluvial-diluvial piedmont belt, composed of a series of composite fans and cones deposited by streams originating in the Central Range, flanks this eastern meridional depression on its western margins. Within the Kitakami drainage system, a series of multiple-purpose dams, whose primary purpose is flood control, is at present under construction.

A chain of eight to ten detritus-choked basins morphologically defines the meridional depressed zone, designated the Median Groove, which separates the Central from the Western Highlands. Along their eastern margins is an alluvial-diluvial piedmont belt similar to that noted for the eastern depressed zone. Such accumulations of stream-deposited materials flanking the Central Range on both east and west reflect the greater elevation of these mountains and the heavy orographic precipitation which they generate. In addition, the abundance of unconsolidated volcanic materials within the Central Range provides a plentiful and easily acquired load for its rivers descending on both flanks.

A series of three or four depositional plains border the Japan Sea side of Tohoku. On their sea margins these western basins have developed wide belts of beach ridges and dunes, which, by obstructing drainage, cause the lowlands back of them to be unusually wet. It is geographically significant that in Tohoku a larger proportion of the plains are interior in location, with no frontage upon the sea, than is true for Japan as a whole.

The climate of Tohoku is emphatically transitional between the cold-winter (Db) climates of Hokkaido and the mild-winter (C) climates

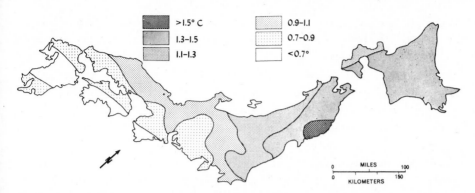

Fig. 10-2.—Above, standard deviation of mean August temperature in Centigrade degrees. After map in *Nippon no Chiri,* Vol. 2. *Below,* areas suffering from spells of abnormally low temperatures in summer. After map by Ogasawara in *Nippon no Tochiryo.*

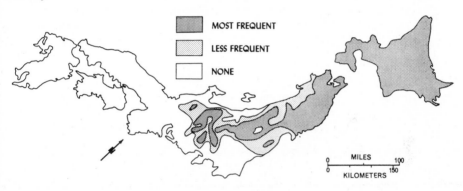

of subtropical Japan. Since the freezing isotherm (32° F. or 0° C.) for January, which has been employed as the boundary between *D* and *C* climates, is in about the latitude of Sendai (38°) in eastern Tohoku and from there loops southward to include the Central Highlands, it is clear that all of northern Tohoku and much of highland southern Tohoku are included within that climatic division designated as "continental with warm summers" (*Daf*). Only the southeasternmost lowlands may be classified as "humid subtropical" (*Caf*). Hot-month average temperatures, unlike those of Hokkaido, are above rather than below 70°. Average August temperatures range from 72° and 73° in northern Tohoku to 76° along its southern margins. As a consequence of the cool Oyashio Current which parallels the east coast, warm-month temperatures are a few degrees cooler on the Pacific than on the Japan Sea side and summer fogs are more common, 20–35 days with fog, nearly all of them in summer, being characteristic. Occasional crop failures, more especially of rice, occur during abnormally cool and foggy summers. A frost-free season of 160–200 days is normal. This represents a month or two more of growing weather than most of Hokkaido has but is, on the other hand, a month or two shorter than the frostless season in southwestern Japan. January average temperatures range from about 27° in northern Tohoku to 33° in the south. In the United States the nearest temperature counterpart to Tohoku's climate is to be found in southern New England and the Middle Atlantic states, or in the Mississippi Valley from southern Wisconsin to about St. Louis.

Marked humidity contrasts between the eastern and western sides of northern Honshu have been discussed earlier, in Chapter 2, with attention to such features as clouds, amount and seasonal distribution of precipitation, snowfall and depth of snow cover.

It is in Tohoku that the temperate forest belt of Japan, characterized by broadleaf deciduous trees and mixed stands of broadleaves and conifers, is most extensively developed. In fact, at low elevations the southern boundary of this broadleaf and mixed-forest zone coincides fairly well with the southern boundary of Tohoku. At greater altitudes in northern Honshu there prevail those conifers and deciduous trees which are representative of the boreal forests.

POPULATION AND SETTLEMENTS

Population density in Tohoku is also definitely intermediate between those of the regions to the north and south. Thus, its density of about 148 per square kilometer is approximately two to three times that of Hokkaido, but less than that of subtropical Japan. Even within Tohoku itself there is an increase in population toward the south; for while the average density in the three northernmost prefectures is nearly 120 per square kilometer, in the three southern ones it rises to 177.

As is true elsewhere in Japan, the lowlands are the conspicuous areas of dense settlement, so that each plain of Tohoku stands out as a cluster of unusually dense population compared with the surrounding hills and mountains. Since many of the lowlands of northern Honshu are interior basins, population in this region is to a greater degree than elsewhere concentrated on lowlands having no sea frontage. Even those plains that do face on salt water are poorly endowed with useful ports, because of the nature of their coastlines; so that the people of Tohoku are unusually dependent upon rail transport. On the whole life is cruder and harsher than it is farther south, partly because of the more severe climate, partly because the region was settled later, being farther removed from the old original culture centers in southwestern Japan from which settlement moved progressively northward.

Birth rates in Tohoku are distinctly above the national average, while death rates are very close to the average, so that the rate of natural increase is well above that for the whole country. But in spite of a relatively high rate of natural increase, the region lost slightly in total population during the period 1955–1960, indicating an important net out-migration, chiefly of people in the younger age groups seeking employment in the great industrial areas farther south.

Tohoku's population is strongly oriented toward the primary industries, especially agriculture. No other large subdivision of the country has such a sizable proportion of its population engaged in the primary industries, with about 55–60 per cent employed in agriculture alone. On the other hand, the percentage employed in the secondary industries (manufacturing and construction) is far below the country average and is the lowest for any large subdivision of Japan. Manufacturing alone employs only about half as large a percentage of Tohoku's population as it does of the whole country's. Obviously the population is more strongly rural in Tohoku than in the rest of the nation.

While nearly 64 per cent of Japan's people live in official *shi,* or cities, the comparable figure for Tohoku (1960) is only 47 per cent. According to the more sophisticated index of urban residence used elsewhere in this book, the DID population for all Japan amounts to 43.7 per cent; and for Hokkaido, 42 per cent; but for Tohoku it is only about 24 per cent. There are only five DID cities of over 100,000 population (1960), four of these in the 100,000–200,000 class, and one, the metropolis of Sendai, with 336,000. Most of the cities are primarily market centers serving local areas. Specialized industrial cities are rare, Kamaishi, the iron and steel center on the Pacific coast in Iwate Prefecture, being a major exception. In all Tohoku there is no port city of consequence. Aomori, the ferry port at the northern extremity is the most important of the few that are open to foreign trade, and even it is chiefly engaged in domestic shipping.

AGRICULTURE AND OTHER
PRIMARY INDUSTRIES

By any yardstick, agriculture, together with its auxiliary primary occupations of forestry and fishing, is dominant in Tohoku. Percentage of the land in forest resembles the national average, but still Tohoku does not produce its proportional share of the lumber, although the reverse is true of charcoal. And although the forest industries are widespread, there is a degree of concentration of lumber production in Akita Prefecture in the northwest, and a still stronger concentration of charcoal output in Iwate Prefecture in the northeast. Charcoal production in Tohoku exceeds that of any other large subdivision of Japan, representing over one-quarter of the national output.

With marine products, as with lumber, the region does not produce its proportional share, but that is attributable to the small importance of Japan Sea fisheries, for actually the Pacific coast of Tohoku is nationally renowned for the value of fish landed. Sardines, mackerel, and swordfish are the chief varieties taken in the coastal waters, but in addition there are more than half a dozen ports along the east side which are important bases for high-sea fisheries.[1]

Mineral production scarcely distinguishes Tohoku, although parts are not without some fame. Some 6–8 per cent of the nation's coal is produced, chiefly in the Joban coastal field in the extreme southeast, while 50–55 per cent of Japan's meager domestic oil output is derived from the Akita-Yamagata fields on the Japan Sea side. About one-third of the nation's pyrite ore and copper are obtained from scattered mines in Tohoku, as is nearly one-quarter of the lead ore. About 60 per cent of the iron sand taken from terrace and beach deposits originates in northeastern Honshu in Aomori and Iwate prefectures, while the Kamaishi mine in Iwate Prefecture is the single most important source for domestic iron ore. Potential water power is comparatively abundant, representing one-third of the national total, but development has lagged, so that normal output is under 15 per cent. Chief center by far, of both potential and developed hydroelectric power, is the Agano River basin, much of whose upper parts, more especially the Tadami tributary, lie within Fukushima Prefecture in southernmost Tohoku, and it is here also that large-scale expansion programs are in progress.

Agriculture in Tohoku differs from that of older subtropical Japan in a number of respects, most of them related to climate. In Tohoku the subtropical climate characteristic of the south and central parts of the country gives way to a more severe continental type, which acts to eliminate certain subtropical crops and to generally reduce agricultural output per unit area. To compensate for lowered production there is an increase in the average size of farms, so that instead of the 0.6 to 1.0 hectares characteristic of subtropical Japan, here in the north farms average 1.2 to 1.6 hectares. This is intermediate in size between those of Hokkaido farther north and subtropical Japan to the south. But unlike Hokkaido farms, a representative Tohoku farm is not a single compact piece of

land, but instead is composed of scattered fields, with the farmstead ordinarily located in a village. Since most of Tohoku lies beyond the northern limit of winter cropping of the paddy lands, a great majority of the
rice fields are allowed to lie fallow in the colder months; the winters are
too severe and long and the growing season too short to permit a dry
crop to mature between the fall harvest and spring planting of rice.
Only about 3 per cent of the paddy area is planted to winter crops, and
much the largest part of this is a green manure crop such as Chinese
vetch, or *genge*. Since the boundary line marking the approximate
northern limit of winter-cropped paddies is located at about latitude
38°, only southeastern Tohoku has a significant amount of fall-sown rice
fields. There, in Fukushima Prefecture, nearly 9 per cent of the paddy
area bears a second crop. By contrast, fall sowing of upland fields is
much more widespread, roughly 45 per cent of the ordinary fields being
so utilized. But here it is the contrast between east and west sides that
is striking, for while about 56 per cent are double cropped in the eastern prefectures of Tohoku, in those facing the Sea of Japan, where the
snow cover is both deep and long continued and the plains are poorly
drained, the comparable figure is only 25 per cent. The regional discrepancy is far greater if green-manure crops are omitted and only fall-sown
cereals are considered. In extreme northeastern Tohoku, as in Hokkaido,
small grains are planted both in spring and fall, but over the region as a
whole cereals other than rice are fall sown.

Tohoku's transitional character between north and south is emphasized by the number of boundaries marking the northern limits of certain crops or agricultural practices which are located within this region.
Two such boundaries—one marking the northern limit of important winter cropping of paddy fields and another separating an eastern region
where 50–60 per cent of the upland fields are winter cropped from a
western region where the practice is much less common—were mentioned
in the preceding paragraph. Other instances are numerous. For example, the sweet potato, which is a mainstay among food crops in subtropical Japan and is still important in southern Tohoku, has become a
very minor crop in northern Tohoku and practically disappears in
Aomori Prefecture. About latitude 39° may be accepted as the approximate northern limit of important sweet potato cultivation. This same
latitude is also close to the northern limit of bamboo, while north of parallel 38° or 39° mulberry-growing and raw silk production become of
minor importance, although southern Tohoku, by contrast, is a region
of silk specialization. Southernmost Tohoku fairly coincides with the
northern limit of citrus culture, and tea likewise is almost wholly absent.

Compensating somewhat for the absence of citrus is the greater emphasis upon deciduous fruits, especially apples. A further contrast is the fact that the horse is a much more common draft animal on the farms of Tohoku, while draft cattle are more prevalent farther south.

As applied to agriculture the name "Tohoku type" until very recently signified an above average prevalence of antiquated farming systems. It is a type associated with a generally backward region, the character of whose farming has been handicapped by a relatively harsh climate, including low temperatures and heavy snowfall, by remoteness from the great urban markets, and by the greater strength and persistence of the feudal system with its prevalent landlordism, so that until recently there existed a great majority of poor tenant farmers and a minority of rich farmers. Accordingly, social stratification and relationships reminiscent of feudal times continued longer here than elsewhere.

Among the characteristic elements of what has been called the Tohoku Type of agriculture are the following: Compared with other regions of Old Japan, farming is more extensive, the average acreage of arable land operated by a farm household amounting to nearly twice that in subtropical Kinki. One crop a year is the rule, and the rate of arable-land utilization (113 per cent in 1960) is the lowest for any large regional subdivision, excepting Hokkaido. Farming is relatively simple, with emphasis on annual crops of cereals and vegetables, while livestock and perennial bush and tree crops are slighted. Subsistence farming is fairly common, less attention being given to special commercial crops. Paddy rice occupies a larger proportion of the cultivated area, and the practice of using permanent seedbeds still persists. Farm equipment is inferior, and production both per unit of area and per unit of labor is relatively low.[2]

Since the war the tempo of agricultural modernization in Tohoku has accelerated as a consequence of the national Agrarian Land Reform laws and the price stabilization of rice and other food cereals resulting from the national government's rice-collection quota and staple-food distribution system. Small farms have been increasing in number and farming is becoming more intensive. With the abolishment of absolute landlordism, a vast majority of the nation's farmers have come to own all or a part of the land they cultivate, with all the benefits which this implies; and in Tohoku, which originally had an abnormally high ratio of poor tenant farmers, the beneficial effects of land reform and price stabilization have been more marked than elsewhere. Moreover, Tohoku, which earlier emphasized subsistence farming, since the war has developed a higher ratio of commercial farming and a lower percentage of self-sufficient farming than the national average. In addition, the employment of day and seasonal labor has been increasing as rice farming has become more intensive, so that the percentage of farm households employing day laborers and also annual laborers is now considerably higher for Tohoku than for Japan as a whole. This in part, of course, reflects the larger size of farms. There has also been a remarkable increase in the productivity of cultivated land, and especially of paddy land, so that at present Tohoku is above the national average in rice production per unit area. Significantly, too, the income per worker in primary industry has risen more rapidly than has the national average, while the income per farm household is well above the national average, and is one of the two highest of the ten large subdivisions of Old Japan.[3] Thus, Tohoku's farming is undergoing important change, although it still suffers unavoidably from harsh climatic conditions and a retarding inheritance from the feudal past.

Rice is indigenous to the wet tropics and subtropics, so that when it is found to be far and away the foremost crop in a region of such severe climate as Tohoku, where it occupies 60–65 per cent of the cultivated land and furnishes 80 per cent of the farm income, it is a matter for wonder. Although Tohoku represents only 18 per cent of the nation's area, it accounts for 19 per cent of its rice acreage and 21 per cent of its rice yield. This pushing of rice northward into what are, for it, marginal climates has inevitably brought it very close to its critical climatic limits, so that it becomes susceptible to damage from cold disasters involving chiefly low summer temperatures and reduced sunshine, resulting in partial or even complete failure of the crop. These cold disasters come to northern Japan about once a decade, on an average. Short spells, as well as longer periods, of low summer temperature in northern Tohoku are associated with a cold northeast wind known as the *yamase*. Some yamase effect is felt every summer, but disaster strikes when it prevails for a long period or comes with great frequency. The unseasonable summer cold appears to occur when lower-than-average water temperatures prevail in the Okhotsk Sea and a strong Okhotsk Anticyclone is positioned to the north and east of Japan with a front over its median latitudes. Such a situation causes a cold northeasterly flow of air over northern Honshu, which in turn intensifies the flow of cold water southward along the northeast coast.[4] Center of the cold damage is northeastern Tohoku, which receives the full brunt of the cool northeast winds, while the protective influence of the north-south mountains diminishes the effects rapidly westward and also gradually southward (Fig. 10-2).

Successful rice specialization—so successful as to make Tohoku the principal region of surplus for domestic export—has been made possible partly by the breeding of new varieties which are quick-maturing and cold resistant, and partly also by the improvement of cultivation methods. Recently greater emphasis has been placed upon varieties which, in addition to possessing the qualities mentioned, are also large yielders, such as Fujisaka No. 5 and Norin No. 17.[5] In the past, the unfavorable climate caused rice yields per hectare in Tohoku to be below the national average, but lately the Tohoku yield, amounting to 4.4 tons per hectare, has exceeded the country-wide average of 3.9 tons, so that this inhospitable northeastern region now boasts the highest unit-area yields of rice for any of the country's large subdivisions. In other words, their obsession with rice has caused the Japanese to bend their efforts to adapting this favorite crop to an unfavorable environment rather than changing their basic agricultural pattern and choosing other crops better suited to the climatic environment.

Improved methods of rice cultivation, also a factor in the increased yields in Tohoku, involve such features as better seedbeds which allow for earlier sowing and transplanting of rice, the overcoming of various kinds of blights, and the improving of paddy fields, especially of their soils and their drainage and irrigation systems. Until recently rice seedbeds in Tohoku remained fallow after the seedlings were removed for transplanting, but increasingly now the beds are replanted to other summer crops. Since the war, also, a new type of rice seedbed has been more widely adopted in Tohoku, the improved type involving the use of oil paper or vinyl covers in order to raise the bed temperature and force the growth of the seedlings. By so doing, the sowing time and transplanting time are made earlier, and sturdier seedlings capable of resisting cold weather are produced. The new semi-hotbed type has found most favor in the northern and northeastern parts of Tohoku where cold disasters are more frequent and more severe.[6] Not a few farmers now also allow the irrigation water to rest for a period in a warming pond before it enters the paddy field, so that its temperature is increased.

In order to step up yields, farmers are inclined to increase their application of chemical fertilizers year after year, but this in turn has had the effect of augmenting the damage due to different kinds of blights, especially the rice-wilting disease. Recently the effects of rice wilt have become equally disastrous, or even more so, compared with damage due to cold. As a consequence, varieties that are resistant to blight resulting from the cumulative effects of chemical fertilizers are being favored.

Modest evidence is accumulating, also, pointing to a slight weakening of the strangle-hold monopoly long held by cereals in Tohoku. Dairy and beef cattle are increasing in numbers, and milk production more and more supplements the farm income. By contrast, horse-raising, which earlier was a specialized development in some of the upland areas, especially the piedmont alluvial-diluvial belts on either side of the Central Range, has declined. As these upland areas of wild pasture have been increasingly reclaimed for cultivation, fewer horses have been pastured on the natural grasslands, a larger percentage being raised on farms. Tanabe[7] finds that in Japan a numerous horse population is characteristic of regions with a less advanced type of agriculture, while on the other hand cattle are indicative of a higher stage of agricultural development. At present in Tohoku the proportion of farm households keeping one or more dairy cows is less than 10 per cent, although more than one-quarter have draft-beef cattle, whose total number is over twice the number of milk cows. Dairy cattle are only 60–70 per cent as numerous in Tohoku as in Hokkaido. Altogether, livestock are still a minor feature of Tohoku's agriculture, but the trend in numbers is upward.

The growing of mulberry and raising of silkworms are important adjuncts of general agriculture in the southern basins of Tohoku, especially in Fukushima and Yamagata prefectures, but this profitable side line has waned markedly here and everywhere in Japan in recent decades. Thus, between 1929 and 1955, the acreage of mulberry declined over 40 per cent in Fukushima Prefecture and over 30 per cent in Yamagata. Still, southern Tohoku remains one of the nation's more specialized areas of sericulture. At present mulberry is almost absent in Aomori and Akita prefectures (in extreme north and northwest), is modestly important in Iwate and Miyagi (farther south and east), more so in Yamagata, and truly specialized in Fukushima (in the extreme southeast). Thus, Fukushima ranks third among all Japanese prefectures in mulberry acreage, while Yamagata is sixth. As the mulberry acreage has waned, it has been replaced in part by an expansion of orchards and vegetable crops, both of which provide additional cash income.

MANUFACTURING

Only a meager 7 per cent of Tohoku's labor force is engaged in manufacturing, which bespeaks the modest development of that economy. Traditional manufactures handed down from the premodern period, such as textiles, paper, and pottery, are widespread, but large-scale modern industries have made slower progress here than in most other regions, and no genuinely important manufacturing district, or even single center, has developed. Three local industrial areas, each containing a few dispersed small centers, may be mentioned: (1) Sendai-Shiogama-Ishinomaki, which in addition to the ubiquitous food industries has several rubber-products establishments, a steel plant, and a number of plants processing marine products; (2) Fukushima-Koriyama, with textile and machine industries; and (3) Yonezawa-Yamagata, specialized in textiles. A few individual cities, such as Kamaishi, Hachinohe, and Akita, likewise represent modest concentrations of manufacturing. Tohoku suffers from a variety of industrial handicaps: most of it is well removed from the flourishing metropolitan regions of central and southwestern Japan; raw materials and developed power resources are only modest; ports are underdeveloped; land communications are inefficient; and the accumulation of local capital is meager. Altogether, Tohoku somewhat resembles Hokkaido, in that it has been accorded a semicolonial status within the nation's economy.

Hasegawa[8] has summarized the main characteristics of Tohoku's manufacturing as follows: (1) It has the slowest rate of development of any of the large regions of Japan. (2) It is unbalanced, in that producer's goods lag well behind consumer's goods. (3) Both wages and productivity of industrial workers are low. (4) There is great heterogeneity of manufactures, together with a prevalence of small enterprises and a paucity of large factories belonging to well-known companies. (5) Premodern social relations still survive in the industries. These characteristics are reflected in the composition of Tohoku's manufacturing as shown in Table 10-1. Compared with the over-all national pattern, Tohoku is weighted heavily in terms of foods, lumber, and wood products, while it is markedly deficient in textiles and machine industries of all kinds, including shipbuilding.

T A B L E   10-1

Composition of manufacturing in Tohoku, 1960, in per cent of value added

Foods	17.8
Chemicals	14.5
Machinery (all kinds)	11.3
Lumber and wood products	10.3
Iron and steel	9.3
Ceramic, stone, and clay products	6.8
Non-ferrous metals	6.1
Pulp and paper	5.6
Textiles	5.4

Source: Ministry of International Trade and Industry, *Census of Manufactures, 1960: Report by Cities, Towns, and Villages* (1962).

TABLE 10-2

Composition of manufacturing in the more important industrial cities ot Tohoku, 1960, in terms of value added (unit: 10,000 yen)

Industry group	Sendai	Akita	Kamaishi	Hachinohe
Food	195,000	80,000	38,000	154,000
Pulp and paper	....	205,000	....	....
Publishing	141,000	53,000	....	....
Chemical	....	260,000	....	200,000
Rubber products	102,000	....	....	....
Ceramic, stone, and clay products	....	....	....	72,000
Iron and steel	....	....	597,000	135,000
Non-ferrous metals	....	48,000	....	....
Fabricated metals	243,000	....	....	....
Machinery	209,000	32,000	....	23,000
TOTAL	1,132,543	793,529	674,497	628,234

Source: Ministry of International Trade and Industry, *Census of Manufactures, 1960: Report by Cities, Towns, and Villages* (1962).

Tohoku's relative poverty and underdeveloped nature are suggested by the fact that personal income here is only three-quarters that of the national average, while bank deposits are only 3.5 per cent of the nation's total, even though its population amounts to 10–11 per cent.[9] The national government has taken cognizance of this retarded condition in northeast Honshu and has initiated multipurpose land development projects to hasten exploitation of the area. These projects look toward an increase in agricultural and forestry production, further exploitation of mineral resources, expansion of hydroelectric generation, and prevention of natural damages. Four regions for special attention have been designated within Tohoku, viz., Tadami, Mogami, Kitakami, Ani-Tazawa. Good progress has been made in the development of hydroelectric power in the Tadami special area, while in the Kitakami drainage area there has been fair progress in the construction of multipurpose dams designed to prevent flood damage and to assure the supply of irrigation water. In the other two areas for special attention progress has been unsatisfactory.[10]

GEOGRAPHICAL SUBDIVISIONS

As was done for Hokkaido, special emphasis will be given to the plains areas, which are the centers of occupance. Only such hill and mountain areas as are particularly distinctive or important will be singled out for individual treatment.

The eastern meridional lowland and its basins

This greatly elongated depressed zone extends beyond Tohoku, northward into Hokkaido, where it coincides with the Ishikari-Yufutsu Lowland, while its southward prolongation can be traced in Kanto Plain and Tokyo Bay.

The northern parts of Shimokita Peninsula, composed of hard-rock hills and volcanoes, are joined to the mainland by what is largely an upland plain—the *Sambongi*—some 1600 square kilometers in area, only about one-fifth of which is alluvial lowland. Sambongi is composed of elevated fluviatile and marine sediments of diluvial age, which on their land side merge with and slope down from the ash aprons of several volcanoes of the Central Range. Rivers have incised wide alluvium-floored valleys into the diluvium-ash upland, while other fragments of lowland are located back of the belt of beach ridges and dunes that form the smooth crescentic coastline. Much of the new alluvium was originally poorly drained, and large areas still remain as swampy wasteland, although a considerable proportion has been reclaimed and is in rice. In such locations the paddy landscape is monotonously uniform, for the rural villages tend to seek drier sites along the margins of the wet valleys. This whole area suffers seriously from cool *yamase* winds in summer, occasionally experiencing a 40–50 per cent decrease in the rice harvest as a consequence.

On the undulating to rolling diluvium-ash upland surface, cropped areas—chiefly dry fields, but also some paddies—are intermingled with moor and woodland. Chief upland crops are barnyard millet, wheat, buckwheat, beans, soybeans, and potatoes grown under a rotation system of three crops in two years.[11] Apple orchards become conspicuous in southernmost Sambongi close to the Iwate prefectural boundary. In this area of relatively new frontier settlement the rectangular pattern of roads and land subdivision resembles that of Hokkaido, although the dispersed farmsteads are largely lacking.

Because of the relatively inhospitable environment, combining summers of a coolness that is occasionally disastrous and low grade Ando soils derived from volcanic ash, the Sambongi Upland was late in developing and was used instead as natural pasture within a specialized horse-breeding area. Some private reclamation for individual farms began as early as 1885, but large-scale reclamation under government supervision did not occur until after 1938. As a result of the latter some 22,900 acres were newly brought under cultivation (14,600 of dry fields

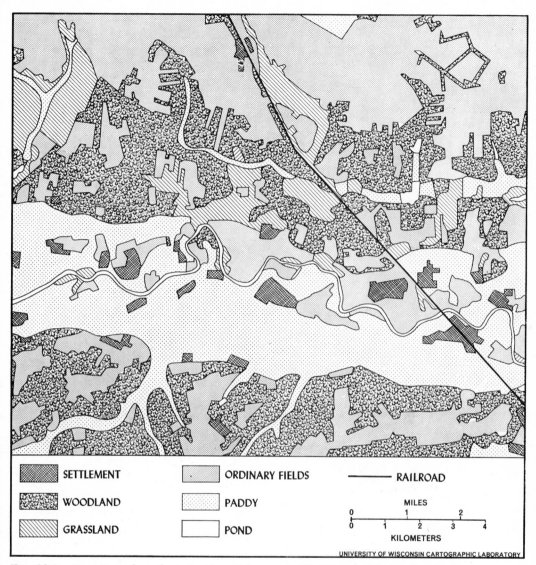

Fig. 10-3.—A portion of southern Sambongi, between 40° 30′ and 40° 40′ N., where land reclamation is well advanced. From Japanese land-use map, 1:50,000, in color.

and 8,300 of paddy) and over 5000 families, most of them farmers, have been settled in the area of new colonization.[12] Reclamation and settlement activity was at its peak shortly after World War II, when large numbers of repatriates were seeking employment. But since the termination, in 1957, of the Central Government's reclamation project, new agricultural settlement has greatly slowed down and this decreased tempo

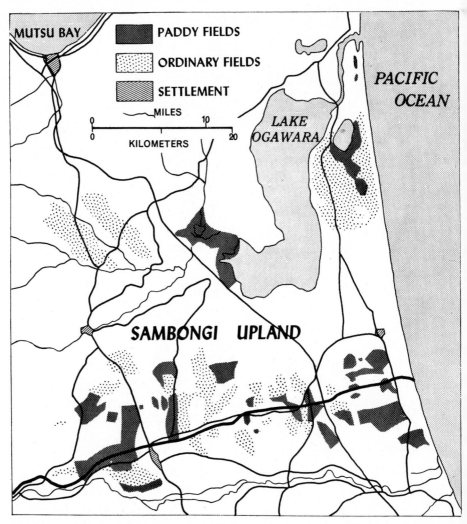

Fig. 10-4.—Distribution of reclaimed lands on Sambongi Upland. After map in *Regional Geography of Japan*, No. 2, *Tohoku Guidebook*.

will likely continue. Compared to what is usual for plains in Old Japan, population density is low in Sambongi, but on the other hand it is markedly higher than on similar upland plains of Hokkaido.

Metropolis of the Sambongi area and one of the leading industrial cities of Tohoku is Hachinohe (104,000 population), located in the southeastern part near the mouth of the Mabechi River. It is an old castle town which is composed of several parts—the business section, Hachinohe, situated a few kilometers inland; the industrial sections of Minato and Konakano, closer to the sea; and the ocean port of Same. As well as being

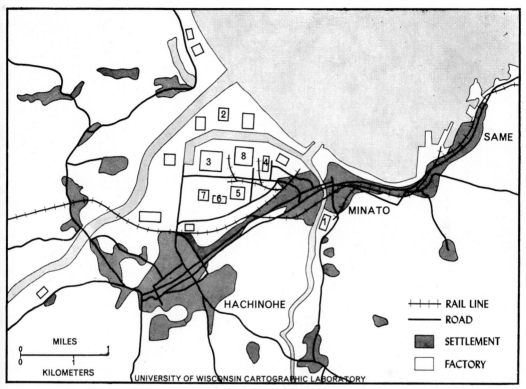

Fig. 10-5.—Hachinohe city, showing port and distribution of main factories: (1) cement, (2, 3, 5, 6, 7) iron and steel, (4) ferro-alloys, (8) chemicals. After map in *Regional Geography of Japan*, No. 2, *Tohoku Guidebook.*

a small port and a major fishing center, it held fourth rank in manufacturing among the urban places of Tohoku in 1960. Seventeen of its factories are of more than average size, including five small pig iron factories, four processing ferro-alloys, one making cement, two canning marine products, one each making *sake*, cans, and chemical fertilizer (Fig. 10-5). In addition there is an electric generating plant and one manufacturing gas from coal.[13] Based upon value added by manufacturing, the ranking industries, in order of importance, are chemicals, food products, iron and steel, and ceramic, stone, and clay products. The metallurgical industries benefit from local supplies of iron sand, while the necessary coal for fuel is obtained from Hokkaido via Muroran.

The *Kitakami Basin*, drained by the southward flowing Kitakami River, is a meridional tectonic depression about ninety kilometers long from Morioka at the northern end to Ichinoseki close to the Miyagi prefectural boundary in the south where the river enters a gorge. It is con-

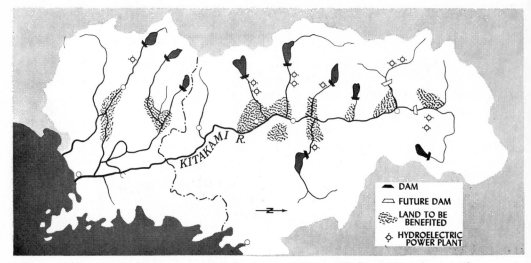

Fig. 10-6.—Multipurpose development project in Kitakami River basin. After map in *Nippon no Chiri*, Vol. 2.   No scale given.

tinued northward beyond Morioka in the form of the relatively narrow valley of the northward-draining Mabechi River. In this latter section ash and lava deposits have so constricted the lowland that continuous settlement has been discouraged, and what exists is located chiefly on the river terraces. Upland fields predominate, for rice is grown by only 10 per cent of the farmers. Noteworthy commercial crops are apple orchards, tobacco, and cabbage, with the latter occupying about 10 per cent of the upland fields.[14]

Only south of Morioka does the Kitakami Valley take on basin dimensions and expand into a lowland some 2800 square kilometers in area. Because this basin has been so frequently damaged by floods, the national government has set aside an area of about 12,000 square kilometers in western Iwate and northern Miyagi prefectures as a special development area, in which chief attention is paid to the prevention of floods, but supplementary plans have been made to organize the development of various resources, looking toward the expansion of mining, agriculture, and manufacturing (Fig. 10-6).

Terrain features of the Kitakami Basin are complicated. East of the fault line which marks its western margins, and occupying one-half to two-thirds of the basin, is an extensive alluvial-diluvial piedmont belt, in which fan configuration and multiple terrace levels are conspicuous.[15] These bulky piedmont deposits have acted to shift the master stream to the eastern side of its basin where it flows in a floodplain rarely more than a few kilometers wide which abuts against the foothills of the Kitakami Highland. The eastern side of the basin is nearly devoid of diluvial and fan deposits, this asymmetry of the alluvial piedmont having been mentioned in an earlier section.

Land use in the basin is varied and complicated. On the new alluvium of the floodplain and its terraces, although paddy rice is dominant, dry crops are far from being absent. In the slightly elevated and uneven riverine zone, emphasis is upon such upland crops as vegetables, mulberry, orchards, and winter grains. Since the paddies are rarely sown to winter cereals in these latitudes, rice can be both planted and harvested fairly early, so that it may be already ripening by mid-August, which is considerably sooner than in those parts of southwestern Japan where double cropping of paddies is practiced. Although dissected by streams, the elevated alluvial-diluvial piedmont belt has considerable areas of flattish to rolling upland surface, large parts of which are under cultivation. Rice as well as the usual upland crops of vegetables and cereals is grown, the artificially terraced paddies occupying a variety of levels and receiving their irrigation water in part from numerous man-made ponds. Mulberry is of small importance and apple orchards are numerous. Iwate Prefecture grows about 5 per cent of the apples of the country, with the principal concentration in the Kitakami Basin and northward along the Mabechi Valley. Most apples are raised in conjunction with general farming, although there are some specialized apple farms. Hard-rock foothills, riverine belt, and alluvial piedmont all provide sites for orchards. On the alluvial piedmont, apple-growing is most concentrated in those parts where paddy rice, rather than dry crops, dominates. For example, on the Toyosawa Fan, which is mainly planted to dry crops, apple orchards are relatively few.[16] Patches of woodland, scattered plots of orchard and mulberry, and numerous isolated farmsteads surrounded by hedges and windbreaks give the alluvial piedmont belt a confused and cluttered appearance, so that extensive panoramic views are difficult to obtain. Dispersed rural settlement appears to be common on the piedmont, and although it is less so on the floodplain, where compact rural villages are the rule, even there it is by no means absent.

Toward the southern end of the basin, south of about Maezawa, the belt of alluvial piedmont and diluvial terrace disappears and the basin terminates in an even-crested Tertiary tableland, possibly an erosion surface, which is dissected by a labyrinthine system of valleys. This dissected upland, known as the Iwai Tableland, separates the Kita-

kami Basin from the Senpoku-Sendai Plain just to the south. The Kitakami River negotiates this barrier by means of a gorge too narrow for
the recurring large volumes of flood water, with the result that flooding
of lowlands is common in the southern part of the basin. Only by means
of a series of short tunnels is the main north-south rail line able to negotiate the Iwai barrier.

In the gap between the two segments of the eastern highlands, Kitakami and Abukuma, the eastern meridional lowland reaches down to
tidewater, and there the extensive *Senpoku-Sendai Plain,* the product
of the southward flowing Kitakami, the northward flowing Abukuma,
and several other rivers, has developed. On it is located the largest population cluster of entire eastern Tohoku. The whole plain may be subdivided into two parts separated by the Matsushima Hills, the larger northern part being known as Senpoku and the smaller southern part as the
Sendai Plain.

Senpoku resembles somewhat the southernmost part of the Kitakami Basin in that it is a series of extensive wet floodplains separated by
irregular masses of hills, so that the whole has a complicated and labyrinthine character. Much of the alluvium is poorly drained and subject
to flood, and there is even some peat bog. Stream channels on the plain
are elevated and flanked by natural and artificial levees. Because the alluvial interfluves are so poorly drained and also are subject to inundation, unirrigated crops are scarce and the whole region is a monocultural
paddy area with monotonously rectangular fields. For the most part,
villages of the cultivators are concentrated on dry sites, such as levees,
or at the base of the adjacent hills (Figs. 10-7, 10-8). Linear-type villages are numerous. Riverine belts of dry crops form conspicuous variations from the paddy landscape.

Smaller Sendai Plain, located south of the Matsushima Block, consists of a narrow belt of alluvium backed by a hilly upland, the latter
complicated in origin and materials, but having some aspects of an erosion surface with a veneer of gravels. Diluvium is present but it is limited.

Senpoku-Sendai Plain terminates seaward in a prevailingly smooth
coast with beach ridges, except where interrupted by the cliffed headlands of the submerged Matsushima Block which encloses shallow island-
studded Matsushima Bay. The picturesque Matsushima archipelago of
irregular pine-clad islands is acclaimed one of the "Three Great Sights"
of Japan, famous in art and literature, and is a resort area as well. On
the parallel beach ridges and dunes, both north and south of Matsushima,
dry crops prevail, although the outermost ridge usually supports a long

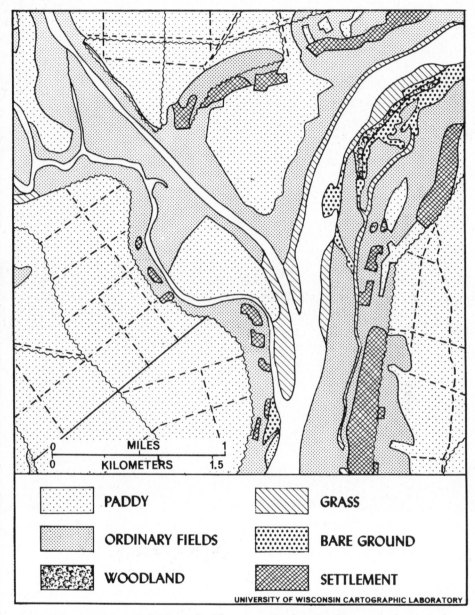

Fig. 10-7.—A portion of the Senpoku Plain near Sendai. On this poorly drained plain, largely planted to rice, settlements and unirrigated crops tend to concentrate on the slightly higher ground near the stream channels. From Japanese topographic sheet, 1:50,000.

Fig. 10-8.—The poorly drained Senpoku Plain in eastern Tohoku, in the vicinity of Sendai. Settlements are concentrated in the somewhat more elevated and drier zone close to the stream. Courtesy of Geographical Survey Bureau, Japan.

wall of conifers which acts as a windbreak and a barrier to drifting sand.

Sendai city (336,000 population), the metropolis of Tohoku, is likewise this northeastern region's prime cultural, economic, political, and manufacturing center. It is an old castle town, a prefectural capital, and a university center, but its outstanding function is that of a wholesale-retail center, serving not only the eastern meridional lowland, but in a general way the whole of northern Honshu. Seriously damaged by bombing and burning during the war, it has been rebuilt, with wide streets and numerous imposing six- to nine-story Western-style buildings in the downtown section, where it has a genuinely metropolitan appearance. Although having no distinctive reputation as a manufacturing city, still it far outdistances any other Tohoku processing center. Mainstays of Sendai's diversified industrial structure are fabricated metals (21 per cent of value-added), machinery (17 per cent), foods (17 per cent), publishing (12 per cent), and rubber products (9 per cent).

Shiogama, situated on a shallow drowned valley on Matsushima Bay, besides being an important fishing base, is a minor commercial port serving Sendai and the tributary Senpoku Plain. A dredged channel about eight kilometers long permits small ships of a few thousand tons to enter and discharge such bulky cargoes as coal, petroleum, and cement.

Although it is the valley of a single northward-draining river, the *Abukuma Lowland* is not quite continuous, for a low divide separates the northern Fukushima Basin from the Koriyama Basin farther south. Still farther south this same tectonic depression is continued in the northern arm of the Kanto Plain.

A conspicuous fault scarp marks the western and northern side of the *Fukushima Basin,* but such a tectonic line is more difficult to trace on the eastern and southern side. Along the base of the northwestern scarp there has developed a piedmont belt of alluvial-diluvial fans showing conspicuous terrace levels, thereby resembling Kitakami (Fig. 10-9). In the southwestern part the piedmont belt is so extensive that the river is pushed eastward against the bordering eastern hills and only a very limited amount of floodplain exists. In the northeastern part fans are much less conspicuous, although broad terraces are still present, and the river here flows in a wide meander-scarred floodplain almost midway in the basin between the mountain borders.[17]

Natural and cultural features resemble in many ways those in the mid-portion of the Kitakami Basin, except that in this more southerly depression winters are milder and snowfall less, mulberry occupies a greater area, population is somewhat less dispersed, and more paddy fields are winter cropped. Not only on the hill lands surrounding the basin and on the diluvial piedmont, but also on the recent floodplain, mulberry is an important crop, while filatures and cocoon warehouses in the villages testify to a specialized sericulture industry, for the Fukushima-Koriyama depression is one of the country's more important raw-silk areas.[18] To be sure, it is much less specialized in mulberry and sericulture now than was the case several decades back, for like other silk regions Fukushima has suffered a large decline in mulberry acreage, with orchard crops

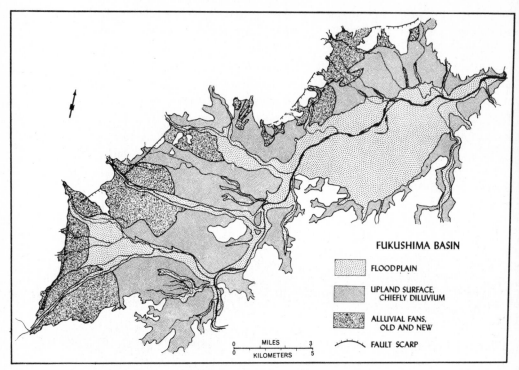

Fig. 10-9.—Terrain types in the Fukushima Basin. After map by Kenzo Fujiwara.

(apples, peaches, pears) and vegetables in a measure having taken its place.[19]

Fukushima city (74,000 population), prefectural capital and metropolis, and a local marketing and manufacturing center, has a diversified industrial composition in which some degree of emphasis is given to ceramic-stone-clay products, foods, steel, and publishing.

In the broader lineaments of its geography there is little to distinguish the *Koriyama Basin* from its smaller northern neighbor. Diluvial piedmont deposits occupy much the larger part of this southern basin. Until about 1873, virtually the only areas of the Koriyama Basin that were importantly used for agriculture were the floodplains of recent alluvium, for the western diluvial piedmont was still largely in its natural state. But shortly thereafter reclamation and settlement of the diluvial uplands were begun, and this was accelerated when water for irrigation was brought in from Lake Inawashiro just to the west. Koriyama city (74,000 population), the basin's metropolis and market center, formerly had some fame as a silk-spinning town but as that industry waned the manufacturing structure has become more diversified, with chemical and machine industries replacing filatures, although textiles do continue to be important.

The Median Groove and its basins

Lying between the Central and Western ranges of Tohoku is a meridional depressed zone, less continuous than that to the east, occupied by a series of six to ten basins, whose number can vary depending on how the several subdivisions are counted. In size they vary from about 200 to 900 square kilometers. Because of interruptions by highland masses the Median Groove is not followed by a continuous rail line from north to south as is the eastern lowland. Most of the basins are occupied by the headwaters of streams crossing the Western Range in antecedent valleys, which have become the routes of rail lines. As a consequence, these inland settlement areas are chiefly tributary to, and the hinterlands of, the delta cities and ports along the Japan Sea. This is less true of the southernmost basins, however, which have fair access to the great Kanto cities. All the basins within the Median Groove have larger or smaller areas of floodplain floor, but piedmont belts of alluvial and diluvial fan deposits are also conspicuous features, as they are in the eastern lowlands. Characteristically the major piedmont deposits have been formed by streams descending from the Central Range and so are best developed along the eastern margins of the depressions.

The climate of these median basins as a group is difficult to generalize because of the localisms that prevail, stemming from contrasts in site and exposure. Conditions differ from one basin to another and even between stations within the same basin. Average cold-month temperatures do not appear to be significantly different, however, from those in the eastern meridional basins, although both interior locations are a few degrees lower than along the western littoral. Protected as they are by the Central Range, these western basins are somewhat less affected than those to the east by the cold *yamase* winds of summer. Total annual precipitation varies greatly, for among twenty-six stations located within the Median Groove it ranges from a low of 1160 millimeters to a high of 2065 millimeters. A similar diversity is to be observed in the annual march of precipitation, for of the same twenty-six stations, eleven exhibited a cold-month maximum, and fourteen a warm-month, while in one case they were equal. What becomes obvious is that for the region as a whole a typical monsoon type of seasonal precipitation distribution is absent, for winter is equally as wet as summer. Emphatically spring is the driest season. In the annual march of precipitation, therefore, the

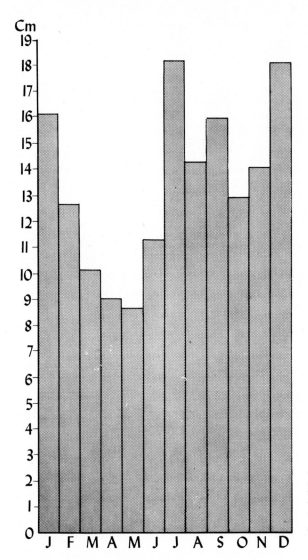

Fig. 10-10.—Annual march of precipitation in basins of Median Groove, Tohoku. Composite of 28 stations; average total precipitation, 1617 mm.

Median Groove resembles the Japan Sea coast and is unlike the region to the east of the Central Range, where the warm season is also much the wettest season. The same regional alignment prevails when days with snowfall and depth of snow cover are considered. During January and February almost every day has snowfall, and snow lies several feet deep on the ground, much deeper than to the east of the Central Range, and also probably deeper than on the plains facing the Japan Sea. Dark, gloomy winter weather is the rule. In these snowy regions of western Tohoku, despite numerous snowsheds, rail traffic is frequently suspended for days at a time. In some villages where the snow is excessively deep, people are forced to live in the upper stories of their homes if they wish to enjoy much daylight. Special forms of architecture are common, such as wide eaves and covered sidewalks, called *gangi*, which when fitted with temporary outer walls in winter form corridors through which pedestrians may walk, protected from the weather.

The basins of the Median Groove contrast somewhat with the eastern depressions in that they show a further shrinking of what even in the east was only modest winter cropping, not only of rice fields, but of upland fields as well. North of about the 39° parallel almost no fall-sown cereals are grown even on upland fields, and even to the south of 39° these crops are not of much importance. Because of the long-continued and deep snow cover, any fall-sown crop would have to be planted so early in fall and harvested so late in spring that the period of growth for summer crops would be dangerously shortened. Mulberry, and sericulture in general, are greatly reduced in importance to the west of the Central Range, but as in the east, there is a distinct increase toward the south, so that the Median Basins of southernmost Yamagata Prefecture are modestly specialized in mulberry.

Each of the small basins of the Median Groove does have individuality, to be sure, yet space in this book does not permit the kind of detailed analysis required to make the aggregate of qualities belonging to each

TABLE 10-3

Mean maximum depth of snow on the ground at selected stations in Tohoku (in centimeters)

Station	Nov.	Dec.	Jan.	Feb.	Mar.	Apr.
West Coast						
Akita	11	20	38	45	36	–
Sakata	1	27	45	51	34	3
Tsuruoka	3	45	72	92	74	6
Median Groove						
Kosaka	2	32	70	93	73	18
Odate	2	28	46	55	41	36
Yokote	4	42	75	81	109	14
Yamagata	2	30	48	57	49	3
Yonezawa	5	77	135	163	135	36
Wakamatsu	3	40	58	37	4	–
Eastern Longitudinal Lowland						
Morioka	3	19	34	40	30	4
Fukushima	–	14	18	17	9	2

Source: Snow Association of Japan, *The Climatography of Snow in Japan* (Tokyo, 1949).

stand forth clearly. One valid generalization seems to be that the northern basins are not so well populated and appear more backward in their economies than those farther south, with less complete and intensive use of agricultural land, less diversification of crops, and a greater lag in the development of the secondary industries. No doubt climatic contrasts play some part in this differentiation, but of even greater significance is the degree of remoteness from the great Kanto metropolitan area.

The *Aomori Plain* at the extreme northern end of Honshu, the only one of the western intermontane lowlands fronting on the sea, consists of a narrow crescentic strip of paddy-covered alluvial lowland which is bordered on its land side by a dissected diluvial bench where woodland, staple upland crops, and apple orchards intermingle. Aomori city (133,000 population) is primarily a ferry and fishing port and the northern terminus of the Honshu rail lines. Frequent boat service, carrying both passengers and freight, connects it with Hakodate and Muroran in Hokkaido. Industries are essentially of local importance.

Next to the south, the *Hanawa-Odate Basin*, which occupies the upper portion of the Noshiro drainage basin, is largely composed of diluvium, portions of it smoothly sloping upland, and other parts in the form of bench terraces. Poor drainage is characteristic both of the floodplain and of flattish parts of the diluvial upland surface. To avoid periodic inundation, rural settlements tend to concentrate on the lower terrace levels. Paddy fields on the terraces depend on numerous ponds for their irrigation water. The higher diluvial surfaces are not well utilized agriculturally, and continue to have extensive areas in woodland. What little industry exists chiefly utilizes the abundant local forest resource and is specialized in lumber and wood products, such as barrels, boxes, furniture, and skis.

Yokote, largest of the western intermontane basins, occupies the upper drainage basin of the Omono River. High volcanic masses separate it from the northern depressions, hence only a roundabout rail connection with them is made by way of the coastal lowlands along the Japan Sea. An eastern diluvial piedmont zone shows varying degrees of utilization, with the gravelly upper levels still retaining extensive woodlands, while the lower elevations are largely given over to dry-crop agriculture, with the settlers living in dispersed farmsteads. On the new alluvium, where rice predominates, the denser population is congregated in compact as well as semidispersed settlements. Definite riverine belts, with dry fields, parallel some of the streams. The very numerous small settlement units, the riverine zones, and the maze of roads and irrigation channels, many of which are lined with trees, give the plain a confused and cluttered appearance.

Occupying middle and upper portions of the Mogami River Valley are the *Shinjo, Yamagata,* and *Yonezawa* basins. Shinjo, one of the smallest and least populated of all, is chiefly composed of diluvial piedmont deposits.[20] Its minimal areas of low-lying new alluvium suffer seriously from floods, a damaging flood having occurred on the average of once every three years between 1900 and 1954. Only 8.5 per cent of the basin area is under cultivation, a low figure, indeed, even for Tohoku. Considering the predominance of diluvial upland, it is surprising to find one crop, paddy rice, occupying about 70 per cent of the cultivated land, a situation which is truly monocultural. The former specialization in horse-raising, with the animals chiefly pastured on the wild grasslands of the higher fans and terraces, has declined as more of the upland surface has been reclaimed and brought under cultivation.

Farther upstream is the larger and more compact *Yamagata Basin,* whose rural population density is the highest anywhere within the Median Groove. Three large fans form a conspicuous piedmont belt on the east side, which long remained uncultivated because of its stony infertile soils and deficiency of water for irrigation.[21] A greater part of the surface of one of the eastern fans is now occupied by the city of Yamagata, while extensive areas on the other two have been reclaimed for upland fields, including apple orchards and mulberry. As the acreage of the latter crop has declined in recent decades, it has been replaced in part by cash crops such as apple orchards and vineyards.[22] Paddy fields largely monopolize the extensive area of low-lying new alluvium except in the slightly roughened and higher riverine belts where orchards and mulberry are both conspicuous. It is worthy of note that while the Yamagata Basin was, and still is in a diminished degree, a specialized silk region with a large acreage of mulberry, in Shinjo Basin just to the north mulberry is greatly reduced in importance. Seemingly a significant cultural boundary is located in the Shinjo-Yamagata vicinity. Yamagata city (99,000 population), an old castle town and the prefectural capital, formerly had fame as a filature center, as did other towns of the basin, but all aspects of sericulture have declined in importance, while metal and machine industries have multiplied; so that currently these sectors represent about half of the total value added by manufacturing. Small processing establishments greatly predominate.

Still farther upriver in the Mogami drainage basin is the *Yonezawa Basin,*[23] which has some distinction owing to the fact that here the depositional piedmont zone lies on the western side of the lowland and not at the base of the Central Range, as is usually the case. Many of the general features of land utilization noted for the Yamagata Basin are repeated here but with local variations. Certain restricted localities are highly specialized in such commercial crops as tobacco, apples, grapes, and pears, some of these specializations having an environmental basis, but others not. Yonezawa city (44,000 population) has some fame as a local textile center, with silk, cotton, and rayon all represented.

Farthest south, and located within the drainage of the Agano River are the *Aizu* and *Inawashiro* basins, whose composite geographic features do not greatly distinguish them from Yamagata and Yonezawa just to the north within the Mogami drainage system, from which they are effectively separated by formidable highlands. Most of the Inawashiro Basin is occupied by an extensive lake resulting from blocked drainage by a mud flow from an adjacent volcano. Delta fans along the lake margins support a narrow interrupted belt of cultivated land.

The western plains

Along the Japan Sea littoral is a series of three or four aggradational plains which occupy portions of insinking kettle-like depressions associated with volcanic activity. In two of them, Iwaki and Shonai, the basins are partially filled with huge symmetrical ash-and-lava cones, which tower above the adjacent lowlands. Each of the plains is at the sea end of an antecedent river whose sediment has been deposited in the shallow waters of lagoons back of wide belts of dune-capped beach ridges or bars. Drainage is generally poor, and in some of the plains large areas of swamp and shallow lake persist. Strong waves and currents generated by the boisterous winter monsoons have tended to smooth the coast, developing wave-cut cliffs where hard-rock highlands reach the sea, and barrier beaches and bars along the alluvial portions. Natural harbors are few, and the two or three ports that have developed are of only minor importance.

Like all of western Tohoku, these plains are afflicted with dark, stormy, snowy winter weather. The snow cover is deep and continuous, though perhaps less so than in the Median Groove or on the plains of Hokuriku farther south, where the highland barrier to the rear of the coast is higher. Winter winds are strong, frequently blowing with gale force. At Akita the wind velocity during three winter months is twice that at Tokyo on the Pacific side of the country. It is common practice in coastal settlements to weight the shingle roofs with boulders, while piles of fuel wood are lashed down to prevent their dislodgment. The unsubstantial dwell-

Fig. 10-11.—Covered sidewalks, called *gangi,* are characteristic of some rural settlements along the snowy Japan Sea side of Tohoku and in Hokuriku farther south.

ings along the shore and on the dunes seem poorly equipped to withstand this onslaught of the winds, despite the windbreaks in the form of hedges, walls of trees, and lattices filled with brushwood or moss. Fishing is less important along this Japan Sea coast than on the Pacific side of Tohoku, partly because of the rougher seas and lack of harbors.

The poorly drained alluvial lowlands are almost exclusively planted to rice, which is one of the important domestic exports of the region. In Akita Prefecture 80 per cent of the cultivated land is in paddy fields. Because climatic handicaps discourage winter cropping, small grains, which are moderately important to the east of the Central Range, are unimportant on these western plains. Unirrigated crops occupy cleared patches on the diluvial terraces, riverine strips along the principal streams, and wide belts on the series of parallel beach ridges which border the coasts.

Iwaki (Hirosaki) Basin,[24] northernmost of the western basins and also the largest (920 sq. km.) and most populous, well represents their common characteristics, except for its high degree of specialization in apples. It is an alluvium-floored structural depression whose sea end is blocked by a wide belt of dune-capped beach material. No port has developed along its exposed, smooth, harborless coastline; Aomori, only thirty to forty kilometers distant, serves in that capacity. Even fishing settlements are almost completely lacking. The lower northern end of the basin back of the belt of dunes is wet, and here the paddy areas are interspersed with patches of shallow lake and peat bog. Completely

Fig. 10-12.—Along the west coast of Tohoku and Hokuriku where winter winds are strong, house roofs are often weighted down with boulders.

surrounding the alluvial floor, except at the sea end, is a border of low diluvial upland, variable in width and covered with a veneer of volcanic ash. Back of the diluvial bench rise the hills and low mountains which enclose the depression and force the railroads to pass the eastern and southern barriers by means of tunnels. Above the western margin of the valley towers the symmetrical ash cone of Mt. Iwaki (1625 meters), which helps to break the force of the winter gales that fill the basin with snow during the colder months.

Population decreases toward the wetter, more exposed northern end of the basin. Except for the wide eaves and covered sidewalks, or *gangi,* there is little evidence that house construction and architecture have been especially adapted to the severe winter weather. Metropolis of Iwaki is Hirosaki (69,000 population), an old castle city which occupies a strategic position on a high prong of diluvial terrace and an adjacent portion of the alluvial lowland, where several valleys converge in the southern end of the basin. It is a market center with no industries of consequence.

The floor of the depression is largely devoted to rice paddy, except in the sandier riverine belts of variable widths along the main streams, where the terrain surface is somewhat uneven, being scarred with relict forms left by migrating stream channels. Here a variety of unirrigated crops are raised, notably apples and potatoes, grown as commercial crops, and rapeseed, some of which appears to be fall sown. In the lower and wetter northern end of the basin a rice monoculture prevails; elsewhere there is a slightly greater variety of crops. The diluvial terraces with

Fig. 10-13.—Iwaki Basin in northernmost Tohoku. Paddy fields occupy the new alluvium on the western, or left-hand, side. Apple orchards are numerous on the diluvial upland and foothills in the central and eastern parts. Settlements are concentrated among the zone of contact between alluvium and diluvium. Courtesy of Geographical Survey Bureau, Japan.

their poor ash soils are in part covered with scrubby trees or a moor-like vegetation, although large areas, particularly in the south end of the valley, are utilized for orchards and other unirrigated upland crops (Figs. 10-13, 10-14). On these sites apple orchards are most extensively developed and irrigation ponds are numerous.

About 48 per cent of the apple orchard acreage of Japan is in Aomori Prefecture, and the larger part of this is concentrated in Iwaki, giving that basin a degree of national fame, for it is the largest and most spe-

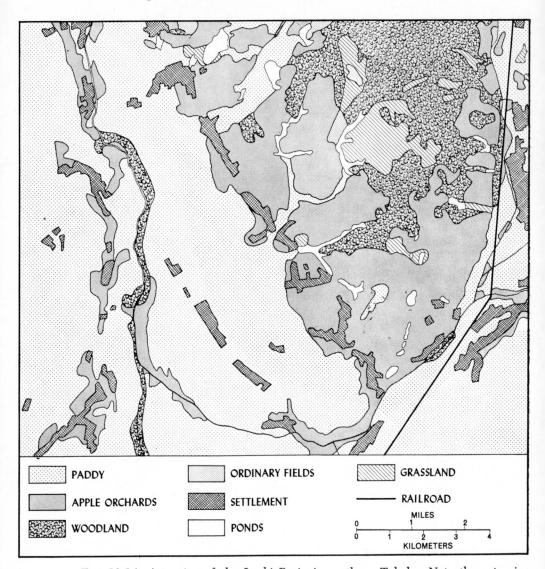

PADDY	ORDINARY FIELDS	GRASSLAND
APPLE ORCHARDS	SETTLEMENT	——— RAILROAD
WOODLAND	PONDS	

Fig. 10-14.—A section of the Iwaki Basin in northern Tohoku. Note the extensive areas of apple orchards, chiefly on the diluvial uplands bordering the alluvial lowland. A smaller concentration of orchards is on the riverine belt on the lowland.

cialized apple-growing area in Japan. Still, full-time apple farmers comprise only 10 per cent of the total, and 72 per cent of the orchards are smaller than 1.2 acres.[25] Orchards occupy three kinds of sites: the lower slopes of the mountain foothills, diluvial uplands, and riverine belts. Lowest yields and shortest productive longevity characterize the slope orchards.[26] For long it has been the practice, using hand labor, to in-

:ase each young apple in a bag made from newspaper to protect the fruit
from fungus and insect blemish, but with the more widespread use of
chemical insecticides this very time-consuming practice is declining.

The *Noshiro-Omono Plain* is composed of two small aggradational
plains located at the mouths of the Noshiro and Omono rivers, together
with narrow strips of coastal sediments which enclose Hachiro Lagoon
and connect hilly Oga Peninsula with the mainland. It is thus a composite plain and lacks compactness. The northern part, or Noshiro Plain,
is composed largely of low diluvial terrace, part of it uncultivated. Its
seaward margins have the usual belt of beach ridges and dunes, which
are sparsely occupied and meagerly utilized. A few of its depressions
have pond-irrigated rice, and there is a little dry cropping, fields of vegetables and Japanese pears being most numerous, especially along the
inner margins of the dune belt, which is also the site of a highway and
a number of villages. Between the coastal zone of beach ridges and the
landward belt of diluvial terrace is a restricted area of alluvium, which
is devoted to rice, only the riverine zone being excepted. Noshiro city
(33,000 population), at the mouth of the Noshiro River, is an anchorage
port for coastwise steamers, which call chiefly for cargoes of logs and
lumber. The city is a sawmill center, with over 200 small establishments
turning out a variety of wood products, including window frames, doors,
and barrels.[27]

The *Omono*, or *Akita, Delta Plain* resembles in many respects that of
the Noshiro River just to the north. Akita city (124,000 population), the
metropolis, prefectural capital, and an old castle town, is situated inland
about four kilometers, behind the protective belt of pine-capped beach
ridges and dunes. Akita port, formerly known as Tsuchisaki, lies nearer
the coast on the Akita Canal, about eight kilometers to the northwest.
Its harbor is so shallow that only small ships of up to 3000 tons can be
accommodated, and for the five or six months of the winter monsoon the
exposed harbor is practically unused. Cargo is chiefly imported Hokkaido coal and exported petroleum products, for from Akita's oil fields
are derived 50–55 per cent (1961) of the domestic petroleum of Japan.
The principal oil field, Yabase, stretches north-south between Akita
and its port of Tsuchisaki. At Akita city is one of the few noteworthy
concentrations of industry within Tohoku. Two distinct centers exist,
one in Akita proper and the other in the port area. In the latter location
is a large plant making railroad equipment, a small oil refinery processing chiefly local oil, and a number of other smaller establishments
making machines, rope, and oil drums. Within Akita proper are two relatively large plants manufacturing fertilizer and wood pulp, and in ad-

dition smaller ones refining ores and making machines (Fig. 10-15).[28] Two manufacturing groups—chemicals, representing 33 per cent of the total value-added, and pulp-paper, with 26 per cent—stand out sharply. Favorable locational factors for industry are an ample supply of industrial water and an abundance of hydroelectric power.

The two sandy strips of beach ridges and dunes joining hilly Oga Peninsula to the mainland and enclosing between them Hachiro Lagoon have a marked linearity and parallelism of both their natural and cultural features. Some of the inner ridges show a considerable amount of dry-crop cultivation, with rice occupying the intervening swales. Villages following the ridge tops have greatly elongated dimensions. A branch line of railroad follows the southern belt of beach ridges from Akita to Oga town, thirty kilometers distant on the southern side of Oga Peninsula. The latter is a supplementary port which is used chiefly in winter when the unprotected river-mouth harbor of Akita is unfit for anchorage. As with Akita port, Oga's cargo is chiefly Hokkaido coal and locally supplied petroleum.

The largest land reclamation and agricultural settlement project now in progress in Japan involves diking and draining shallow Hachiro Lagoon, with an area of 223 square kilometers. Target date for completion is 1965. About two-thirds of the lake's area will be converted into arable land while one-third will be a reservoir for impounding flood waters and from which irrigation water may be drawn. Of special importance is the fact that the new land is preponderantly for paddy; over 13,000 hectares are to be so used, while only 940 hectares are designated for ordinary fields. Some 700 new agricultural households will be settled on the reclaimed Hachiro lands, and in addition over 7000 neighboring farm families already there will have their holdings enlarged.[29]

The *Shonai Plain,* product of the Mogami River, is a compact lowland composed almost entirely of new alluvium, except for the extensive belt of sandy beach ridges and dunes along its seaward margins. In this dune area numerous newly reclaimed farm areas are shielded from the strong sea winds and drifting sand by long rows of protection forest. On the lowland proper, rice dominates except in the riverine zones. Sakata, a minor river-mouth port, has all the disadvantages of an open roadstead harbor located on a stormy coast. Shallow waters permit its use by only small ships. Tsuruoka, an inland city with a feudal castle, has long been famed as a silk-weaving center.

On the northern margin of the Shonai Plain is the lordly volcanic cone of Chokai (2230 meters), the lower and flatter slopes of whose ash apron are partially under cultivation. Along the plain's northwest margins,

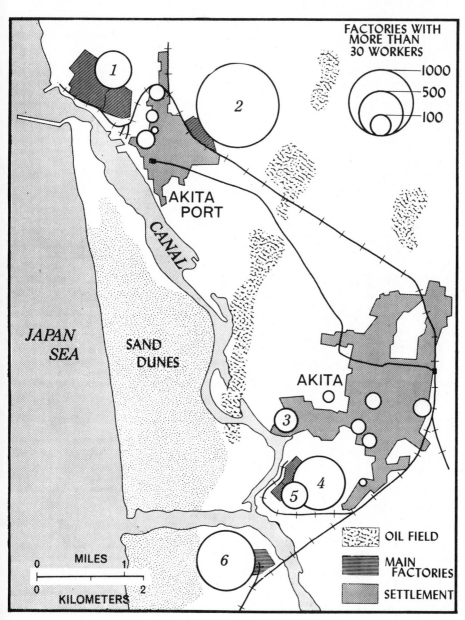

Fig. 10-15.—Akita city and environs, showing locations of factories: (1) oil refinery, (2) railroad equipment, (3) machinery, (4) fertilizer, (5) metals, (6) pulp and paper. After map in *Regional Geography of Japan*, No. 2, *Tohoku Guidebook.*

dissection is so far advanced that a labyrinthine valley pattern has resulted, the floors of which are covered with paddy fields.

Some highland areas

Compared with the plains, only a few of the Tohoku highlands are sufficiently distinctive and important geographically to warrant individual comment. Among these are the two elliptical shaped hill masses of Kitakami and Abukuma in eastern Tohoku. *Kitakami,* in the north, composed of complicated old sedimentary rocks and ancient intrusives, is an incompletely peneplaned surface which later experienced elevation and subsequently was reduced to slopes by normal stream erosion. Remnants of the old erosion surface still persist at about 1000 meters elevation, and monadnock peaks rise several hundred meters higher. The more elevated interior portion is best described as subdued mountains; elsewhere it is rugged hill country. The northern two-fifths of the coast, having experienced emergence, is smooth in contour and is characterized by multiple marine terraces with abrupt wave-cut fronts. Along parts of this coast iron sands are mined from the terrace deposits. By contrast, the southern Kitakami coast has undergone subsidence resulting in a deeply indented ria-type coastline where fishing is a main occupation of the modest population clustered on the tiny bay-head delta plains.

Population is sparse in the Kitakami hill land, and agriculture is largely of the subsistence upland type with some local emphasis on the pasturing of livestock, especially horses. It has the reputation of being one of Japan's poorest and most backward regions, and this is reflected in the forlorn and shabby appearance of the semidispersed agricultural settlements.[30] Cool summer weather, which so greatly reduces the importance of rice, stimulates the growing of such climatically resistant crops as buckwheat, barnyard grass (*hie*), soybeans, and millet. *Hie*, a type of millet which is a dependable producer even in the coolest summers, is the key crop.

Backward Kitakami does not strike one as a likely area for large-scale industry, yet along its barren coast at Kamaishi, situated at the head of a long indentation, is one of the nation's centers of pig iron and steel. Moreover, this was the first Western-style blast furnace–steel mill center in Japan, established in 1874 under direct government management. Isolated even for Tohoku and well removed from the nation's important metropolitan market areas, the Kamaishi plant owes its location to the proximity of the single most important domestic supply of iron ore. The ore body is located in hilly country about twenty kilometers inland from the port of Kamaishi, with the ore beds outcropping along the flanks of the hills some 400–500 meters above the nearest drainage level. But large amounts of foreign ore importantly supplement the local supply, while coal and coke are brought by ship from Muroran, in Hokkaido, and from outside the country as well. Relatively, Kamaishi has declined in importance during the postwar period, as market-oriented steel centers in central and western Japan have burgeoned. Like other steel plants in Japan, this one at Kamaishi is also in a port location, but oddly its site is about one and a half kilometers removed from the harbor.

Like Kitakami, the more southerly *Abukuma* hill land is an uplifted and dissected peneplain of complex structure. It differs from Kitakami in that (1) it is composed chiefly of granite rather than old sedimentaries; (2) its elevations average only about half as high; (3) it is bordered on its sea side by a belt of upland plain; and (4) it contains many fault valleys. Average elevation of Abukuma is only about 400 meters, although the highest points reach nearly 1000 meters. The upland surface is strongly rolling but scarcely to be described as rugged. Peneplanation having been more complete than in Kitakami, there is greater uniformity of upland levels, and larger remnants of the erosion surface remain.

Reflecting its lower altitude and latitude, its more moderate slopes, wider valleys, and its comparative proximity to Kanto, the density of population in Abukuma is nearly twice that of Kitakami. Along its western margins are several small depressions where agricultural settlement distinctly thickens. At Hitachi in southernmost Abukuma is one of the country's seven most important mines, its output including copper, silver, gold, and sulfur. Not far removed from the Hitachi mine is a smelter and refining plant.

Along its sea side and lying east of the abrupt fault scarp which defines its eastern margins, Abukuma is bordered by a belt of low upland plain some eight to ten kilometers wide, and usually under 150 meters elevation. The several terrace levels probably owe their origin to both wave and river erosion.[31] Dissection by streams has opened up numerous wide valleys in the diluvial-Tertiary platform and made uneven the terrace surfaces. In places the upland terminates in sea cliffs along the coast, but at the mouths of the valleys there are belts of beach ridges and dunes, capped with windbreaks of picturesque conifers. Paddy landscape typifies the floodplains, while a variety of dry crops occupy both the sandy sea margins of the valleys and the least roughened parts of the interstream upland surfaces. Woodland prevails on the steeper slopes. Numerous artificial ponds provide supplementary irrigation water for the lowland paddies. A major highway and railway follow this coastal belt, connecting the Kanto cities with Sendai, and Aomori farther north.

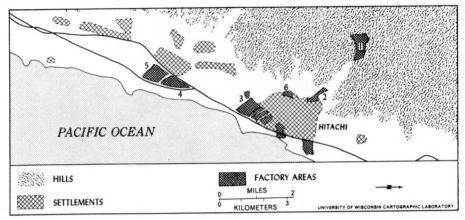

Fig. 10-16.—The Hitachi industrial district: (1) ore refinery, (2, 3, 4, 5) electrical equipment and machinery, (6) electric wire. After map in *Nippon no Chiri*, Vol. 3.

The northern and more rural part of the Abukuma coastal belt, known as the *Hamadori Plain,* has a population density only one-quarter that of the southern half which contains the Joban coal field and its associated industrial development. At Joban some 4 million tons of low-grade bituminous coal are mined annually, or about 8 per cent of the country's total. In dimensions the coal field extends eighty kilometers from north to south and varies in width from four to twenty kilometers. While some seventy collieries are in operation, only three have an annual output exceeding 300,000 tons.[32] Northern Joban is much the more important part, accounting for 90 per cent of the total output, containing all of the larger mines, and employing close to 90 per cent of the labor force. Four large collieries employ 61 per cent of all the workers.[33] Located less than 200 kilometers northeast of Tokyo and connected with it by a double-tracked rail line, the Joban coal field finds the chief market for its output in the Keihin industrial center.

But closer at hand is the smaller Joban-Hitachi industrial district, which is likewise powered by Joban coal. Here there exists a zone of dispersed smaller industrial centers, by all odds the most important of which is Hitachi, in the extreme southernmost part of the coastal belt and hence in the northeastern part of Ibaraki Prefecture, which more precisely belongs to eastern Kanto rather than to Tohoku. While this local Joban manufacturing district is not without variety in industrial composition, since it includes coal-based chemical industries, smelters and refineries of non-ferrous ores, cement plants, and those making general machinery, its genuine specializaton is in things electrical, including wire, appliances, machinery, and equipment, much of it housed in large, modern factories (Fig. 10-16).

South and west of Tohoku is old subtropical Japan, the Japan of legend and tradition, of ancient occupance and mellowed living. In contrast, Hokkaido and, to a less degree, Tohoku are the "provinces," where many of the features considered typical of Japan either are lacking or are present only in modified form. This subtropical Japan, the land of warmth and climatic bounty, of tiny farms, terraced fields, and extensive multiple cropping; of tea gardens, bamboo, orange groves, and cherry blossoms, lies principally south of the 36° or 37° parallels.

The generalizations and descriptions for Japan as a whole which have been provided in Part I apply more often and more specifically to subtropical Japan than to the two northern subdivisions, which represent departures from the base level or standard for things Japanese. To avoid repetition, therefore, no over-all summary like those in the chapters on Hokkaido and Tohoku is provided here for subtropical Japan. As a convenience for organizing description, the region is here separated into two principal subdivisions—Central and Southwestern—and they in turn are further subdivided into smaller sectional units. But these Central and Southwestern subdivisions have no such regional distinctiveness as do Hokkaido and Tohoku, so that for them, also, any general description is abbreviated in character. It will be recognized immediately that not a few of the sectional subdivisions are determined largely by terrain character. Within each of them there may be numerous local areas, many possessing national importance and striking individuality, and it is with these local areas, principally lowlands, that the remainder of the book chiefly deals. As a rule the highland subdivisions, of sparse population, receive little or no attention.

CENTRAL JAPAN

In Central Japan (98,759 sq. km.) there is attained the maximum breadth and maximum altitude of Honshu Island. Watanabe recognizes this broad, high, and rugged area of Central Japan as one of the major landform divisions of the country, and what degree of unity the region may have is largely due to its terrain character. In turn, the terrain character, to an important degree, results from the fact that this is not only the junction point of the two subdivisions, Northeast and Southwest, of the main Japan Arc, but also the juncture of the Japan Arc as a whole with the Bonin or Shichito-Mariana Arc entering from the south, which is represented chiefly by a line of Pacific islands. Reflecting these complex tectonic structures are numerous lesser terrain features, in addition to those of breadth and altitude noted earlier. Not only altitude and land relief are great, but also adjacent ocean-basin relief, for Suruga Bay on the Pacific side and Toyama Bay along the Japan Sea exhibit unusual depths of more than 2000 meters at no great distance from the coast. Such pronounced total land-ocean relief of over 5000 meters within relatively short horizontal distances reflects the complex and strong orogenic forces operating within this region.

Most conspicuous topographic boundary within Central Japan is the Itoigawa-Shizuoka Tectonic Line, with its associated fault scarps, which extends in a nearly north-south direction through the mid-section from Mt. Fuji and Suruga Bay on the Pacific side to a point somewhat east of Toyama Bay on the Japan Sea. This line and its associated fault scarps separates southwestern Japan with Mesozoic and older rocks from northeastern Japan where mountains are chiefly of Tertiary rocks and volcanoes are numerous. West of this tectonic line and its great scarps, there rise in echelon the Hida, Kiso, and Akaishi mountain ranges, together designated the Japanese Alps, whose highest peaks rise to 3000 meters. Of the three ranges forming the Japanese Alps, the Akaishi is distinguished from the other two in that it is an eastward extension of the Outer Zone of Southwest Japan and like it is composed of folded sedimentaries. Geological boundary between the Akaishi and Kiso Mountains is the Median Dislocation Line, also marked by fault scarps, and

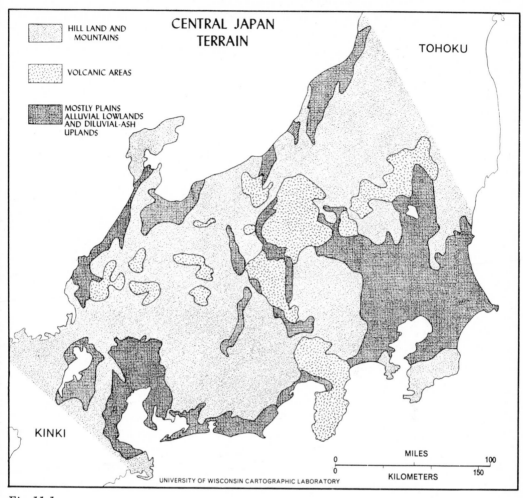

CENTRAL JAPAN
TERRAIN

HILL LAND AND
MOUNTAINS

VOLCANIC AREAS

MOSTLY PLAINS
ALLUVIAL LOWLANDS
AND DILUVIAL-ASH
UPLANDS

TOHOKU

KINKI

MILES
0   100
0   150
KILOMETERS

UNIVERSITY OF WISCONSIN CARTOGRAPHIC LABORATORY

Fig. 11-1.

the elongated Ina Graben. Within the Japanese Alps, cirques of Pleistocene glaciation are found on peaks above 2500 meters.

East of the great Itoigawa-Shizuoka tectonic and fault line runs the
Fossa Magna Depressed Zone, extending in a north-south direction from
the Pacific to the Japan Sea. Although geologically a continuous depressed zone, morphologically it is far from being an uninterrupted
lowland, for most of the depression is filled with recent Tertiary and
volcanic masses, the southernmost of which is the Fuji Volcanic Group.
All of the important fault basins of central Honshu, with the exception of the Ina Graben mentioned previously, lie within the Fossa Magna

Zone. Alluvial cones and fans in the form of piedmont belts occupy much of these basins.

Three great north-south embayments—Tokyo-Sagami, Suruga, and Ise —indent the Pacific side of Central Japan. Alluvial filling at the heads of Tokyo and Ise bays has resulted in the formation of two of the largest and most important plains of Japan—Kanto and Nobi. At the head of Suruga Bay the great volcanic masses of the Fuji Volcanic Group come down to the sea margins, so that a bay-head lowland is absent there. A series of smaller alluvial and diluvial plains fringe the littoral between these larger lowlands. Along the Japan Sea coast is the extensive Echigo (Niigata) Plain and the smaller Toyama, Kaga, and Fukui plains.

Four major subdivisions of Central Japan are usually recognized: (1) Tosan (Nagano, Gifu, and Yamanashi prefectures), embracing much of the highland core and including the Japanese Alps and the Fossa Magna and its basins; (2) Kanto, with seven prefectures, located along the eastern Pacific margins; (3) Tokai, comprising three prefectures along the western Pacific margins; and (4) Hokuriku, including four prefectures, facing the Japan Sea. Important differences exist between the Hokuriku and Pacific sides. Thus, climatically Hokuriku with its overcast, windy, snowy winter weather stands in contrast to the Pacific side where winter is sunnier and drier and there is no durable snow cover. The Pacific side is a part of modern industrial Japan, characterized by high population densities, great urban and industrial development— including two of the nation's three greatest industrial nodes—and a strongly commercialized agriculture. By contrast the Hokuriku side is less modernized and more provincial, with a slower tempo and somewhat lower standard of living. It is often referred to as the "back door" of Japan, and in many ways it resembles western Tohoku. Population densities are lower, no great urban-industrial centers or port cities have developed, manufacturing is less, foreign trade is insignificant, agriculture is almost monocultural in rice, and commercial crops are few. In winter, heavy snows, low visibility, and stormy seas seriously handicap all forms of transport.

TOSAN AND ITS BASINS

Tosan, the roof area of Central Japan, is a region of prevailingly mountainous relief and steep slopes. Various sections of the Tosan Mountains differ from each other, to be sure, and they vary in their degree and kind of occupance, but population and production are very much concentrated in a half dozen detritus-choked fault basins. A map

Fig. 11-2.—Alpine scenery and picturesque thatched roofs in a mountain basin of Tosan. Courtesy of Consulate General of Japan, New York.

of population distribution clearly defines the locations and boundaries of the six basins. Consequently, an analysis of the Tosan ecumene is essentially an analysis of these basins, each a few hundred square kilometers in area and containing a few hundred thousand inhabitants.

Conspicuously sloping alluvial and diluvial piedmont belts, most of them composed of coarse materials, make over half of the basin floors unsuitable for rice cultivation, so that home-grown rice is insufficient for local needs. Nevertheless, rice output per unit area is here the highest in Japan, a seeming anomaly considering that the altitude induces relatively cool summers and a short growing season. High rice yields are attributed in large measure to the intensive farming methods, involving a large labor input and a generous use of fertilizer, but partly also the size of the yields is related to the abundant summer sunshine, high day temperatures, and a large and reliable supply of irrigation water from the surrounding highlands.[1]

Tosan is one of the two silk centers of Japan, its three prefectures accounting for 26 per cent both of the nation's mulberry acreage and of its output of cocoons. Mulberry fields, which occupy large areas on the steeper alluvial and diluvial fans, the river levees and terraces, and the lower hill slopes, represent 16–18 per cent of the crop land in Nagano and Yamanashi prefectures. Formerly the acreage of mulberry was much larger and the emphasis on silk greater, but the impact of the Pacific War and the great decline in the overseas markets for raw silk have been accompanied by a marked reduction in mulberry acreage and in silk-reeling operations. Within the three Tosan prefectures mulberry acreage in 1960 was only about one-third what it was in 1930. As mulberry declined, its place has been taken by such crops as wheat, potatoes, and orchard fruits; filatures have been supplanted by a great variety of light industries, including precision instruments, small machines, and light metals. Despite the changing situation, however, mulberry and silk are still highly important in the region and the main source of income for Tosan farmers is from the sale of cocoons and fruit, but because of the small size of the farm unit (0.4 to 0.8 hectares, or 1 to 2 acres) farmers are relatively poor in spite of the diversified commercial farming. And while the rise in substitute manufactures, such as machines and precision instruments, has partially offset losses in employment and income from a decline in filatures, there has not been complete compensation.

Five of the Tosan basins are located within the Fossa Magna Depressed Zone, while the sixth, the Ina Graben, lies farther west in the Kiso Mountains, associated with the Median Dislocation Line and its fault scarps.

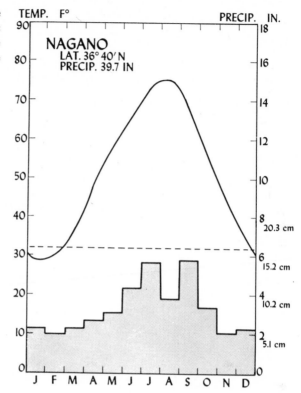

Fig. 11-3.—Temperatures are lower and precipitation less in the elevated basins of the Fossa Magna than on the coastal lowlands.

Three of the basins of the Fossa Magna—from north to south, Matsumoto, Suwa, and Kofu—lie at or near the base of the great fault scarp which defines the western margins of the depressed zone. Two others, Nagano and Ueda, lie close to the eastern margins of the Fossa Magna, also in close proximity to fault scarps, but of lesser magnitude. The basins are separated from each other by hill and mountain masses, chiefly volcanic and Tertiary.

These intermontane basins, varying in elevation from 250 to 800 meters, have many characteristics in common, both natural and man-induced.[2] Generally their climates are more continental than those of the coastal lowlands, with temperatures lower and precipitation less abundant. In most of the basins the average January temperature is a degree or two below freezing; so that a humid continental *Da* climate prevails. August temperatures are three to five degrees Fahrenheit below those of Tokyo and Osaka, and the growing season of 180–200 days is 25–50 days shorter. Precipitation averages 40–50 inches, which is modest for subtropical Japan. To be sure, there is a fair amount of climatic varia-

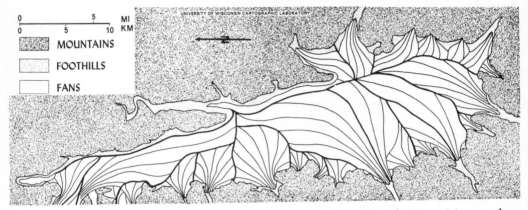

Fig. 11-4.—Matsumoto Basin, representative of those within the Fossa Magna, where steeply inclined fans are characteristic. After map in *Nippon Chiri-Fuzoka Taikei.*

tion between different basins, and even between individual stations within the same basin. One feature common to all is the large amount of alluvial and diluvial piedmont material, often coarse in texture and perceptibly sloping, which comprises much of their areas. Varying amounts of this detritus are diluvium, carved by streams into terraces at sundry levels, but fan and cone configuration is prominent in both recent and old river deposits (Fig. 11-4). Soil materials tend to be coarse or even stony, a feature which, in conjunction with the sloping piedmont surfaces, the cooler, shorter summers, and the smaller amount of rainfall, somewhat handicaps rice cultivation, although it by no means excludes it. On some of the steeper fans the paddy fields are artificially terraced, their outer retaining walls being composed of smooth, waterworn boulders excavated from the fan materials. Normally the rice fields produce only one harvest during the annual cycle, although many of the dry upland fields are planted to wheat and barley in autumn after summer vegetables and other warm-season crops have been harvested. But of notable importance is the mulberry crop, for these mountain basins of the Fossa Magna are the very heart of Japan's silk-producing region. Mountainous Nagano Prefecture, in which all but one of these grabens are located, is easily the second ranking prefecture in sericulture in Japan, having 15 per cent of the country's mulberry acreage in 1960 and a similar proportion of its cocoon crop.

Tosan's specialization in sericulture reflects in part the region's relative inaccessibility and the predominance of slope land. Raw silk, being a relatively compact and highly valuable commodity, can absorb higher transportation costs than most products. Moreover, since mulberry is soil tolerant and does not require irrigation, it is well adapted to the stony soils of hill slopes and steeper fans. The two most common mulberry locations are (1) a zone several hundred feet wide along the foothills of the bordering mountains and (2) the higher and stonier parts of the alluvial-diluvial fans, where irrigation is difficult. Because of the shorter growing season in these high-altitude basins an early-budding, quick-maturing dwarf variety of mulberry bush is commonly grown, which yields less per unit area and from which only one picking is possible. Since late spring frosts are a hazard to mulberry at these altitudes, there is greater emphasis upon summer and autumn cocoons than upon the spring crop, which for the country as a whole is more important. Partly because of the unusually high degree of specialization in silk, and the fact that fewer substitute crops can be grown here successfully, there has been less decline in mulberry acreage in these mountain basins than in the country at large.

The interior basins of Tosan, like those of Tohoku, are very conspicuous on a population map, their dense settlement standing out in contrast to the meagerly inhabited surrounding highlands. And while the Fossa Magna does not represent an easily traversed morphologic corridor with easy gradients, still portions of it are followed by transverse rail lines connecting Pacific and Japan Sea coasts via the fault basins.

Matsumoto Basin, northernmost of the three situated at the foot of the western fault scarp of the Fossa Magna (although the narrow Hime-kawa Valley does continue on to the Japan Sea coast), in most respects resembles the general description of the Tosan basins just given. Extensive alluvial-diluvial piedmont deposits, derived chiefly from streams entering from the high mountains to the west, occupy extensive areas; so that flattish floodplain is limited in extent. The metropolis, Matsumoto city (75,000 population), is an old castle town whose chief fame had been that of a filature center, but which is now diversified in industry, with machinery ranking first, followed by food products and paper.

Suwa Basin, next to the south of Matsumoto, is the smallest and highest of the five fault basins of the Fossa Magna, the altitude of its floor ranging from 760 to 800 meters above sea level. Steep fault scarps flank it on both the northeast and southwest sides. Its central and lowest section is occupied by Suwa Lake, upon whose southern and northwestern margins delta-fans are encroaching. The one on the north has a slope sufficiently steep in its upper parts to warrant being called a cone; that at the south end is much flatter. The Tenryu River, one of the largest in Japan, has its source in Suwa Lake and drains southward via the Ina Graben, reaching the Pacific in western Shizuoka Prefecture.

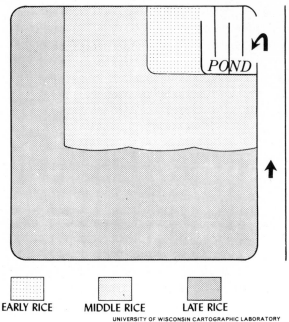

Fig. 11-5.—In mountain basins, where the irrigation water is cold, the water is circulated in a warming pond before entering the rice fields. Rice with different maturing periods may be planted in locations with water-temperature contrasts. After sketch in *Regional Geography of Japan,* No. 4, *Chubu Guidebook.*

EARLY RICE   MIDDLE RICE   LATE RICE

UNIVERSITY OF WISCONSIN CARTOGRAPHIC LABORATORY

A few decades back Suwa Basin was nationally famous as the greatest silk-reeling area in all Japan and probably in the world, for it contained in addition to first rank Okaya city two other filature centers, holding fourth and fifth rank in the country. Such a concentration of raw silk in Suwa has its roots in the past. Lying between what were earlier the cotton-producing plains along the Pacific coast and the non-cotton-producing region to the west and northwest, at a natural sag in the dividing range and near the heads of a number of radial valleys which ascend from the Pacific coast, the Suwa region after the middle of the eighteenth century became a center for the relaying, distributing, and cleaning of cotton. In the feudal period three of Japan's famous national roads converged here. After the country was opened to world trade, cotton was imported instead of grown locally, so that Suwa lost its position as a cotton distributing and processing center, and gradually it turned to another fiber, raw silk. Within Okaya city the filatures were concentrated along the Tenryu River, where the large water supply required was easily obtained and direct water power, antedating hydro-electricity, was available. Suwa city, a castle town on the lake, is famous as a resort and boasts several excellent inns. Hot sulfur springs, skiing and skating facilities in winter, and the scenery of mountains and lake all combine to make the basin attractive as a resort center.

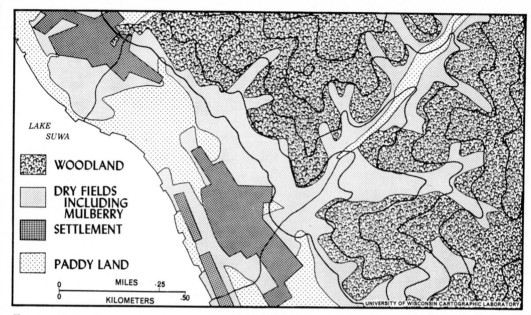

LAKE
SUWA

WOODLAND

DRY FIELDS
INCLUDING
MULBERRY

SETTLEMENT

PADDY LAND

MILES .25

KILOMETERS .50

UNIVERSITY OF WISCONSIN CARTOGRAPHIC LABORATORY

Fig. 11-6.—Land use along the eastern margins of Lake Suwa, in Suwa Basin. Paddy fields occupy the lowest land, along the lake shore and in the valley. Mulberry and other dry crops occupy the foothills and the fans at the mouth of the valley.

One unfavorable feature of the basin is that it is visited every few years by floods which do extensive damage to the rice fields on the basin floor, especially those on the flatter delta-fan at the southern end of the lake. This flooding has given rise to an unusual practice known as *tuka-dukuri,* or mound cultivation.[3] Many small circular mounds of earth, about a foot high or more and four to seven feet in diameter, dot the paddy fields. On each is planted a single fruit tree, usually quince, though some-times apple or pear; grapes and vegetables, too, may be grown on the mounds. Tuka-dukuri crops furnish the farmers food and income even on occasions when floods seriously damage the rice crop.

Principal locations of cultivated land within the basin are (1) the two large delta-fans, together with additional smaller fragments of low-land along the lake's other margins, and (2) the lower slopes of the en-circling hills. Paddy land dominates on the flatter southern delta and also on the lower parts of the northern delta-fan. Ordinary fields are rele-gated chiefly to the hill slopes and to the steeper upper part of the north-ern fan. A relatively cool summer and short growing season discourage important winter cropping of paddy lands, but much less so of the upland fields.

Like other silk-specialized areas, Suwa has suffered serious dislocations both in its mulberry-oriented agriculture and its raw-silk-oriented industry, involving a drastic reduction in mulberry acreage and a conversion of former mulberry fields largely to food crops of summer vegetables and winter grains. Since mulberry had been a crop chiefly of dry sites, such as hill slopes and the steep upper part of the northern fan, it was there that the change-over was most radical. For example, on the northern fan, 85 per cent of the 1929 mulberry acreage had been eliminated by 1960.[4] Unlike southern Tohoku, and also the Nagano Basin, Suwa did not turn in an important way to orchard crops as a replacement for mulberry.

Industrial Suwa also has suffered derangement, the silk-reeling capacity within Suwa *gun* having declined by over 90 per cent between 1928 and 1960. During World War II numerous filatures were converted to plants turning out shells, airplane parts, and optical instruments; and further change and diversification has taken place since the war. Within Suwa *gun* in 1929, silk reeling employed nearly 97 per cent of the laborers in industry and accounted for 95 per cent of the industrial output, while in 1960 the comparable figures were about 15 per cent and 12 per cent. Compensating for this decline, there was a marked increase in machines, precision instruments, and metals, from practically nothing in 1929 to 48.3 per cent of the total value of output in 1960, precision instruments alone accounting for 31.3 per cent.[5] Food products likewise showed a remarkable increase, so that at present machines, precision instruments, and foods all exceed textiles. Typical products from Suwa's machine-instrument industries are watches, cameras and their parts, auto cycles, piston rings, valves, and electronic equipment.[6] As a consequence of this type of specialization, Suwa is sometimes referred to as the Switzerland of Japan.

Many of the factories are subcontract units of larger establishments in Tokyo. Within Suwa Basin the chief centers of all industry, including silk reeling, are at Okaya and Suwa city. It is noteworthy that silk processing, the industry that declined, employed chiefly female labor drawn from the farms and so was rurally based, while the newer machine and instrument factories, developed during and since the war, are much less dependent on farm women for labor. Most obvious of Suwa's attractions for new industry is the availability from the surrounding overcrowded farms of relatively cheap but efficient labor, plus the amenities which the basin offers in the form of splendid scenery and ample opportunities for both summer and winter sports.[7]

Kofu Basin is separated from Suwa by the Yatsuga volcanic mass, so that the connection between the two basins is reduced to a relatively constricted valley. Still, there is sufficient occupance of the ash slopes and river terraces bordering the connecting valley to constitute an attenuated settlement corridor which is evident on a detailed population map. Kofu, most southern and least elevated of the Fossa Magna basins, is warmer, with less frost and snow than the others; Kofu city has only 11.7 days with snow compared to 46.5 for Matsumoto and 82 for Nagano, while its growing season is 194 days as against their 157 and 167. The basin's terrain consists of a central core of low alluvium, encircled by a depositional piedmont belt of well-defined fans which has developed at the base of the surrounding fault scarps.

Paddy rice occupies much of the lowest central part of the basin and part of the alluvial piedmont in addition, the well-preserved remnants of the Jori land division system on the lower land testifying to the early development of paddy in this area. Up until about the fifteenth century the alluvial piedmont and the adjacent volcanic slopes were left un-

TABLE 11-1

Value of products by type of industry, Suwa gun, 1960

	Value	
Type of industry	1000 yen	%
Food and kindred products	6,432,023	18.0
Textile mill products, including silk reeling	4,359,874	12.2
Apparel and other finished products	20,916	0.1
Lumber and wood products	995,301	2.8
Furniture and fixtures	401,977	1.1
Pulp, paper, and paper-worked products	179,043	0.5
Publishing, printing, and allied industries	126,265	0.4
Chemical and allied products	1,545,264	4.3
Ceramic, stone, and clay products	32,740	0.1
Leather and leather products	8,000	0.0
Iron and steel	101,000	0.3
Non-ferrous metals and products	24,496	0.1
Fabricated metal products	571,928	1.6
Machinery	3,051,465	8.6
Electrical machinery, equipment, and supplies	1,775,622	5.0
Transportation equipment	590,044	1.7
Precision instruments	11,155,727	31.3
Others	1,995,350	5.6
TOTAL	35,619,035	100.0
(Data discrepancy)	(2,252,000)	(6.3)

Source: Richard Fairchild Hough, "The Impact of the Drastic Decline in Raw Silk upon Land Use and Industry in Selected Areas of Sericultural Specialization in Japan" (Ph.D. dissertation, University of Wisconsin, 1963).

cultivated and used as horse pasture. Later, with the construction of more extensive irrigation canals, the piedmont was gradually brought under cultivation as well. Mulberry became a specialized crop on the slope land, and silk reeling evolved as the most important industry of the basin, while vineyards were of more than ordinary importance.

Changes growing out of the decline of the silk market have been important. With the marked shrinking of the mulberry area there has been an important replacement with orchards, while vineyards have recovered from their wartime slump and are expanding, so that at present one of the most distinctive features of agriculture within the basin is the prevalence of vineyards and orchards.[8] While fruit is grown both on slope sites, including fans and lower hill slopes, and on the floodplain, it is the former which are most specialized and where the recent expansion has been greatest, for there the mulberry acreage had earlier been concentrated.[9] Of unusual importance is the expansion of peach orchards, so that their current acreage equals that of vineyards. Of somewhat less importance are cherries and persimmons, and there is also a degree of specialization in the growing of melons and vegetables for the Kanto market. About one-quarter of the nation's grape production is from Kofu Basin. Vineyards usually are small, the average size of the operating unit, even in the most specialized areas, being only about 0.5 hectares, or 1.2 acres.[10] Grapes are grown for table use as well as for wine. Orchard-vineyard-mulberry-vegetable combinations are numerous and complicated, and their distribution patterns differ in various parts of the basin.

Silk reeling and wine making are the only industries of more than local significance, so that Kofu is underdeveloped industrially. Kofu city (117,000 population), the metropolis of the basin and an administrative and wholesale commercial center, serves in the capacity of middleman linking the Kanto industrial area and numerous retail outlets in Kofu towns and villages. Traditional light industries predominate, foremost among them being foods and textiles, with emphasis on silk reeling and weaving, but with other plants making wood products, glassware, wine, leather goods, and electrical appliances.[11]

Along the eastern margins of the Fossa Magna, at the base of less conspicuous fault scarps, is a north-south corridor, constricted to valley dimensions in some parts, but containing two wider basin-like sections, *Nagano* and *Ueda*. Both basins have been highly specialized in various aspects of the silk industry, but as elsewhere in Tosan, this has suffered great decline. Nagano Basin has experienced an 80 per cent reduction in mulberry acreage since the peak period of the early 1930's and a still greater loss in silk reeling.[12] As in Kofu, an important part of what was formerly mulberry land has subsequently been converted to orchards, especially apples and peaches, and to vegetable fields. Apples have become the outstanding commercial crop since the decline of mulberry. While they are grown as a subsidiary crop in many parts of the basin, as a main commercial crop they are concentrated in the riverine levee zone, on the diluvial uplands of the western and northern parts, and on the fans of the eastern part.[13] Next to Iwaki Basin in Aomori Prefecture, Nagano Basin at present is the most important apple area in Japan.

In the industry of the area, small plants manufacturing such items as electrical equipment and telephone parts have only partially filled the gap left by abandoned filatures.[14] Within Nagano city (99,000 population), metropolis of the northern basin, the ranking industries are publishing, foods, machinery, textiles, and fabricated metals; and at Ueda (39,000 population), chief city of the southern basin, machinery far outranks all other sectors, followed by foods and textiles.

Ina Trench, a much elongated graben, 72 kilometers long by 12 kilometers wide, at most, occupies the upper part of the Tenryu Valley. Fault scarps set the limits of the basin on both the east and the west. An extensive piedmont belt of confluent fans, increasing in size toward the south, has developed at the foot of the western scarp, thereby shifting the Tenryu River to the eastern side of the basin. Most of the alluvial piedmont is covered several meters deep with young volcanic ash. Rejuvenated tributary rivers have cut deep lateral valleys into the fans, while the Tenryu has eroded their eastern margins to create a series of terrace levels, as well as a floodplain. Less conspicuous terrace fragments are also found on the eastern side of the graben.[15]

Like the other basins of Tosan, Ina is strongly oriented toward the various aspects of the silk industry, with the mulberry acreage almost equaling that of rice. Expectedly, rice takes chiefly those locations most susceptible to irrigation, while mulberry is relegated to the fans and less accessible terraces. Orchards, principally pear and apple, occupy about 10 per cent of the cultivated land and are gradually replacing mulberry. As elsewhere, Ina has suffered serious decline in all phases of sericulture, which is only about one-third as important there now as in the heyday of raw silk, but the sale of cocoons remains the single largest source of income for most farmers.[16] Hatching and care of young silkworms are phases of sericulture usually carried out in community nurseries located in the rural villages. At the end of about ten days the young larvae are distributed to the farm households where they are tended for an additional twenty to thirty days before they begin to spin cocoons.

Normally a farm family raises three crops of silkworms per year. Significantly, in this locality of important sericultural specialization, most phases of the silk industry are under the direction of farmers' cooperatives. In Iida and Ina cities are located two of the country's largest filatures, each employing about 1000 workers, and both owned and operated by farmers' cooperatives.[17]

Prior to the last few decades silk reeling was the single important form of manufacturing in Ina. But paralleling the history of Suwa, as the overseas silk markets shrank and the effects of the war in the Pacific intruded, filatures declined and new industries—chiefly metals, machines, and precision instruments—multiplied. More specifically, the newly developed industry, which is mainly housed in small plants, includes the manufacture of watches, microscopes, sewing machines, and parts for radios, cameras, and electrical machinery.[18]

THE HOKURIKU LOWLANDS BORDERING THE JAPAN SEA

Between much indented Wakasa Bay to the south and the area where the Echigo Mountains reach the coast between parallels 38° and 39° in Niigata Prefecture, is a series of coastal alluvial plains whose immediate hinterlands are well dissected Tertiary hill country, back of which rise higher mountains. Sado Island and Noto Peninsula are a part of this Tertiary hill land. From north to south the five Hokuriku plains are the Echigo or Niigata Plain, one of the largest in Japan; the Takada Plain; and the Toyama or Etchu Plain, all three to the north and east of Noto Peninsula; and, to the south and west of Noto, the Kaga Plain and the Fukui or Echizen Plain. Of these lowlands, those to the northeast of Noto occupy fault depressions that have been partially filled with river sediments. All of the plains except Toyama are bordered on their sea sides by extensive belts of parallel beach ridges and sand dunes built up by the strong winds and waves of winter. In part, therefore, the lowlands behind these coastal barriers are filled lagoons, in which drainage is deficient and flooding frequent.

In climate, as in many other respects, Hokuriku bears resemblance to western Tohoku just to the north. Being somewhat farther south it has slightly less severe winter temperatures, with midwinter-month temperatures averaging a few degrees above freezing, rather than below as in Tohoku. The January mean at Niigata is 35°, at Takada 33°, at Aikawa (Sado Island) 36°, and at Kanazawa 37°. In the United States similar January means are to be found in Kentucky, Virginia, and Maryland.

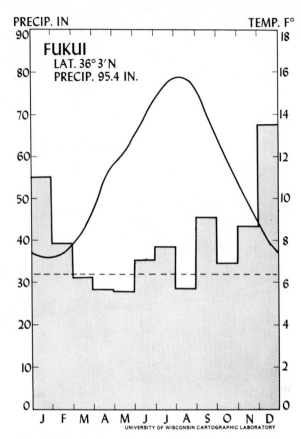

Fig. 11-7.—Stations in Hokuriku, facing the Japan Sea, characteristically show a winter maximum of precipitation, with a heavy snowfall.

August mean temperatures average 77° or 78°, so the summers are distinctly hot. Hokuriku has some of the heaviest precipitation in Japan, the mean for the year generally being between 80 and 100 inches. The winter half-year usually has more than the summer half, and fall is wetter than spring. Lying almost at right angles to the strong winter monsoon and backed by relatively high mountains, Hokuriku is the snowiest of all the lowland regions of Japan. Rail traffic is impeded by heavy snows to such an extent that in places the tracks are laid on elevated roadbeds, and snow fences and snowsheds are common. As pointed out in an earlier chapter, this part of Japan is climatically unique among the subtropical regions of the earth by reason of its deep snow cover. A considerable part of the winter precipitation falls as rain, so that the number of days with snowfall, averaging fifteen to twenty in the midwinter months, is less

Fig. 11-8.—A snowbound rural village in Hokuriku. Courtesy of Consulate General of Japan, New York.

than that farther north in Tohoku, where the amount of snow is actually less. That part of northern Hokuriku which lies leeward of Sado Island has somewhat less snow than the more exposed regions farther south. Snowfall also increases inland from the coast as the hills and mountains are approached. For example, in January the maximum depth of snow cover at Niigata on the coast is only 36 centimeters, whereas at Nagaoka,

TABLE 11-2

Maximum depth of snow on the ground at selected stations in Hokuriku (in centimeters)

Station	Nov.	Dec.	Jan.	Feb.	Mar.	Apr.
Fukui	1	17	81	69	35	2
Takada	1	69	158	187	154	56
Toyama	2	18	105	96	64	0
Nagaoka	1	46	98	119	86	14

Source: Snow Association of Japan, *The Climatography of Snow in Japan* (Tokyo, 1949).

inland and closer to the mountains, it is 98 centimeters. Thawing of the heavy snow cover in spring creates serious flooding of the rivers which in turn inundate the plains. As might be expected in a region of such abundant precipitation and heavy snowfall, hydroelectric power is plentiful.

Culturally Hokuriku somewhat resembles Tohoku, for in many ways it is a backwater area which lags behind the rest of Old Japan. Separated by mountains and by distance from urbanized industrial Kanto, Chukyo, and Kinki, where westernization is farthest advanced and large domestic markets have developed, it is in an isolated position. In addition, the abundant precipitation, resulting in frequent floods, and the generally poor drainage on the lowlands impose serious limitations upon agriculture and lead to frequent crop losses. The long, gloomy, snowy winters with their strong winds and the attendant boisterous seas, impeding ocean shipping, add further to the disadvantages of life in Hokuriku. Supplementing these physical handicaps is a man-made one, for like Tohoku, Hokuriku suffered longer and more grievously than most regions from an aggravated form of landlordism and land tenancy.

Settlements and houses on the Hokuriku plains have certain distinctive characteristics. Because of the superabundance of water on the lowlands, settlements tend to be concentrated on elevated dry sites, most common of which are the serpentine levees along present or abandoned stream channels, the beach ridges and dune ridges bordering the sea margins of the plains, and the older beach ridges back from the coast, marking ancient strand lines (Fig. 11-9). Since most of the elevated dry sites have linear dimensions, their settlements likewise have shoestring form, frequently consisting of a single row of rather closely spaced houses lining either side of a road along the crest of a levee or beach ridge. Many of these settlements extend unbroken for more than a mile, and a few for more than five miles with only an occasional break. Over 90 per cent of the settlements on the Echigo Plain are of this shoestring form.[19] Dry fields planted to vegetables and other unirrigated crops usually flank these linear villages. As a protection against the strong winter winds and heavy snows, all rural settlements, whether elongated or compact, agglomerated or dispersed, are surrounded by high, dense hedges of trees, so that from a distance they look like wooded islands rising out of the inundated paddy land.

The typical house of stormy Hokuriku is somewhat more substantially constructed than those in most parts of subtropical Japan. Heavy board siding is conspicuous. Thatched roofs are less numerous, being replaced by thick wooden shingles weighted down by cobblestones. Commonly, the

Fig. 11-9.—Serpentine-shaped settlement occupying the levees of an abandoned river meander, a dry site on the poorly drained Niigata Plain. From the *Geographical Review*, published by the American Geographical Society, scale 1:35,000.

roof is asymmetrical with the shorter slope facing the thoroughfare, an arrangement which lessens the danger of large masses of snow sliding into the street. It is also common practice to equip the roof with some kind of barrier or obstruction to anchor the snow. These *nadedome*, as they are called, may take the form of rows of hooked tile on tile roofs, or of long wooden cleats on thatch roofs.[20] On the shingle roofs, cobblestones serve the same purpose. Another fairly common roof feature in this region of heavy snow and much dark winter weather is the *takamado*, a kind of skylight, which rises well above the roof line and serves to bring additional light into the house.[21] Because of the strong westerly winds, frequently accompanied by driving snow, the takamado usually faces east.

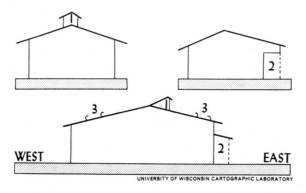

UNIVERSITY OF WISCONSIN CARTOGRAPHIC LABORATORY

Fig. 11-10.—Certain special features of house construction are characteristic of snowy Hokuriku: (1) takamado; (2) gangi; (3) nadedome. After sketches by Hatuo Yasuda.

In many of the older towns in those areas where the snow accumulates to a depth of more than 50 centimeters, the buildings of shopping streets are equipped with *gangi,* the wide board awnings or overhangs which protect pedestrians. This feature seems to be gradually disappearing, however. In this snowbound region where winter farming activity is small, the seasonal labor surplus is attracted to household industries or to seek employment in the cities, both local and distant.[22]

Agriculture in Hokuriku also has certain distinguishing features. Partly because the lowlands are poorly drained, they are more specialized in rice than are the plains in most parts of the country. Thus, whereas in Japan as a whole paddy fields occupy about 55 per cent of the total cultivated area, in Niigata Prefecture 75 per cent is so utilized, while comparable figures for Toyama, Ishikawa, and Fukui prefectures are 86 per cent, 76 per cent, and 92 per cent. This closely approaches a monocultural type of agriculture. In Hokuriku rice is not only the principal food crop but an important cash crop as well; so that a large surplus leaves the region annually for the markets in the great urban industrial areas on the Pacific side of Honshu. While on such naturally inundated plains the problem of irrigation usually is simple, flooding, on the other hand, frequently does great damage to the rice crop. A conspicuous feature of these wet plains that specialize in rice are the rows of *hasagi*—dwarf trees, commonly water oaks, that are planted along the footpaths and the irrigation and drainage canals. The lower branches of these trees have been removed usually, leaving a small crown of foliage at the top, and there the rice is hung to dry after it is harvested, for the heavy fall rains make portions of the plains so wet that ordinary methods of drying are unsuitable.[23] In the most deeply inundated areas only the tallest varieties of rice can be grown, and occasionally the harvesting must be done from boats. Small elevated platforms are built in each paddy on which the rice is stacked as it is cut. Undrained marshy areas are commonly cropped to lotus.

Echigo or *Niigata Plain,* largest single area of compact new alluvium in all Japan, is the product of the Shinano River, the longest in Nippon, and a number of smaller streams fed by the heavy precipitation in the high mountains to the south and east. The plain is coincident with the greatest population cluster north of latitude 37°. Occupying in part a fault basin, and bordered by a belt of parallel beach ridges and dunes one to three miles wide along its sea margins, the lowland is poorly drained and contains a considerable area of swamp and shallow lake. Across its flat surface wander numerous large and small streams, the story of whose shifting courses can be read in the scars of old channels and the oxbow lakes. The plain's long smooth coastline backed by arcuately arranged sand dunes, the outermost reaching heights of fifty meters, has undergone important retrogression, that part near Niigata city having retreated 325.3 meters over the last sixty years, or an average of 5.4 meters per year.[24]

Like a series of giant corrugations, the dozen or more parallel beach ridges and dunes twenty to twenty-five meters high, with their intervening wet, trough-like swales, form the widest belt of wave- and wind-deposited materials anywhere in Hokuriku or western Tohoku (Fig. 11-11). The outer ridges are often dune-capped, and fence-like windbreaks are necessary in some places to prevent the drifting of sand. Large areas are planted with a cover of trees which perform the dual service of anchoring the loose sand and protecting the settlements from strong sea winds. A distinctly linear arrangement of the culture features prevails. Rice and some Japanese-pear orchards occupy the elongated swales; the intervening sandy ridges, where these are not in woods, are the sites of villages and dry crops, vegetables being prominent, especially such sand-tolerant ones as melons, legumes, and tubers. On the ocean side of the outermost ridge the wide expanse of sloping beach is also in vegetables.

Settlements of this coastal subdivision of the Echigo Plain bear the earmarks of poverty. While compact settlements are not lacking, often they are only loose agglomerations of huts without well-defined streets completely hedged in by walls of bamboo trees. In the absence of effective protection against strong sea winds, stone-weighted roofs are common. Among the vegetable patches in many of the coastal villages, wells with great wooden sweeps are conspicuous features, reflecting the necessity for artificial watering of these porous sandy soils.

On the poorly drained lowland back of the belt of dunes and beach ridges is one of the largest and most exclusive rice-growing areas of Japan. Because of the wetness of the plain, yields per unit area are only average or slightly below, and the grain is not of the highest quality, being somewhat soft. Rows of rice-drying poles and *hasagi* line the field margins. Kanbara, the lowest and wettest part of the Niigata Plain, exemplifies the various adjustments that are made to an excess of water. While the earliest settlements sought out dry sites, later they located along the margins of small waterways which served as routes of transport for farm products. Rice sometimes had to be harvested from boats. Since World War II drainage of the Kanbara section has been improved, with resulting increases in crop yields.[25] An early-maturing, glutenous type of rice called *mochi*, which is used for pastry, is extensively planted on Echigo, and by early August the ripening fields of this variety are conspicuous by reason of their yellowish color. Low, isolated remnants

Fig. 11-11.—Air view of a portion of the coastal dune belt of the Niigata Plain in Hokuriku in winter. Note the linearity of the features. Dry fields and settlements occupy the elevated dunes and beach ridges while paddy fields dominate in the intervening swales. Courtesy of the Geographical Survey Bureau, Japan.

of beach ridges rise slightly above the alluvium in some places, these dry sites having become islands of upland crops in the midst of the paddies. Irregularities representing abandoned stream courses, as well as riparian belts of variable width along the present diked streams, are also the sites of unirrigated crops. In the vicinity of villages, small artificially elevated plots of vegetables, fruit, and mulberry are conspicuous. Here on Echigo, where extensive areas of wet land were not reclaimed for rice until the Edo Period, average farm size is relatively large for subtropical Japan, usually one to one and a half hectares. Double cropping of fields is not common, the ratio of crop area to area of cultivated land being only 105 per cent in Niigata Prefecture.

Compact villages, many of them small and tree-enclosed, are the prevailing settlement form on the new alluvium. Linear types are especially common, paralleling highways on the wet plains and occupying dry sites along the dikes of streams, relict levees, and the low remnants of beach ridges. The landscape is deprived of spacious vistas by the numerous villages, with their protecting hedges and trees, and by the rows of trees along rivers, canals, and even some field boundaries.

The city of Niigata (236,000 population), occupying both sides of the Shinano River near its mouth, is located in the midst of the beach ridges. Canals and bridges are numerous. Its hinterland includes not only the immediate Echigo Plain, but the adjacent productive Tertiary hill country and the northern basins of the Fossa Magna as well. The Niigata manufacturing district, one of the few examples of extensive industrialization along the Japan Sea side of the country, comprises a number of dispersed centers, of which Niigata city is the most important. Other lesser centers are Nagaoka, Tsubame, and Sanjo. Industry, varied in character, includes oil refining, electrochemicals, gas and petrochemicals, machinery, primary and fabricated metals, textiles, and the construction of small ships. Niigata city is strongly specialized in chemicals, which account for one-third of the total value added by manufacturing, followed by machinery of all kinds and pulp-paper (Fig. 11-12). Chemical industries based upon local natural gas are among the newer developments. Even more specialized is Nagaoka, where machinery of various kinds is responsible for 47 per cent of the value-added. Sanjo and Tsubame have national fame as centers of cutlery and tableware manufacture. Abundant and cheap hydroelectric power and supplies of limestone have acted to foster the chemical and electrochemical industries.[26]

With a shallow, silting, estuary harbor, exposed to strong winter winds, Niigata is seriously handicapped as a port; still it is probably the most important industrial port on the Japan Sea coast. Larger vessels are

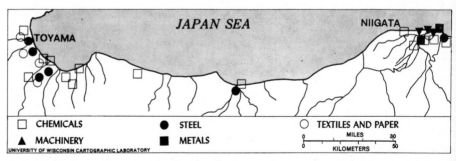

Fig. 11-12.—Distribution of kinds of industry in the Toyama and Niigata industrial districts of Hokuriku. From map in *Nippon no Chiri*, Vol. 4.

obliged to lie at anchor in the open roadstead and load and discharge cargo by means of lighters; small ones may enter the river. Niigata being primarily an import port, its incoming cargo is mainly raw materials for industry, including oil, iron ore, and phosphate ore, while its exports, preponderantly domestic and in processed form, are largely fertilizer and oil. Rail transportation is used to a much greater extent by the region and its industries than are ships.[27]

Although much less extensive, the *Takada Plain* closely resembles Echigo, its northern neighbor. It is included within the general Niigata industrial district previously described.

The *Toyama Plain,* situated just to the east of Noto Peninsula, is distinguished from all other lowlands in Hokuriku and western Tohoku in at least three respects: (1) it is composed chiefly of a series of fairly steep alluvial and diluvial fans which form an extensive piedmont belt, while low delta plain is restricted in area; (2) the usual wide belt of parallel beach ridges and dunes along the seaward margins is lacking; and (3) there is a remarkable development of dispersed rural settlement.

Toyama is a bay-head plain partially occupying a depressed block between Noto Peninsula and Honshu proper. Absence of a marked belt of beach ridges and dunes along the sea margins is likely due both to the depth of water in Toyama Bay and to a weakening of the strong winter winds and waves as a result of the protection offered by Noto Peninsula. It is the high Hida Range immediately to the rear of Toyama, with its heavy precipitation, numerous short vigorous streams, and abundance of fluvial materials, that largely accounts for the well-developed alluvial-diluvial piedmont belt.[28] Subsequent crustal movements have acted to increase the inclination of the fan surfaces.[29] Coarse diluvial sediments in the form of fans and benches at various levels, resting unconformably upon Tertiary prongs, characterize the inner higher margins of the plain. In some places relatively broad and smooth depositional upland surfaces still persist; in others stream erosion has roughened them. Conspicuous cliffs, providing excellent cross-sections of assorted fluvial materials, not infrequently separate the several benches, but gradual transitions from one level to another are more common. Red loams resulting from a weathering of the diluvial sands and gravels cover these older deposits.

The new fans, also composed of coarse stony materials and resting upon eroded diluvium or Tertiary surfaces, merge along their upper slopes with the older formations without a very perceptible disconformity. Broad boulder-strewn stream courses containing vigorous rivers occupy both fan-crests and inter-fan areas. Scars of old stream channels testify to river vagaries. Toward the sea margins the conspicuous cres-

Fig. 11-13.—Dispersed rural settlement characterizes the Toyama alluvial piedmont. Courtesy of the Geographical Survey Bureau, Japan.

centic or scalloped arrangement of features so characteristic of compound fans gives way to a flattish plain.

On the steeper alluvial fans of Toyama the completely dispersed rural settlement type, with the individual isolated farmstead as the unit of occupance, is the rule.[30] This feature is most perfectly developed on the upper part of the western fans in the vicinity of Fukuno, where the individual farmstead is the only form of settlement outside a few fairly large market towns (Fig. 11-13). No system or pattern in the arrangement of the individual farmsteads is apparent. Each is hidden in a tall hedge of conifers which completely surrounds and protects it against the sea winds and also against the downcast *föhns* which arrive from the interior. From an elevated vantage point the fans appear to be dotted with tiny groves, so hidden are the farmsteads in foliage. In such regions of disseminated population it is common for peddlers to carry provisions from the market town to the rural farmstead. Gradually, as the fans flatten out upon approaching the sea, small compact settlements become more numerous at the expense of isolated farmsteads, until finally on the flattish ocean margins of the lowland, rural villages are the rule. Hamlets and villages, like the individual farmsteads, are hedge-enclosed.

As noted in an earlier chapter, dispersed settlement in Japan often indicates relatively recent occupance. Retarded settlement on the steeper, coarser parts of many fans is probably attributable to the stony infertile soils and the flood hazards associated with vagrant rivers. But here again Toyama seems to be the exception. From a study of old maps it has been concluded that the scattered rural residences on Toyama may not be so much indications of recent occupance as they are relic features associated with the ancient Jori system of land division, in which one cho (2.45 acres) was the common unit of landholding. It has become a tradition in the Toyama area for a farmer to cultivate a single block of land in the immediate vicinity of his dwelling. A further suggestion is that the strong *föhn* winds descending from the Hida Range, with the attendant fire hazards, may help to explain the persistence of isolated farm residences.[31]

More exclusively than the terrain features would lead one to expect, rice is the dominant crop in the Toyama region; so much so that large amounts are exported to other parts of Japan. About 86 per cent of the arable land in Toyama Prefecture is in paddy. Not only on the lower, flatter areas are paddies conspicuous, but even on the higher parts of the fans, where, however, mulberry, vegetables, and even woodland, offer competition. While the proportion of the arable land replanted to winter crops is relatively high, amounting to about 40 per cent, which is a much higher figure than that for Niigata, still the area sown to winter cereals is very small. Seemingly the high rate of utilization of arable land is a consequence of the large number of readily drained paddies on the fans which are fall-planted to green manure, soiling, and forage crops. Since stones and boulders are numerous on the fans, many of the farmsteads are surrounded by low walls of smooth, waterworn boulders, and the retaining walls of the terraced paddies are similarly constructed. With the numerous tree-enclosed isolated farmsteads and small villages, and the lines of trees along roads and waterways, the plain has a somewhat disordered appearance and affords no extensive vistas.

The Toyama industrial zone, relatively new, rapidly growing, and recently about on a par with the Niigata zone in importance, has many of the same advantages as the latter. Cheap and abundant hydroelectric power and local limestone deposits have been the principal magnets attracting chemical, electrochemical, electrometallurgical, paper, and textile industries.[32] The greatest collection of factories is around Toyama (128,000 population) and Takaoka (81,000 population) cities, with the single largest center at Higashiiwase harbor, about four miles northwest of the center of Toyama city near the mouth of the Jinzu River, where port facilities exist and canalized flat lands provide ample factory sites. This is the only harbor-port on the Japan Sea coast which is being energetically developed at the present time. Extensive lagoon areas are currently being reclaimed for additional factory sites, with the expectation that oil refineries will be attracted to locate there. Harbor improvements include the construction of an entrance channel 200 meters wide and 12 meters deep, together with a breakwater 1700 meters long.[33] In Toyama city, chemicals and machinery, about on a par in importance, together account for nearly 70 per cent of the value added by manufacturing. Smaller Takaoka is to some extent more diversified and has a somewhat different industrial composition, with metals well in the lead, followed by chemicals, paper-pulp, machinery, and textiles.

The nearly continuous *Kaga* and *Fukui* plains lying west of Noto Peninsula are throughout most of their combined length a relatively narrow strip of alluvium, which broadens to the south in the vicinity of the valley and delta of the Kuzuryu River, where the city of Fukui is located. Essentially these plains are filled lagoons, in some places poorly drained, whose seaward margins, lacking the protection given to Toyama by Noto Peninsula, are characterized by a wide belt of parallel beach ridges and dunes, with a smooth harborless coast line. They therefore resemble more the Niigata and Takada plains and those of western Tohoku than they do Toyama. Still, on the Kaga Plain in the vicinity of the fan of the Tedori River, poorly drained lagoon plain is largely absent, and the belt of coastal dunes is distinctly narrower. Also the general assemblage of cultural features, including those of agricultural land use on dune belt and wet lowland alike, are similar to those noted previously for Niigata.

In industrial composition the Ishikawa-Fukui manufacturing district, which is coincident with these plains, differs from Toyama and Niigata in that there is far greater emphasis on textiles. Since the Edo Period this local area has been nationally renowned as a producer of traditional silk fabrics, while more recently it has expanded into the field of fabrics made from rayon and synthetic fibers, so that at present Fukui is probably the greatest rayon-weaving center in Japan. Supplementary industries include the designing and manufacture of textile machinery, construction machinery, industrial machinery, and automobile bodies, and the processing of chemicals, ceramic-clay products, and paper-pulp. In Fukui city (104,000 population), textiles represent 67 per cent of the value added by manufacturing. In Kanazawa (225,000 population) machinery of all kinds ranks first (30 per cent), followed by textiles (22 per cent).

Noto Peninsula, which forms the principal irregularity of the whole

Japan Sea coast, is an area of low Tertiary hill country, separated into northern and southern parts by a narrow graben valley whose alluvial floor is covered with rice fields. The indented coast of Noto has numerous fishing villages as well as the protected harbor and small port of Nanao, which serves the adjacent plains whose smooth coasts are deficient in natural harbors.

The Kanto Region, embracing seven prefectures in northeastern Central Japan, covers some 32,000 square kilometers and has a total population of about 27 million. Hence, with an average density of 800+ per square kilometer, it is the most densely populated as well as the most populous of the large subdivisions of Japan. This populousness stems from the fact that it contains two of the country's metropolises—Tokyo and Yokohama—as well as the largest plain of Japan. The Kanto Plain covers 40–45 per cent of the area included within the seven prefectures of the Kanto Region, the remainder being hill and mountain land bordering the plain on the north, west, and south. Discussion in this chapter will focus on the Kanto Plain.

PHYSICAL CHARACTER

A depositional plain occupying a tectonic basin, Kanto is composed of unconsolidated fluvial, estuarine, and marine deposits totaling several hundred meters in thickness. From the lowest central part where elevations are scarcely twenty meters above sea level, the lowland rises to elevations of thirty to fifty meters along the seaward eastern margins, and to still greater heights along the western borders where fan configuration is conspicuous. Hilly Miura and Boso (Chiba) peninsulas in south Kanto, flanking Uraga Channel at the entrance to Tokyo Bay, are composed of

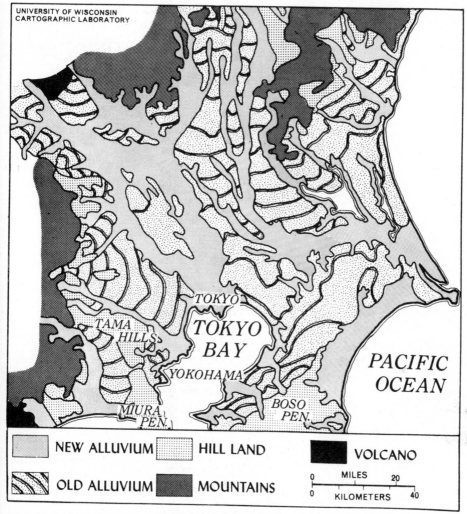

Fig. 12-1.—Terrain types and earth materials of the Kanto Plain. Old alluvium, or diluvium, is chiefly in the form of tabular uplands. Hill land is largely diluvium or Tertiary. From map by Sohei Kaizuka.

faulted and dissected Tertiary rocks. Like the Tokachi Plain in southeastern Hokkaido, Kanto is not a broad and continuous surface, but instead is composed of surfaces at different levels, including low-lying floodplains and coastal plains, as well as extensive marine and fluvial terraces at different elevations, the latter resulting from alternating transgressions and regressions of the sea within the Kanto embayment.[1] It is predominantly an upland plain, for some 60–65 per cent of its total area is diluvial upland, which sets Kanto apart from a great majority of the plains of Old Japan (Fig. 12-1).

The diluvial upland surfaces are of different ages and elevations and have variable surface configurations. Those in the central and eastern parts have flattish crests, and here the tabular upland surfaces are undulating or sometimes rolling. Toward the northern, western, and southern margins, with approach toward the foothills, fan configuration and increased surface gradients are the rule. A small percentage of the diluvium, and more especially the oldest and highest parts, has been so dissected by running water that the upland surface has largely disappeared and the terrain form is that of hill land. An excellent example of hilly diluvium, closely resembling the terrain developed on weak Tertiary rocks at low elevation, is the Tama Hills just north of the Miura Peninsula and inland from Yokohama. Numerous smaller fragments of dissected old diluvium are to be seen also along the plain's western and northern margins close to the mountains. Still, much the larger part of the diluvial surface is tabular in appearance, but with several different terrace levels, each level being terminated by a fairly steep declivity, which characteristically is marked by a sinuous belt of trees.

Overlying the diluvial upland surfaces of Kanto, like a mantle of snow, is an aeolian deposit of acidic volcanic ash. As a rule, the higher and older the terrace, the thicker the ash cover, and in places it attains a depth of forty meters. The reddish-brown ash deposit is known as Kanto Loam.

Streams have incised broad valleys into the diluvial upland and subsequently floored them with deposits of more recent alluvium. The zone of contact between the low-lying newer and the higher older alluvium is a dissected zone that in places forms a narrow belt of hilly land. Minor river terraces in the form of low benches characterize the floodplains. A relatively recent depression of the Kanto Basin transformed the lower ends of the river valleys into broad estuaries, large parts of which subsequently have been filled. But long estuarine remnants of swamp and lake still persist along the lower Tone Valley. Most recent of the crustal movements has been a slight uplift that has produced belts of new coastal plain along the Pacific littoral, both to the north and south of Cape Inubo, marking the mouth and delta of the Tone River, as well as along the head of Sagami Bay. A hilly diluvial-terrace margin overlooks the belts of new coastal plain, in places rising as much as eighty meters above them. Extensive crescentic beaches are backed by linear belts of dunes and beach ridges roughly parallel with the shore, each ridge separated from the next by a shallow swale.

Excepting the immature new alluvium, the soils of Kanto have as their parent material the acidic volcanic ash which mantles the diluvial sediments. Preponderantly the Ando ash soils are ruddy-brown in color, although a smaller proportion is sufficiently high in organic matter to warrant their being described as black.[2] They differ in texture from loams to clays and are inherently infertile, acidity varying from medium to strong, while base minerals are deficient. Because of the presence of active colloidal substances, Ando soils are able to absorb and retain large amounts of water, which in years of excessive rainfall cause them to be wet and soggy, even to the point of injuring crops.

CULTURAL FEATURES

Considering the long period of Japanese history, widespread and intensive occupance of Kanto has been both late and relatively slow. Ten centuries ago, when Japanese culture attained its fruition in Nara and Kyoto, old national capitals in the Kinki region, Kanto was still in large part a wilderness known as Musashino. It was not until the latter part of the sixteenth century, when Edo, now Tokyo, was made the residence of the ruling Tokugawa shoguns, that development of the adjacent extensive lowland was markedly accelerated. The predominance of infertile, ash-covered diluvial upland, whose soil and relief characteristics are not well suited to irrigated rice, is the chief cause for the retarded settlement of this largest plain of Nippon. In addition, the diluvial uplands of Kanto are notoriously deficient in water, which further discouraged their use for agriculture.[3] Even today, although Kanto, as the largest of the lowlands, has the single greatest population cluster in Japan, the general density of rural settlement is probably no more than half as great as that of other important plains along the Pacific coast, which are predominantly of low, fertile, easily irrigated new alluvium. Large areas of Kanto's flattish diluvial uplands continue to have a cover of planted woodland or wild grasses. It is especially in northern and eastern Kanto, with its predominance of poor ash-covered diluvial upland and poorly drained lowlands of newer alluvium, that the densities are relatively low, which in turn tends to reduce the average for the whole plain (Fig. 12-2).

But the Kanto Plain is by no means a homogeneous unit. Thus, in a geographic regionalization of the country, Ishida separates Kanto into North Kanto, containing the three northern prefectures, and South Kanto, containing the four southern ones. Northern Kanto is not unlike Tohoku in some ways, especially southern Tohoku, for this more rural northern part contrasts with the strongly urban southern part in having a much lower per capita income—in fact, well below the country average—a much larger proportion of its population in primary industry and a smaller

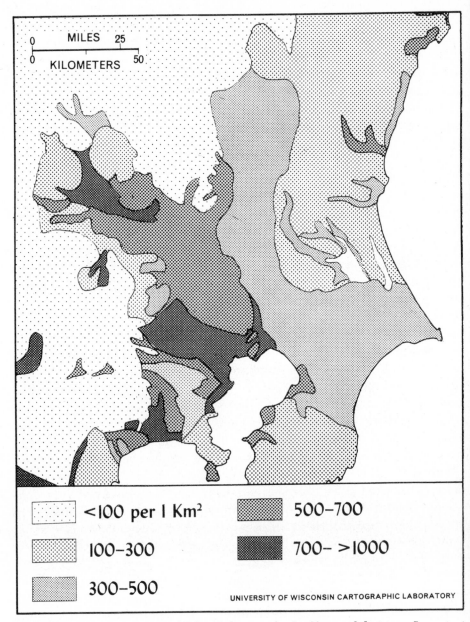

Fig. 12-2.—Population density of the Kanto area by landform subdivisions. Important cities are omitted in making density computations. From map, *Population Density by Landform Subdivisions,* scale 1:800,000 (Geographical Survey Institute of Japan, 1955).

part in secondary industry, an industrial income per capita which is less than one-third what it is in South Kanto, and a yearly net population loss, compared with a gain in the south which is the greatest for any large subdivision of the country. (See Table 9-1.)

In his map of agricultural land-use regions, Ogasawara[4] places western Kanto in his central zone of most intensive and advanced land utilization, while eastern Kanto, like Tohoku, is located within the frontier zone of less complete and intensive land utilization. In western Kanto, land reclamation is well advanced, and due to the nature of the terrain, upland fields make up 70 per cent of the total cultivated area. Stimulated by the great Tokyo-Yokohama market, the raising of commercial vegetables, fruit, flowers, and tree crops is here well developed. Around the great cities there is a clearly observable zoning of agricultural production. The proportion of double-cropped paddies is relatively high in western Kanto, per unit-area yields large, and commercial farming notably advanced. Specialization in mulberry and cocoon production is especially strong.

By contrast, eastern Kanto, with its lower population density, is more retarded in agricultural land use.[5] Considerable areas of marsh, pond, and lake are still unreclaimed. The proportion of the diluvial uplands under cultivation amounts to only 30–50 per cent as against 40–60 per cent in west Kanto, and slope lands also are much less completely utilized, while more area remains in planted woodland and genya. Yields of rice are lower in east Kanto, and the percentage of double-cropped paddies is much less. Though cash crops such as fruit, vegetables, mulberry, and tobacco are raised to some extent, as influenced by the Tokyo market, subsistence farming remains more important than in the west. This relatively retarded nature of the agriculture in east Kanto, in spite of the proximity of Tokyo, reflects the smaller area covered by fertile new alluvium, plus the marshy and frequently inundated tracts along the Tone River which make communication with the Tokyo market difficult. Eastern Kanto is also less industrialized and has fewer cities.

Rural land utilization

Low-level occupance of floodplains and high-level occupance of diluvial uplands, marked by contrasting land-use features, is distinctive of Kanto. It is on the low floodplains and deltas with their superior soils and easy access to abundant irrigation water that the highest population densities are found. In such locations on Kanto the ratio of rural people to area is similar to that on other crowded lowlands. As noted previously, the western half of Kanto, which contains the larger share of the new alluvium, is the most crowded part, with the greatest concentration of population coincident with a belt extending in a northwesterly direction from Tokyo along the complicated floodplain nets of a number of closely spaced rivers.[6] Settlements on the new alluvium tend to be compact village units. Because of the wet nature of the floodplains the rural villages are generally concentrated on dry sites, either at the base of the diluvial uplands where they make contact with the floodplains or along river levees and elevated highways. The location at the contact zone between alluvium and diluvium is advantageous not only because of the better drainage, but also because it gives the farmers in the village easy access to both the paddy fields below and the dry fields above.

Aside from the riverine belts where elevation is slightly greater, surface irregularities more pronounced, and soils sandier, the low-level floodplains are largely devoted to irrigated rice, except in the peri-urban belt adjacent to Tokyo and Yokohama where there are numerous artificially elevated dry fields devoted to market-garden crops. A considerable part of the paddy land of Kanto remains fallow after the rice harvest, or has only a green manure or forage crop (Fig. 12-3). In part this reflects the poorly drained nature of much of the alluvium, especially that in eastern Kanto. It is also a result of intensive market gardening, which leaves the farmers little time for fall draining and planting of their rice fields. In some of the recently reclaimed lands along the seaward margins of the larger deltas there are typical polder tracts where master dikes enclose extensive areas of wet land devoted almost exclusively to rice. In such parts the rectangular fields laid out in grid patterns suggest the recency of reclamation. Where drainage is especially poor, the rural villages, characteristically situated on the dikes, are of shoestring form; or single isolated farmsteads may occupy artificially elevated plots.[7]

The higher-level or second-story type of occupance, characteristic of the diluvial uplands, is in contrast to that of the floodplains. Settlements are of both the dispersed and the village type. The isolated farmstead is the older form and was a natural development, whereas the farm villages are of later origin and were planned and fostered by the government of the shoguns.[8] Since scarcity of drinking water offered a major problem to the settlers on the diluvial uplands, the early occupants located their farmsteads along the margins of the terraces, where shallow wells could provide the necessary water. In the interiors of the tablelands, where water was farther below the surface, government officials

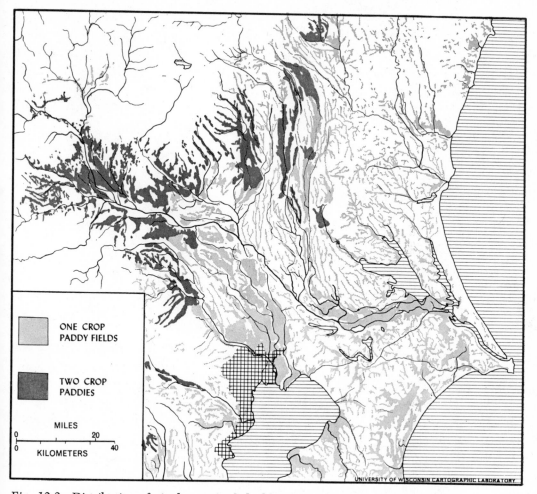

Fig. 12-3.—Distribution of single-crop and double-crop paddies on the Kanto Plain.

charged with village planning were forced to select sites where underground water was most accessible.[9]

A striking feature of the farm villages on the diluvial uplands is the large number of long, narrow shoestring settlements, composed of a row of closely spaced dwellings on either side of, and paralleling, a highway (Fig. 12-5). One comes to associate this form of settlement with elongated dry sites such as river levees, beach ridges, and elevated highways on wet paddy lands, but on the Kanto uplands they represent relatively new, or *shinden,* settlements, resulting from the active land reclamation for agriculture that took place during the Tokugawa Shogunate (Fig. 12-6). In some instances the linear village form appears to have its

Fig. 12-4.—Illustrating the typical two-level type of land use on the Kanto Plain. Small irregular-shaped paddy fields characterize the poorly drained new alluvium. Here settlements are few. Dry fields characterize the diluvial upland, with settlements concentrated along the zone of contact between alluvium and diluvium. The dissected margins of the diluvial upland often have a woodland cover. Courtesy of the Geographical Survey Bureau, Japan.

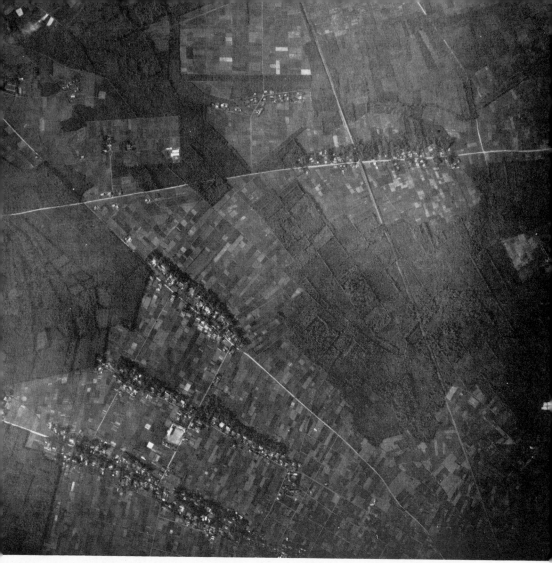

Fig. 12-5.—Air view of linear-shaped *shinden* settlements on the diluvial upland of the western Kanto Plain. Note also the extensive areas in planted woodland. Courtesy of the Geographical Survey Bureau, Japan.

origin in the difficulties associated with obtaining adequate water supplies. Since subsurface water was commonly deficient on the diluvial uplands, canals were built which carried underground water from areas on the uplands where wells were relatively abundant, to those sections that were deficient in water. The canals tend to parallel the highways, with the result that settlements have developed a linear form. Often farms and fields of such villages are also linear, with their long dimension at right angles to the highway and the village.[10] It is common for such farms to show a definite belted arrangement of land use, mulberry occupying those fields closest to the village, with dry fields of vegetables and grains lying beyond the mulberry, and woods or wasteland located at still greater distances. Concentrating the mulberry closest to the farm

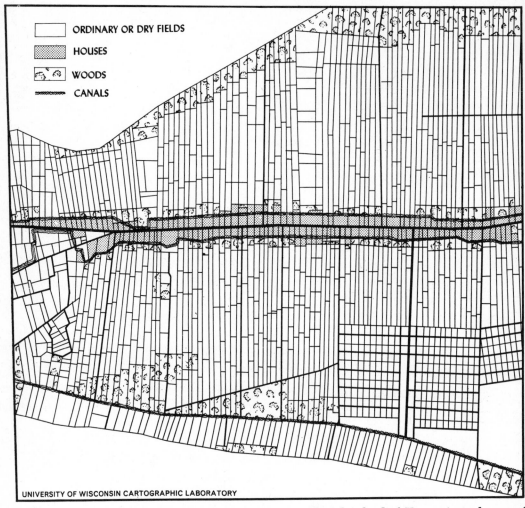

ORDINARY OR DRY FIELDS

HOUSES

WOODS

CANALS

UNIVERSITY OF WISCONSIN CARTOGRAPHIC LABORATORY

Fig. 12-6.—A shoestring settlement on the diluvial upland of Kanto. A single row of houses fronts upon either side of the highway. At the rear of the houses pass the canals supplying the village with water. A hedge of trees flanks the village on the north and south. Beyond are the long fields with their long dimension at right angles to the highway. After map by Nikiti Yazima.   No scale given.

home makes picking and carrying the bulky fresh mulberry leaves a less time-consuming job. A conspicuous feature of settlements on many parts of Kanto is the dense hedge of trees planted around them, giving village or isolated farmstead the appearance of larger or smaller groves. In some sections near Tokyo as many as 70 per cent of the farmhouses have tree hedges, usually surrounding them on all sides, with the densest planting on the north and west, which is the direction of the strong, cold winter winds.[11]

On the flattish to undulating younger and lower surfaces of diluvium land use is varied. Considerable areas remain uncropped, having instead a cover of planted woodland or even of low-grade wild grasses and brush. Paddy fields are scarcely absent, but on the other hand they are exceptional, for the dearth of water and the difficulties involved in lifting it from the floodplain level handicap a crop requiring long-continued inundation. Hence dry crops greatly predominate. These represent the usual variety of annuals characteristic of subtropical Japan, including a great diversity of summer-sown vegetables—sweet potatoes being unusually important—plus pulses, millet, and buckwheat, followed by fall-sown wheat, barley, and naked barley. About 80 per cent of the fresh vegetables for the Tokyo-Yokohama market come from a market-gardening zone lying mostly within a fifty kilometer radius of the cities.[12] Here a most intensive rotation of crops using inter-tillage methods is practiced, so that three or more harvests of vegetables are obtained from the same field. Even within the market-garden belt there is a degree of zonation of the various crops, with the more perishable ones grown closest to the city. Chiefly the market gardens are located on the diluvium, although they are not absent even on the floodplains. To some extent the vegetable zone is still fertilized by night soil from Tokyo, but this practice has declined markedly as commercial fertilizers have rapidly taken over. To protect against frost and cold northerly winds which can do serious injury to vegetable crops in early spring and in fall, farmers have devised ingenious methods of shielding the plants against cold and wind without cutting out the sunlight. One common practice is to plant vegetables on the southern sides of rows of barley which offer some protection against wind. Hedges of tea bushes may serve a similar purpose.

Certain parts of the Kanto uplands are specialized in commercial crops other than vegetables, among them tea, tobacco, fruit, and mulberry. The latter is by far the most important and is highly concentrated in western Kanto in Gumma and Saitama prefectures. The combined acreage of these two prefectures represents nearly one-quarter of the nation's total area in mulberry, making northwestern Kanto one of the most specialized and important regions of silk production. Like all such regions, this one too has suffered a remarkable decline, the mulberry acreage having dropped by more than one-half between 1930 and 1960.[13] The relative decline appears to have been more serious in the hilly borderlands, however, and somewhat less on the diluvial areas of the adjacent plain.

Remarkable to a foreigner is the fact, mentioned earlier, that extensive areas on the Kanto diluvial tablelands should still remain out of cultivation and given over to woodland uses or to wild grasses and shrubs.

This is an old story in Japan, reflecting the low esteem in which land unsuitable for paddy is held. Even the food stringencies associated with World War II seemingly did not bring about a wholesale expansion of the crop area on the diluvial uplands of Kanto. But in the postwar period admittedly there has been an important encroachment of the cultivated area on some of these tableland sites, one of the best examples of such conversion being that of Sagamihara, a diluvial upland surface about thirty kilometers long by four to eight kilometers wide, just inland from Yokohama.[14] Even as late as the 1920's this upland had extensive tracts where the proportion of cultivated land was less than 50 per cent. Before, during, and even for some time after the war, Sagamihara was the site of numerous military installations of an extensive character. But more recently dams have been built and the impounded water has been led on to this upland surface where it will provide supplementary irrigation for 2700 hectares of upland fields, this being something of an innovation since irrigation is almost exclusively reserved for paddy.

On those more restricted areas of diluvium, usually older and higher, where stream erosion has proceeded so far that only disconnected remnants of the upland crests remain and the terrain features are those of the low hill country, population distribution and occupance features are in contrast to those of the smoother uplands. Because of the relatively large area of intricately branched valley floodplain, rice is a very important crop, and in consequence population density may equal or exceed that on the tablelands. There is also some cultivation of the slope lands in the dissected diluvium, but much of it remains in woodland and shrub. The Tama Hills in southwestern Kanto constitute the largest area of such dissected diluvium.

On the more recently emerged strips of coastal lowland bordering the plain proper, settlements seek out the dry sites on the parallel lines of low beach ridges. Because of the considerable amount of low wet land, rice is an important crop, characteristically occupying the troughs between the sandy beach ridges. The latter in turn support crops of vegetables and mulberry, or they are crowned with rows of trees which act as windbreaks and prevent the blowing of the sandy soil.

KANTO CITIES AND THEIR FUNCTIONS

Not only is the Kanto Plain coincident with what is by far the largest single compact cluster of total population in Japan, but even more emphatically it is the greatest concentration of urban population. Within the seven prefectures comprising the Kanto Region, where the total population in 1960 was nearly 27 million, close to 14.9 million were in Densely Inhabited Districts with over 5000 inhabitants, while 17.6 millions were in official *shi*. Thus the Kanto Region, which accounts for about 29 per cent of the nation's total population has nearly 30 per cent of its *shi* inhabitants and 36–37 per cent of its DID population. To be sure, the above figures are for all seven prefectures comprising the whole Kanto Region, and hence contain much hill and mountain land in addition to the Kanto Plain, so that they are somewhat too large if applied to the plain only. But the rough lands contain few cities, with the consequence that the area within a fifty kilometer radius of Tokyo Station, equaling less than half the Kanto Plain, has a DID population of 12.37 million. This includes 8.1 million in the Tokyo DID, the capital; 1.1 million in the Yokohoma DID, one of the greatest ports; and 3 million in scores of lesser cities. Tokyo and Yokohama, together with Kawasaki, which lies between them, all fronting on tidewater, form a great conurbation of close to 10 million people (as of 1960).

An outgrowth of recent, and expected future, burgeoning of population and industry within this Tokyo-centered region has been the creation by the national government of the so-called Capital Region Development Plan, whose authority embraces an area with a radius of about one hundred kilometers centering on Tokyo Station. The Capital Region planning area includes not only the Kanto Plain, but some adjacent hill lands as well, especially those to the south in closest proximity to the great cities, so that the total area covers close to 25,000 square kilometers, or 6.5 per cent of all Japan. In 1955 the population of the included area was about 19.8 million, or more than 22 per cent of the nation's total, and it is estimated that 6.8 million will be added in the two subsequent decades, thereby creating a cluster of 26.6 million inhabitants by 1975.

The Capital Region has been zoned into three areas. First of these is the already built-up section, or Inner Urban Zone, which is almost exclusively the combined DID areas of Tokyo, Kawasaki, and Yokohama, bordering Tokyo Bay on the north and west (Fig. 12-7). In the absence of planning it is estimated this congested core area would acquire an additional 5.6 million inhabitants by 1975, a figure which the plan hopes to reduce to 2.9 million by shunting the remainder into some thirty industrial satellite cities lying outside.[15] Within the built-up inner urban area permission is required before additional population-attracting features such as factories and educational institutions may be built. A second Suburban or Green-Belt Zone, roughly ten kilometers wide, borders the Inner Urban Zone on its land sides. This is a region of combined urban and rural living in which efforts will be directed toward preserving scenic

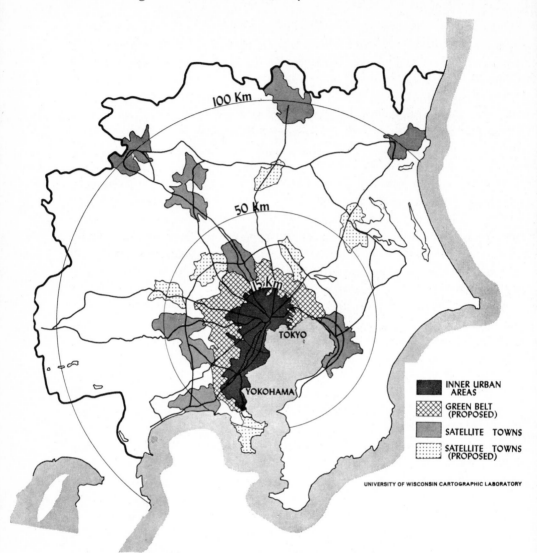

*Fig. 12-7.—*Capital region development plan. From *City Planning, Tokyo* (1961).

areas and providing parks and other open spaces for recreational uses, features which occupy such a minimal area in the Inner Urban Zone. It is within this second zone that universities, research institutes, hospitals, and the like will be encouraged to locate. A third or Peripheral Zone includes the remainder of the Capital Region. Here it is planned to locate much of the new factory industry that might otherwise crowd still more the Inner Zone, and here also should be absorbed a large part of the in-migrating population flowing toward the Tokyo labor market. Accordingly, it is planned to foster the development of about thirty industrial satellite districts within the Peripheral Zone, each including one or more cities, ten of which have already been designated and six additional ones proposed.[16] Nearly all of these currently are, or shortly will be, connected with the core area by arterial highways. But these sixteen designated and proposed satellite districts by no means include all of the DID centers of Kanto, or even the twenty-four existing non-metropolitan DID with over 50,000 population.

It needs to be emphasized that although they are separate political entities, Tokyo and Yokohama (together with intermediate Kawasaki) comprise—and function as—a distinct conurbation of about 10 million people which goes by the name of Keihin. Each of the political units of Keihin is complementary to and dependent upon the others. In this respect Tokyo-Yokohama closely resembles that other great binuclear conurbation in Kinki, composed of Osaka and Kobe. In each combination the larger city functions principally as the great business and industrial center, while the other serves as its deepwater port. Tokyo, however, is more of a political and educational center than is Osaka. Definitely, the Kanto metropolises have as their service area all of northern Japan.

Tokyo

For reasons previously stated, Tokyo, formerly Edo, remained an insignificant fishing village for centuries after Kyoto and Osaka, the great cities of the southwest, had reached maturity. It became a daimyo's headquarters in the mid-fifteenth century when a strongly fortified castle was erected on the site of the present Imperial Palace. But it was not until late in the sixteenth century, when Edo became the capital of the usurping Tokugawa shoguns, who ruled the country for two and a half centuries, that rapid development of the city began. During this period Edo was a vast military encampment, with the nearly impregnable shogun's castle on a diluvial bluff overlooking the bay, and the fortress-like dwellings of the daimyos, or feudal lords, spread out on the floodplain below it. In its most prosperous days under the shoguns, population of the capital city exceeded a million.

Following the shogunate's fall, Edo's threatened decadence was averted when the restored Mikado made it his "East Capital" and renamed it Tokyo. Today it lays claim to being the world's most populous city, and unquestionably it is the primate city of Japan, no other even approaching it in size and importance. Indubitably, also, it is the advantages accruing from its position as the national capital of a highly centralized state that have made it so much the focus of wealth, education, and economic power. All leading banks, major business concerns, and pre-eminent institutions have their head offices in Tokyo. This primate character acts like a great lodestone to attract such a stream of immigrants that the 1955–1960 population increase within Tokyo-to was nearly 40 per cent of that in the whole nation. As a consequence of Tokyo's large-scale in-migration the city's population characteristics differ in important respects from those of the country as a whole. Thus, in 1960, nearly 40 per cent of the resident population was born outside of Tokyo-to; the percentage of young productive population in the age groups of fifteen through twenty-nine is unusually large; and the ratio of males to females is 107.4 compared with a male deficiency of 96.5 for the nation. For example, Tokyo has 37.6 per cent of its population in the fifteen to twenty-nine age groups compared with 27.5 per cent for the country average, while the comparable figures for young males are 40.5 per cent and 28 per cent.

The capital city is located near the head of Tokyo Bay where a number of streams debouch, thereby shallowing the waters. The Tone, or master stream of Kanto, formerly entered at this point also, but in the mid-seventeenth century, in order to lessen the flood hazard in Edo, the lower Tone was diverted eastward through a prepared channel to empty into the Pacific. Other smaller stream channels traversing the city have been improved and diked, thereby reducing, but not eliminating, the flood hazard. Tokyo is a two-storied city, with much of the eastern part, where business, commercial, and industrial functions are concentrated, being low-lying delta and floodplain, while the western part, more exclusively residential, occupies the undulating to rolling surface and dissected margins of a diluvial upland, whose crests are twenty to forty meters above sea level. Relatively steep grades frequently mark the transitions from one level to another. Considerable areas of recently reclaimed land border the bay. Portions of eastern Tokyo along the lower parts of the Arakawa Drain are less than one meter in elevation and parts are even slightly below sea level (Fig. 12-8). Recent land subsidence resulting from the large-scale pumping of water for industrial purposes from the underlying sediments has aggravated the flood hazard in these low-lying districts, requiring the construction of protective coastal embankments which have been extended inland along the rivers.[17]

The moat-and-wall-enclosed Imperial Palace, strategically located on the extreme eastern margin of the diluvial upland, has been the core around which accretion has taken place and even yet it is fairly near the geographical center of the modern city. The Imperial Palace is also the focal center of a cobweb pattern of urban transport lines. Even beyond

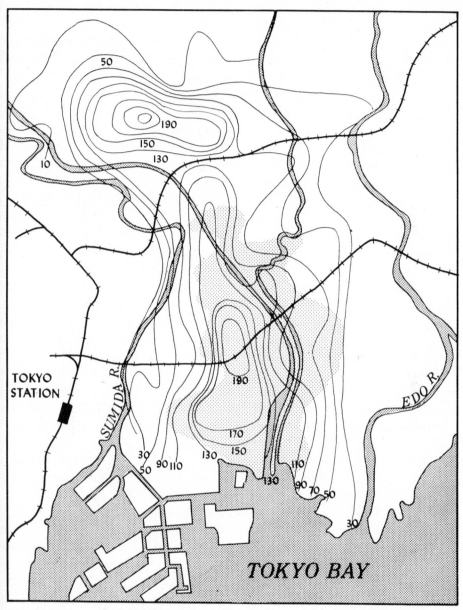

Fig. 12-8.—Amount of subsidence in millimeters in the eastern lowland of Tokyo, from March 1951 to January 1955. Shaded area is at about sea level or below. From map in *Geography of Tokyo and Its Planning*, IGU Regional Conference in Japan, 1957. No scale given.

the city limits these radial highways localize urban development in the form of linear settlements, so that octopus form is conspicuous. Throughout the commercial and industrial sections, which occupy flood-plain and delta sites, the city plan is somewhat more rectangular than in other parts. Canals and canalized rivers thoroughly intersect the lowland district, making hundreds of bridges necessary.

Two twentieth century disasters—the earthquake and fire of 1923, and the air raids of World War II—have resulted in great losses of life, wholesale destruction of property, and prodigious temporary declines in population. In each instance the reconstruction of the city has permitted the partial modernization of its street system and the construction of a host of new, Western, fireproof buildings in the city's main business sections. As a consequence Tokyo's appearance is the most modern and Western of any of Japan's cities. Still, the ratio of street area to city area is unbelievably small, Tokyo's 11 comparing most unfavorably with Washington's 42, New York's 35, and London's 23. The result is a traffic maelstrom unknown elsewhere.

A fairly well developed regional differentiation of functions can be recognized within Tokyo (Fig. 12-9). The central business district, including offices, stores, wholesale establishments, and government buildings, occupies the westernmost part of the alluvial lowland, located between the Sumida River and the eastern margins of the diluvial upland. Within the extensive central business district there is a moderate degree of segregation of such subfunctions as government, business agencies and offices, retail shops, and wholesale firms. It is especially in the government and business-office sections in and around Otemachi, Marunouchi, and Nihonbashi that new six- to nine-story Western-style buildings, forming solid blocks, are concentrated. Most of the retail sections are less impressive, and the areas of wholesaling distinctly less so. Secondary, but large, isolated shopping centers, notably at Shinjuku, Shibuya, and Ikebukuro, have developed along the loop rail line farther west on the diluvial upland, where they are imbedded in the extensive general residential area.[18] Indeed, throughout the whole residential area there is a dense network of shopping streets, a great majority of whose shops and buildings are small and of Japanese style.

Industrial Tokyo occupies chiefly alluvial lowlands in the eastern part of the city, beyond the Sumida River, and coastal lowlands along the northern and western side of the bay. Three general manufacturing districts may be recognized. First and foremost is the one which occupies the eastern lowland between the Sumida River and the Arakawa Drain, extending far inland to the northwest along these waterways. Fronting

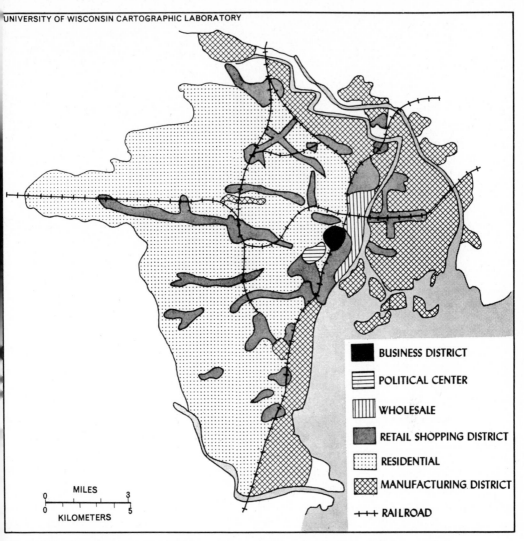

UNIVERSITY OF WISCONSIN CARTOGRAPHIC LABORATORY

BUSINESS DISTRICT

POLITICAL CENTER

WHOLESALE

RETAIL SHOPPING DISTRICT

RESIDENTIAL

MANUFACTURING DISTRICT

+++ RAILROAD

MILES
KILOMETERS

Fig. 12-9.—Functional areas of Tokyo. From map in *Geography of Tokyo and Its Planning,* IGU Regional Conference in Japan, 1957.

upon the bay, and intersected by a network of canals inland, this region's prime advantage is the services provided by water transportation. Variety is the keynote of its industrial structure, but chemical, metal, and machine industries predominate, these being somewhat segregated in a number of specialized localities. In its western parts small and medium-sized plants predominate, but toward the east the number of new, large, and modern factories increases. A second district specialized in manufacturing is a relatively narrow belt extending from Tokyo's cen-

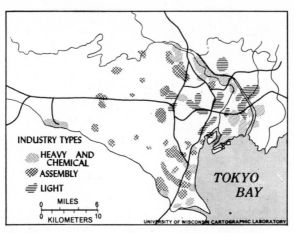

Fig. 12-10.—Regional specialization of manufacturing in Tokyo. After map by Yoshiro Tsujimoto and others.

tral business district southward along the coast until it makes juncture with a similarly located factory belt in Kawasaki and Yokohama. Here some of the industrial plants occupy sites on land reclaimed from the sea. Metal and machine industries dominate along the coast, with assembly industry more prevalent in the shallow valleys of the adjacent diluvial upland. While many of the coastal factories are served by marine transportation, since the protected Keihin Sea Canal does not extend northward beyond Kawasaki, large industrial plants served by their own company piers are much less common along the Tokyo water route than farther down the bay in Kawasaki and Yokohama. Unlike the other two, Tokyo's third manufactural district lacks compactness, but instead is made up of a number of scattered areas located inland on the diluvial upland. Many of these represent relatively new developments which are specialized in the making of parts for assembly industries.[19]

Tokyo, the principal unit of the Keihin Industrial District, is characterized by small and medium-sized factories, over 50,000 of them having fewer than thirty employees and only 6500–7000 having more than that number. Mainly these plants are engaged in the processing of durable and non-durable consumer's goods of sundry character. Variety of industrial composition prevails, but the concentration rate in steel, non-ferrous metals, machinery of all kinds, precision instruments, chemicals, leather and rubber goods, and printing is high, while with coal and petroleum products, textiles, and lumber and wood products it is low.[20] Heavy metal and machine industries and chemicals form the core of the city's manufactural structure, but even these engage mainly in making parts for assembly industries which turn out consumer's goods. Large steel and chemical plants are less dominant here than in either Kawasaki or Yokohama, so that there is less striving for tidewater location. A majority of Tokyo's small establishments depend upon manual labor, and these are locally grouped according to the products produced. Typical of the production organization is the local grouping of numerous small plants making parts around a large parent assembly plant, or they may be organized with wholesalers as the center of a grouping.[21] Thus, most Tokyo factories are integral parts of a larger complex production organization and cannot exist independent of that organization.

From Table 12-1 it may be observed that three-fifths to two-thirds of the employees in manufacturing are engaged in the heavy, chemical, and machine-assembly industries, while less than two-fifths are in the light industries. Metals, machinery, and chemicals are concentrated in

TABLE 12-1

Composition of manufacturing in Tokyo, 1960, based on number of employees

Industry group	% of industrial employees	
Iron and steel	3.5	
Non-ferrous metals	2.5	
Fabricated metals	9.1	
Machinery	10.1	
Chemicals	6.7	
Petroleum and coal	0.5	
Rubber	1.5	
Ceramics, stone, and clay	2.8	
Total of heavy and chemical industry		36.7
Electrical machinery	14.4	
Transport equipment	6.1	
Precision instruments	5.8	
Total of machine-assembly industry		26.3
Food and kindred products	6.4	
Textile mill products	3.8	
Apparel and other finished textile products	2.8	
Lumber and wood	1.5	
Furniture	2.0	
Pulp and paper	3.3	
Publishing and printing	9.2	
Leather	1.5	
Miscellaneous	6.5	
Total of light industry		37.0

Source: 1960 Establishment Census of Japan, 3.

the eastern industrial district, most strongly on the delta of the Sumida River, but also extending inland along the Arakawa Drain, and in the southernmost part of the coastal belt around the mouth of the Tama River. The assembly industry is strongly concentrated in the southern part of the coastal belt, and to a smaller extent inland along the Arakawa-Sumida floodplain and in scattered areas on the diluvial upland. Light industries are widely distributed and localized according to the articles produced. To a greater degree than is true of the others, they tend to concentrate in the central areas of the metropolis.[22]

It was noted earlier that the more exclusively residential parts of Tokyo are to be found in the western half of the city, chiefly on the diluvial upland. A smaller residential area is located east and north of the Arakawa Drain on what is floodplain. To be sure, many dwellings are found in the industrial and commercial sections as well, and throughout the residential areas streets with shops are very numerous. It is estimated on the basis of population that there is a 20 per cent housing shortage in Tokyo, a situation of such seriousness that it has led to the widespread development of multiple-dwelling-unit housing estates, both publicly and privately financed. At least six of these have over 1000 dwelling units each, and there are two to three times as many with over 500 units.[23] Some of the larger housing estates are equipped with public facilities in the form of schools, parks, playgrounds, a shopping center, and a public hall.

It was because of the shallow silted nature of Tokyo's harbor, at the head of Tokyo Bay, that following the Restoration the new deepwater port of Yokohama was developed, thirty to forty kilometers down-bay. Accordingly, a large proportion of Kanto's overseas trade is done through Yokohama, even though much of the cargo originates in, or is destined for, the metropolis. There remains the inconvenience, therefore, or having to use barges, trucks, and railroads to join Tokyo with Yokohama, its outer port. But gradually the harbor and port of Tokyo have been improved so that at present it is the fifth-ranking port in Japan, although it is used chiefly by smaller freighters. Only since 1953 has Tokyo been able to accommodate vessels of 10,000 or more gross tons, and harbor and port improvements are now in progress which will allow still larger vessels to dock. The sea approach to Tokyo is by means of a dredged fairway 9 meters deep and 150–200 meters wide. For the most part, ships anchor at buoys and load and discharge cargo with the aid of lighters and barges. Domestic trade, which is three to four times larger than the foreign, is preponderantly imports, a fact that reflects the importance of the local market. Coal and cement are the chief domestic imports.

Most important classes of foreign cargo unloaded at Tokyo are raw materials for industry, food, and semiprocessed goods, with greatest emphasis on rice, wheat, sugar, lumber, and scrap iron. Chief foreign exports are general cargo, steel, and machinery.

Yokohama

Yokohama is one of those few Japanese cities which had little or no contact with the feudal period, since it was called into being by the exigencies of the modern era. Although only a small fishing village in 1859 when it was declared an open port, at present it is a metropolis of over a million, and one of the nation's two really great deepwater ports. Preeminently it is a port city, and its growth has been closely associated with its expanding services as a foreign-trade port, not only for Tokyo and the Kanto region, but for all of northern Honshu as well. Lying in a small indentation along the west side of Tokyo Bay, it has adequate depth of water and docking facilities to permit the largest trans-Pacific ships to anchor alongside the piers. Most freighters, however, anchor at buoys and load and unload with the aid of lighters. Because of the highly congested condition of the harbor, which is cluttered with numerous barges and small boats, ship collisions are not uncommon. Foreign and domestic trade are about equal in tonnage, and foreign trade value is about equally divided between imports and exports, although the tonnage of incoming cargo is five times that which is loaded. The latter situation reflects the industrial nature of the Kanto hinterland, which imports chiefly bulky raw materials destined for manufacture and exports largely processed goods. (See Table 12-2.) It is both a commercial and an industrial port. Compared with Kobe, the only port which exceeds

TABLE 12-2

Important items in the foreign trade of Yokohama, 1960

Exports	Value (1000 dollars)	Ratio to all exports	Imports	Value (1000 dollars)	Ratio to all imports
Marine products	108,447	12%	Petroleum	97,122	11%
Radios	69,489	8	Wheat	54,381	9
Clothing	62,461	7	Non-ferrous metals	54,128	6
Iron and steel	52,242	6	Soybeans	39,036	4
Optical instruments	43,639	5	Wool	23,731	3
Vessels	43,403	5	Scrap iron	19,564	2
Silk fabrics	40,025	4	Raw cotton	19,122	2
Raw silk	39,897	4	Raw hides	14,358	2

Source: Economic Bureau of Yokohama Municipal Office, *Yokohama Trade and Industry.*

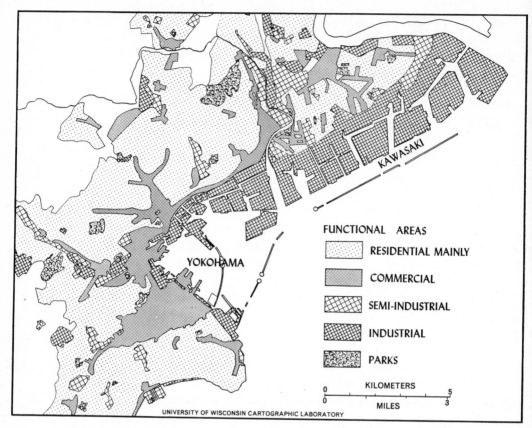

Fig. 12-11.—Land use in Yokohama and Kawasaki. From map prepared by the Kanagawa Prefectural Office, scale 1:50,000, 1962.

it in foreign trade, Yokohama equals its rival in value of imports, but exports only 60 per cent as much—and this in spite of the fact that Keihin exceeds Hanshin in industrial output. On the other hand, the hinterland that Yokohama serves outside of Kanto, chiefly northern Japan, is not nearly so industrialized as is southwestern Japan which is tributary to Kobe.

The city occupies three kinds of sites: (1) in complete and compact fashion, three small wedge-shaped delta-plains opening out upon Yokohama Bay; (2) spurs of diluvial upland forty to sixty meters high which separate the three delta-plains, these being much less completely urbanized; and (3) a coastal strip, a considerable part of which has been recently reclaimed from the sea, extending northward from the harbor along the coast (Fig. 12-11). The small streams which have formed the delta-plains occupied by the city do not carry enough load to seriously silt the harbor. On the other hand, they provide a splendid network of waterways throughout the lowland portions of the city, offering exceptional transport advantages to business and industry. The heart of Yokohama, containing its commercial core, is on the southernmost of the three small alluvial plains, immediately back of the principal piers. Here a goodly number of multistoried Western-style buildings have been erected since the bombing and burning associated with the war, but they are more scattered than in Tokyo, and consequently Yokohama does not have the strongly Western and metropolitan appearance of its larger northern neighbor.

Although the low delta-plains and the diluvial spurs are both sites for dwellings, it is the flanks and crests of the uplands that are most exclusively residential and have the most attractive living quarters. From the crests of the diluvial upland splendid views may be had of the lowland city and its harbor.

Throughout most of its history Yokohama has been chiefly a port city and manufacturing has been somewhat overshadowed by trade. One reason for the slower development of industry has been the restricted area of level land suitable for factory sites; for the local plains are small, congested, and hemmed in by diluvial uplands with steep seaward margins; so that the expansion inland of large-factory industries has been seriously handicapped. As a consequence most modern industry is coastal in location. Three factory areas may be recognized. First and foremost is the strand area, occupying a strip of reclaimed land along the bay north of the main harbor but still within the outer breakwaters that protect the Keihin Sea Canal. Since 1945, 1.3 million square meters of land for factories have been reclaimed from the sea, and plans are under way for still further expansion. Here large modern factories, many of them with their own company piers served by deep canals and spur lines of railway, are the rule (Fig. 12-12). The main commodities produced in this section are iron and steel, vessels, automobiles, machinery and tools, metal goods, chemicals, and foodstuffs, whose combined output is about 80 per cent of the entire industrial production of Yokohama. A second and less exclusive industrial area lies in the southern section of the city, on the deltas, where textile products in the form of shawls, scarfs, kimonos, handkerchiefs, and the like, made from silk and rayon fabrics, are processed in small plants of workshop size. And finally, there are scattered inland centers with a variety of industries, chief of which are light electric and chemical manufacturing. South and west of Yokohama, where hills reach almost to the sea margins, isolated modern factories, processing chiefly consumer's goods, are concentrated along the route of

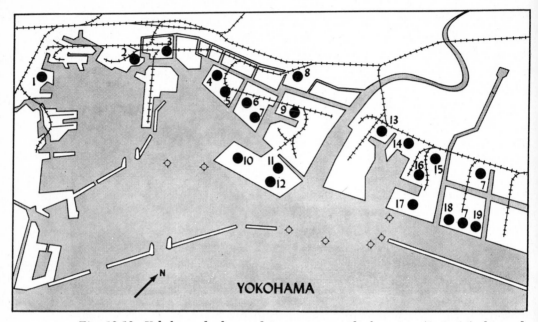

Fig. 12-12.—Yokohama harbor and important seaside factories: (1, 2, 14) shipyard, (3) flour mill, (4) elevators and grain warehouses, (5, 9, 11) chemicals, (6) motorcars, (7, 10, 18) oil refinery, (8) brewery, (12) thermoelectric power plant, (13) glass, (15) steel mill, (16) gas plant, (17) electronic machinery, (19) trading company. From map in *Nippon no Chiri*, Vol. 3.   No scale given.

the Tokaido rail line, where they are to be observed from the train windows, their fine appearance and modernity seeming to advertise the burgeoning industry of Japan.

Industrial Kawasaki (566,000 population), filling the gap along the coast between Tokyo and Yokohama, although a separate municipality, is only the intermediate segment of a continuous conurbation. Its seaside industrial belt is a continuation of those of its larger southern and northern neighbors, while in composition its industries resemble those of Yokohama's coastal belt. Because of the heavy type of industry, housed in large plants, which predominates in both Kawasaki and Yokohama, there is an unusual striving for tidewater location where the individual plants can be served by ocean-going ships. Almost exclusively an industrial port handling bulky raw materials, the tonnage ratio of foreign imports to exports is about fifty to one in Kawasaki.

Also included within the Keihin industrial zone bordering Tokyo Bay is the newly developing Chiba zone adjoining Tokyo on the east and extending along the northeastern shore for as much as thirty kilometers (Fig. 12-14). As one of the country's recently emerging tidewater zones of heavy industry, it directly reflects the necessity within the Kanto area

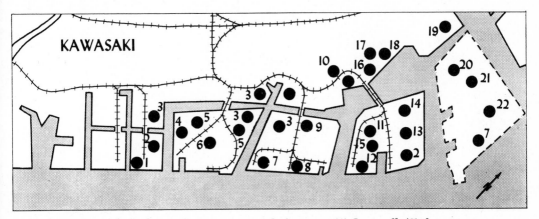

Fig. 12-13.—Kawasaki harbor and important seaside factories. (1) flour mill, (2) thermo-electric plant, (3) steel, (4, 14, 15, 16, 17, 18, 22) chemicals, (5, 6, 7, 8, 21) oil refinery, (9) sugar refinery, (10) metals, (11, 13) petrochemicals, (12) canned fish, (19) motorcars, (20) electronic equipment. From map in *Nippon no Chiri*, Vol. 3. Scale not given.

for providing new seaside sites for the soaring heavy and chemical industries. This one had its beginnings as late as 1953. Already by 1960 nearly 13,500 hectares of new tidewater factory land had been reclaimed from the sea; almost four times that amount is scheduled to be added by 1967, and another 46,000 hectares by 1975.[24] In this region the planners contemplate a duplication of the Kawasaki-Yokohama tidewater belt of heavy and chemical industries. Very recently, there have been built a steam-electric plant and a large and modern integrated iron and steel

TABLE 12-3

Composition of manufacturing in Yokohama and Kawasaki, 1960, in terms of value added (unit: 10,000 yen)

Main industry groups	Yokohama	Kawasaki
Food	1,254,352	1,992,203
Chemicals	1,020,745	1,486,862
Petroleum and coal products	702,368	1,042,549
Ceramic, clay, and stone products	1,174,872	558,370
Iron and steel	849,249	4,151,770
Non-ferrous metals	697,709	499,085
Fabricated metals	688,184	576,213
Machinery (general)	1,798,829	1,124,008
Electrical machinery	2,476,580	7,008,656
Transportation equipment	5,498,321	2,010,765
TOTAL, including all others	17,263,991	21,108,522

Source: Ministry of International Trade and Industry, *Census of Manufactures, 1960: Report by Cities, Towns, and Villages* (1962).

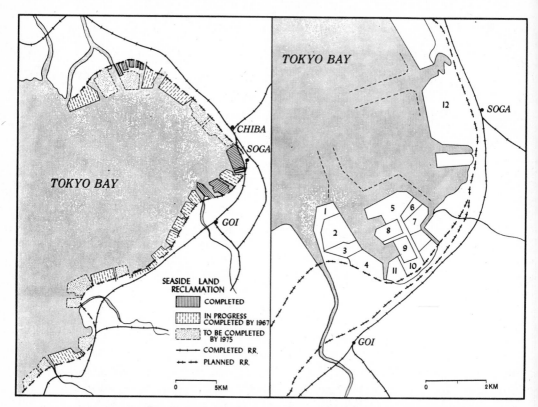

Fig. 12-14.—Development in the new Chiba industrial area. The left side shows the progress of area planning of seaside land reclamation for industrial sites. The right side indicates the factories already established in the vicinity of the Chiba-Goi section of the left side of the figure: (1, 7, 8, 9) thermoelectric plant, (2) oil refinery and petrochemicals, (3, 10) chemicals, (4) glass, (5) shipyard, (6) metals, (11) steel (from iron sand), (12) Kawasaki steel plant. After maps by Takeo Arisue.

plant in the vicinity of Chiba city, while a few kilometers farther south in the Goi-Ichihara area there are in operation, or in process of construction, a very extensive oil refinery–petrochemical combine and in addition upwards of ten other relatively large plants turning out glass, chemicals, ships, fabricated metals, and machinery.[25] This Chiba zone not only is one of the most rapidly growing, but seemingly has one of the highest potentials, of any of the new manufacturing localities in the Keihin area. No commercial ports of consequence exist, so that new industrial port facilities must be developed contemporaneously with the establishment of new factories. Already two dredged channels, each twelve meters deep, have been constructed to allow cargo ships to approach the iron and steel plant near Chiba city and the collection of plants in the Goi-Ichihara district.[26] Ninety-nine per cent of the foreign cargo tonnage is imports.

Cities	DID pop.	Food	Textile	Chemical	Iron and steel	Non-ferrous metals	Fabricated metals	Machinery	Ceramic, clay products	Rubber products	Total value added (10,000 yen)
Tokyo-to											
Musashino	120,000	....	....	16.2%	....	....	....	85.1%	....	....	511,381
Hachioji	93,000	....	51.1%	....	....	....	....	18.6	....	....	709,823
Mitaka	84,000	....	....	....	....	....	....	83.7	....	....	1,102,121
Tachikawa	66,000	....	....	....	....	....	....	68.0	12.5%	....	245,139
Fuchu	54,000	....	....	....	....	....	....	84.6	....	....	868,120
Kanagawa Pref.											
Yukosuka	220,000	11.6%	....	....	....	....	....	78.6	....	....	1,404,229
Fujisawa	78,000	11.8	....	....	....	....	11.5%	59.6	....	....	682,166
Hiratsuka	74,000	13.2	....	....	....	....	....	38.2	....	18.5%	556,320
Odawara	72,000	11.2	14.3	15.8	....	....	....	33.5	....	....	672,612
Kamakura	64,000	....	....	....	....	....	....	57.0	....	....	361,813
Chiba Pref.											
Chiba	167,000	7.3	....	....	84.3%	....	....	....	....	....	2,048,478
Ichikawa	130,000	26.5	13.0	11.8	8.1	....	....	17.9	....	....	663,602
Funabashi	79,000	14.1	....	6.9	....	7.4%	....	27.8	29.2	....	701,444
Choshi	56,000	90.0	....	....	....	....	....	5.2	....	....	511,297
Saitama Pref.											
Kawaguchi	131,000	....	5.0	....	33.6	....	7.6	30.7	5.3	....	2,798,082
Urawa	130,000	....	....	8.7	....	....	....	48.4	....	5.1	679,539
Omiya	115,000	....	12.5	9.0	....	....	8.3	52.5	....	....	605,391
Gumma Pref.											
Maebashi	107,000	25.9	21.1	....	....	....	....	12.3	....	....	537,781
Kiryu	81,000	....	55.2	....	....	....	....	22.8	....	....	860,064
Takasaki	80,000	19.0	....	....	8.0	....	....	35.1	....	....	802,742
Tochigi Pref.											
Utsonomiya	133,000	25.0	....	....	....	....	....	36.1	....	....	533,793
Ashikaga	57,000	....	58.2	....	....	....	....	....	....	8.5	785,900
Ibaraki Pref.											
Hitachi	100,000	....	....	....	....	33.5	....	50.2	7.4	....	2,630,672
Mito	78,000	45.2	....	....	....	....	....	18.1	....	....	283,854

Source: Ministry of International Trade and Industry, *Census of Manufactures, 1960: Report by Cities, Towns, and Villages* (1962).

The whole Keihin industrial zone fronting on Tokyo Bay is suffering from acute growing pains requiring large-scale planning. It is estimated that by 1970 the Kanto ports will be obliged to handle three to four times the cargo they did in 1959, an expansion that will demand much in the way of port development and improvement.[27] To meet the urgent demands for new land for factories, much of it with water-front location, it is planned to reclaim a total of 67–70 million square meters of land, some 37 million of this intended for industrial use in the Tokyo-Chiba area, and another 8–9 million along Negishi Bay near Yokohama harbor. Such a remarkable expansion in the seaside industrial area bordering on Tokyo Bay will require construction of a new coastal railroad and highway linking Yokohama to Chiba via Tokyo.

Outside of Keihin (chiefly Tokyo, Yokohama, Kawasaki, and more recently the Chiba area) the more extensive Kanto industrial zone includes in addition more than two-score other cities, located inland, with over 5000 workers each, where textiles, clothing, small metal products, foods, and other light industries predominate. Benefiting by proximity to rapidly growing Keihin where most of their processed wares are sent for consumption, domestic redistribution, or export, these dispersed large and small centers of Kanto may be expected to develop rapidly in the ensuing decades as new factories have increasing difficulty finding space within congested Keihin.[28] But the type of manufacturing in interior Kanto is likely to continue in its orientation toward light industries, whereas tidewater Kanto will increasingly emphasize heavy and chemical manufacturing.

On the extreme western margins of the southern Kanto Plain and extending into the adjacent foothills in the vicinity of Hachioji, is one of the country's important local textile areas. Here concentrated in such cities and towns as Hachioji, Ome, Ogawa, Tokorozawa, and Chichibu are upwards of 10,000 plants engaged in various aspects of the silk and rayon textile industry.[29] It is a well integrated textile area, including spinning factories, dye works, and weaving and finishing plants, together with numerous merchants engaged in wholesaling the merchandise. Many of the silk factories originated during the Tokugawa Period when such industry was encouraged by the feudal lords. Over a long period it has profited by proximity to a large local raw silk supply in the Gumma-Saitama sericulture region.

The industrial city of Hitachi in extreme northeastern Kanto, most important single manufacturing center within the Joban industrial district, has already been commented on in the discussion of the Abukuma Region within Tohoku.

13 · CENTRAL JAPAN: TOKAI

THE SUN-EN COASTAL STRIP

Along the Tokai littoral from Izu Peninsula to about Mikawa Bay,[1] the high mountains of central Honshu come close to the sea, so that extensive plains are absent, and the small fragments of coastal alluvium that are present are separated from each other by spurs of highland. Essentially this is the littoral belt of Shizuoka Prefecture (Figs. 13-1, 13-2). But in spite of its configuration, this Sun-en district has had an importance in Japanese life far out of proportion to its restricted area of level land. Its earlier consequence and fame largely derive from its having been the corridor through which, in Tokugawa days, passed Japan's most famous road, the Tokaido, connecting Tokyo (then Edo) and the old capital at Kyoto. Because of the several highland barriers separating the plains, as well as the errant streams, the route was not an easy one. Situated as it is between the great Kanto cities to the northeast and Nagoya and the Kinki cities to the southwest, it still retains its importance as a transit route, for Sun-en is traversed by two of Japan's most modern and efficient thoroughfares, the double-tracked Tokaido railway and the Tokaido motor highways. It is this superior location with respect to markets that has stimulated the modern development of a diversified commercial type of agriculture based on tea, fruit, and truck gardening.

A number of relatively important rivers descending from the central mountain core have built a series of small delta-fans which have become the focal points for population, agriculture, and industry. The Oi Plain

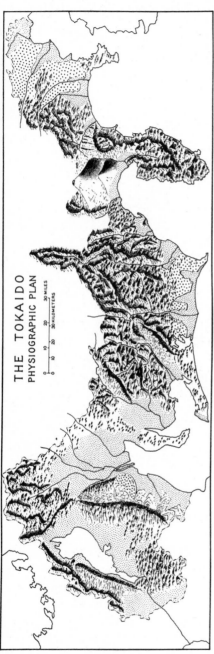

Fig. 13-1.—Physiographic diagram of the Tokaido District. Map by R. B. Hall, from the *Geographical Review*, published by the American Geographical Society.

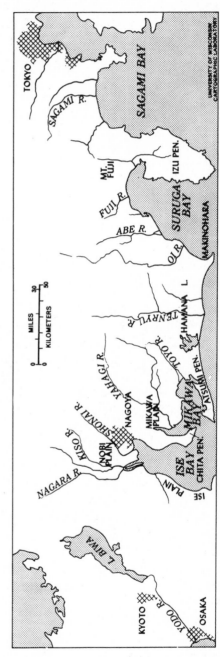

Fig. 13-2.—Location map to accompany Figure 13-1.

is distinctive because of its striking fan configuration, and even more so because its rural population resides mostly in isolated farmsteads rather than in rural villages. On alluvial plains in Old Japan the clustered type of rural living is so characteristic that the exceptional Oi Plain with its dispersed farm dwellings has been the subject of considerable discussion among Japanese geographers.[2] There seems to be agreement that this settlement feature is in some way associated with the vigorous and errant Oi River and its floodings, for the present scattered settlements appear to date from the great inundations of 1604 and 1627 which laid waste the Oi Plain. It is possible that the advantages offered by isolated dry sites, or the scattered nature of fertile spots, also had some effect in dispersing the homes of the new settlers on a plain made desolate by water and gravel.

The delta of the Tenryu River in western Shizuoka Prefecture is rather exceptional because of the unusual number of elevated dry fields which rise like tiny rectangular buttes two to four feet above the lower inundated paddy fields.[3] These thousands of elevated plots owe their origin to the fact that the Tenryu has refused to be contained within its dikes. The sand and gravel spread over the plain in times of flood were first scraped off and piled up into heaps, and at subsequent periods leveled off and converted into dry fields. Succeeding inundations, which ruined the rice crop in the paddy fields but spared the crops on the raised fields, proved the positive value of elevated sites. Currently the delta of the Tenryu has the appearance of a huge checkerboard, on which great numbers of the irregular squares have been pushed up above the surface. In some parts the elevated plots appear to occupy nearly as much area as the paddies. A two-storied type of land utilization prevails, for the raised fields are devoted to an intensive rotation system involving summer vegetables and fall-sown crops of grain and rapeseed.

Along the coasts of those alluvial plains fronting upon the open ocean west of Cape Omae, wide belts of sandy beach ridges and dunes are characteristic, coastal settlements are few, and fishing is meagerly developed. Such formidable belts of dunes are absent along the protected coasts facing on Suruga Bay, and here the lower beach ridges are important sites for settlements and for specialized truck-and-market-gardening developments. Fast rail service carries the perishable products to the great urban centers lying both to the northeast and to the southwest. Particularly famous as truck-gardening centers for out-of-season vegetables are the narrow plain at the base of Kuno Mountain, southeast of Shizuoka city, and Miho Spit, a little farther to the east, which forms the protected harbor of Shimizu (Fig. 13-3).[4]

Fig. 13-3.—Out-of-season strawberry culture in the truck gardening zone at the base of Kuno Mountain near Shizuoka in central Tokai.

High and fairly extensive diluvial uplands, composed of thick sand and gravel strata resting upon a base of Tertiary rocks, whose flattish to rolling crests are covered with a veneer of volcanic ash, are conspicuous features of the Sun-en littoral, especially its western half. Some of these uplands rise by strikingly steep slopes for 150–300 meters above the surrounding lowlands. Sun-en is nationally famous as the area of greatest specialization in citrus and tea, Shizuoka Prefecture having about 53 per cent of the nation's tea acreage and 20 per cent of its area of mandarin oranges. It is both on the crests and flanks of the diluvial uplands and on the hill slopes bordering the alluvial lowlands that these two perennial commercial crops are largely concentrated. While both crops are grown on the two types of upland sites just mentioned, tea is more a specialized crop of the diluvial surfaces and citrus of the hill slopes, with a high degree of citrus concentration on those protected slope lands with southerly exposure in eastern Shizuoka bordering Suruga Bay. Shizuoka city has become the primary center of the nation's tea organization, including processing and merchandising.[5] Both tea and oranges are important foreign exports through the local port of Shimizu.

Included within the nation's industrial belt, though not forming an important node, Sun-en contains at least four separate local centers of manufacturing which warrant individual brief comment. All of them

Fig. 13-4.—Location of three local industrial districts and their main cities at the head of Suruga Bay in eastern Tokai, Shizuoka Prefecture.

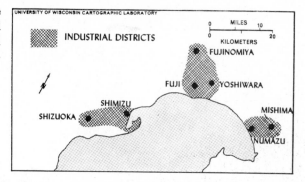

profit by location on the Tokaido railway and highway, and only one has direct access to ocean transport. Two of the four, the Numazu-Mishima and the Gakunan centers, are located in easternmost Shizuoka at the head of Suruga Bay and consequently close to the base of the Fuji volcanic group where abundant hydroelectric power is an important asset (Fig. 13-4). Numazu-Mishima, easternmost and smallest of the four, has an industrial composition which is varied in character, including machinery, non-ferrous metals, fabricated metals, foods, rubber, and chemicals as important items; but all-machinery leads in importance by a wide margin, representing over 40 per cent of the total value added by manufacturing. The Gakunan district, located at the mouth of the Fuji River and including such individual industrial towns as Fuji, Yoshiwara, and the Fujinomiya, has two to three times the industrial output of the Numazu center. Earlier its fame was closely associated with silk reeling, but this has largely disappeared and at present it has a slightly more diversified industrial structure, with pulp and paper well in the lead, however, providing 50 per cent of the value-added, followed by machinery of all kinds with about 37 per cent (Fig. 13-5).[6] This is one of the foremost pulp and paper centers of Japan.

Nearly on a par with Gakunan in terms of value-added, the third manufacturing district, Shizuoka-Shimizu, is the only one having adjacent port facilities. Its manufacturing composition shows great variety, but machinery of all kinds takes first rank, providing about 23 per cent of the total value-added. At the same time foods, chemicals, non-ferrous metals, paper and pulp, and lumber and wood products all are noteworthy items. Shizuoka city (230,000 population) reflects the local hinterland in that it is the nation's foremost tea processing and packing center, while Shimizu (106,000 population) does the same with its orange canning. Distinction also attaches to Shimizu by reason of its shipyards and oil refinery–petrochemical combine. In the latter city, man-

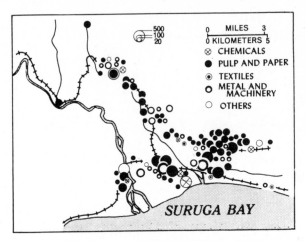

Fig. 13-5.—Types of factories and their distribution in the Gakunan manufactural district, in eastern Tokai, Shizuoka Prefecture. After map by Isamu Ota.

ufacturing plants are chiefly concentrated along the water front.⁷ Shimizu, although a lighterage port, is one of the more important secondary ports of call in Japan, serving not only the local industrial district, but those of eastern Shizuoka as well. Principal foreign exports are canned fish, canned oranges, composition board, tea, and machinery, most of the items reflecting the local agricultural and industrial hinterland. Imports are chiefly industrial raw materials such as oil, ores, and lumber, together with foodstuffs.

Hamamatsu, westernmost of Sun-en's four manufacturing districts and unlike the others in that it includes only one urban center, ranks just slightly below Gakunan and Shizuoka-Shimizu in importance. But on the other hand, Hamamatsu (180,000 population) is Sun-en's foremost manufacturing city. Like the other industrial districts, its manufactural composition is varied, but unlike them its fame until recently has been chiefly in cotton textiles. In this respect Hamamatsu resembles more the Chukyo or Nagoya area, just to the west, than it does the remainder of Sun-en. However, in recent decades Hamamatsu's industrial structure has been changing, for while in terms of number of employees textiles still hold first position, in terms of value-added all-machinery (44 per cent) ranks well above textiles (25 per cent). Foremost single industry is the manufacture of motorcycles, while perhaps the most unusual specialization is that in musical instruments, more specifically pianos and organs. All aspects of the textile industry—spinning, weaving, dyeing—are here represented, with the spinning being done in large plants, and the other processes in thousands of inconspicuous establishments.

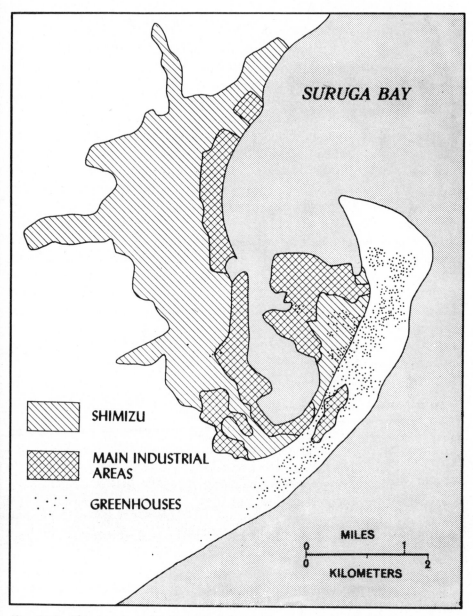

Fig. 13-6.—Shimizu port city, its main industrial areas, and the distribution of green-houses on Miho Spit, a specialized truck gardening area. After maps by Shozo Yamamoto and others.

THE PLAINS BORDERING ISE
AND MIKAWA BAYS

Westward from Sun-en, and between it and Kinki, are the plains ad-joining Ise Bay and tributary Mikawa (Atsumi) Bay. Parts of three pre-fectures—Aichi, Gifu, Mie—are included within the region. The entire plain's area, composed of both alluvium and diluvium, exceeds 4000 square kilometers, which makes it, next to Kanto, the most extensive plain of Japan. Less compact than Kanto, it includes as its core area the Nobi or Nagoya Lowland (1600 sq. km.) at the head of Ise Bay, and in addition other lowlands bordering that water body on the east and west, and still others along the margins of Mikawa Bay. It is difficult to compute the population of this somewhat arbitrary area, but within a circle having a fifty-kilometer radius and centered on the Nagoya City Office—a circle whose circumference embraces an area which is roughly what is here designated as the Ise Bay borderlands—there were in 1960 about 5.4 million inhabitants, of whom 49 per cent lived in DID or ur-ban settlements. The latter figure includes the metropolis of Nagoya with nearly a million and a half inhabitants as well as seven other cities of 100,000–200,000 people. Of the three great metropolitan areas, this is the smallest in population, and its percentage of urban people is the low-est. These Ise Bay lowlands are roughly coincident with the Chukyo manufacturing node, described in an earlier chapter, which ranks third in the country, after Kanto and Kinki.

The Nobi Plain

Largely a lowland of new alluvium, but with a fringe of diluvial ter-race, chiefly on its eastern margins, Nobi geomorphically may be divided into two parts, a higher eastern and a lower western. The eastern subdi-vision, where the main rivers, including Kiso, the master stream, enter the lowland, is not only higher, but it is also characterized by fan con-figuration, terraces, natural levees, more extensive fragments of dilu-vial upland, and altogether by greater surface irregularity and local re-lief. The lower western section, the receptacle of the major drainage lines, is mostly back and tidal marshes, threaded by conspicuous levees which are the highest and driest sites.[8] This contrast between eastern and western Nobi has a twofold explanation: (1) The Kiso and other rivers entering from the northeast carry a far greater load of material than do the smaller streams entering from the west and northwest. (2) There has been a differential crustal movement, involving rising toward the east and subsidence in the west.[9] Low western Nobi has had a long history of constant battle against floods and inundations, many of which

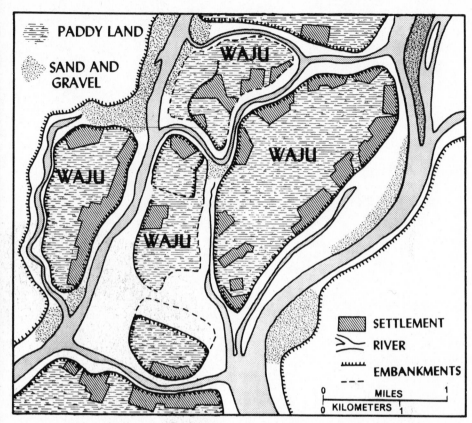

Fig. 13-7.—An example of *waju* lands in the Kiso Delta of the Nagoya Plain. Note the location of settlements along the embanked margins of the several *waju,* or polders. After map in *Regional Geography of Japan,* No. 4, *Chubu Guidebook.*

were accompanied by serious losses of life and property. This is *waju,* or polder, country, with immense dikes as high as ten meters along the streams, enclosing waju of varying sizes.[10] Construction of the dikes was begun about a thousand years ago, some of the original work having been under the supervision of Buddhist temples which were in possession of large areas of the wet lowland. Individual polders of considerable extent are sectioned into subdivisions by cross dikes, so that different kinds of agricultural land use can be carried out within the same general polder (Figs. 13-7, 13-8). Conspicuous throughout Nobi are the scars of abandoned stream channels and levees testifying to the capricious migrations of swollen rivers in time of flood. Historical records reveal a total of 110 damaging floods over a period of 150 years.

Nobi has had a long history of human occupance; the presence of relics of the Jori land division system indicates that paddy fields were devel-

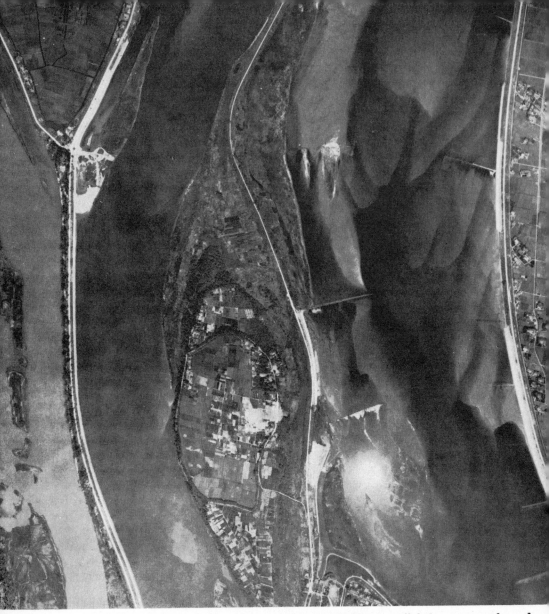

Fig. 13-8.—Waju country in western Nobi. High dikes parallel the rivers with roads and settlements commonly concentrated along the dikes. Courtesy of the Geographical Survey Bureau, Japan.

oped there as far back as the seventh century. But the absence of Jori features on the Kiso alluvial fan in the northeast, and in the low marshy areas of the southwest, denotes that in these sections rice fields were a later development. In fact, large parts of the swampy southwest, including the tidal flats, were not reclaimed until between the seventeenth and nineteenth centuries, during which period the coastline near the mouth of the Kiso River was pushed seaward more than ten kilometers.[11] The reclamation of the tidal lowlands continued throughout the Meiji Period and even down to the present time, for a project by the national government to reclaim several hundred hectares at the mouth of the Kiso River for agricultural use is now in progress.

Population of this plain is around 3 million, of which nearly one-half are in Nagoya city. About 80 per cent are urban. Because of its multiplying factories, Nobi is a region of strong net in-migration, although the stream of migrants is not so large as that entering either the Kanto or the Kinki industrial areas. Within the plain, population densities are greater in the higher, drier, and more industrial eastern section, where market gardening is also specialized, than in the wetter southwest where industry is meager and a rice monoculture prevails.

Rural Nobi is largely planted to rice, the ratio of paddy fields to dry fields being about 7:3.[12] But although rice fields are widely distributed over the plain, that crop is most exclusively grown in the low waju lands of the southern and western sections, where it occupies as much as 90 per cent of the cultivated area. By contrast, in the higher and drier east, with its somewhat more uneven surface, where irrigation is less easy and a more varied and complicated cropping system prevails, paddy fields and dry fields divide the arable area about equally. Until recently mulberry was an important crop in eastern Nobi, for the Ise region was one of Japan's specialized areas of sericulture, but the decline has been drastic, Aichi Prefecture's mulberry acreage in 1960 being only 12 per cent of what it was three decades earlier. As the mulberry area has shrunk, market gardens have proven a profitable substitute in those parts close to Nagoya, while in more distant sections the usual upland food crops have taken over.

In the polder country where summer rice so completely dominates, drainage is so poor that about half the paddy area remains fallow in winter in spite of recent land improvement. Chief winter crops in the paddies are green-manure vetches, vegetables, and rapeseed. On the broad levees and dikes dry crops are planted. Rural settlements in the waju lands are of two kinds, those occupying dry sites on the dikes and those situated within the polders. Most dike villages have linear dimensions, whereas the others are more compact. Houses in the waju villages are built upon artificially elevated earthen platforms as a protection against inundation, and special *mizu-ya,* or flood-houses, are erected upon foundations some three meters high to serve as storage places for articles of value, or even foodstuffs, in times of flood. With the completion of new embankments in 1900, the danger of floods has been greatly reduced, rendering mizu-ya less necessary. Each waju village is enveloped in a dense screen of trees whose present function is to protect against winter winds, although formerly the trees also provided a temporary refuge during serious floods.[13] There are about seventy waju of various sizes, the largest having a circumference of seventy kilometers.[14] Only by means of continued pumping are the waju made sufficiently dry for cropping, and even so, water areas—including ponds, marshes, and canals used for drainage, irrigation, and transport—may constitute as much as 20 per cent of the total area of some waju.

Cities of Nobi

On the Nobi Plain there are at least five cities of over 50,000 DID population—Nagoya, 1,465,000; Gifu, 204,000; Ichinomiya, 82,000; Ogaki, 63,000; and Seto, 75,000. Several of these larger cities, and more of the smaller ones, are specialized in one or more phases of the textile industry, for the Ise Bay area is one of Japan's two principal textile concentrations. Cotton spinning and weaving, the manufacture of woolen yarn and cloth, and the processing of rayon thread as well as rayon and rayon-mixtures cloth—all of these subdivisions of the textile industry are importantly represented.

Ichinomiya is very much a one-industry city, for textiles are responsible for 96 per cent of the total value added by manufacturing. Machinery, next in rank, provides less than 2 per cent. Here at Ichinomiya and in surrounding smaller cities and towns (Bisai, Kasamatsu, Tsushima, Kisogawa) is centered the wool manufacturing of central Nobi and the wholesaling of wool manufactures as well, the wholesale establishments numbering 150 to 200. Weaving and yarn-twisting mills are of small size, but the spinning and dyeing-finishing plants are large. This is Japan's principal focus of wool manufacturing, and many large exporting firms with headquarters in Osaka and Tokyo have representatives or branches in Ichinomiya so as to deal directly with the mills.[15] This same area likewise specializes in the making of Western-style woolen clothes.[16]

Gifu, an old castle city and a prefectural capital, is not highly industrialized as are the other large cities of Nobi, and what fame it has as a processing center is related to indigenous craft industries making Japanese umbrellas, fans, paper lanterns, and textiles for home consumption, rather than to modern manufacturing. Still, there are several fairly large textile mills and a great number of small plants, so that their combined production represents some 27 per cent of the total value added by manufacturing, and this rises to 36 per cent if apparel is included. Machinery ranks next in order, with about 16 per cent of the value-added. As a result of proximity to the concentrated textile area around Ichino-

TABLE 13-1

Composition of manufacturing in important industrial cities within the three subdivisions of Ise-Mikawa, 1960, based on number of employees

Industry group	Nobi cities					
	Nagoya	Ichinomiya	Bisai	Seto	Ogaki	Gifu
Food	23,497	662	117	214	1,240	2,337
Textiles and apparel	33,885	52,615	24,763	......	10,972	12,139
Lumber, wood products, and furniture	31,575	467	82	233	462	2,499
Chemicals	14,448	102	......	44	2,572	286
Ceramic and clay products	18,870	133	25	16,323	817	740
Iron, steel, non-ferrous metals	21,866	96	......	......	402	884
Fabricated metals	24,846	74	......	306	404	1,124
Machines (general and electric)	55,512	979	259	788	2,399	3,508
Transportation equipment	31,787	112	......	......	1,148	1,170
Precision instruments	6,966	......	......	......	121	68

Industry group	Mikawa cities					
	Okazaki	Handa	Kariya	Toyota	Gamagori	Toyohashi
Food	2,009	1,804	365	487	436	5,378
Textiles and apparel	12,584	7,010	......	431	11,591	10,475
Lumber, wood products, and furniture	1,057	446	604	323	247	5,174
Chemicals	3,592	171	91	......	332	301
Ceramic and clay products	1,102	729	1,008	453	52	442
Iron, steel, non-ferrous metals	592	141	701	......	......	991
Fabricated metals	296	268	930	612	102	1,090
Machines (general and electric)	1,444	1,999	9,382	273	317	1,633
Transportation equipment	909	664	7,879	13,332	428	1,154
Precision instruments	732	......	......	......	818	242

Industry group	Ise cities				
	Tsu	Yokkaichi	Ise	Kuwana	Suzuka
Food	1,236	1,746	819	438	448
Textiles and apparel	5,313	11,105	1,271	1,763	3,675
Lumber, wood products, furniture	837	605	759	798	161
Chemicals	26	5,539	51	63	604
Ceramic and clay products	1,119	8,560	115	490	368
Iron, steel, non-ferrous metals	84	1,777	84	5,030	107
Fabricated metals	146	1,014	262	784	.....
Machines (general and electric)	1,494	5,087	1,713	4,393	362
Transportation equipment	.....	221	1,171	226	.....
Precision instruments	16	34	.....	323	6

Sources: Statistical Yearbook of Aichi Prefecture, 1962; and Ministry of International Trade and Industry, *Census of Manufactures, 1960: Report by Cities, Towns, and Villages* (1962).

miya, Gifu has developed into an important wholesale garment center, with hundreds of wholesale establishments. The city also boasts a tourist attraction: the unique and primitive cormorant fishing in the nearby Nagara River, which annually draws large numbers of sight-seers.

Seto, like Ichinomiya, is a one-industry city, but in this instance the specialty emphasized is pottery or chinaware, which contributes 70 per cent of the total value added by manufacturing. Seto has had a long history as a famed pottery center, so much so that chinaware products in Japan are often referred to as Seto-mono or Seto-ware. In the Tertiary hills along the eastern side of the Nobi Plain pottery clay of good quality is obtainable, which resource gave rise to pottery manufacturing even centuries ago. Of the hundreds of ceramics factories in the city, 90 per cent have fewer than ten workers. A high degree of specialization exists among the small plants, both in the different steps of the industry and in the individual products made. Some plants make saucers only, others cups, etc. They are not equipped to produce the finished wares for export, so from Seto and other towns in the hills the semifinished products are sent to Nagoya for finishing and decorating. Over half of the export product goes to North America.

Present-day *Nagoya,* Japan's third largest city, has resulted from the merging of two former cities, old Nagoya, developed around a feudal castle situated several kilometers inland, and Atsuta, a famous shrine city, located on the adjacent coast, whose annexation in 1907 provided Nagoya with a seaport. Nagoya's advantages include its important local hinterland, the Ise Bay plains, and also its intermediate location between Kanto and Kinki, with which it is connected by the Tokaido rail line and highway.

The core of the city is situated about eight kilometers from tidewater, and as late as 1930 its connection with the harbor was by means of a relatively narrow and incompletely urbanized zone, so that the whole city had the shape of a funnel tapering to the south (Fig. 13-9). Nagoya's original site, and still the site of much of the central part of the presently expanded city, is a very low diluvial upland whose smooth surface is only five to fifteen meters above sea level. The earlier funnel shape approximately coincided with the slightly elevated site, which in turn operated to protect the city from the floods that in the past scourged the adjacent alluvial lowlands. Yet the diluvium was not high enough to entirely preclude the development of canals within the city, although they are fewer than in either Tokyo or Osaka. Over the past three or four decades, however, there has been a rapid expansion of the Nagoya urbanized area, especially southward, with a consequent spread of the city

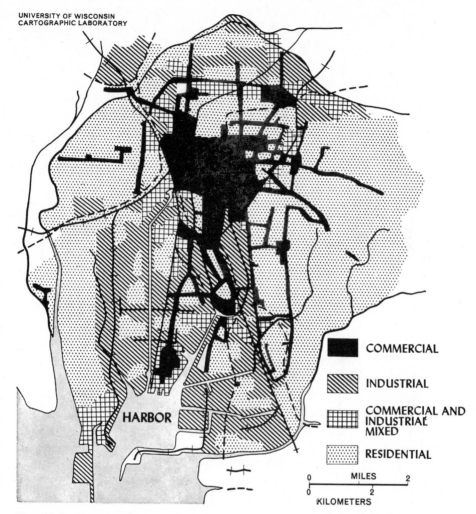

Fig. 13-9.—Functional areas in Nagoya. After map in *Regional Geography of Japan,* No. 4, *Chubu Guidebook.*

on to lower alluvial lands in the direction of the harbor, where inundation is a hazard.

The central business area is situated in the north central part, just south of Nagoya Castle, a modern replica of the one destroyed by fire during the war. In this section modernization, in the form of widened streets and new multiple-storied Western-style buildings, gives parts a distinctly metropolitan appearance. Here are concentrated many public buildings and of course retail and wholesale establishments, the latter comprising about one-fifth of the total. Important wholesale functions are relatively new to Nagoya, but at present the wholesale business accounts for about 90 per cent of the value of all commercial transactions, indicating the role of the city as middleman for manufacturers in the smaller cities of the Ise area.[17] Nagoya's tributary wholesale area is somewhat limited by competition both from Osaka on the west and from Tokyo on the east, but still it represents a population of 6 to 7 million. The city's most exclusively residential area lies on the eastern side, and less exclusively to the north and west, of the business district.

Partly because of its later development in industry, no doubt, manufacturing functions appear to be less areally segregated than in most of the other metropolises. Newest and most exclusive of the industrial areas is that in the southern part of the city, located on low alluvial land between the business center and the harbor, where canals are relatively numerous. Within this area there is a noteworthy concentration in the coastal zone, chiefly of large modern factories specializing in heavy and chemical industries, together with lumber mills, the tidewater location being particularly attractive to industries requiring extensive factory sites and ready access to imported bulky raw materials (Fig. 13-10). Some of the present tidewater factories are located on land reclaimed from the sea, and with future industrial expansion in mind, extensive additional areas are in process of being reclaimed south and west of the present harbor, while plans have been formulated for subsequent increments along the east side of Ise Bay.

Other less exclusively industrial areas within Nagoya form what is usually designated as the inland industrial zone. It is not one area, however, but several, and these are peripherally located with respect to the main business section, the most important being to the north on the alluvial plain of the Shonai River. Large factories are not absent, to be sure, but dominantly the inland areas are characterized by small and medium-size plants specializing in textiles, ceramics, foods, and similar consumer's goods. Several large establishments engaged in various kinds of machine industries are also situated inland. Most of Nagoya's ceramic plants are occupied in finishing and decorating the semifinished products received from Seto and other pottery centers in eastern Nobi.[18]

Although Nagoya is pre-eminently an industrial city, with a higher proportion of its employed population engaged in manufacturing than is true of most of the other metropolises, its present accelerated development is somewhat in the nature of a compensation for an earlier retardation which resulted in part from inferior water transport facilities. Like Tokyo and Osaka, Nagoya is located on a bay-head plain where silting rivers have shallowed the coastal waters so that originally ocean-going vessels were unable to reach the city; but unlike Tokyo and Osaka, Nagoya had no adjacent deepwater port such as Kobe or Yokohama. Manu-

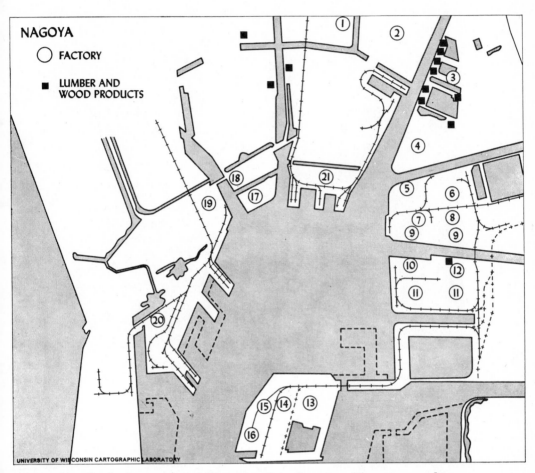

Fig. 13-10.—The harbor section of Nagoya and its factories. (1) sewing machines, (2) light metals, (3) composition board, (4, 12) steel, (5, 13, 19) thermoelectric plant, (6) rayon, (7) railway cars, (8) aircraft, (9) heavy industry, (10, 18) shipyard, (11, 17) chemicals, (14, 15, 16) oil refinery, (20) concrete. After map in *Nippon no Chiri*, Vol. 4. No scale given.

facturing industries first became important in those parts of Japan provided with adequate marine transport. Not only was the original city nucleus of Nagoya located some eight kilometers inland from shallow tidewater, but within the city itself the network of canals was not comparable to that of either Osaka or Tokyo, so movement of bulky goods was made more difficult. Consequently, the city was largely dependent upon rail transport, and it was not until after the development of efficient rail service that Nagoya's modern expansion began. Today it is an important rail center on the main Tokaido Line. Its trade territory is primarily the borderlands of Ise Bay, with their five to six million in-

Fig. 13-11.—The harbor industrial zone of Nagoya, showing extensive areas of seaside land reclamation for factory sites in various stages of completion. Courtesy of Nagoya Municipal Office.

habitants, which comprise a major settlement area and consuming center. As mentioned earlier, the outer hinterland of Nagoya is somewhat limited by the proximity of Osaka and Kobe. However, its influence has been extended (1) toward the Matsumoto and Suwa basins of Tosan, by a rail line following the Kiso Valley, and (2) toward the Japan Sea coast, by way of the Biwa Depression.

Until very recently Nagoya's industry, like that of the larger Chukyo Industrial Zone of which it is the principal unit, was heavily weighted in favor of light consumer's goods, more so than was the case with the other metropolises of Japan. For example, in 1955, about half the value of industrial shipments from Nagoya belonged in the light-industry class. Machinery was the only one of the heavy industries in which, percentage-wise, Nagoya measured up to the other great cities; in metals and chemicals it lagged well behind. Factors accounting for the poor showing in primary metals and chemicals are (1) the longer distance from coal fields, (2) the late development of modern port facilities, and (3) the inadequate water transportation facilities within the industrial parts of the city.[19] These same handicaps did not have such an adverse effect on those industries which profited from the abundance of hydroelectric power, such as electric steel, electrochemicals, and synthetic fibers.

But the five-year period 1955–1960 witnessed a remarkable change in Nagoya's industrial composition, as metals rose in value of output from 11.6 per cent to 17.5 per cent; machinery, from 22.7 per cent to 28.3 per cent; and chemicals, from 10.4 per cent to 13.8 per cent; so that heavy industry and chemicals as a class were lifted from 44.7 per cent to 59.6 per cent. In the same period textiles dropped from 16.9 per cent to 8.8 per cent; and all light industry, from 50.4 per cent to 35.8 per cent. As a consequence of these recent shifts in emphasis, Nagoya now much more resembles the other great cities in industrial composition, and this trend toward heavy and chemical industries no doubt will continue.

Although ranking among the first five commercial ports of Japan in foreign trade, Nagoya is scarcely in a class with Kobe and Yokohama, but on the other hand it does compare favorably with Osaka and Tokyo, which are similarly handicapped by location on shallow water at the head of silting bays. Not opened to foreign trade until as late as 1907, this man-made harbor and its port, in spite of natural handicaps, have grown importantly as the gateway to the expanding Chukyo industrial zone. But even yet a considerable amount of the cargo in and out of Chukyo passes through the Kinki ports, both because of their better facilities and services and because of the fact that Kinki business and financial interests are importantly involved in Chukyo industry and its commercial transactions. Approach is made to Nagoya harbor by means of a dredged fairway 9.1 meters deep, and since the turning space within

TABLE 13-2

Composition of manufacturing in Nagoya city, 1955 and 1960, in percentage of value added

Industry group	1955	1960
Heavy industry	34.3%	45.8%
Metals	(11.6)	(17.5)
Machines	(22.7)	(28.3)
Chemicals	10.4	13.8
Pottery	4.9	4.7
Light industry	50.4	35.8
Food products	(15.6)	(12.0)
Textiles	(16.9)	(8.8)
Wood products	(9.4)	(7.2)
Others	(8.5)	(7.8)

Source: Nagoya Municipal Office, *Industrial Nagoya* (1962).

the inner harbor is restricted, great congestion prevails. Most ships anchor at buoys and load and discharge cargo with the aid of lighters, but commodious new piers with depths-alongside reaching 9 meters are able to accommodate vessels of fair size. The recent invasion of the Nagoya area by modern integrated iron and steel plants, as well as petroleum refining and petrochemical industries, has necessitated extensive harbor and port improvements. It is estimated that the cargo to be handled at Nagoya port in 1970 will be three times the 1959 total. As a consequence the outer harbor fairway is being increased to 300 meters in width and 12 meters in depth; and the inner harbor lane, to 220 meters in width and 10 to 12 meters in depth. Depths at the industrial piers will be increased to 12 to 19 meters.[20] In the light of the disastrous consequences of the typhoon of September, 1959, upon the harbor area, a new anti-high-tide breakwater, 8300 meters long, is under construction.

Domestic trade is usually double the tonnage of foreign trade, and imports are several times the magnitude of exports, this latter inequality reflecting the bulky nature of the industrial raw-materials imports. Chief foreign imports, by value, are textile raw materials (53 per cent), chiefly wool and cotton, and foodstuffs (13 per cent), while exports are in the form of processed goods—ceramic wares (19 per cent), machinery (25 per cent), textiles and clothing (8 per cent). Among the exports, textiles and ceramics are declining proportionally, while machinery is increasing. Arriving cargo is largely from the United States and Oceania, while outgoing cargo is destined chiefly for North African and Asian countries.

The Ise Plain west of Ise Bay

Long, narrow Ise Plain, a southwestern extension of Nobi, whose area is roughly 1100 square kilometers, is to a considerable extent composed of low diluvial uplands averaging some ten meters in elevation above sea level. Shallow valleys, floored with new alluvium, have been cut into the diluvium, and these in turn are connected at their sea ends by a narrow and not completely continuous coastal lowland, for in a number of places spurs of the upland reach even to the sea coast.[21] Most of the new alluvium is paddy land, while the diluvial surfaces are largely given over to dry fields. Mulberry, which was formerly so important here, has greatly declined, especially in the vicinity of the coastal industrial cities where the opportunities for full, or part-time, employment are good.[22] Commercial vegetable farming, stimulated by the nearby Osaka market, has importantly supplanted mulberry, while tea gardens have operated in a similar way, especially in the southern parts.

Altogether, the trend in the region is toward a greater diversification of agriculture.

The Ise Plain, as one part of the larger Chukyo industrial node, has many of the latter's characteristics. Textiles, both cotton and wool, rank high in importance, and large wool and cotton spinning mills are widely distributed. But recent expansion in heavy and chemical manufacturing is causing much greater diversification than formerly prevailed, so that textiles, machinery, and chemicals have become the major triad. Tsu city (73,000 population) has remained highly specialized in textiles (54 per cent of total value-added), as have smaller places like Ise and Suzuka, but Yokkaichi, the Mie prefectural metropolis (109,000 population), an ocean port and also a textile city of renown, has burgeoned recently as a heavy and chemical industries center of high importance. Almost spectacular is the growth of oil refining and the petrochemical industry; and with the new developments that are now under way, Yokkaichi bids fair to become the nation's foremost petrochemical center. At the same time, however, textile and machine industries continue as important elements of the city's industrial structure. As of 1960, chemicals represented nearly 36 per cent of the total value added by manufacturing, followed by textiles with close to 18 per cent; ceramic and clay products, 14 per cent; all-machinery, 12 per cent; and petroleum products, nearly 12 per cent. Large factories prevail, most of them concentrated in the harbor area where imported raw materials are readily accessible (Fig. 13-12). Smaller Kuwana, with its emphasis on machinery and steel, somewhat resembles Yokkaichi.

TABLE 13-3

Foreign trade of Nagoya port, 1960 (unit: $1000)

Exports	Value	Imports	Value
Ceramic wares	62,703	Raw cotton	108,522
Automobiles	50,417	Wool	80,853
Textile machinery	18,217	Timber	30,344
Plywood	16,189	Wheat	25,769
Woolen textiles	14,819	Corn	18,612
Toys	14,319	Iron and steel scrap	11,456
Sewing machines	12,039	Hemp	9,561
Iron and steel	9,619	Coal	3,822
Clothing	7,878	Crude rubber	3,347
Cotton textiles	4,992	Soybeans	3,153
Others	99,183	Others	82,917
Total	310,375	Total	378,356

Source: Aichi Prefectural Office, Nagoya.

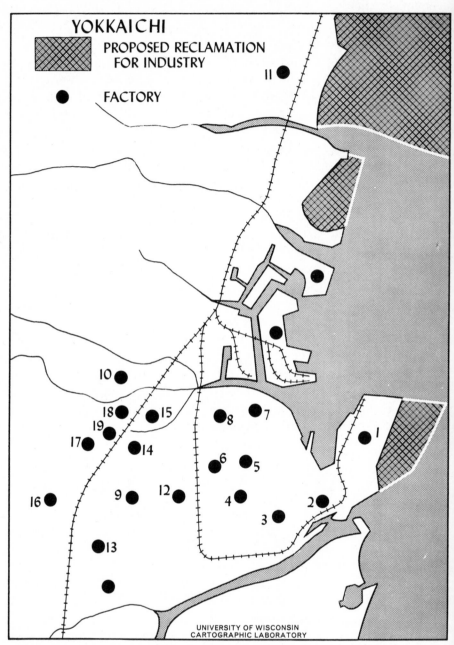

Fig. 13-12.–Kinds and locations of factories in the port city of Yokkaichi: (1, 4, 6, 7, 8, 14, 18, 19) chemicals, (2) thermoelectric plant, (3, 10) oil refinery, (5) petro-chemicals, (9) glass, (11, 15) electrical machinery, (12, 16, 17) textiles, (13) synthetic rubber. After map in *Nippon no Chiri,* Vol. 4. No scale given.

Yokkaichi has a down-bay location on Ise Bay similar to that of Yokohama and Kobe on their respective bays, but it has never approached the importance they have as commercial ports serving industrial hinterlands. This is partially because Yokkaichi does not fit the description of a naturally deepwater port, its offshore water being relatively shallow. Moreover, the slower industrial development of the Chukyo area, together with the good shipping services it received from the nearby Kinki ports, made a down-bay port less essential. In recent years Yokkaichi has held sixth or seventh rank among Japan's ports. At present the fairway leading to the harbor has been dredged to a depth of 12 meters, so that relatively large vessels can approach the industrial piers, but much additional harbor and port improvement is essential in order to serve adequately the swelling industry. Overwhelmingly an import port serving an industrial region, Yokkaichi's unloaded cargoes are chiefly wool, raw cotton, and petroleum.

The Mikawa plains east of Ise Bay

Along the east side of Ise Bay, and bordering Mikawa Bay, are several plains belonging to the general Ise lowlands, whose combined areas are over 1200 square kilometers. On the northwest they merge with the Nobi Plain, and on the east they may be considered as terminating at about Hamana Lake in westernmost Shizuoka Prefecture. Taken together they do not form quite a continuous plain, for in the vicinity of Gamagori city a spur of highland approaches almost to the coast, providing a convenient boundary separating the whole into eastern and western parts. The western subdivision, bordering Ise Bay, is usually designated as the West Mikawa, or Okazaki, Plain, while its eastern counterpart is the East Mikawa Plain, which has Toyohashi city as its metropolis. The two large peninsulas, Chita to the west and north, and Atsumi to the south and east, like giant claws almost enclose Mikawa Bay. While Chita is a rolling upland, chiefly of diluvium, Atsumi is a southwestern extension of the Akaishi Mountains, and its isolated higher parts are composed of Paleozoic rocks, but these in turn are tied together by diluvial and alluvial deposits to form the complex peninsula.

The West Mikawa Plain is composed of the alluvial lowlands of the Sakai and Yahagi rivers at the northwestern and southeastern extremities, separated by the low Hekkai diluvial upland which rises only three to seven meters above the adjacent alluvial lowlands. Diluvial Chita Peninsula may be considered a southward extension of West Mikawa. As might be expected, rice is the dominant crop in the alluvial areas, but less expected is the fact that the relatively large Hekkai diluvial upland is

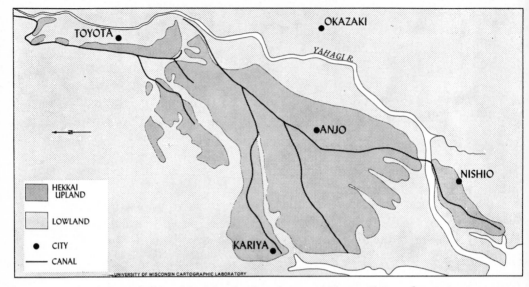

Fig. 13-13.—Hekkai diluvial upland in the west Mikawa Plain and its irrigation system. After map in *Nippon no Chiri*, Vol. 4.   No scale given.

also chiefly a paddy area, about 70 per cent of its cropland being in rice. This has been made possible by the extensive Meiji Irrigation Canal system which pours water from the Yahagi River through over 300 kilometers of trunk and branch canals.[23] Although primarily a region of grain farming, West Mikawa has diversified its agriculture through the addition of livestock and the cultivation of vegetables and orchard crops (Fig. 13-13). Agriculture on the rolling upland surface of Chita Peninsula has been greatly improved as a consequence of the extension, along its whole length, of the Aichi Aqueduct, providing water for irrigation and for urban and industrial uses as well. A similar development is planned for the Atsumi Peninsula. Both of these peninsulas are renowned as specialized areas of flower and vegetable culture.

West Mikawa is an integral and important part of the larger Chukyo industrial zone and partakes of many of the latter's characteristics. Textiles, ceramic and clay products, and food processing are specializations of long standing, but machinery is a relative newcomer. Dispersed and relatively small industrial centers are characteristic, among the more significant ones being Okazaki, Anjo, and Nishio, all specialized in textiles but importantly supplemented by machinery, chemicals and metals; Hekinan, emphasizing machinery, steel, and ceramics; Kariya, making machinery, textiles, steel, and ceramics; and Toyota, specializing in one form of machinery, viz., motor vehicles. Toyota's automobile plant, one of the country's largest, represents almost the only industry in this small city of 13,000 population. Throughout West Mikawa, large modern factories are numerous, although hardly dominant.

East Mikawa Plain, comprising the Toyo River Valley and Atsumi Peninsula, has been handicapped in its agricultural development by a deficiency of irrigation water, a situation which is in the process of being corrected by the Toyokawa Irrigation Canal Project which involves the construction of a dam and reservoir at the head of Toyo River, and canals on the coastal lowlands and diluvial uplands which will irrigate 22,000 hectares of cropland.[24] A somewhat diversified agriculture, but with considerable emphasis on commercial vegetables and fruit destined for the large urban markets, is characteristic. Toyohashi (110,000 population), the metropolis of East Mikawa, was formerly highly specialized in silk reeling, raw silk representing 70 per cent of its manufactures, for the Ise Bay region had been a principal concentration of mulberry culture. Especially since World War II, Toyohashi's manufacturing has diversified, so that at present cotton textiles, food processing, machinery, lumber and wood products, and metals are all importantly represented. Light consumer industries strongly predominate. Both along the eastern end of Atsumi Bay near Toyohashi, and along narrow Kinu Bay just east of Chita Peninsula, important land reclamation from the sea is planned, or is in progress, which will provide extensive factory sites for new coastal industrial zones, with associated industrial port facilities.

Southwest Japan is that part which extends west and south from the narrow waist of the country where it is constricted by the two major water indentations of Wakasa Bay on the north and Ise Bay on the south, with the waist appearing all the narrower because so much of its area is occupied by Lake Biwa. To recapitulate earlier descriptions, it is that part of Japan, included within the Southwestern Arc, which at its westernmost extremity is intersected by the Ryukyu Arc, resulting in the great volcanic masses of central and southern Kyushu. Elsewhere in the region volcanoes are few, and an extensive development of young strata is lacking.[1] Morphologically, Southwest Japan is divided into two subregions, an Inner Zone and an Outer Zone, separated by the Median Dislocation Line and its great northward-facing fault scarp. Rugged folded mountains comprise the Outer Zone, the three representative areas being southern Kyushu, southern Shikoku, and Kii Peninsula. Plains are few and restricted in area. By contrast, the Inner Zone, including southwestern Honshu, northern Shikoku, and northern Kyushu, is a region of complex and variable landforms, characterized by a dense network of faults, associated with which are numerous horst-like hill masses and fault depressions. Accordingly, Southwest Japan comprises the whole of the two islands, Shikoku and Kyushu, as well as the Chugoku and Kinki regions of western Honshu.

All lowland parts of Southwest Japan belong climatically to the humid

494

subtropics (*Caf*). Temperature contrasts between the several parts are not great, average temperatures for July varying between 75° and 78°, and those for January oscillating around 40°. Contrasts in precipitation are more striking, the Inland Sea margins having the lowest annual totals and the least cloud. Summer rainfall is especially heavy on the slopes of the mountains facing the Pacific, while the Japan Sea littoral has more winter precipitation, so that there snow may lie on the ground several centimeters deep in the colder months.

Although it was the region of earliest settlement by Yamato peoples, Southwest Japan today exhibits remarkable contrasts. Those parts looking inward upon the Inland Sea are usually thought of as the larger heartland of Old Japan, containing the ancient capitals of Nara and Kyoto, which from 710 to 1192 were the centers of national government and which in their palaces, shrines, and temples continue to bear witness to an earlier Imperial occupance. Here within this core region is located the western half of the country's manufactural belt, with the attendant features of numerous cities, dense population, intensive land use, and commercialized agriculture. But in those peripheral areas of Southwest Japan, bordering both the Pacific Ocean and the Sea of Japan, the pulse of economic life beats more slowly.

Kinki, situated at the eastern end of the Inland Sea, is the more restricted cultural heartland of premodern as well as modern Japan. No other part has as high rural population densities on the occupied land; no other region has the same degree of maturity, richness of culture, or perfected land use. The name, Kinki, meaning "Neighborhood of the Imperial Palace," evokes images of antiquity and grandeur, promising a wealth of historic landmarks and cherished works of art. But in addition to its cultural and spiritual endowment Kinki is one of the great industrial and commercial centers of Japan, also, containing three of the country's six metropolises of over a million people, and two of its greatest foreign trade ports. Only Kanto exceeds it in industrial and total population. Six prefectures (Osaka, Hyogo, Kyoto, Nara, Shiga, Wakayama) are included, the whole having an area of about 2700 square kilometers, with 14 million people, so that its over-all population density exceeds 500 per square kilometer.

Based upon terrain character, Kinki may be divided into three sections: (1) a northern upland, (2) a central lowland, sometimes called Central Kinki, and (3) Kii Peninsula to the south.

First of these, the *Tamba Upland*, is merely the eastward projection of the Chugoku hill land, where sedimentary rocks predominate, rather than granites, and where the fault net is less dense than that in Chugoku, so that occupance is more confined to stream valleys and to a few small detritus-floored fault basins such as Kameoka, Tamba, Sasayama, and Fukuchiyama. The wide interfluves, whose rolling summits average about 600 meters in elevation, are largely forested, for settlement is meager. Tamba reaches the sea at Wakasa Bay, which represents one of the principal irregularities along the entire Japan Sea coast. A consequence of down-faulting, fault scarps bound the bay on its eastern and western margins, while fault-line valleys are similarly responsible for its deeply indented ria coastline. Occupying little alluvial patches at the heads of several of the narrow bays are three or four small port cities, all of very minor importance except Maizuru (58,000 population), formerly a fortified naval base, which has developed into a shipbuilding and steel center of local consequence, and Tsuruga, a commercial port, which shows some evidences of a revived trade with the U.S.S.R.

Kii Peninsula, one unit of the folded sedimentary mountains comprising the Outer Zone of Southwest Japan, is higher and more rugged than the Tamba Upland, its highest peak reaching nearly 2000 meters. Inland Kii is one of Honshu's more unoccupied areas. Settlement is strikingly peripheral, but even there it is modest, for characteristically the hills come down to the coast where they terminate in wave-cut cliffs. Settlements of the fishermen-farmers are chiefly in spots where weaker Tertiary rocks or wave-cut platforms afford slightly expanded coastal areas of less precipitous land. A new and more direct rail connection with Osaka has permitted recently an active exploitation of the considerable forest resources of interior highland Kii.

It is the *Central Lowland,* however, which is the economic heart of Kinki. Here the terrain pattern is one of alluvium-filled fault basins separated by hill masses of horst origin. Principal lowlands are the Kyoto, Nara, and Omi (Biwa) depressions, all located inland, and the littoral plains around the head of Osaka Bay. Diluvial uplands of different ages and levels are widespread (Fig. 14-1). A striking feature of the lowlands is the elevated and diked stream channels rising five meters or more above the surrounding plain. Rapid silting of the channels, a result of the abundant loads provided the rivers from the denuded granite slopes flanking the basins, necessitates repeated raising of the levees. The origin of these denuded hill slopes is disputed: some regard the denudation as contingent upon the granite materials; others are inclined to attribute it chiefly to the overcutting of the common forests by the dense population.[2]

The Keihanshin manufacturing node spreads over the lowland parts of Kinki, but is most intensively developed on the littoral plains in western Osaka and southern Hyogo prefectures and in the Yamashiro Basin

MILES
0 — 5
0 — 8
KILOMETERS

ALLUVIUM
YOUNGER DILUVIUM
OLDER DILUVIUM
HARD ROCK AREAS OF
HILLS AND MOUNTAINS

LAKE BIWA
OTSU
KYOTO
NARA
OSAKA
AMAGASAKI
NISHINOMIYA
ASHIYA
KOBE
SAKAI

UNIVERSITY OF WISCONSIN CARTOGRAPHIC LABORATORY

Fig. 14-1.—Terrain types in central Kinki.

of Kyoto Prefecture, all three prefectures boasting cities of over a million population. From Table 14-1 it becomes obvious that Nara Prefecture, and more realistically Nara Basin, makes the least contribution to Keihanshin industrial output, while Wakayama and Shiga are likewise relatively weak members of the Kinki group. As noted in an earlier chapter on manufacturing, Keihanshin during the postwar period of rapid industrial expansion has not kept pace with its rival, Kanto. While, to be sure, Keihanshin is characterized by variety in its manufactural composition, nevertheless, it does contain a relatively high proportion of the traditional, labor-intensive types of industry. Therefore, accelerated manufactural expansion can be achieved only by a stronger infusion of industry of a modern character, making greater use of automation and labor-saving devices.

NARA (YAMATO) BASIN

Nara Basin,[3] oblong in shape and about 300 square kilometers in area, is surrounded by hard-rock hills except on the north, where a low belt of dissected old diluvium (possibly Tertiary) separates it from the Kyoto Basin. Along the margins of the basin there are belts of lower young, and higher old, diluvial terrace, the latter in places reaching elevations of fifty to sixty meters.

Agricultural occupance in Nara is very ancient, going back to Yayoi times. It is also very intensive, with farms averaging only about half a hectare, which situation compels a large number of farmers to seek supplementary employment in rural village industries, or to commute to and from Osaka and Kyoto. Rice is the main crop, of course, and half of the tilled land bears a secondary crop, while even three crops are not

TABLE 14-1

Relative importance of manufacturing in the six prefectures comprising Kinki and the Keihanshin industrial zone, 1960

Prefecture	Number of employees	Value added (1,000,000 yen)
Osaka	902,763	664,565
Hyogo	453,747	356,189
Kyoto	196,204	121,512
Shiga	64,628	38,806
Wakayama	69,330	34,160
Nara	35,457	15,883

Source: Ministry of International Trade and Industry, *Census of Manufactures, 1960: Report by Cities, Towns, and Villages* (1962).

uncommon. Truck crops in great variety, marketed chiefly in nearby Osaka, have become increasingly important in Nara Basin, so much so that they have become involved in a rotation system with rice, and have even displaced rice in parts. In addition to the more staple vegetables, specialty crops such as watermelons, tomatoes, strawberries, and flowers are emphasized. Some sensitive truck crops are raised under vinyl covers.

An almost unique feature of Nara Basin is the strong dependence upon pond irrigation for the paddy fields, a feature whose origin goes back to the seventh or eighth centuries. Sixty per cent of the paddy area is watered from ponds, compared with a national average of 20 per cent. Most of the ponds are rectangular, perhaps influenced by the early Jori land subdivision system. The great dependence upon these tiny reservoirs reflects in part the moderateness of the rainfall, but even more so the fact that no large river flows through the basin. Interesting village customs and rites are associated with the irrigation ponds, more than 6000 of which are to be found in Nara, or an average of four per hectare. For example, in some villages there are *mizuoya* (water parents) who are in charge of the ponds and decide the dates and hours when water shall be applied to, and drained off, the paddies. These same pond guardians direct the ceremonies held on the first day of each crop year when the ponds begin to operate. In some villages owners of paddy land auction off any surplus water allotted to them by the pond guardian.[4]

But in spite of its deficiency in river water for irrigation, the Nara Basin is nonetheless subject to floods, so that all communities lying below 100 meters elevation have suffered flood damage. This has led to the development of warehouses for protection against inundation and also to the widespread construction of flood embankments in the lower lands,[5] such embankments being maintained by strict community regulations.

Nara Basin provides a classic example of the effects of the Jori land partition system which was imposed on the region in the seventh century, for road pattern, fields, villages, canals, and even ponds, on the basin floor are strongly rectangular (Fig. 14-2). Rural settlements are predominantly of the nucleated or compact type, rather regularly spaced, a feature which also seems to be associated with Jori planning and its grid development of waterways and canals. Over eighty villages in the basin are surrounded by defensive moats which may date back to the sixteenth century when the rural areas were in great turmoil.

Although life in the Nara Basin is strongly influenced by proximity to the great metropolises of Kyoto, Osaka, and Kobe, strangely enough there remain vestiges of older patterns of community life that are completely non-urban in character. For example, villages maintain their an-

Fig. 14-2.—Nara Basin, showing evidences of Jori in its compact rural villages and rectangular systems of ponds, roads, and land subdivision. Courtesy of the Geographical Survey Institute, Japan.

cient grouping into communities, based upon devotion to a common tutelary deity. Such a community of villages has a strong tendency to be exclusive and clannish, regarding other villages as outsiders. Each group has a common shrine, whose appointed guardian performs divine services which mainly consist of ceremonies associated with the agricultural calendar.[6]

Nara Basin is not well developed in modern industry, and genuine industrial cities are absent, although cottage industries are locally important. Nara city is a town of ancient glory which by reason of its famous temples and shrines and fine scenery has been revived as one of the country's main tourist centers. It attracts an estimated 5 million visitors annually. Over half its area is occupied by parks and temple grounds. Nara was the first permanent capital of Japan (710–784 A.D.), but the Todaiji Temple, with its huge Buddha, and the Kasuga Shrine are the only remaining relics of the Imperial past.

KYOTO (YAMASHIRO) BASIN

Oblong in shape, with dimensions of about 36 by 8 kilometers, Kyoto Basin has an area of nearly 300 square kilometers. Four main streams empty into the basin, where they join to form the Yodo River which, after crossing Osaka Plain, debouches into Osaka Bay at Osaka Harbor. Where the four rivers join within the basin is the lowest part of its alluvial floor and there until recently a shallow lake existed. Its drainage was begun in 1939, and the newly reclaimed land was subsequently converted into a monocultural paddy landscape. But unusually heavy rains occasionally still result in temporary standing water. Around the margins of the basin, at the base of the surrounding foothills, are diluvial terrace uplands, those in the vicinity of Uji having national fame, not only for the extensiveness of their tea gardens, but even more so for the quality of the tea grown. Uji ranks second to Shizuoka in amount of tea produced but is first in quality. Commonly the higher and more dissected diluvial levels, where soils are coarse and slopes steep, are not importantly used for agriculture. While much of the basin floor is in rice fields, especially the lower, wetter parts, large areas in the vicinity of Kyoto city have half or more of the agricultural land devoted to market gardening.

Kyoto (1,168,000 population), fourth-ranking city of Japan, is one of a triad of million-cities in the Kinki metropolitan area, which has a total DID population of 7.8 million. Being located inland, Kyoto is the only one of the six great metropolises which is not a port. Its growth, therefore, has stemmed from conditions somewhat different from those of the other five, whose industrial importance is closely linked with littoral locaton. Kyoto has the further distinction of having been the national capital for over a millenium, until after the Meiji Restoration, when the seat of government was moved to Tokyo. The history of Kyoto in large measure epitomizes the history of Japan, and even today the city harmoniously combines the East and the West, the best of past and present. Aptly called the Florence of the East and considering itself the cultural center of Japan, Kyoto possesses a multitude of national treasures in the form of temples and shrines by the hundreds, royal villas, the former Imperial Palace, and, derived from these sources, a matchless store of antiquities, including paintings, sculptures, manuscripts, ceremonial robes, furniture, and military equipment.

It follows that one of Kyoto's large sources of income is from its tourist industry. Over 10 million sightseers and tourists visited Kyoto in 1960, about one-third of them in conducted-tour parties. Although tourist patronage continues throughout the year, it is relatively concentrated in the spring months of March-April-May, with nearly 40 per cent of the annual total, and in the two fall months of October-November, with close to 23 per cent.[7] Nearly a thousand hotels and inns in Kyoto are supported chiefly by tourists, while some 5 per cent of the inhabitants are connected directly or indirectly with tourism. As noted above, there are multiple attactions to lure the visitors—historical sites and treasures, including the Nijo Palace, residence of the Tokugawa shoguns, the pre-Meiji Imperial Palace, the Katsura and Shugakuin Detached Palaces, famous Shinto shrines and Buddhist temples in abundance, remarkable gardens, scenic Mt. Hiei with its ancient temples and magnificent panoramic views of Lake Biwa and its borderlands, and renowned Arashiyama Park bordering the rapids of the Oi River.

Much the greater part of the city is situated on gently southward-sloping alluvial sediments at the extreme northeastern part of Kyoto Basin. To the eye, the site appears flat, but in a distance of eight or ten kilometers from the northern to the southern ends of the city there is a decline in elevation of about fifty meters, so that drainage is good. On the east, the city creeps up onto a narrow diluvial bench and the hard-rock foothills behind it. Like Nara's, Kyoto's street pattern exhibits an almost perfect checkerboard design, with streets oriented with the cardinal directions, the whole pattern resembling those of the ancient Chinese capitals.

Functional areas within Kyoto are neither as specialized nor as segregated as they are in the other metropolises, for it is primarily a political, religious, educational, and cultural center. Within the commercial core, situated in the east-central part of the city, are a fair number of multistoried steel-and-concrete buildings of European style, but such structures are distinctly fewer than in Tokyo and Osaka, and they are not continuous even along the main thoroughfares, where the façade is interrupted by small Japanese structures, usually of stucco or wood.

Even in downtown Kyoto the Japanese imprint is still strong, and the appearance is not particularly metropolitan.

Without doubt, areas that are exclusively or even predominantly industrial are much less conspicuous in Kyoto than in the other major cities. Here workshops and small factories mingle with residences and commercial establishments. But this should not be interpreted to mean that in Kyoto manufacturing is a much less important sector of the economy than it is in the other great cities, for the proportion (16 per cent) of the DID population engaged in manufacturing in Kyoto is about the same as in the industrial port cities of Kobe and Yokohama, which are similar in size. It is somewhat below the comparable figure for Tokyo (19 per cent), and even lower compared with Nagoya (22.5 per cent) and Osaka (24.3 per cent). Still, manufacturing represents the single most important form of employment in Kyoto. The fact that exclusive manufacturing districts are less characteristic here probably reflects the weaker emphasis on heavy and chemical industries, and the greater prevalence of small and medium-sized establishments. So, while the number of employees in manufacturing does not greatly differ as among Kyoto, Yokohama, and Kobe, yet Kyoto has three to four times as many manufacturing establishments, a very large number of them being of workshop size.[8] But insofar as is consistent with the situation just described, a large proportion of the processing plants are located south and west of the commercial core of the city. Here the only apparent advantage is the greater availability of less expensive flat land for factories, since hills flank the city to the east and north.

Doubtless Kyoto's lack of frontage on tidewater handicaps it greatly in attracting heavy and chemical industries such as primary iron and steel and petrochemicals, which require the import of heavy, bulky raw materials, much from abroad. Table 14-2 shows the varied structure of Kyoto's manufacturing but also the strong emphasis upon textiles and machinery, the former representing nearly one-third of the total output by value, and machinery close to one-quarter. The combined outputs of machinery and metals nearly equal that of textiles. Of the six major cities, Kyoto has not only the largest number of employees in textiles, but the largest proportionate number (41 per cent of all manufacturing employees).

Since Kyoto's national and international reputation in manufacturing is that of a craft and workshop center, it is somewhat surprising to note that machines-metals-chemicals represent 38 per cent of the total output by value, and it is these same industries which are expanding most rapidly while the craft industries grow more slowly. Moreover, the advertising by the city administration plays up the craft industries and

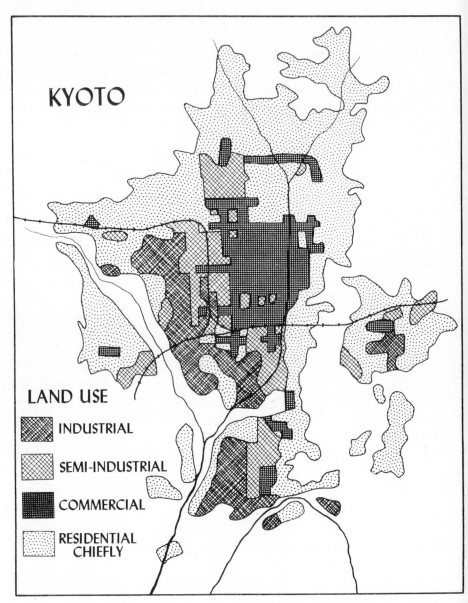

Fig. 14-3.—Generalized functional areas in Kyoto. After map by Takeo Tanioka. No scale given.

plays down the less picturesque ones such as machinery, metals, and chemicals. This is understandable in the light of Kyoto's function as a tourist center, since visitors are much more attracted by the arts and crafts than by machines and chemicals. Moreover, the latter products are ordinarily marketed through Osaka, while the craft products are sold in, or marketed through, Kyoto, and hence add more to the city's reputation.

Products of Kyoto's machinery industries are characteristically not large and bulky, but instead include such items as small gasoline engines for hand tractors, textile machinery, electrical appliances, scales, small motors and dynamos, electrical equipment such as switches and transformers, and precision machines and instruments. Similarly, the chemical industries specialize largely in consumer's goods such as medicines, drugs, soaps, and batteries, while the metals are non-ferrous, with emphasis on copper, brass, and bronze, which are used in some of the craft and tourist-products industries. The machinery, metal, and chemical establishments are relatively modern, for the most part, and are larger than those making craft products. Within the city such factories tend to be particularly concentrated in the less congested western and southern parts. Tall smokestacks characteristically mark their sites.

A host of varied traditional craft and workshop products—fine pottery, lacquer ware, dolls, fans, silk fabrics—bespeak the fame of Kyoto as an artisan center. Typical of the traditional workshop industries is that of the Nishijin silk fabrics, originally started by Chinese immigrants in the fifth century. Early specialization was in fine silk fabrics for the imperial family and its court. By the early eighteenth century more than 5000 weavers were employed, and Nishijin in Kyoto was one of the most flourishing textile centers. At present Nishijin processing employs about 15,000 weavers, 80 per cent of the looms used being of the hand variety.[9] An important product of the industry is brocade, using both silk and synthetic fibers. Nishijin is concentrated in the northwestern part of the city close to the site of an earlier Imperial Palace, and is housed in frame workshops that have the appearance of residences. Other than Nishijin,

TABLE 14-2

Composition of manufacturing in Kyoto, 1960, by percentage of value of output

Food	12.5
Textiles	32.8
Lumber and wood products	4.1
Pulp and paper	3.2
Printing and publishing	4.4
Chemicals	6.3
Ceramics	1.8
Metals	8.3
Machinery	23.7
Others	2.9

Source: Kyoto Municipal Office.

the Kyoto textile industry includes also the weaving of cotton and wool, the spinning of thread, the dyeing of cloth, and the making of garments, but dyeing and weaving are the most important of these.[10] Nearly 9000 establishments scattered throughout the city are engaged in the various aspects of textile manufacturing. A special technique of dyeing, inherited from ancient times, involves hand-washing the newly dyed fabrics in the streams of the city and subsequently spreading them out on the river banks to dry, both processes providing a colorful sight.

THE OMI (BIWA) BASIN

This largest of the several Kinki fault basins, containing the most extensive fresh-water lake in Nippon, coincides with a narrow land section and provides the most complete break in the Honshu mountain barrier separating the Pacific from the Japan Sea coast. A narrow belt of hills some fourteen to sixteen kilometers wide, cut by numerous north-south fault valleys, is all that divides the Omi Basin from the deep tectonic indentations of Wakasa Bay on the Japan Sea coast. In ancient days this hilly belt was the site of a barrier gate comparable to the one at Hakone Pass on the Tokaido. A still narrower and lower mass of hills, three to five kilometers wide, forms the southern divide separating the Omi from the Yamashiro (Kyoto) Basin, while another narrow zone of hill land is all that separates the east Biwa Plain from the Ise Bay lowlands. From ancient times down to the present this corridor between north and south coasts has been an important transit route used for both military and commercial purposes. It is followed today by the Hokuriku, and in part by the Tokaido, railroad lines, which negotiate the northern and southern barriers by means of tunnels. It is by this natural lowland route through the Biwa Basin that the Japan Sea coast of Hokuriku is made tributary to the industrial port cities of Osaka, Kobe, and Nagoya.

The asymmetrical basin occupied by Lake Biwa has higher and more precipitous walls on the west than on the east, with the result that the alluvial and diluvial fringe surrounding the lake is much narrower and the individual fans steeper on the western margin, where the waters are also deeper. The shore-line is conspicuously scalloped by the numerous advancing fronts of the delta-fans which the radial drainage lines have produced. On the eastern margins a few hard-rock outliers rise above the alluvium, or out of the shallow water of the lake. Within the Biwa area, three distinct depositional terrain forms are recognized: (1) the delta-fans of new alluvium, those along the eastern margins of the lake being broader and flatter, (2) the lower and younger diluvial terraces,

Fig. 14-4.—In the foreground a section of the lower diluvial upland surface in Kinki on which rice fields and dry fields are intermingled. In the background is the higher diluvial terrace (1), chiefly in woodland or wasteland, although the valley bottoms are given over to paddy land. Still farther in the background are the higher granitic mountains (2). Scene near the southern end of Lake Biwa.

and (3) the higher and older diluvial terraces.[11] The last are composed chiefly of coarse fluvial deposits with some intervening clay beds, the total thickness often being over 100 meters. Where gravels predominate, the surfaces are likely to be rugged, exhibiting badland characteristics, although level to undulating crests are also represented. Unlike diluvial terraces in many parts of Japan, these of Omi have no mantle of volcanic ash. The lower terraces, which represent a second stage of uplift, have smoother surfaces and finer, better mixed soils, with less sand and gravel. Elevated stream channels (*tenjo-gawa*), flanked by high dikes, are characteristic of the delta-fans. In one instance a railroad and highway are carried under an elevated river bed by tunnels rather than over it by bridges. Thick groves of bamboo characteristically parallel the river courses, their roots serving to hold the levees in time of flood. Ruthless cutting of timber from the surrounding granite hills to supply wood for rebuilding the Buddhist temples of Mt. Hiei and for the great urban centers of Kinki led, in times past, to disastrous floods which deposited

Fig. 14-5.—A section of the relatively barren higher diluvium of Kinki.

sand and gravel on the fertile plains below. It was here in the Biwa Basin that some of the first work of torrent correction was begun, under the direction of a Dutch engineer in 1871, yet even now the whitish scars of bare granite slopes are conspicuous.

Climatically the basin subdivides into two parts, a northern half, down to about the Echi River, which is similar to the Hokuriku coast, and a southern half which resembles the Inland Sea regions. In winter especially the contrast is marked, for at that time the strong northwest monsoon blows over the low range of intervening hills into the northern end of the Omi Depression, giving gray, overcast, snowy weather, typical of Nippon's windward Japan Sea coast in that season. During the three winter months precipitation near the southern end of the lake is not much more than half what it is along the northern margins, only sixty to seventy kilometers distant, where snow lies fairly deep on the ground.

Such a climatic contrast leads to major differences in land use as well, for while in the southern half of the Omi Basin most of the rice fields are winter cropped, being resown to wheat, barley, and rapeseed in fall, those in the snowier north mostly remain fallow after the rice harvest. In northern Biwa, as on the Japan Sea coast, are to be seen along the paddy margins rows of *hasagi*, bare upright poles or slim almost branchless trees, upon which the rice is hung to dry after harvest. While Biwa is one of the most representative paddy areas in Kinki, yet its proximity to large urban markets has produced a degree of specialization in truck farming. That agricultural occupance is very ancient is indicated by the rectangular pattern of land subdivision, a product of the Jori system. The low diluvial terraces are somewhat less completely utilized than is the new alluvium, but nevertheless they bear important crops of rice, mulberry, summer vegetables, and winter grain, as well as some tea. In such loca-

Fig. 14-6.—Paddy fields occupy the intricate valley systems of the higher diluvium.

tions pond irrigation is common. The ruddy-colored high terraces, as well as the steeper, stonier upper parts of the alluvial fans, have much wasteland and woodland. Where the high terraces have been much dissected, a complicated dendritic pattern of paddy land characteristically coincides with the intricate valley systems. Bamboo is relatively abundant, and in favorable locations some dry crops are cultivated, even on the crests of these higher uplands.

The lake itself, together with its rugged surroundings and the legend, history, and art which are associated with its classic shores, makes the Biwa region attractive to resorters, tourists, and sightseeing groups from the adjacent metropolitan centers. Two funicular railways carry tourists to the top of Mt. Hiei, where there is the attraction of ancient Buddhist temples set in groves of magnificent cryptomeria. From Hiei's elevation of 848 meters one can behold a panorama of almost the entire Biwa Basin and of Kyoto as well. Small tourist boats, carrying some freight, ply between various towns along the lake shore. Biwa is well stocked with fish, and although most of the catch is consumed by the local population, some is sent to the adjacent cities.

Biwa water has long been used to irrigate the lake margins of the bordering plains, but until recently the primitive methods of lifting—by well sweeps and tread mills—placed a definite limit on the height, and hence area, of land that could be supplied with lake water. As motorized lifting has replaced the older methods, however, the area irrigated from the lake has greatly expanded, for water can now be raised to the higher levels back from the immediate shore. Most of this expansion has taken place on the flatter plains to the east of Lake Biwa. Introduction of the motor for irrigation purposes has likewise promoted a more intensive type of farming as represented by commercial vegetable growing.[12] Biwa's waters find their only natural outlet at the south end by way of the Seta River, which furnishes the chief volume of the Yodo River that flows southward through the Osaka Plain, providing the principal water supply for the Hanshin industrial area. By means of a tunnel through the intervening hills, Biwa water is likewise conducted to the Kyoto metropolis. Plans are afoot to tap the lake for a much larger volume of water for the Kinki metropolises, perhaps as much as a 50 per cent increase by 1975. Such an increase could result in much larger fluctuations in the water level of Biwa than now exist, and this in turn would adversely affect recreation and tourist establishments concentrated on the shores of the narrow southern arm of the lake, which represents only about 10 per cent of the total lake area. It is planned, therefore, to construct an embankment across the lake at its waist in order to maintain a nearly constant water level in the narrow southern part and thus protect the riparian interests there concentrated.[13]

During the feudal period, Omi Basin, which was at the juncture of three major national highways—Tokaido, Nakasendo, and Hokurikudo—was of considerable industrial importance as a region processing crepe silk, linen mosquito netting, linen fabrics, and porcelain products, including *hibachis* and table chinaware. But handicapped by its inland location, Omi has remained relatively underdeveloped in modern industry, despite its proximity to Hanshin, the nation's second most important manufacturing center. It has been forecast that with the completion of the new Tokaido Super-Express Rail Line and the new super-highway linking Nagoya with Osaka-Kobe, Omi Basin will become a more attractive location for modern industry of certain types. Land and water for factories are both available here in abundance, features that are deficient in most parts of Kinki.

Significantly, nearly half the employees in manufacturing in Shiga

Prefecture are in relatively large factories with over 300 workers. This proportion is greater than that for any other prefecture of Kinki excepting Hyogo, which contains the city of Kobe. Among the larger plants are two cement factories, two cotton spinning mills, a rayon factory, and one making diesel engines. Over one-third of the factory employees are in textiles, which branch of manufacturing overshadows all others, for only 15–16 per cent are engaged in machine and instrument manufacturing, and 15 per cent in chemicals. The single largest manufacturing center is Otsu (74,000 population) at the southern end of Lake Biwa, where there is a concentration of chemical fiber plants attracted by the abundance of clean, soft water available. Accordingly, chemicals provide 43 per cent of the value added by manufacturing in Otsu, but machinery of all kinds is a close second with 40 per cent. Large supplies of limestone have made this same general location a desirable one for cement factories.

THE OSAKA LITTORAL PLAIN

Osaka Plain (760 sq. km.) is more than twice as extensive as either the Yamato or the Yamashiro basins, but like them it is essentially an alluvial lowland with bordering areas of both younger and older diluvial terraces. All that separates it from the Yamashiro Basin is a low and narrow divide in the form of diluvial or Tertiary hills, while a similar low terrain barrier divides the Yamashiro from the Nara Basin. Unlike the other three Kinki lowlands previously described, Osaka Plain has the immense advantage of fronting on tidewater—an advantage which has been a critical factor favoring its development of industrial and commercial functions on a scale not known to the others. Here on the Osaka Plain is the great Hanshin industrial node, next to Keihin the nation's greatest, and also the core of the more extensive Keihanshin industrial zone embracing most of lowland Kinki.

Essentially the Osaka Plain is the advancing bay-head delta of the broad and diked Yodo River and a number of smaller streams, with amphibious ocean margins and shallow water offshore. Large areas of back marsh along the littoral have been reclaimed only during the last two or three centuries. Population is exceedingly dense, so that the rural landscape of paddy and vegetable fields is greatly interrupted by the numerous settlements of various sizes and by other non-rural cultural features. Although Jori influence is discernible in settlement forms and land subdivisions, it is not dominant. Myriads of irrigation ponds dot the diluvial borderlands, and there are some on the alluvial floor of the plain as well, a feature which reflects the relative scarcity of running water for irrigation purposes, for the rivers are short and have small catchment basins, so that their volumes are restricted.[14] Some 48 per cent of the irrigated land in Osaka Prefecture derives its water from artificial ponds or reservoirs.[15]

A relatively wide and continuous belt of diluvial-Tertiary terrace borders Osaka Bay on the south and east, and in places the terrace reaches the water's edge and terminates in low, wave-cut cliffs. Toward the north, approaching Osaka, as the diluvial bench gradually recedes from the coast, there is a progressively wider strip of sandy coastal plain, principally devoted to market gardens and rice, between the sea and the terrace. Along the seacoast of this southern arm of the plain is a series of small industrial cities, satellites of Osaka, which are specialized in textiles. Much of the smooth, rolling surface of the low diluvium is in rice, vegetables, and other dry crops, while the belt of higher and more dissected diluvium (perhaps Tertiary) farther back from the coast is a specialized orange-growing region, in which pond-irrigated paddy fields occupy the maze of intricately dendritic valleys. To be sure, woodland and wasteland comprise the most extensive type of cover on the higher diluvium, but cultivation seems to be more general here than on most of the older diluvium of Kinki.

Much the larger part of the Osaka Plain is included within Osaka Prefecture although its northwestern part, bordering the bay, falls within southern Hyogo Prefecture. Within the Hyogo section are located the metropolis of Kobe and five or six other cities specialized in manufacturing, each having 50,000 or more people. On that larger part of the Osaka Plain contained within Osaka Prefecture, besides the metropolis of Osaka city, there are at least seven other industrial urban places of more than 50,000. Diminutive Osaka Plain, including its Osaka and Hyogo sections, is approximately coincident with the Hanshin industrial concentration, and it epitomizes modern industrial Japan, for its factories employ about three-quarters of all the workers engaged in manufacturing within the more extensive Kinki or Keihanshin industrial zone (Fig. 14-7).

The Hanshin industrial node

The Hanshin industrial node has participated in Japan's postwar manufacturing boom, although not to the same degree as Keihin. Thus, in 1935, Kinki exceeded Kanto, and Osaka Prefecture exceeded Tokyo Prefecture, in value of industrial output. But in the postwar period, even by 1950, the scales had tipped slightly in Tokyo's favor, and this trend has strengthened in later years.[16] Economic power gradually has been mov-

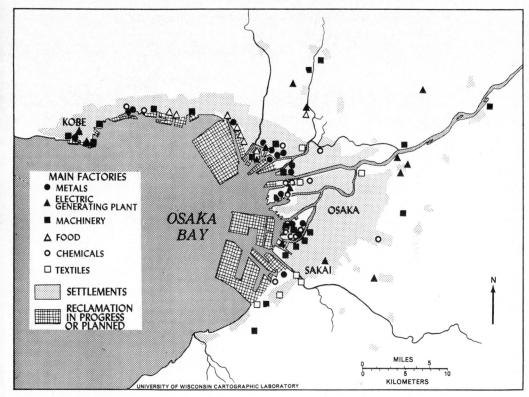

Fig. 14-7.—The Hanshin industrial district and the distribution of its main factories. After map in *Nippon no Chiri*, Vol. 5.

ing toward the national center of political power and the country's most populous market. In Japan, therefore, the situation rather resembles that in Britain, where industry gradually is shifting southeastward and especially toward the environs of London. Since 1955 the index of industrial production, both in the larger Kanto district and in the smaller Tokyo metropolitan area, has exceeded the country average, while in the whole Kinki district it has been below the country average, and in Osaka Prefecture has just equaled the country average.[17] Reasons why Hanshin has not kept pace in industrial expansion with Keihin, or the Tokyo metropolitan area, are several. Of some importance is the fact that there was a strong centralization of the Allied Occupation's military and governmental functions in the Tokyo area after the Pacific War, which resulted in a more rapid reconstruction of the Tokyo urban complex and in a truly remarkable concentration of industrial and commercial functions near the seat of political power. As a consequence, developmental momentum was more rapid there than in the country's other centers.[18]

A further explanation of Hanshin's lag is the fact that following the war technological developments and reorganization of the world economy hastened the creation of a new national mixture of industries, a number of which had been only modestly represented in the Hanshin area, which was traditionally a great textile and clothing center and a processor of general merchandise, the establishments of which were commonly small or of medium size. But for the country as a whole textiles have been a sector of manufacturing that has not experienced the rapid growth characteristic of metals, machines, and chemicals, a fact which has affected Hanshin more than Keihin where textiles are much less important. To be sure, Osaka remains the nation's greatest center of head offices of major textile manufacturing companies, even though their factories are now widely scattered, but on the whole the textile industry in Hanshin is declining in relative importance. Many of the new industries have tended to locate elsewhere, especially in Keihin.

Industrial composition, which is notably varied in both regions, does not differ markedly between Hanshin and Keihin. Still, certain disparities, beyond the difference in importance of textiles, do exist. In Hanshin there is considerably less emphasis on machine industries, which for the nation as a whole are one of the fastest growing sectors of manufacturing. Moreover, Hanshin's main machine industries are shipbuilding, railway equipment and rolling stock, textile machines, engines for agricultural and machine use, and home electrical appliances. By contrast those of Keihin emphasize motor vehicles, electrical machinery for industrial use, machinery for the generation and transmission of electricity, electrical instruments for metrical uses, machinery for construction and mining purposes, optical and other precision instruments and machinery, including radio, television, and other electronic equipment. In general, the machine industries of Keihin are newer, more modern, represent a more advanced technology, are science-based to a greater degree, require more skilled manpower, are housed in larger and more modern factories, and to a greater extent are included within that group designated as "promising industry." Similarly, while in both regions the chemical industries are well represented, those of Hanshin are weighted in the direction of inorganic chemicals—paints, fertilizer, drugs, toilet goods, and chemicals for textile processing—which have shown no remarkable expansion, while the organic chemicals, especially the booming new petrochemicals, are less well represented in Hanshin.[19]

A third factor contributing to slower postwar expansion was Hanshin's traditional role as the chief center for Asian trade in Japan. The loss of Korea, Formosa, and Manchuria, the weakening or severing of trade ties with other parts of Asia following the war, the unsettled conditions in Southeast Asia, and the reorientation of Japan's trade to a trans-Pacific pattern—all dealt blows to Hanshin's trade and industrial structure, the shock of which was felt most keenly in its industrial port cities.

But some of these postwar handicaps are lifting, at least in part. Trade with Asia, particularly Southeast and South Asia, is staging a comeback, so that the trans-Pacific orientation of Japan's commerce is less dominant than it was. In addition, Osaka, and Hanshin in general, have a marked advantage over Keihin in potential availability of large quantities of industrial and potable water, as provided by extensive Lake Biwa only forty-five kilometers northeast of Osaka. Biwa is the principal source of water for the Yodo River, which at present provides almost all of the metropolitan water supply for Osaka, about half of Kobe's, and a considerable proportion of the requirements of the cities to the north and northwest of Osaka. The total amount of water available from the Yodo-Biwa system is about 5 billion tons a year, which is ample for the needs of the Hanshin area if it is efficiently utilized.[20]

If Hanshin is to be expanded industrially in proportion to the rest of the country, capital-intensive manufacturing, characterized by a high input of capital per worker and employing relatively small numbers of highly skilled personnel, must be attracted more strongly to the region. Such industries are of two groups: those associated with the highly automated processing of bulky raw materials and those concerned with the production of intricate machinery by assembly-line methods. Traditionally these have not been so well represented in Hanshin. The first group includes industries such as petrochemicals and ore-and-metal processing and manufacturing. The second includes electronic equipment, machine tools and dies, and other producer's goods, together with the so-called "research and development" industries, which pioneer in the development of extremely complex products or production techniques.[21]

Some of the desirable conditions for attracting the raw-material processing industries are extensive coastal factory sites and large quantities of industrial water, both of which are in short actual supply in parts of Hanshin at present. The machinery and data-processing industries are not to the same degree tidewater-oriented, but some do require extensive sites. For these it may be wise to develop large and relatively new industrial districts, some within Hanshin, such as the coastal lands south of Osaka, or Sennan, and along the Yodo corridor connecting Osaka and Kyoto, but others outside, as for example Harima, lying westward of Kobe, the Nara Basin, and the borderlands of Lake Biwa.[22]

The Hanshin industrial region may be divided into three subregions:

Osaka subregion, centrally located on the broader plain at the head of the bay; Kobe subregion, on the narrower Hyogo part to the north and west; and Sennan subregion, on the narrow strip to the south and west. The Osaka subdivision consists of Osaka city and such satellite cities of Osaka Prefecture as Sakai (260,000 population), Fuse (194,000), Toyonaka (173,000), Moriguchi (93,000), Suita (90,000), Yao (73,000), Takatsuki (45,000), Kawachi (33,000), Hiraoka (32,000), and other smaller ones. Industries are varied in character, and typically the metal and machine industries employ 45–60 per cent of all factory workers, with textiles next in rank, but well below. (See Table 14-3.)

The Sennan subregion, occupying that part of the coastal strip located south and west of Osaka city, contains no metropolis, but is composed of such lesser cities, all in Osaka Prefecture, as Kishiwada (73,000 popula-

TABLE 14-3

Industrial composition of important cities in the three Hanshin manufacturing subregions, 1960 (in percentage of each city's industrial workers)

Area	Total workers in industry	Percentage of workers in		
		Textiles and clothing	Metals and machinery	Chemicals
Osaka subregion				
Osaka	563,234	10.7	48.0	7.8
Sakai	59,692	20.4	54.2	4.4
Fuse	40,229	8.1	64.8	2.9
Toyonaka	15,050	5.9	66.8	11.3
Takatsuki	14,016	2.6	60.6	11.6
Yao	15,875	18.4	36.8	8.0
Moriguchi	12,081	19.1	55.3	0.9
Kawachi	10,044	11.7	59.7	1.1
Hiraoka	8,792	6.9	75.7	4.4
Kobe subregion				
Kobe	140,827	5.2	54.3	2.4
Amagasaki	83,118	4.1	65.0	9.8
Nishinomiya	19,621	2.5	39.8	7.2
Akashi	23,424	7.4	52.8	1.7
Itami	15,770	17.0	50.7	3.7
Sennan subregion				
Kishiwada	16,506	70.0	15.3	0.4
Izumi	15,104	83.6	....	...
Izumiotsu	15,754	86.7	0.7	...
Izumisano	13,562	61.7	28.4	...
Kaizuka	19,105	73.8	12.9	...

Source: Osaka and Hyogo prefectural offices.

tion), Izumi (24,000), Izumiotsu (33,000), Kaizuka (28,000), and Izumisano (22,000). This group is highly specialized in textiles (spinning, weaving, dyeing, and bleaching), which usually account for from 60 per cent to over 85 per cent of the laborers in manufacturing, while only 10–20 per cent are in metals and machinery. Chemicals are negligible. Factories occupy sites along the coastal lowland and also in the valleys that lead back into the diluvial and Tertiary hills to the rear. There is some degree of concentration of large cotton mills along the coastal tramline. Considering Osaka Prefecture as a whole—that is, including the subregions of Osaka and Sennan, as well as other less specialized and less urbanized areas—some 20 per cent of the industrial employees are in textile and clothing establishments, a little over 8 per cent in chemicals, 16.7 per cent in metals, and 26 per cent in machinery industries, or nearly 43 per cent in metal and machinery industries combined.

The westernmost or Kobe subregion, in Hyogo Prefecture, resembles the Osaka subregion in its emphasis on metal and machine industry, but

TABLE 14-4

Osaka prefecture, industrial composition, 1960, in establishments of all sizes

Types of manufacturing	Number of persons engaged	Per cent
Food	58,111	5.3
Textiles	166,898	15.3 ⎫
Apparel	55,834	5.1 ⎬ 20.4
Lumber and wood products	21,374	
Furniture and fixtures	18,635	
Pulp and paper	36,212	2.5
Publishing and printing	46,784	4.3
Chemicals	88,920	8.2
Petroleum and coal products	3,054	
Rubber products	18,090	
Leather	9,386	
Ceramic and stone	33,316	3.1
Iron and steel	52,411	4.8 ⎫
Non-ferrous metals	20,490	1.9 ⎬ 16.7 ⎫
Fabricated metals	109,300	10.0 ⎭ ⎬ 42.7
Machinery (except electrical)	131,075	12.0 ⎫ ⎭
Electrical machinery and equipment	88,089	8.1 ⎪
Transportation equipment	55,496	5.1 ⎬ 26.0
Precision instruments	8,971	0.8 ⎭
Miscellaneous	55,204	
TOTAL	1,089,808	

Source: 1960 Establishment Census of Japan, 3.

on the other hand it has distinctly smaller percentages employed in textiles and chemicals. Here the single chief focus of manufacturing is the
port city of Kobe, but lesser centers, all coastal, include Amagasaki
(382,000 population), Nishinomiya (239,000), Itami (49,000), and
Akashi (81,000). Relatively large plants, chiefly engaged in iron and steel,
shipbuilding, and various types of machine industry, are more characteristic here than elsewhere. Since Kobe has fewer small factories which are
functionally related to the larger ones, it has to rely for such services on
smaller plants in Osaka and adjacent cities.

Osaka city

Osaka, the primate city of Hanshin, Kinki, and all of southwestern
Japan, had in 1960 2.97 million people within its densely inhabited district. This is to be compared with 8.11 million for Tokyo.

Growing out of its low deltaic site, certain serious problems confront
the city of Osaka. As in Tokyo, so here there has been significant ground
subsidence, notably in the city's lower western part. Most devastating
damage associated with ground subsidence occurs with unusually high
tides and heavy rainfalls accompanying typhoons, when as much as 40
to 60 million square meters of the lowest land may be inundated (Fig.
14-8). Subsidence is attributable to the shrinking of clay layers when the
groundwater table is lowered by excessive pumping for industrial and
other urban uses. As a means of protection cement sea walls have been
constructed along fifty kilometers of river and canal within Osaka.

Osaka is located on the Yodogawa Delta at the head of shallow Osaka
Bay, where a score or more small streams debouch along a coastal strip
of twenty kilometers, with the main delta's projecting front creating
the principal irregularity of the otherwise smoothly crescentic littoral.
Nearly three-quarters of the city is built on flat delta sediments only
slightly above sea level, so that large modern buildings must be supported on piles. Throughout this lowland section Osaka is intersected by
a remarkable network of rivers and canals crossed by over 1300 bridges
—a veritable Venice, but a smoky, grimy one. Only along the city's extreme eastern margins are the site conditions different, and here the best
and most exclusively residential areas occupy a north-south spur of diluvial upland, ten to twenty meters high and one to two and a half kilometers
wide. Commercial thoroughfares are frequent even here, however. At
the upland's northern tip, from which eminence one commands a view
of the entire city and its environs, there stands the impressive replica
of one of Japan's greatest feudal castles, with its wall-and-moat-enclosed

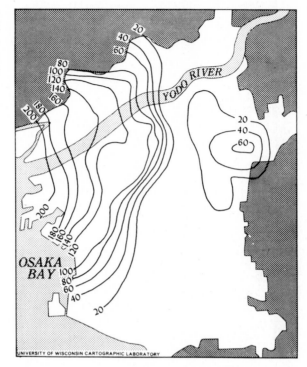

Fig. 14-8.—Ground subsidence in Osaka city between 1935 and 1959, in centimeters. From map in *Nippon no Chiri,* Vol. 5. Scale not given.

grounds. From this high northeast corner of the city there is an abrupt descent to the lowland.

The commercial core of the city is situated just southwest of Osaka Castle and about five kilometers inland from the water front (Fig. 14-9). Here the flat alluvial site has permitted the development of a grid pattern of intersecting north-south and east-west streets that rivals that of Kyoto. But unlike Kyoto, the grid is not so conspicuous beyond the core area. Large multistoried Western-style buildings are numerous in downtown Osaka, so that it looks more metropolitan than any other Japanese city, excepting Tokyo. Retail stores, banks, theaters, and wholesale houses here vie for space, for Osaka is nationally famous as a banking and wholesale center. In normal years, the city's wholesale transactions amount to 25–30 per cent of the nation's total by value, with textiles and clothing contributing most (46 per cent of the city total) followed by metals-machines-tools with 31 per cent.

Osaka is not only the great financial and wholesale center of Kinki and Keihanshin, but their manufactural center as well. Within the whole Kinki region, embracing six prefectures, nearly 36 per cent of the industrial workers are employed in Osaka city, while within Osaka Prefecture,

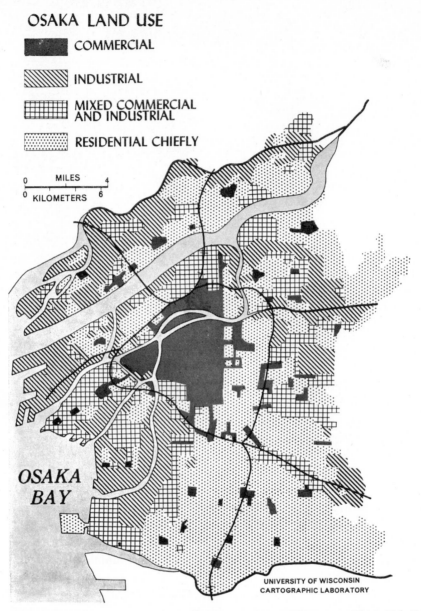

OSAKA LAND USE

■ COMMERCIAL

▨ INDUSTRIAL

▦ MIXED COMMERCIAL AND INDUSTRIAL

⣿ RESIDENTIAL CHIEFLY

MILES 0 — 4

KILOMETERS 0 — 6

OSAKA BAY

UNIVERSITY OF WISCONSIN
CARTOGRAPHIC LABORATORY

Fig. 14-9.—Functional areas in Osaka. From map in *Nippon no Chiri*, Vol. 5.

approximately three-fifths of the factory workers and a similar proportion of the value of manufactures shipped and the value added by manufacturing are accounted for by the metropolis. In Japan only Tokyo outranks it as a manufacturing center.

That its manufactures are varied in character is indicated by Table

14-5. Some 47 per cent of the city's workers in manufacturing are employed in metals and machinery, indicating a relatively strong specialization in the heavy industries. Textiles, including clothing, account for about 11 per cent of the industrial employment; and chemicals, including products of coal and oil, about 8 per cent. Compared with Osaka Prefecture where over 20 per cent of the workers are employed in textiles, Osaka city is less specialized in this light industry.

Manufactural functions in Osaka are not so areally concentrated as they are in some of the other metropolises, for many Osaka factories are small or only moderate in size, and such plants are often intermingled with residences and commercial establishments. The heavy industries, especially, seek out tidewater locations and hence are concentrated along the immediate coast or along major waterways leading inland. Much the larger part of the city's manufactural area lies seaward from the commercial core, although there are lesser concentrations well inland, more especially to the east of the castle.

As it applies particularly to Osaka city, but with some extension beyond its borders, a number of discrete industrial areas may be recognized.

TABLE 14-5

Composition of manufacturing in Osaka city, 1960
(in factories with 4 or more workers)

Industry group	Number of workers	Per cent
TOTAL, all industries	563,234	100.0
Food	35,239	6.3
Textiles	35,002	6.3
Clothing	27,451	4.9
Lumber and wood products	14,218	
Furniture	13,082	
Pulp and paper	26,207	
Publishing and printing	39,674	
Chemicals	42,757	7.6
Petroleum and coal products	1,424	
Rubber products	9,200	
Leather and its products	6,098	
Ceramics, stone, and clay products	19,094	
Iron and steel	36,280	6.5
Non-ferrous metals	13,564	2.4
Fabricated metals	71,684	12.7
Machinery (general)	74,792	13.3 ⎫
Electrical machinery	40,320	7.2 ⎪ 25.7
Transportation equipment	28,478	4.3 ⎬
Precision instruments	5,276	0.9 ⎭
Miscellaneous	25,394	

Source: Osaka Municipal Office. *Osaka City, 1961* (in Japanese).

The coastal industrial belt in Osaka continues across the city boundaries into Sakai on the south and Amagasaki on the north and west and, with some modifications, even beyond the latter to Nishinomiya and Kobe. Shallow water is a handicap to maximum use of tidewater location in much of this area, while within seaside Osaka ground subsidence has been so serious that a number of factory sites have had to be abandoned. Sakai and Amagasaki ports are especially shallow, so that boats of only a few hundred tons can come alongside the factory sites. In the port section of Amagasaki the seaside was developed for heavy industry from the beginning, so that even at present it gives the appearance of an exclusive and planned industrial area, and the same is true of parts of the Osaka water front, but in other sections the composition of manufacturing and the degree of planning and of industrial exclusiveness show wide variations.[23]

In order to provide much needed sites for heavy and chemical industries, an extensive seaside reclamation project is in progress which will allow for a greatly expanded coastal industrial zone. The plan calls for a reclamation by dredging and filling of nearly 21 million square meters at South Harbor of Osaka city and at Sakai Harbor which joins Osaka on the south. Parts of the reclamation project are completed, and a new factory of the Yawata Iron and Steel Company is already under construction. Depths of water alongside the newly reclaimed lands vary between twelve and sixteen meters. Consideration is being given to the development of a great combination Hanshin port, including Osaka-Kobe and other smaller coastal cities, behind a breakwater to be constructed across the head of the bay from Kobe to Sakai.[24]

Inland from the coastal strip, and between it and the city center, where factories tend to be concentrated along a number of canalized distributaries, is a manufacturing district where industrial planning has not prevailed, so that factories are randomly located and there is a confused pattern of land use. Accessibility to barge services makes riverside location attractive to the variety of industries there concentrated, including machinery, electrical appliances, metals, textiles, pulp and paper, and chemicals. Small and medium-sized plants prevail.

The so-called urban industrial areas are those that surround the old city core. There, establishments usually are not large, diversity of industry prevails, and land-use functions are mixed. Large numbers of individual structures combine residential and industrial functions, signifying that home industries are numerous. Around the commercial cores of Sakai and Amagasaki similar zones of mixed industry are common.

Beyond the city proper, and therefore in the suburban areas and along the national highways, are scattered factories within a periurban zone where urban and agricultural land use are both present.

Modern Osaka harbor is entirely of artificial construction. Two converging moles extending seaward some three kilometers, with a narrow entrance between their terminals, enclose a harbor of modest dimensions. Approach to the harbor is by means of a dredged fairway 9 meters deep and 230 meters wide. Shallow water at the head of the bay where Osaka is situated makes it difficult for deep-draft ships to enter, so that most of the shipping entering the port is in the form of small and medium-sized vessels. To be sure, the harbor has been greatly improved, but its remaining handicaps make it unlikely that Osaka can ever be the equal of Kobe as a deepwater port, for not only is the harbor shallower, but also it is less well protected from strong winds and high tides. At the tidewater industrial sites on the newly reclaimed lands, depths of thirteen to sixteen meters are provided. Within the mole-protected harbor are more than thirty anchor buoys at which vessels of 10,000 tons and over can be berthed. Osaka is primarily a cargo port, and much of the cargo is loaded and unloaded by means of lighters. Using canalized rivers, cargo unloaded into lighters from vessels moored by buoys could formerly be transported directly to factories located along the waterways well within the city, while products manufactured at these same places could be taken by barges to the harbor for shipment. Wharfs and piers with a total length of 9266 meters and with average water depth alongside of 9.6 to 10 meters, provide alternative means of loading and unloading cargo. As a rule, the larger vessels anchor at buoys. In 1961, 20.7 million gross tons of shipping used the wharves and piers, and 11.2 million tons used the buoys. Large vessels on the American and European runs do not make so much use of Osaka port; instead it is chiefly the 6000-ton to 10,000-ton vessels, many of them operating on shorter runs, that enter this shallower harbor.

While Osaka is primarily a cargo port, its location at the eastern end of the scenic Inland Sea has likewise given it some fame as the eastern terminus of numerous domestic sightseeing ships. Nearly 1.9 million passengers used Osaka port in 1960.

In terms of total tonnage of sea-borne commerce through the five major ports of Japan in 1961, Osaka ranked next to Yokohama and only very slightly below it. Eighty per cent of its cargo tonnage was imports and only 20 per cent exports. This disparity reflects the proximity of the great Kinki market and the industrial nature of the hinterland which emphasizes the import of bulky raw materials. Foreign trade amounts to only 37 per cent of the domestic. Foreign exports are very small, under a million tons, and domestic shipments represent 84 per cent of the total. Cargo receipts are divided, about 70 per cent domestic and 30 per cent foreign. Dominantly the domestic trade is with Inland Sea ports, including those of northern Kyushu (89 per cent of exports and 80 per cent of imports), the principal receipts being coal, petroleum products, iron ore, stone, and cement. Foreign imports emphasize logs, coal and coke, scrap iron, iron ore, phosphate, foods, non-ferrous minerals, fertilizer, raw cotton, and salt. Thirty-five per cent of the foreign imports are from the United States, and 25–30 per cent are from south and east Asia. Foreign exports are largely to Asia.

Osaka's commercial and industrial fame is of long standing. During much of the fourth century A.D. it was the capital city of Japan and consequently the political, as well as the economic center. During the long Nara-Kyoto period of political pre-eminence Osaka, then called Naniwa, was their principal port and commercial and financial center. Very early it became the port of entry for official envoys from Korea and China. In the sixteenth century Hideyoshi, predecessor of the Tokugawa shoguns and a great tycoon of the feudal period, chose the city for his residence, built Osaka Castle, and induced the merchants of Sakai and Fushimi to locate there. Throughout the whole feudal period, even after Edo (Tokyo) became the Tokugawa capital, Osaka continued to grow as a commercial center, the daimyos erecting huge warehouses there for the storage of the products from their respective fiefs. Osaka merchants acted as middlemen and bankers in the distribution of these products so

TABLE 14-6

Sea-borne trade of Osaka port, 1960

Foreign				Domestic			
Exports (895,141 tons)		Imports (6,181,837 tons)		Exports (4,695,280 tons)		Imports (14,639,696 tons)	
Metal products	14%	Coal	15%	Coal	15%	Coal	24%
Cement	12	Logs	15	Petroleum		Petroleum	
Iron	10	Scrap iron	15	products	9	products	20
Machinery	5	Iron ore	13	Coke	9	Iron	16
Vehicles	3	Phosphate	5	Phosphate	5	Limestone	9
Fertilizer	3	Sugar	4	Cement	3	Cement	6
		Petroleum		Scrap iron	3		
		products	4	Fertilizer	3		
		Wheat	4	Iron ore	3		
				Salt	3		

Source: Osaka Municipal Harbor Bureau, *Port of Osaka* (1962).

that their city became the commercial hub and they the merchant princes of the country.

Kyoto, site of the Imperial residence, and a city of perhaps half a million throughout much of the feudal period, required tremendous quantities of water-borne supplies, all of which passed through the port of Osaka, only forty or fifty kilometers distant. Thus at the time Japan was opened to foreign trade about a century ago, Osaka was the greatest trading mart in the country, but it was all domestic trade carried on in shallow, clumsy boats which required little in the way of port facilities. In 1868 the city was opened to foreign trade, but the large ocean-going foreign ships could not enter its shallow and generally unimproved harbor, and consequently Osaka's business suffered. It was at this time that Kobe rapidly grew into prominence as the transshipping port for foreign cargoes.

Kobe

Kobe (1,006,000 population), third most populous of the triad of major cities within Kinki and Keihanshin, and second within Hanshin, is similar to Yokohama in the locational and functional relationship it bears to its area. Located some twenty-five to thirty kilometers down-bay from Osaka and beyond the area of active delta growth, so that waters are deeper, Kobe is a relatively new city, having been brought into being by the shipping demands of the modern era following the Restoration in 1868. At the time Japan was opened to foreign trade, after the middle of the last century, the central and main part of present Kobe port was the site only of insignificant fishing villages, although, to be sure, Hyogo port, now included within westward-expanded Kobe city, was already a town of 20,000 population. Hyogo had profited by its proximity to the flourishing metropolis of Osaka. But since Osaka's shallow, silted harbor was unable to accommodate the larger foreign vessels from the United States and Europe, the need arose for an adjacent deepwater port to serve the general Kinki region, and this need was satisfied by the development of present Kobe port on a minor coastal indentation known as Kobe Bay, slightly north and east of older Hyogo.

The site of the city, a narrow coastal and alluvial-piedmont strip backed by a high and precipitous granite horst, has determined its linear shape —twelve kilometers long by one and a half to two kilometers broad. Its immediate hinterland is relatively barren of settlement, which again serves to emphasize the fact that Kobe's *raison d'être* is populous industrial Kinki. Industrial and commercial sections of the city occupy the flatter land along the waterfront, while more exclusively residential Kobe monopolizes the piedmont slopes to the rear. The central business district, some of it emphatically Occidental in appearance, occupies a mid-position along the coast just back of the principal piers (Fig. 14-10). Portions of the city along the water front are built upon reclaimed land, with the larger buildings resting upon piles. Since all parts of the business and industrial section are so near salt water, the lack of an extensive system of canals such as serves most of the great Japanese cities is not a serious handicap.

Primarily a port, and somewhat overshadowed industrially by the adjacent metropolis of Osaka, Kobe is nevertheless one of the nation's principal manufacturing cities. It lies at the northwestern extremity of the great Hanshin industrial node and represents the second most important point of manufactural concentration within that node. Employment in its manufacturing plants is only about 23 per cent that of Osaka's but only slightly less than that of Kyoto's.

Noteworthy of Kobe's manufacturing, and distinguishing it from either Osaka's or Kyoto's, is the greater importance of large factories, for establishments with 1000 or more employees represent 39 per cent of the manufacturing employment in Kobe, but only 19 per cent in Osaka. This feature of larger establishments is related to the greater emphasis on heavy industries in Kobe, where iron and steel, shipbuilding, rubber products, and machinery are dominant. On the other hand, textiles and clothing, more commonly housed in small and medium-sized plants, are less important in Kobe, where they employ only 5 per cent of the industrial workers as compared with 11 per cent for Osaka. By contrast, 57 per cent of Kobe's factory employees are in metals and machinery, as compared with 47 per cent for Osaka. Kobe is especially important as a shipbuilding center, containing the Kawasaki Dockyard and the Mitsubishi Kobe Shipyard, two of the giants of Japan. And while Osaka likewise is a shipbuilding center, Kobe, because of its deeper water, is able to build and launch larger ships.

The port section of Kobe, together with the central business district located just to the rear of the main Shinko Piers where the foreign trade facilities are concentrated, divides the industrial belt into an eastern and a western section. In both there is a strong orientation of factories toward the coast, with the larger plants engaged in heavy industry profiting by actual tidewater location where they can be serviced from their own piers. Considerable areas of the industrial belt are on land reclaimed by dredging and filling. Western industrial Kobe consists of four principal subdistricts, three of them centering upon the two large shipbuilding and ship-repair yards and one rolling-stock plant. The

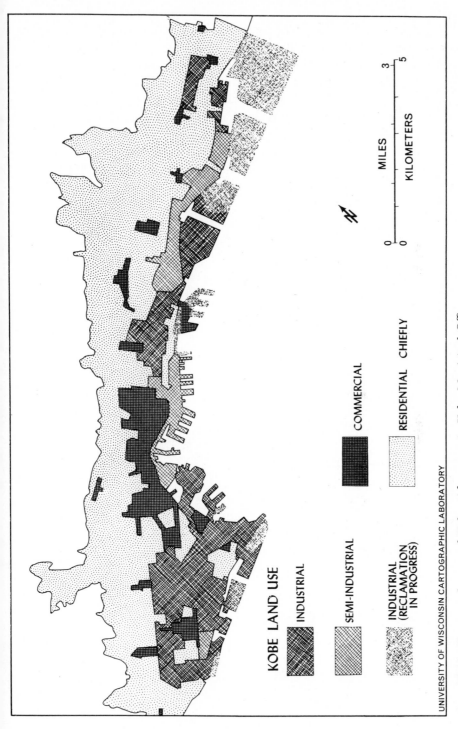

KOBE LAND USE

INDUSTRIAL

SEMI-INDUSTRIAL

INDUSTRIAL
(RECLAMATION
IN PROGRESS)

COMMERCIAL

RESIDENTIAL CHIEFLY

MILES

KILOMETERS

Fig. 14-10.—Functional areas of Kobe. After map in Kobe Municipal Office.

latter is not on the immediate coast but instead is situated on the Hyogo canal just slightly inland. Smaller factories, mixed in with residential and commercial features, occupy the spaces between all three of the giant plants. Along both sides of the four canals in western Kobe are varied industries, including foundries, flour mills, sugar refineries, lumber and woodworking plants, and electrical appliance factories.[25] In the Nagata section, located just beyond and to the west of the main industrial nucleus of western Kobe, is still a fourth fairly distinctive factory area where small and medium-sized plants engaged chiefly in manufacturing rubber products are crowded in with groups of row houses to form a very confused and congested district. The water fronts of both eastern and western Kobe bristle with well over a hundred large oil storage tanks.

Eastern industrial Kobe is dominated by two large iron and steel plants, both on tidewater. Eastward of the steel plants, and still coastal in location, but with the plants not necessarily on tidewater, is a mixed industrial area specialized in *sake* brewing and various food industries.

Kobe has two very serious handicaps to further expansion of manu-

TABLE 14-7

Composition of Kobe industry, 1961

Industry group	Number of workers	Output (million yen)
Foods	14,557	87,234
Textile mill products	4,528	7,467
Apparel and other similar products	2,486	4,160
Lumber and wood products	2,272	3,575
Furniture	1,611	2,152
Pulp and paper	1,648	2,442
Publishing and printing	4,765	7,106
Chemicals	3,528	18,649
Petroleum and coal products	304	1,014
Rubber products	22,810	46,195
Leather and leather products	387	450
Ceramic, stone, and clay products	3,075	7,874
Iron and steel	20,009	99,945
Non-ferrous metal	1,080	1,756
Fabricated metal products	6,198	12,309
Machinery	11,235	18,891
Electrical machinery and equipment	12,389	23,211
Transport equipment	31,781	80,261
Precision instruments	283	270
Miscellaneous industries	3,437	6,806
TOTAL	148,383	431,678

Source: Kobe Municipal Office, *Kobe City, 1962.*

facturing: lack of suitable land for factory sites and a deficiency of industrial water. Both have acted to retard manufacturing expansion in recent years but plans are in operation to improve the situation. The high hills terminating Kobe's narrow plain to the rear and to the west make factory expansion in those directions out of the question. Consequently there has been a rapid filling in of the coastal lands between Kobe and Osaka to the east, resulting in the growth of such in-between cities as Amagasaki and Nishinomiya. But to break the space strangle hold on Kobe itself, that city is now engaged in filling in areas along its coast, whereby over 6 million square meters of new seaside industrial land are expected to be made available by 1970. This new land is located at both the eastern and the western extremities of the city, but chiefly the former (Fig. 14-11). The fill, amounting to 67 million cubic meters, is being obtained from excavations in the hills behind the city, from which location it is carried by conveyor belts and dumptrucks to the water front. Of the new lands reclaimed by fill, a million square meters have been allocated for iron and steel plants: 860,000 square meters for machine and metal industries; 400,000, for oil storage; 200,000, for food processing; and 300,000, for shipbuilding.[26] More than seventy firms have already contracted for space.

Kobe's position as the leading port of Japan in value of its sea-borne commerce reflects not only its superior harbor but even more so its advantageous location. Its trade hinterland is not restricted to the populous industrial Keihanshin area, the second most important in Japan; it serves a much more extensive region, also, including the Chukyo or Nagoya industrial center to the east and north, and Chugoku, Shikoku, and Kyushu to the west and south—all regions lacking in first-class ports specialized in foreign trade. Indeed, Kobe has the most populous and most highly industrialized general hinterland in Japan, so that it is a truly national port. Of further significance is its location at the eastern extremity of the important Inland Sea route, which is followed by both domestic and foreign shipping.

Kobe's harbor in a number of respects resembles Yokohama's and is dissimilar to those of Osaka, Nagoya, and Tokyo, all three of which are on advancing deltas and are handicapped by shallow water. No important rivers enter the sea in the vicinity of Kobe, for its narrow plain is not primarily deltaic in origin. As a consequence, natural depths of ten meters are to be found at a distance of one to three kilometers seaward from Kobe, so that no lengthy dredged fairway is required to reach the inner harbor. Located in a natural coastal indentation, the relatively commodious harbor (55 sq. km.) faces south and east, enclosed by a series of

Fig. 14-11.—A part of the coastal industrial zone in Kobe west of the harbor. Shipyard on extreme right. In the foreground is an extensive area of seaside fill where factories will be erected. Courtesy of Kobe Municipal Office.

thirteen breakwaters, seven kilometers in total length. High hills to the rear of the city protect the harbor from the strong northerly and westerly winds of the winter monsoon, and there is likewise greater protection than at Osaka from violent hurricane winds. Shipping at Kobe is therefore much less exposed than at Osaka, and inundation by high tides, especially at times of strong winds from the south and west, is not felt to the same degree. Although the port is equipped with multiple large piers, six of them with depths-alongside of ten meters or more, probably about 80 per cent of the cargo is still handled by means of lighters. Large line ships, carrying passengers and mail as well as cargo, use the piers more exclusively, while cargo vessels more often anchor at mooring buoys, of which the harbor has forty-four.

In Kobe, as in most Japanese ports, there is a dangerous congestion of traffic, with ships generally waiting to use buoys and piers. To increase port efficiency, several new piers are under construction, including Hyogo Pier No. 3, Shinko Pier No. 8, and Maya Piers 1, 2, 3, and 4. In addition, the new industrial sites in eastern and western Kobe will have their own wharves.

Kobe's sea-borne trade is divided between foreign and domestic tonnage in the ratio of about 53 to 47. Domestic cargo is preponderantly petroleum and metal products, followed by ores, coal and coke, fertilizers, wood, and raw cotton. The large domestic trade reflects Kobe's importance as a port of transshipment.

About one-quarter of the nation's total foreign trade, by value, is handled by Kobe, giving it first rank among the ports of Japan. In value the ratio of foreign exports to imports is about 59 to 41, but in tonnage the ratio is reversed and is 31 to 69. These figures reflect the contrasting nature of Kobe's incoming and outgoing cargo, for imports are largely bulky raw materials for manufacture, while exports are more exclusively processed goods. Both types of cargo are indicative of a strongly industrialized hinterland and the general dearth of domestic raw materials in Japan.

Far in the lead among foreign imports (20–25 per cent of total value) is raw cotton, chiefly from the United States, Mexico, and India. This is followed in order by crude rubber, scrap iron, non-ferrous ores, and flax. The relatively great importance of textile raw materials, in spite of Kobe's very modest development of the textile industry, demonstrates this port's service to an extensive hinterland where the processing of fibers holds high rank. Similar large-scale foreign imports of scrap iron, crude rubber, and ores testify to the importance of steel and rubber processing within Kobe itself, as well as within Hanshin in general. It appears worthy of re-emphasis that Kobe serves as the deepwater port not only for nearby Osaka and all Keihanshin, but also for more distant Chukyo and for numerous local industrial centers within the Inland Sea basin and northern Kyushu. By truck, rail, and small ship the imported raw materials are distributed from Kobe to points both near and far, while the finished goods from a similar extensive area find their way to market via Kobe. Foreign exports, like imports, are strongly oriented toward textiles, which account for 39 per cent of the total, and this in spite of the fact that Kobe is not a textile city. The second major group of exports, consisting of steel, ships, and metal goods, more closely reflects Kobe's own specialization, as well as that of its hinterland.

The question may well arise as to why Kobe should greatly exceed Yokohama in foreign exports (36–37 per cent of the national total for Kobe, 22 per cent for Yokohama) although the two cities handle similar percentages of the country's imports (19–20 per cent). Both ports serve populous industrial hinterlands of a local character, with Kanto somewhat outranking Keihanshin. But Kanto has, in the metropolis of Tokyo and the other Kanto cities, a greater consuming market than is provided locally by Keihanshin, so that more of Kanto's manufactures are consumed locally. In addition, Kobe serves as the outlet for a more extensive industrial region than does Yokohama; for Chukyo, Hokuriku, Chugoku, Shikoku, and Kyushu all use Kobe and Osaka as their outlets. And while it is true that 38 per cent of Kobe's imports and 22–23 per cent of its exports are with the United States, still it has a very large export trade in textiles and other consumer's goods with the countries of southern and southeastern Asia, and it is in such items that Hanshin and the general Kobe hinterland specialize. Without doubt the recent revival of trade with southern and southeastern Asia has considerably benefited the Kinki ports. It has been suggested, also, that the great use made of Yokohama's port facilities by the Occupation authorities after World War II may have operated to divert more of the commercial sea trade to Kobe.

Kino Valley, just south of the Osaka Plain, is a narrow, spear-shaped, sediment-filled lowland lying between Izumi Horst on the north and the Kii Folded Mountains on the south and coinciding with the great fracture zone and morphological fault which separates the Inner and Outer Zones of Southwest Japan. Although somewhat isolated from Kinki by a range of hills, in most respects it has closer attachment to that region than to Kii. Diluvial deposits in the form of several terrace levels, representing different stages of uplift, fill the northern part of the valley. As a consequence, the river and its narrow alluvial floodplain are crowded against the hard-rock hills that bound the valley on the south. Rice predominates on the alluvial floor. Both lower smooth, and higher dissected, terraces are represented, and on these elevated sites the usual dry crops of cereals, together with mandarin orange trees, mulberry fields, and vegetables, occupy the greater share of the cultivated land, although pond-irrigated rice also competes for space. On the hill

TABLE 14-8

Foreign trade of Kobe, 1961

Exports	Value (millions of dollars)	Imports	Value (millions of dollars)
Cotton fabrics	288.9	Raw cotton	227.7
Clothing	112.2	Scrap iron	48.7
Staple fiber	95.9	Crude rubber	41.9
Iron and steel	61.1	Non-ferrous ores	33.0
Ships	52.2	Flax	28.5
Metal goods	49.4	Hides and skins	22.5
Rayon fabrics	47.1	Petroleum	21.2
Radios	38.2	Soybeans	20.3
Woolen fabrics	24.0	Wheat	19.2
Rayon yarn	23.0	Wool	17.8
Others	727.7	Others	575.4
TOTAL	1,519.7	TOTAL	1,056.2

Source: Kobe Chamber of Commerce.

slopes bordering the valley there is even a greater concentration of orange groves than on the diluvial uplands, for Kino is one unit of that specialized Kinki orange district lying along the south side of Osaka Bay and bordering Kii Channel.

Wakayama city (176,000 population), at the mouth of the valley, is in the nature of an outlier of the Keihanshin industrial region. Until lately it was strongly specialized in textiles and consequently resembled the Sennan district just to the north along Osaka Bay. But recent construction of a large seaside iron and steel plant has fomented diversification in manufactures, so that in 1960, although textiles were still the leading sector, representing 31.6 per cent of the total value added by manufacturing, next in line were steel (16.2 per cent), chemicals (12 per cent), and machinery (11.5 per cent). Significantly the metal and machine industries, combined, outranked textiles.

15 · SOUTHWEST JAPAN:
CENTRAL SETOUCHI IN CHUGOKU
AND SHIKOKU

Westward from Kinki, southwestern Japan comprises three discrete land areas: the separate islands of Shikoku and Kyushu, and Chugoku Peninsula, the narrow western extremity of Honshu. Together these encircle the justly famed Inland Sea, known as Setouchi.[1] A logical method for organizing the description of this western part of Japan would be to focus attention on each of the three land areas separately, but to follow the promptings of logic would have the effect of diminishing the strong element of geographic unity which dominates the central Inland Sea and its borderlands in Chugoku and Shikoku. Thus, the southern coast-lands of Chugoku and the northern ones of Shikoku, facing each other across Setouchi, have much more in common than either does with its own transmontane coast. Consequently principal attention in this chapter will focus upon the central Inland Sea borderlands, including southern Chugoku and northern Shikoku, a strongly centripetal region. The highlands of both Chugoku and Shikoku, together with their coast-lands along either the Pacific or the Sea of Japan, are less important regions and therefore will receive less attention. Volcanic northeastern Kyushu's coast-lands on the Inland Sea are sufficiently different from those ʳ Chugoku and Shikoku to earn a separate treatment for Kyushu Island.

CENTRAL SETOUCHI IN GENERAL

In the various schemes employed by different Japanese geographers for subdividing their country, Central Setouchi is almost invariably recog-

534

nized as a unit. It is Japan's Mediterranean. Translated literally the word "Setouchi" means "within the channels," referring to the several straits which connect the Inland Sea with the ocean. Morphologically the sea and its borderlands are the lowest part of a subsiding, down-faulted land area, which prior to subsidence had been a dissected erosion surface cut by a complicated net of faults. Recent drowning of extensive areas has resulted from the rise of sea level following the Ice Age. What was originally a very irregular coastline resulting from submergence of a thoroughly dissected land surface has been somewhat smoothed by the later deposition of alluvium, although the coast of Chugoku continues to be relatively irregular. Alluvium is not confined to tidewater location, for it also fills the intricate network of channels and valleys that extend inland a few miles from the coast. Alluvial plains are almost everywhere small, and in many parts the hills reach down to tidewater.

The waters of the Inland Sea are so shallow that an uplift of fifty meters or a little more would expose a land surface whose essential features would resemble those of the present coast-lands. Before submergence the region was a series of five distinct basins separated by dissected horsts, a pattern quite similar to that of Kinki. These basins or bays are now the more open portions of Setouchi, called *nadas*. The archipelagic portions, composed of hundreds of islands arranged in roughly parallel rows, are the exposed crests of the intervening dissected horsts, and the narrow channels between the islands mark the fault lines. At present the deepest water is found not in the open nadas, but in the constricted interisland channels, where depths are maintained by tidal scouring, tidal races being so vigorous in some places as to make navigation difficult, if not actually hazardous.

The dominant granitic and Paleozoic rocks in the Setouchi borderlands show significant landscape differences. Thus, the drainage density net is finer in the granitic areas and the vegetation cover is sparser. Tanaka ascribes these differences to the coarser grain size of the granitic soils and their lower water-holding capacity, as well as to the great susceptibility of the granites to erosion.[2]

A very conspicuous feature of the Setouchi borderlands is the large amount of bare, whitish slope land which has been denuded of its soil and vegetation cover. It is not at higher elevations that bare slopes are most common, but rather on the foothills adjacent to densely populated plains. The causes seem to be both of natural and of human origin, associated in part with the large amount of granite which, as noted elsewhere, is readily weathered and eroded and produces a coarse droughty soil, and in part with the lower than average rainfall which makes reforestation more difficult. Chiba[3] is inclined to think that human causes are even more important, and that the forests were seriously denuded in early Meiji times by a dense population which intensively gathered fuel and fertilizer materials from the common forests.

Climatic features are likewise distinctive within the Inland Sea district, for annual precipitation is less, number of rainy days fewer, humidity lower, and amount of sunshine greater than in most other parts of subtropical Japan. Extensive areas of Central Setouchi receive less than 1200 millimeters (48 in.) of rainfall annually, and in restricted areas it drops below 1000 millimeters (40 in.). Climatic and weather contrasts are particularly striking in winter between the brighter and quieter Inland Sea littorals and the coast facing the Japan Sea, across the Chugoku Peninsula, where weather is much more tumultuous, skies more lowering, and precipitation, including snowfall, heavier.[4]

The shores of Setouchi comprise a region of ancient occupance coeval with the Christian Era in the West. With the importation of bronze and iron from eastern Asia and the migration of people as well, nearly two millenniums ago, the Yayoi culture developed in Japan, with its two main centers at the eastern and western extremities of the Inland Sea, in Kinki and northern Kyushu. Connection between the two centers gradually developed by way of the Inland Sea, and in about the fourth or fifth century the Yamato Court in Kinki placed the northern Kyushu center under its command. Because of its proximity to the continent, which was the source of danger from invasion, northern Kyushu was developed as an advanced military and political stronghold against such potential Asiatic invasion, so that the Inland Sea corridor route connecting Yamato and Kyushu was furthered in its early development by its importance as a line of communication. Much later, when the manors were obliged to pay a tribute in kind, chiefly in rice, to the Emperor, shipment was made from ports on the Inland Sea to Kyoto, the capital, thereby fostering the development of Setouchi port cities.

Even down to the present Setouchi Basin has continued to function as an important corridor of marine and land transportation. Along the Sanyo coast of southern Chugoku run main rail and highway lines, while the sea itself is a veritable thoroughfare for shipping.

Influential factors in furthering Setouchi's development as an important commercial sea are its protected waters and numerous natural harbors, together with the dense population which occupies its margins. Not only is it followed throughout its entire length by ocean-going and local steamers, but it is also crossed from north to south by multitudes of small intercoastal and interisland boats engaged in an important domestic trade. Since this is a segment of the nation's well-populated industrial belt, cities are relatively numerous, although none of them is of first rank. Almost all the larger cities have coastal locations and are maritime in aspect, each being an important trade center for a local hinterland and linked by frequent boat service with neighboring ports.

At the extreme eastern and western ends of Setouchi are, respectively, the important twin ports of Kobe-Osaka and the Kammon ports of northern Kyushu. Between these, and much more engaged in domestic trade, are nearly a score of lesser ports, most of them located in the central portion of Setouchi where well-populated islands are most numerous and where they do not compete with the great "end ports." Some of the leading ones are located opposite constrictions in the sea or on important channels. All of them have vital shipping connections with the great Kinki ports of Kobe and Osaka, upon whom they are highly dependent for marketing and port services. The main traffic is in raw materials and fuels from northern Kyushu and Shikoku destined for Hanshin where they are processed or consumed, and subsequently the distribution of Hanshin's finished products to the different parts of the Kinki region. Boats plying only between the local ports tend in general to follow north-south routes connecting opposite cities on the Chugoku and Shikoku coasts and touching at some islands between. Characteristically, the water fronts of these minor ports suggest a forest of masts, so crowded together are the small craft of various sorts. Small boats employing both engines and sails are unusually important carriers on the quiet waters and short hauls characteristic of the Inland Sea, for this *Kihansen* service accounts for over one-half of the volume of goods transported.[5] Such small cargo-carriers have a number of important advantages over larger steamships, including (1) adaptability to the congested traffic conditions in the Inland Sea, (2) reduced expense associated with loading and unloading, (3) the cheapness of short-distance hauls, (4) the facility with which small cargoes can be handled, and (5) adaptability to inefficient, antiquated port facilities. Much Inland Sea cargo is bulky and cheap and consists of coal and coke (from North Kyushu and Ube), pig iron, cement, limestone, salt, gravel, fertilizer, and raw cotton.

Although this is a part of the nation's industrial belt, it is an attenuated part and not a major node or center like Hanshin or Chukyo. Rather, it consists of a number of small isolated local districts strung out along both the Chugoku and Shikoku margins of Setouchi (Fig. 15-1). Excluding North Kyushu, the combined output of the scattered manufacturing cities of Central Setouchi represents some 6 or 7 per cent of the nation's manufacturing. Thus far, industrial expansion has not been sufficiently rapid to absorb the current population increase, so that the In-

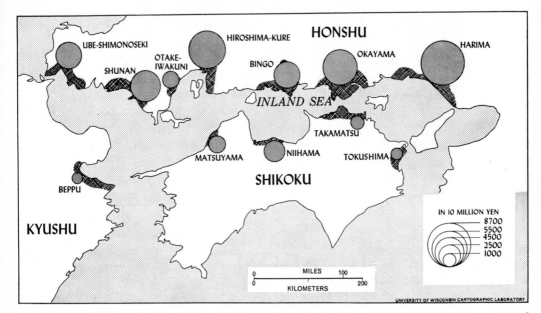

Fig. 15-1.—Local industrial districts along the borders of the Inland Sea. Proportional circles indicate the value added by manufacturing in each district, 1960.

land Sea prefectures are either declining in population or growing at less than the country average.

Although different local districts show contrasting specializations in manufacturing, up until the Pacific War the region as a whole emphasized textiles. During the last decade or two, however, while textiles have lagged, the chemical and heavy industries have soared, so that at present chemicals and machinery of all kinds are the sectors which are far in the lead, and even metals now exceed textiles in value added by manufacturing. Great industrial combines involving oil refining and petrochemicals have grown especially fast in Setouchi, with the result that at present there are three or four noteworthy districts so specialized. With the shift toward heavy industry and chemicals has come a shift also in the location of factories toward tidewater sites. This in turn has stimulated active reclamation of land from the sea by dredging and filling at numerous locations along the margins of the Inland Sea (Fig. 15-2).

Agriculture is unusually intensive, and the area cultivated per farm family (0.6 hectares in southern Chugoku, 0.5 hectares in northeastern Shikoku) is below the national average of 0.8 hectares. Such diminutive farms result in numerous part-time farmers, who in turn represent a relatively large and stable labor pool available to the rapidly multiplying factories. Similarly reflecting intensity of land use is the remarkable

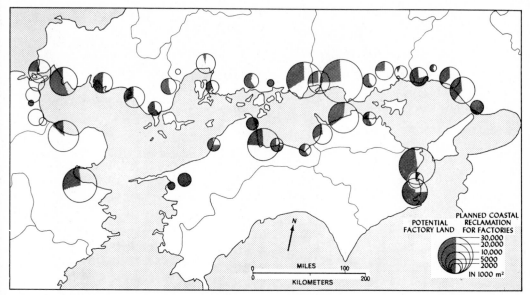

Fig. 15-2.—New land presently available for industrial expansion, or to be made available by seaside reclamation, along the margins of the Inland Sea. After map in *10 Nen Go No Setonaikai Chiiki, 1960.*

exploitation of the hill slopes, in the form of terraced fields. In few areas of Japan is terraced cultivation of steep slopes so extensively practiced. But in spite of intensive cultivation, yields per unit area are not as high as might be expected, especially in paddy rice, for the prevailing soils, derived as they are from granites, are coarse in texture and shy in the alkaline minerals, which have been depleted by long use. As a consequence, their absorption of commercial fertilizers is inefficient, so that it is not uncommon for the rice plants to wilt suddenly just before the grain is about to ripen. In addition to the usual staple crops of rice, wheat, barley, sweet potatoes, and vegetables, this general region has special and national fame as a center of production both of citrus, chiefly mandarin oranges grown on hill slopes, and of Igusa, an aquatic reed from which are made tatami mats, the nearly universal floor covering.

Some fame likewise attaches to the borderlands of Setouchi by reason of the fact that over 90 per cent of the nation's salt fields are located there. The process whereby table salt is obtained from sea water, through natural evaporation, is a government monopoly in Japan. Its concentration in the Setouchi area reflects the greater prevalence of hot sunny days with low relative humidity here than elsewhere. Other favorable elements have been (1) the moderate tidal range, averaging about three meters, which facilitates the periodic filling and draining of the canals

Fig. 15-3.—A flow-down type of salt field along the shores of the Inland Sea. Photograph by Setsutaro Murakami.

in the salt fields; (2) numerous small protected alluvial areas with extensive clean sandy beaches and quiet waters off shore; and (3) the ready availability of cheap coal, brought from Ube and North Kyushu by small boat, for use as fuel in boiling the brine.

Since about 1950 there has been a remarkable change in the prevailing type of salt field in the Setouchi region. The older type, known as *irihama enden,* is developed on a sandy beach at the head of a small bay. It is encircled by embankments and crossed by ditches in geometric pattern, by means of which sea water is allowed to enter the field. Through natural evaporation processes the salt from the sea water crystallizes on the surface sand, which is then scraped up and put into a percolator established at the center of each evaporation flat. There the salt is washed out of the sand by water in the percolator, the brine collecting at the bottom and subsequently being carried off by underground pipes to a boiling station in one corner of the salt field. The geometric pattern characteristic of the surfaces of the salt fields has been a very conspicuous landscape feature of the Inland Sea borderlands.

In recent years the so-called "flow-down system" has rapidly replaced the irihama method of brine accumulation, so that at present well over 90 per cent of the salt fields have undergone conversion. In the new system a gently inclined evaporator plot is floored with angular pebbles coated with black tar. Sea water is allowed to flow slowly over the irregular blackened surface whose high temperature speeds up evaporation and concentrates the brine. It is then further concentrated through evaporation by allowing it to trickle down over pyramid-shaped bamboo screens about five meters tall, after which the highly concentrated brine collected at the lower end of the bamboo screen is sent by pipes to a boiling station. Under the new system the yield of salt per hectare is half again as much as that under the irihama type, while the number of employees required is markedly decreased.[6]

Fortunate are the Japanese who live along the shores of Setouchi, for the gentle charm of its landscape, both natural and man-made, makes it one of the most attractive parts of the country. Nor has the beauty of the Inland Sea region been lost upon the Japanese, judging by the large numbers of people who for centuries have crowded onto its dimunitive plains and adjacent hill slopes. Protected by highlands both to north and to south, Setouchi basks in a climate which is softer and brighter than the normal for Japan. Blue skies, calm bright water, studded with islands and enlivened by picturesque boats and sails, shining sandy beaches of disintegrated granite, exceedingly intricate coastlines with coves sheltered by pine-crested hills, whose terraced slopes are pleasingly geometric in pattern and variegated in color—these are a few of the elements which combine to make the region unforgettably beautiful.

THE ISLANDS

The thousands of islands, large and small, which dot the surface of Setouchi are a characteristic feature of this Japanese Mediterranean. An opportunity to observe the landscape features of the islands is afforded by a voyage on one of the small intercoastal steamers, which call at numerous island settlements, not putting in alongside a dock at any of the ports, but always being met by a sculled lighter. Such was the case, at any rate, on one particular crossing, in a thousand-ton boat, from Hiroshima in Chugoku to Imabari in Shikoku.

Most of the islands are several hundred feet high, with steep, forested slopes. Those composed of granite have somewhat softer contours with much bare whitish rock showing through the sparse woodland cover. Slopes frequently terminate abruptly in the form of low wave-cut cliffs at the water's edge. Hidden away in little coves are tiny alluvial accumulations on which small agricultural-fishing villages have grown up, although many of the deltas are so small that expansion onto the adjacent lower hill slopes and onto reclaimed land along the sea margin has been necessary. The little fishing boats belonging to the island villagers

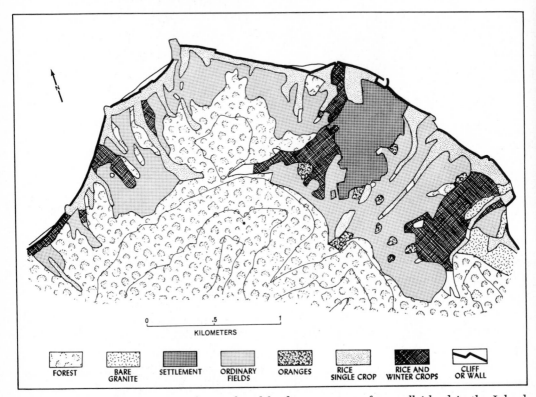

KILOMETERS

FOREST | BARE GRANITE | SETTLEMENT | ORDINARY FIELDS | ORANGES | RICE SINGLE CROP | RICE AND WINTER CROPS | CLIFF OR WALL

Fig. 15-4.—Features of agricultural land use on part of a small island in the Inland Sea. After map by I. Watanuki.

anchor behind the protection of low stone breakwaters, which enclose a trifling bit of sea. Many small islands adjacent to a larger mother island have no settlements. Those that are used exclusively for rice cultivation are less likely to have permanent habitations than those on which dry fields prevail, for the latter require more constant attention. Since grass is much in demand for forage and fertilizer, it is the exclusive resource on some islands. Because island farms are reached more often by boat than by road, the cultivated spots are distributed in a most irregular manner and are likely to be found wherever access to and from the sea is least difficult.

Shrines, often dedicated to *Benten*, sea goddess and favorite deity of the island people, occupy conspicuous promontory sites and in pre-lighthouse days served as landmarks for the navigators.[7] Wondrous to a foreigner is the widespread cultivation of steep slopes by means of artificial terracing. By this laborious method cultivation is often carried up the steep hillsides for as much as 150 to 200 meters. In such a slope environment unirrigated crops naturally predominate, although rice is grown wherever possible. Mandarin oranges are one of the most widespread crops of the terraced slopes.[8] On at least two islands of the Inland Sea —Naoshima (Kagawa Prefecture), near the coastal smelter town of Tamano in Okayawa Prefecture, and also Shisaka, northwest of Niihama in Shikoku—are located important smelters processing copper and other ores. In such island locations the smelter fumes can do least damage to surrounding vegetation and crops.

CHUGOKU

The Sanyo coastal district of southern Chugoku

The Inland Sea margins of Chugoku are known as Sanyo, which translated means "sunny side," as contrasted to the northern coast, facing the Sea of Japan, which is labeled Sanin, meaning "shady side."

The irregular south coast of Chugoku has resulted from the sinking of a hilly land whose original presubsidence features were caused by faulting and stream erosion. Along certain stretches islands are abundant, but where *nadas* (presubsidence basins) prevail they are largely lacking. Many former irregular arms of the sea, or archipelagic channels, have been filled with detritus, forming a complicated and seemingly almost patternless system of alluvial valleys and basins, or small plains studded with hard-rock hills. Most of the region's dense population is confined to the intricate network of fault valleys which interlace the seaward margins of the slope lands.

Scarcely any of the delta-plains are sufficiently large and distinctive to warrant their consideration as separate units. Alluvial deposits accumulate here with surprising rapidity because of the rapid weathering of the rounded and often bare granitic hills. A relatively dense population has kept the vegetation cover of these slopes impoverished for centuries, and reforestation is therefore extremely difficult. Hills come down to the water's edge so frequently that many stretches of the Sanyo trunkline railway, paralleling this coast and connecting Kobe and Kitakyushu, are located inland at variable distances from the shore and separated from it by hills. Diluvium is not extensive along the coast of Chugoku except toward its extreme eastern end between Himeji and Kobe. There in the basin of the Kako River is a relatively large area of old diluvium and Tertiary deposits which has been so thoroughly dissected that almost all upland surfaces have been obliterated. Associated with the labyrinthine pattern of alluvium-floored valleys there has developed a relatively dense dispersed rural population, whose floodplain paddies are watered by numerous artificial ponds located in the nearby hills.

Fig. 15-5.—Within the Inland Sea Basin, mandarin oranges are characteristically grown on terraced steep hillsides. Photograph by Setsutaro Murakami.

While arable land is completely occupied, there are no large, compact clusters of population, since there are no extensive plains. Various forms of rural settlement are represented, dispersed and compact, as well as intermediate amorphous types. An unusual feature is that in such a long-settled region there should be so many isolated farm dwellings, such dispersion actually prevailing on the hilly lands, but also far from uncommon on the more open plains. In no other part of rural Japan are there to be observed so many prosperous looking and attractive residences of more than average size. Whitewashed, with roofs of bronze-colored tile, many of them look more Mediterranean than Japanese. These do not predominate, to be sure, but they are sufficiently numerous, especially in those parts westward of Onomichi, to add brightness, color, and charm to the countryside. On the extreme outer margins of some of the larger deltas, where new land has recently been reclaimed for agricultural uses, the isolated rural dwellings and the rectangular patterns of land subdivision and lines of communication remind one of Hokkaido.

Certain features of the agricultural landscape are distinctive, as for example the already noted prevalence of terraced fields on slopes. A further resemblance to the islands of Setouchi is the specialization of

Fig. 15-6.—Harvesting *Igusa* reed. Photograph by Setsutaro Murakami.

these slope fields, particularly in western Sanyo, in the mandarin orange, for close to 9 per cent of the nation's acreage of this crop is concentrated in southern Chugoku, largely in Hiroshima and Yamaguchi prefectures. Grapes, pears, peaches, apricots, and persimmons also are locally important crops. Still more widespread are the fields of summer vegetables, many of which are replanted to cereals in the fall. High-grade rice is produced on the relatively coarse alluvial soils, which have been weathered from the adjacent granite hills. Winter cropping of both the paddies and the upland fields is very common, so that the region has become one of the country's largest producers of naked barley and wheat, both fall-sown crops which are used for human as well as animal food. The cattle population is relatively dense for Japan, many farmers raising beef-draft animals for sale, thereby increasing their meager cash incomes. This specialization in cattle in southern Chugoku seems to date from the Sino-Japanese War of 1894–1895 and the Russo-Japanese War of 1904, when army headquarters were located at Hiroshima and the region supplied large amounts of beef to the military. A number of towns, both coastal and interior, have become important as cattle markets. The animals are kept chiefly at the farmsteads and fed on grass cut from the hill slopes and on crops grown on the tiny farms.

An industrial cash crop of some importance is the Igusa reed, a dark green aquatic plant three or four feet tall whose small cylindrical stem is used chiefly in making *tatami*, but also as the raw material for rugs, floor mats, place mats, and curtains. Close to 60 per cent of the nation's acreage of this crop is concentrated in Okayama and Hiroshima prefectures. Igusa grows as a winter crop in paddy fields, to which it is transplanted from seedbeds in December and early January.[9] Following the harvest of Igusa in July, rice seedlings are planted in the same fields. After cutting, the reed stems are dipped in a water and white clay mixture which aids in bleaching and in an even drying. Subsequently they are laid out on roads, paths, and hill slopes to dry. In Okayama Prefecture, where the crop is highly concentrated, Igusa occupies about one-ninth of the total paddy area. Some of the reed crop is hand woven in the farm homes, while a part is processed by power looms in the villages.

No part of Chugoku, including Sanyo, is an important producer of minerals. In western Sanyo is the Ube coal field whose normal annual yield is a little over 3 million tons of generally low-grade bituminous coal, or about 6 per cent of the nation's total. However, the 28 per cent of the Ube yield classed as anthracite makes this field the largest anthracite-producer in Japan. Coal from Ube serves as a power and raw-materials base for a local center of industry in the vicinity of the field, and it also powers industry in various cities throughout the Inland Sea Basin. Iron pyrites mined at Yanahara near Okayama provide the basis for a local sulfuric acid industry of national importance.[10]

Harima, or *Banshu, industrial district* in easternmost Sanyo, which may logically be classified with either Central Setouchi or Kinki, is an old textile district which is in the process of being transformed into a center of modern heavy industry. (See Fig. 15-1.) Himeji (161,000 population), the main city, is the fourth largest in Hyogo Prefecture and the only large one in that prefecture west of Akashi Strait, which separates large Awaji Island from the mainland. Much of Harima's sea-borne trade takes place through the nearby deepwater port of Kobe, although a number of shallow local ports are able to handle transshipped cargo carried in small vessels. Banshu is the smaller of the two main centers of manufacturing in Hyogo Prefecture, the first, of course, being the Kobe part of Hanshin in eastern coastal Hyogo. Still, Harima employs about one-fourth of all the factory workers in industrial Hyogo Prefecture and accounts for about the same proportion of the output of goods, but this equals less than half the employment and only one-third the value of output of the Kobe center. Measured in terms of output value, iron and steel rank first, followed by machinery (including transport equipment), chemicals, textiles, and food. As one might expect from the emphasis on heavy industry and chemicals, a great proportion of the factories are located on, or close to, tidewater where they can be served by water transport; so that the industrial area forms a coastal belt extending from the city of Akashi on the east to Ako on the west, with Himeji located near the center (Fig. 15-7). Actively planning for an anticipated large-scale expansion of industry in the district, the prefectural authorities are using dredge-and-fill methods to reclaim extensive areas of water-front land, a considerable part of which has already been allocated for specific factory construction.

An older industrial subcenter specialized in textiles, sometimes called *Harima North,* is located inland in the vicinity of Nishiwaki. The moderately extensive plain in the Harima area is in part a dissected diluvial surface, twenty to thirty meters in elevation.

Next west of Harima is the *South Okayama industrial district,* located chiefly on the plain of the same name and with Okayama city (155,000 population) as the metropolis. (See Fig. 15-1.) After the Harima Plain, this is the largest in Sanyo, with an area of nearly 600 square kilometers. A low flat delta, with a few isolated hills projecting above its surface, it consists of two parts, an older inner section which, as evidenced by the Jori imprint, has long been occupied, and an outer newer plain, reclaimed from Kojima Bay, parts of which are below sea level. Kojima represents one of the largest coastal reclamation projects anywhere in Japan. Important reclamation began in the seventeenth century, continued through the eighteenth and nineteenth, with large-scale additions being made during this century, so that Kojima Bay has had a reclamation history of over three hundred years (Fig. 15-8). In 1624 the bay had an area of 12,000 hectares, but the newly reclaimed lands had shrunk it to 7000 hectares by 1884 and to only 2500 hectares at the present time.[11] One of the serious and not completely solved problems on the low, flat reclaimed land is the insufficiency of irrigation water.

TABLE 15-1

Composition of manufacturing in the Harima district, 1960 (major industries only)

Industry group	Number of workers	Output (million yen)
Textile products	14,994	24,533
Chemicals	10,119	31,840
Iron and steel	17,324	66,406
Machinery (all kinds)	28,787	41,488
Food	10,200	19,442

Source: Hyogo Prefectural Office.

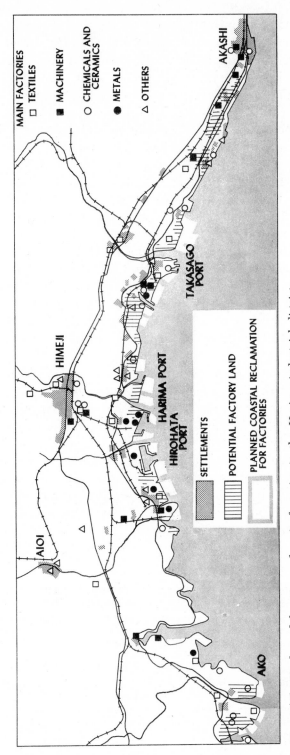

Fig. 15-7.—Kinds and locations of main factories in the Harima industrial district. After map in *Nippon no Chiri*, Vol. 5. No scale given.

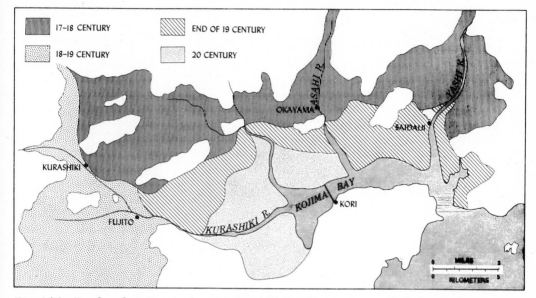

Fig. 15-8.—Land reclamation in the vicinity of Kojima Bay over a period of several centuries. Compiled from maps in *Regional Geography of Japan*, No. 5, *Inland Sea and Kyushu Guidebook*.

Rather striking contrasts exist between the ancient plain, known as the *Koji,* and the modern reclaimed parts, or *shinden,* where settlement is new. In the former, a compact village type of rural settlement prevails, while dispersed settlement is the most common form on the new parts, although different shinden areas representing various periods and methods of reclamation may vary somewhat. On the older plain agriculture is very intensive, farms are small, and orchards are plentiful; but on the shinden lands agriculture is more extensive, and larger rectangular farms of about 1.5 hectares are the rule. On these farms small-scale mechanization is well advanced and seasonal labor is employed on a large scale. Igusa reed is an important cash crop.

The South Okayama industrial district has had modest fame as a local center of manufacturing for some time, with textile fibers and other light industries paramount in and around Okayama, Saidaiji, Kurashiki, Tamashima, and Kojima cities, while Tamano city on the coast of Kojima Peninsula has nearby a large shipbuilding plant and a smelter for copper and other ores, neither of which is new.[12] But beginning with the Pacific War, and more especially since the war, spectacular planned developments of factories on new sites along tidewater have taken place within the South Okayama area. Emphasis is upon chemical and heavy metal and machinery industries, which of course makes seaside location es-

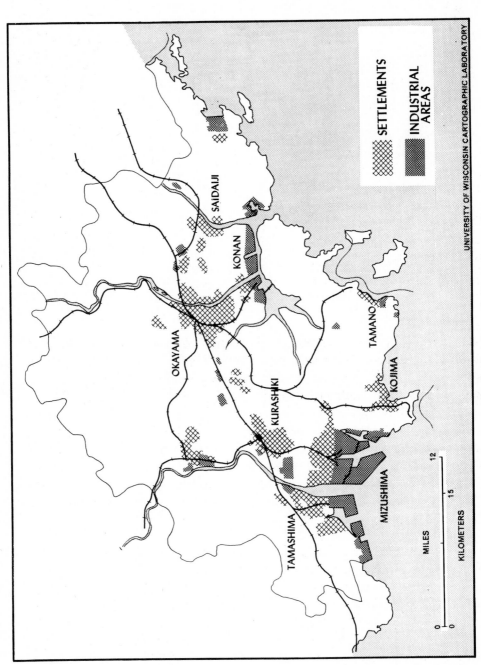

SETTLEMENTS

INDUSTRIAL
AREAS

SAIDAIJI

KONAN

OKAYAMA

TAMANO

KOJIMA

KURASHIKI

MIZUSHIMA

TAMASHIMA

MILES

12

15

KILOMETERS

0

0

Fig. 15-9.—Industrial areas within the South Okayama industrial district.

sential. Particularly noteworthy is the fact that plans have been developed for the reclamation by dredging and filling of 48 million square meters of new coastal lands for industry. Some new factory sites on this made land will have water depths reaching sixteen meters alongside the wharves, which will accommodate oil tankers of 100,000-ton capacity.

Already large littoral areas have been reclaimed and the new lands partially occupied by factories in two centers: (1) the *Mizushima center,* west of Kojima Peninsula and just seaward of Kurashiki and Tamashima cities, and (2) the *Konan center,* along the northern side of Kojima Bay just south of Okayama and Saidaiji cities. The former has had the more spectacular growth, and the plans for its future expansion are likewise more comprehensive. One gets the impression that Mizushima may have one of the greatest growth potentials of any industrial center within the Central Setouchi region. Factory lands are in all stages of development —some already reclaimed, with factories in operation; others reclaimed but without factories as yet; and others still submarine, or partially so, but with multiple suction dredges rapidly extending the area of made land. Emphasis at Mizushima is upon chemical and heavy industries. Already there have been constructed two large oil refineries, and petrochemical industries are expected to be a principal component of the new industrial complex. In addition there is a small steel and steel-products plant, a cement factory, two concrete-products plants, and other establishments for the manufacture of rayon fiber, plastics, synthetic fibers, diesel engines, motor trucks, and bean oil. As of 1960, transportation equipment and chemicals were far in the lead (Fig. 15-10). Kawasaki Limited has contracted for extensive space on one of the "in-progress" reclamation areas, on which it plans to erect a major iron and steel plant. The new industrial harbor and port of Mizushima now under construction is well protected and the depths of the fairways and channels (sixteen meters) will permit relatively deep-draft vessels to approach alongside the various factory wharves.

Konan industrial area, like Mizushima, is a new development, but it is less impressive both in its accomplishments and in its future plans. In a sense this is the outer port of Okayama city, with which it has connections by rail. Although still very raw and unfinished in appearance, it already has a nucleus of manufactural establishments, including a thermoelectric plant, sugar refinery, rayon plant, one chemical plant, a woodproducts establishment, and an impressive array of oil storage tanks.

Of the five prefectures of Chugoku and Shikoku fronting on the Inland Sea, Hiroshima, just to the west of Okayama, has by far the largest number of workers employed in manufacturing. Heavy industry predominates,

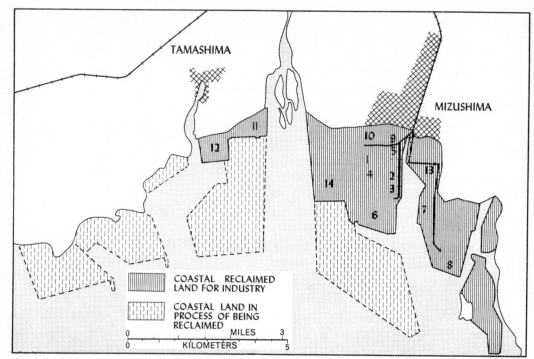

Fig. 15-10.—The Mizushima center within the South Okayama industrial district. Factories are located on coastal lands recently reclaimed from the sea: (1) rayon, (2) cement, (3) soybean oil, (4) chemicals, (5, 10) concrete, (6, 8) oil refining, (9) motor vehicles, (11) rayon fiber, (12) machinery, (13) steel products, (14) iron and steel (planned). After map prepared by the Okayama Prefectural Office.

with transport equipment ranking first in number of employees—ship-building is highly important—followed by machinery in general. Textiles rank third, followed by chemicals, fabricated metals, and iron and steel. Two general districts of industrial concentration within Hiroshima are observable, an east and a west, the former having only about one-third the importance of the latter.

Bingo district, to the east, comprises at least six distinct local centers, including Fukuyama, Mihara, Fuchu, Matsunaga, Innoshima (insular), and Onomichi. Here great variety is characteristic of the industrial com-position, which changes from one local center to another. Thus, Fuku-yama, the leading center of the Bingo district, includes textile, rubber, chemical, and machine tool industries; Onomichi-Matsunaga has some distinction as a *geta, tatami,* and fertilizer-processing center; Mihara is known for cement, rayon, and transportation equipment. At several of these small industrial centers, but more especially around Fukuyama,

Fig. 15-11.—Mizushima harbor and its factory area, the latter on newly reclaimed land. Photograph from Okayama Prefectural Office.

plans are under way for important reclamation of seaside lands for factory sites. Only Fuchu has distinctly interior location, so that the others are all served by water transport, although their port facilities leave much to be desired. Onomichi is a ferry port on a line connecting Sanyo with Shikoku.

The *Hiroshima-Kure district* in western Hiroshima Prefecture consists of two or three centers—the two large ones of Hiroshima city and Kure and the smaller Otake center on the border of Yamaguchi Prefecture (Fig. 15-12). Hiroshima city (407,000, population), the metropolis of

TABLE 15-2

Composition of manufacturing in Hiroshima Prefecture, 1960

Industry group	Factory employees	Output (million yen)
Textiles and clothing	24,087	26,592
Chemicals	12,286	25,268
Iron and steel	10,054	40,901
Fabricated metals	12,245	12,704
Machinery	30,200	46,628
Transport equipment	36,083	98,447

Source: Hiroshima Prefectural Office.

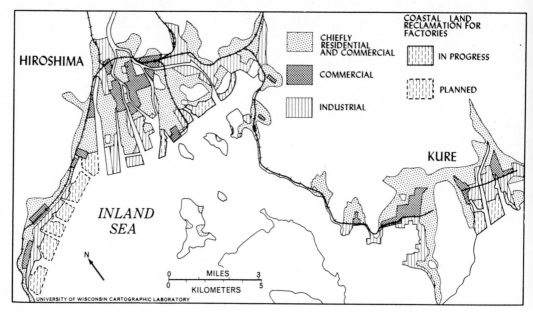

Fig. 15-12.—Land use in the Hiroshima and Kure industrial cities within Hiroshima Prefecture. After maps provided by Hiroshima Prefectural Office.

Central Setouchi, is an old castle town whose modern growth was stimulated by its function as the greatest military and mobilization center in western Japan and the location of Western Japan Army Headquarters.[13] Although the city suffered devastating destruction from the atomic explosion of August 6, 1945, the central parts which were totally destroyed have now been completely rebuilt, so that at present these sections represent one of the most modern appearing of Japan's urban areas, with substantial fireproof buildings and rearranged and widened streets. It is a multifunctional city, being a prefectural capital, a major market place, and the most important manufacturing center between Kobe and northern Kyushu. Strong emphasis is given to machinery, more especially transportation equipment, including motor vehicles and ships, all-machinery being responsible for nearly two-thirds the total value added by manufacturing. Foods rank a very poor second. To a considerable degree the more exclusive industrial areas are coastal in orientation, with an extension inland along the rivers and canals.[14]

The site of Hiroshima city is the small delta of the Ota River with multiple distributory prongs, at whose sea ends are several land-tied islands. Canals are abundant. Along the advancing delta prongs water is so shallow that it has been necessary to develop an outer port of modest importance at Ujina, located close to one of the land-tied islands three or four kilometers from the heart of the city. Large areas of the outer delta have been reclaimed for industrial sites, and additional seaside reclamation is in the planning stage. Hiroshima Bay is nationally famous for its oyster beds.

Until World War II, Kure's fame was that of a fortified naval base and arsenal, the greatest in the Far East. Its large and protected bay and the prevailingly deep water have fostered its postwar development as a shipbuilding center, with auxiliary industries such as iron and steel, machinery, and smelters. Some 30 per cent of the value added by manufacturing is attributable to iron and steel, 24 per cent to machinery of all kinds including transportation equipment, and 10 per cent to fabricated metals. Emphatically it is a center of heavy industry. The great naval dockyards that earlier fabricated the superbattleship *Yamato* are now constructing some of the world's largest tankers. In 1960 it was the third-ranking industrial city of the prefecture, after Hiroshima and Fukuyama. Because of the limited area of flat land, residential Kure has spread up the adjacent hill slopes, but industry is confined to the water front.

Of modest current importance, the *Otake-Iwakuni manufacturing district,* located astride the Hiroshima-Yamaguchi prefectural boundary, has distinction because of the nature of its industrial specialization. Here unusual emphasis is given to chemicals, including chemical fibers, which account for two-thirds of the total value added by manufacturing, with pulp and paper adding another 24 per cent.

Yamaguchi, the westernmost prefecture in Chugoku, like Okayama and Hiroshima has several separate industrial districts of local fame and distinction. (See Fig. 15-1.) Easternmost is the Iwakuni chemical district mentioned above. Farther west is the considerably more important *Shunan manufacturing district,* comprising such individual cities as Kudamatsu, Tokuyama, Hikari, and Hofu, where the combination of oil refining and chemicals (especially petrochemicals), together with steel, appears to be paramount. Under way at present in this Shunan district is the development of a petrochemical combine which bids fair to become one of the largest in the country. Nucleus and pillar of this combine is the Idemitsu Kosan petrochemical establishment in Tokuyama city, which through pipelines will supply nine nearby companies —each specializing in a single group of chemical products—with basic materials such as ethylene and various other petro products.

The *western Yamaguchi industrial district,* close to Shunan in importance, comprises two leading centers—Shimonoseki and Ube—and several minor ones. In reality, Shimonoseki may with equal logic be considered one unit of the highly compact North Kyushu industrial area, even though it is separated from it by a narrow strait. Ube's subbituminous coal[15] has provided a source of power and raw materials which has fostered the development of a chemical complex based on coal and including sulfuric acid, ammonium chloride, caustic soda, and soda ash. Chemicals represent some 38 per cent of the total value added by manufacturing, followed by machinery of all kinds (largely railroad rolling stock) with 26 per cent, and cement with 16 per cent. At the great fishing port of Shimonoseki, the processing of fish products is an industry worthy of mention.

Interior Chugoku

Upland Chugoku, complicated in geological structure and composed of both igneous and sedimentary rocks, shows a multicycle staircase morphology consisting of three levels of erosion surfaces, of which the middle one, the Kibi upland surface, is the most extensive.[16] Remnants of the erosion surfaces in the form of undulating and rolling crests are widespread, but the larger part of the area has been reduced to relatively rugged hill country whose elevations normally do not exceed 1000 meters. Subsequently the region has been fractured into a series of blocks, some rising and others depressed. The principal fault system trends northeast by southwest, but this is intersected roughly at right angles by another series of parallel faults, with the resulting major drainage lines having a marked lattice pattern. Where two fault-line valleys intersect, wider alluvium-floored stretches result. Unlike Kinki, however, Chugoku is not a region of alluvium-floored grabens. Passes and saddles across stream divides, resulting from frequent instances of stream capture, are numerous. Within Chugoku the highest land—the divide between Inland-Sea and Japan-Sea drainage—is asymmetrically located and is much closer to the north coast than to Setouchi.

The complicated terrain features and land-use patterns of this region make broad generalizations on the basis of reconnaissance observation not only difficult but somewhat hazardous. Five rail lines cross Chugoku from north to south, connecting the trunk lines which parallel the Sanin and Sanyo coasts. Except in the higher northern parts where coarser-textured slopes prevail, the hill lands of southern Chugoku, considering the terrain handicaps, are relatively well occupied, both dispersed and semidispersed settlements characterizing the rural areas. The settlement pattern is unusually complicated, coinciding as it does with the intricate system of drainage lines. Densest settlement occurs in two lower, basin-like areas of weak Tertiary rocks in which the local market centers of Miyoshi and Tsuyama are located. Slope cultivation, with terraced hillsides planted in tea, tobacco, orange groves, summer vegetables, and winter grains, is characteristic. Rice occupies all irrigable sites. The region is not thoroughly specialized in any one crop, variety being a distinctive feature.

Some fame is attached to the Chugoku uplands because they are probably the most extensive region specialized in beef-draft cattle. Still, the typical farm household of the hill lands keeps only a cow or two as an adjunct of general farming, but even this permits of some income from the sale of calves. Dealers purchase calves in the hill town markets and drive them to the lowlands where they are sold to the paddy farmers. The lowland farmer, in turn, uses the animal for draft purposes and as a source of manure for two to three years and then sells it for beef.[17] Kinki provides the principal market for these animals.

The hill lands of central and western Chugoku are likewise an important source of forest products. Excluding Hokkaido, Hiroshima Prefecture ranks fifth in the country in volume of saw logs cut and is first in southwestern Japan, both in saw logs and in logs cut for pulp. The other two prefectures of Chugoku also make large contributions to the total log production, with the result that westernmost Honshu has become one of the nation's leading pulpwood sources. This relatively high importance of forest industries is reflected in the stacked bundles of wood and charcoal which are to be seen on the platforms of railway stations throughout the area.

The Sanin or Japan Sea littoral of Chugoku

The name "Sanin," meaning "shady side," as applied to the northern littoral of Chugoku, refers more particularly to the darker, gloomier, stormier winter weather there as compared with Setouchi. But the contrasts between these two regions do not end with climate, for Sanin differs from Sanyo also in its less indented coastline, more limited hinterland, more restricted areas of alluvium, lower population density, fewer cities, more meager development of manufacturing and commerce, and absence of salt manufacture and citrus culture. Sanin is a representative section of peripheral Southwest Japan.

Since Chugoku is asymmetrical in that the drainage divide is closer to the north than to the south coast, the total area of land naturally tributary to Sanin is neither large nor productive. Marginal downwarping, together with deep water offshore, has tended to produce an abrupt coast where wave-cut sea cliffs are conspicuous and large indentations rare. No relatively extensive alluvial accumulations, such as exist in Hokuriku and along the Japan Sea coast of Tohoku are to be found in Sanin. Where small, short streams enter the sea, tiny accumulations of alluvium develop behind outer belts of beach ridges and dunes, but strong waves and currents prevent their seaward extension beyond the protection of the headlands. A single exception to this rather featureless shore line should be noted, where, in mid-Sanin, the Shinji Range, a horst sixty to seventy kilometers long, parallels the coast and is attached to the mainland by a narrow isthmus of Tertiary rock and by river and wave sediments deposited in the intervening graben valley.

Four separate volcanic groups occupy caldron-shaped depressions along the Sanin coast, the mightiest of which is Daisen (1773 meters), just to the east of Matsue, a youthful cone on whose ash and lava slopes cattle and horses are pastured in some numbers. Hyonosen volcanic group (1510 meters), south of Tottori, is so thoroughly dissected that little if any of the original smooth slope remains, and a wild mountainous country is the result. Two smaller and less conspicuous groups of lava domes are found west of Matsue.

Lacking extensive plains, Sanin has no large compact settlement clusters. Small agricultural-fishing villages are strategically located on little alluvial patches at the mouths of short rivers whose valleys provide access to the rugged local hinterlands. Still, considering the nature of the terrain, rural population is relatively dense. Settlements, both compact and dispersed, occupy shallow upland basins as well as valley floors, but there is not the same abundance of open valleys with important linear concentrations of population extending back into the interior such as exist in Sanyo. Artificial terracing for slope fields is common. General subsistence agriculture prevails, though some cash income is derived from the sale of cattle, wood products, and tea.

Since both manufacturing and commerce are relatively unmodernized in Sanin, no large urban centers have developed. Only three cities of over 25,000 population are to be found in the region, the two largest, Matsue and Tottori, having fewer than 60,000 inhabitants. Tottori and Shimane, which include most of Sanin, are two of the nation's prefectures least developed in manufacturing, and significantly the two most important sectors are pulp-paper and foods, both of which depend on local raw materials. Small establishments are the rule. A rail line, constructed with difficulty because of the abrupt and hilly coast, parallels the Sanin littoral. Ocean shipping is relatively undeveloped and ports are rare, there being few natural harbors along the generally smooth coastline. Moreover, during the winter, weather conditions and boisterous seas make navigation difficult.

Shinji Basin, a rift valley lying between Shinji Horst and the main-

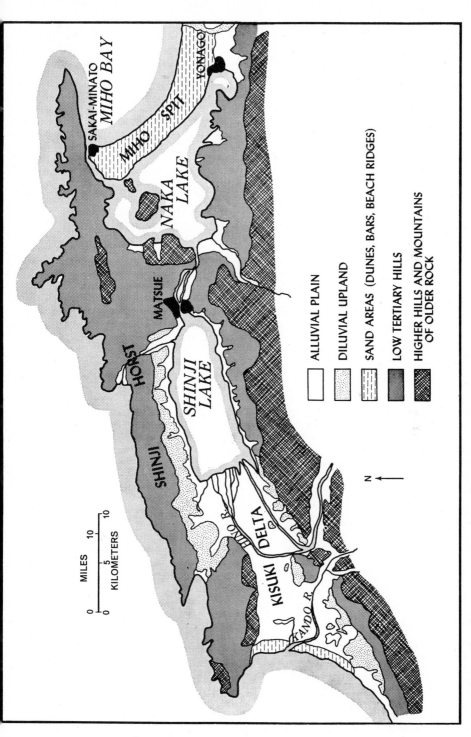

Fig. 15-13.—The Shinji Depression along the Sanin coast.

land, is occupied by Naka Lagoon, Shinji Lake, and the delta-fan of Kisuki, deposited by the Hino River. This elevated and strongly diked river is subject to violent floods, during which periods it has frequently shifted its channel. A wide combination sand spit and bar separates salty Naka Lagoon from Miho Bay, while Naka and Shinji lakes are separated by an isthmus composed of Tertiary rocks and alluvial materials (Fig. 15-13). The western end of Shinji graben is blocked by a bay-mouth bar composed of beach ridges and sand dunes, back of which lies the steeply inclined Kisuki delta-fan of the Hino River. Since the alluvial sediment within the Shinji Depression represents the most extensive lowland area anywhere along the entire Sanin coast, it is not surprising that it is coincident with the single largest population cluster anywhere in northern Chugoku. Density of population is high.

In the older western and central parts of Kisuki, settlements are very ancient, but they become progressively more recent toward the eastern parts of the advancing delta, some along the extreme eastern margins having been in existence less than seventy-five years.[18] Distinctive of this plain is its dispersed rural settlement, each individual farmstead occupying an artificially elevated site as a protection against the flood menace. Equally distinctive is the feature of tree windbreaks twenty to fifty feet high which border the individual farmsteads on their north and west sides, so that from a distance each dwelling has the appearance of a small grove of trees. Usually the windbreak is trimmed squarely on top so that it has an even crestline (Fig. 15-14). Farmers take great pride in the height and perfection of their hedges, which are not simply utilitarian, but are status symbols, their quality and appearance being an indication of the prosperity of the owner. In the almost universal swaled roof crest of the rural houses there is strong resemblance to the architecture of southern Korea. Only here within the old province of Izumo, including the Shinji region, which has had the longest and most intimate contact with Korea, is this unique house type to be found. For purposes of transportation each of the isolated farmsteads has direct access to, and use of, one of the very numerous watercourses that interlace the plain. Each household possesses a boat, which in times of deep flood may become the only means of communication.[19]

The eastern end of the graben back of Shinji is likewise blocked, in this case by a combination spit and bar with conspicuous parallel beach ridges. Here a linear pattern of settlements, communications, and agricultural features is very evident. Dry crops prevail. At the extreme northern tip of the spit and protected from the violent winter winds by the Shinji hills is the local open port of Sakaiminato. From here goods are

Fig. 15-14.—An isolated farmstead with its tall trimmed tree hedge on the Kisuki Delta in the Shinji Depression. Note the thatched saddle roof.

taken by motorboat or by rail to Matsue, the prefectural capital and business center, and by lighter and rail to Yonago, a minor focus of industries, chiefly pulp and paper. Matsue, located at the eastern end of Lake Shinji astride the channel leading to Naka Lake, is an ancient castle town, whose narrow streets and alleys were laid out with defense in mind. It has remained remarkably unaltered to the present day.

INLAND SEA MARGINS OF SHIKOKU

Shikoku has two very dissimilar geographic subdivisions separated from each other by the great Median Dislocation Line and its fault scarp, which both in Shikoku and in Kii Peninsula are associated with elongated rift valleys. To the north of this morphologic boundary is a fractured granite area of hill-land proportions, in many respects like southern Chugoku; to the south is a region of rugged folded mountains, a segment of the Outer Zone of Southwest Japan and hence resembling southern Kyushu and Kii Peninsula. Morphologic and geologic boundaries do not exactly coincide, for a small and lower portion of the Outer Zone of Pacific Folded Mountains lies north of the Yoshino Valley and Matsuyama Plain, which in turn do coincide with the line of dislocation. The landform layout of northern Shikoku, with its two blunt granite peninsulas, Takanawa (west) and Sanuki (east), and broad Hiuchi Nada, or Bay, between them, is the result of differential fault-block movement, the two peninsulas being horsts and the bay a depressed block. It is noticeable that the granite peninsulas are in line with numerous islands in Setouchi, the archipelagoes marking the tops of fractured sunken horsts.

Many of the natural and cultural forms described for Sanyo are duplicated in northern Shikoku with slight modifications. In the unusually intensive use of land, numerous terraced fields, bare granitic slopes, absence of large alluvial plains, specialization in mandarin oranges, numerous salt fields, and the attractions for tourists, both sides of Setouchi have much in common. But at the same time there are significant, if not striking, contrasts. Taken as a whole, the coastline of northern Shikoku is less irregular in detailed outline than that of Sanyo, for the Shikoku coast comprises chiefly the two blunt peninsulas, so that good natural harbors are fewer than along the more indented Setouchi north coast. Also, insular Shikoku must labor under the handicap of having no direct rail connection with the great economic centers of the country. Partly as a consequence of this greater isolation, modern industrial and commercial functions are less well developed and cities are fewer and smaller than in Sanyo.

Low diluvial terraces with smooth or rolling surfaces are more numerous, rivers are shorter and more variable in flow, and fan configuration of the detrital deposits is more conspicuous on the Shikoku borderlands than along Sanyo. Supplementary irrigation from almost innumerable ponds is very common on the alluvial delta-fans as well as on the diluvial uplands, this feature reflecting the shortage of river water. The latter sites, although they are less completely utilized than the more fertile new alluvium, are extensively planted to both rice and dry crops, reflecting long occupance of the region by a relatively dense population.

Occupance patterns associated with the Jori system of land subdivision are very numerous on the plains of north Shikoku. To be sure, evidences of this system are not entirely lacking in Sanyo, but certainly the imprint there is not only fainter but less widespread than in north Shikoku. Even there it is not equally evident in all parts, and local variations in forms and patterns are observable. From the flat top of an andesite hill just north of Takamatsu one can look southward over one of the more extensive lowlands of the island and behold a rural scene where Jori features, expressed in terms of rectangular land subdivision, road pattern, and irrigation ponds, are very conspicuous. But the compact village so well represented in Jori areas of Kinki is not prevalent here; instead, there are numerous isolated farmsteads, many of them located along the grid pattern of highways as in Hokkaido. A similar rectangular pattern of land subdivision and roads is conspicuous on the Matsuyama Plain, but there compact villages are much more numerous and ponds are fewer. On the lowland adjacent to Imabari the mesh of the rectangular grid of roads and farms is considerably finer than in other sections.

Significant contrasts exist, however, between the eastern and western peninsulas. Tanakawa, to the west, is compact and less fractured, contains very little valley alluvium, and presents a relatively smooth coastline, so that settlement is very meager except along the sea margins. On the other hand, Sanuki, the eastern peninsula, resembles Sanyo in being more fractured and fragmented, with considerable areas of alluvium occupying broad valleys and what were formerly interisland channels. As a consequence, over-all settlement is much denser in the eastern peninsula than in the western.

Regional variations

Sanuki, the eastern of the two blunt northern peninsulas, is fairly coincident with Kagawa Prefecture. Its southernmost part is the Sanuki Range, of horst structure, which is a part of the Outer Zone of folded sedimentary rocks lying north of the Median Dislocation Line, in these parts defined by the fault scarp along the northern side of Yoshino Valley. But most of the Sanuki Peninsula is composed of subdued dissected granite hills which gradually decline in elevation to the north where the terrain becomes that of an alluvial plain, above whose surface rise isolated hard-rock hills, so that the whole has the appearance of an archipelago, such as at present actually prevails off the north coast of Sanuki. Three types of contrasting hill masses on the Sanuki Plain can be recognized: (1) a few conical andesite hills which may be remnants of volcanic plugs, (2) gravel-veneered tabular hills, some over 300 meters high, which are possibly remnants of a local peneplain, and (3) a few craggy hills. Mesa-like in profile, the flattish crests and precipitous upper slopes of the tabular hills are coincident with an andesite layer, while the flaring, milder lower slopes are developed on granite. All rivers in the area are short and torrential, descending as they do from the nearby Sanuki Mountains just to the south. Their lower courses are variable in flow and they silt rapidly with abundant detritus collected from the granitic slopes. Most channels are dry except in the rainy season or periods, so that water for irrigation and for industrial uses is limited. Even the alluvial fans have meager water supplies, for they are shallow in depth and underlain by impermeable rocks.

All of the potentially arable land appears to be occupied, and not only the plains but parts of the surrounding hill slopes as well, for terraced hillsides and slopes covered with dry fields are very characteristic. The plain proper is largely in paddy fields made possible by the myriads of irrigation ponds, developed not only in natural sites such as ravine heads and valley junctions, but even on the flat lands where they must be excavated and diked. In Kagawa Prefecture, ponds occupy 13 per cent of the area of all paddy fields, while about 70 per cent of all paddies are irrigated by ponds and only 13 per cent by river water.[20] This is one of Japan's most noteworthy areas of pond irrigation.

Cropping is extremely intensive, as much so as anywhere in the country. In fact, for the period 1946–1959, Kagawa Prefecture had the highest ratio of crop area to cultivated area (174.3) for any prefecture in Japan. Eighty to ninety per cent of the paddies are planted to winter crops, and even triple cropping of paddies, employing a method which involves the transplanting of wheat and naked barley seedlings, is significantly developed. For example, after the rice harvest in fall, wheat is transplanted; and then about a month before the wheat harvest in late spring the wheat stems are bound together and tobacco seedlings, or perhaps those of sweet potatoes or *daikon,* the giant white radish, are set out between the rows of wheat. In the specialized vegetable areas as many as five or six harvests may be obtained. Rice yield is above the national average while the wheat yield is the highest in the nation. A great variety of fruit is grown—mandarin oranges, grapes, peaches, persimmons, apples—the recent expansion of which crops has been stimulated by the introduction of hand tractors, freeing labor formerly spent in arduous spading operations. Farms average only half a hectare, or well below the national average, so that the income per farm family from the land is low in spite of an increased emphasis on non-staple crops and specialized forms of land use such as orchards, tobacco, dairying, and the raising of poultry, edible frogs, and birds. Such specialties are not in the hands of particular farmers, but instead are adjuncts of general farming. More and more farmers are obliged to supplement their income from the land by auxiliary employment in industry.

Compared with Hiroshima and Okayama prefectures in Sanyo, industrial development in Sanuki has been slow. Much of what exists evolved as household industry in the farm villages, so that large-scale modern establishments associated with heavy and chemical industries are few.

Minerals—metallic or fuel—and hydroelectric power are insufficient to attract industry, while most of the industrial raw materials must be obtained from outside. Scarcity of land for large factories and a meager supply of industrial water are further discouraging elements. Abundant labor and marine transportation are the chief attractions Sanuki has to offer. Most manufactures are characterized by small bulk and a high value added by processing.

Noteworthy of the home and workshop industries are those making lacquer ware, knit gloves, shell buttons, imitation pearls, and paper fans. Among the more important modern factory industries are textile mills and those making agricultural implements, measuring instruments, fertilizer, and paper. Lacking, or nearly so, are the genuine boom industries such as petrochemicals, steel, and complicated machinery, for while Kagawa Prefecture's manufactural composition is highly varied the leading items are textiles and clothing, food products, and small machinery of various kinds.

Urban settlements are few and, significantly, all except one very small one have coastal location, their development having been associated with marine transport. Only one city, Takamatsu (120,000 population), has a population of over 30,000, and it rather monopolizes the north and east coast of Sanuki, while several small cities are scattered along the western littoral of the peninsula. Takamatsu is the gateway city for the whole island, for it is the southern terminus of the Japan National Railway Ferry whose scheduled service between Uno, on the Sanyo coast, and Takamatsu—the shortest of the trans-Inland Sea routes—carries 80 per cent of all passengers and freight moving between the mainland and Shikoku. It is likewise a prefectural capital, university town, tourist center, and the principal local area with adequate trading and financial services. But it has been slow in developing manufacturing functions, partly because industrial water and available land for factories are both scarce. Notable among its factories, all of them small-scale enterprises, are those associated with ship repairing, farm machinery, diesel engines, scales, and pulp and paper. Machinery of all kinds contributes about 30 per cent of the total value-added. Among the indigenous workshop industries, lacquer ware and paper parasols rank highest.

Also in eastern Shikoku, although separated from the Sanuki Lowland by the Sanuki Range, is the *Yoshino Valley,* a long narrow lowland associated with the Median Dislocation Line, as is the Matsuyama Plain in western Shikoku. Yoshino, about eighty kilometers long, is shaped like a narrow spear. A conspicuous belt of well-cultivated river terraces and fans, with relatively dense dispersed settlement, occupies the north side of the rift valley. On the valley floor the evidences of floods and lateral stream migration are numerous, and coarse sand and gravel deposits are abundant. Here dry crops compete with paddy rice. In general agricultural characteristics, Yoshino appears to be somewhat transitional between the Inland Sea borderlands and the Outer Zone. Tokushima (123,000 population), the one city of any size and the prefectural capital, is just another local market town, and without any particular distinction as a processing center.

In *Ehime Prefecture,* westward from Sanuki, narrow plains border northern Shikoku, both where it fronts upon Hiuchi Nada and along the margins of blunt Takanawa Peninsula. More compact than its eastern counterpart, the latter lacks anything comparable in size and character to the hill-studded plain of Sanuki previously described. Takanawa is terminated on the south by the Median Dislocation Line, along which there has occurred a slight broadening of the lowland area to form the Dogo or Matsuyama Plain on the west, and the Dozen or Nakayama Plain on the east. The narrow lowlands bordering Hiuchi Nada are also approximately coincident with the same tectonic line.

Compared with Sanuki, western North Shikoku in Ehime Prefecture not only has less lowland, but salt fields are less extensive, water supply from rivers is more abundant and irrigation ponds less numerous, urban development is greater, modern factory industry is better developed, and evidences of the Jori system of land subdivision are largely absent, as is dispersed rural settlement.

Agriculture for the most part does not differ conspicuously between Takanawa and Sanuki. Most noteworthy of comment is the greater attention given to citrus in the west, Ehime Prefecture ranking next to Shizuoka in citrus acreage and having five times the production of Kagawa Prefecture. Citrus is chiefly concentrated on the hill slopes bordering the lowlands, but in the rapid expansion that is now taking place, orchards are beginning to encroach even on the lowland paddies. It is chiefly the more exposed and windier north-facing slopes that are least well suited to citrus.

Although such traditional industries as hand-made paper, lacquer ware, roof tile, china ware, and Japanese wax are well represented, the development of modern industry housed in factories of large and medium size, especially the heavy and chemical industries, has progressed farther in Ehime than in Kagawa, so that the value added by manufacturing is three times as great. Measured in terms of value-added, chemicals lead by a wide margin, with 40 per cent of the total, followed by non-ferrous metals (13 per cent), textiles and clothing (12 per cent),

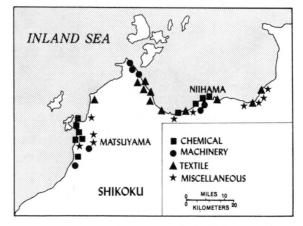

Fig. 15-15.—Important industrial plants in Ehime Prefecture in northwestern Shikoku. After map prepared by Satsutaro Murakami.

and machinery of all kinds (11 per cent). Ehime is favored by a greater abundance of industrial water than Kagawa, and while land for factories is limited, more effort appears to have been made to reclaim coastal lands for industrial development. Most industry has tidewater location (Fig. 15-15).

Ehime has eleven official shi compared to Kagawa's five, and its total DID urban population is twice as large. Matsuyama (142,000 population) is the metropolis, but three other cities exceed 40,000 in DID population. Imabari and Niihama are the only open ports.

Farthest east of the local industrial districts in Ehime, on the shores of Hiuchi Nada, is small Kawanoe-Iwo-Mishima, which is completely specialized in paper and paper products. Besides over two hundred household-industry plants, there are some sixty factories of all sizes, a few of them employing several hundred workers each. About 70 per cent of the raw material comes from outside the prefecture.

Centrally located on the narrow Hiuchi Nada plain is Niihama, leading industrial city of Shikoku and one of the most important in the whole of Central Setouchi, a product of Sumitomo capital and organization. The industrial complex, involving chiefly chemicals and non-ferrous metals but also machinery, had its inception in the exploitation of the Besshi copper (plus silver and gold) mine, located in the rough lands south of Niihama. A large plant for refining the locally produced, as well as imported, ores is located on Shisaka Island, twenty kilometers offshore. From this Besshi mining and ore-dressing center and the Shisaka refining center, Sumitomo capital has expanded into chemical and machine industries, four or five of its establishments employing several thousand workers each, and in addition there are some thirty smaller factories which are subcontractors to the larger plants or otherwise re-

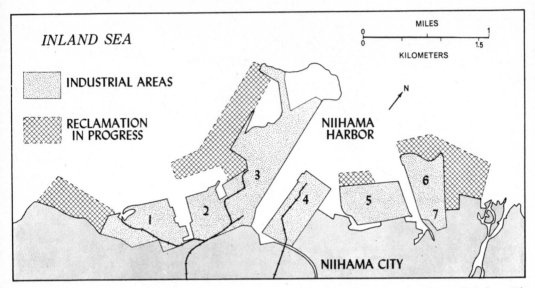

Fig. 15-16.—Kinds and locations of factories in Niihama in northern Shikoku: (1) thermoelectric plant, (2) machinery, (3, 5, 6) chemicals, (4) metals. After map prepared by the Ehime Prefectural Office.

lated to them.[21] Large factories occupy sites on seaside reclaimed land where they can be directly served by marine transport (Fig. 15-16). Plans for the reclamation of extensive new tidewater factory lands are going forward not only in Niihama, but also at Saijo just to the west, and along the coast to the east.

Niihama port serves chiefly the local industries, and its domestic trade is four to five times its foreign trade. Foreign imports are almost exclusively minerals, and exports are 80 per cent fertilizer. Domestic trade is somewhat more varied, but imports still are strongly weighted toward coal and coke, metallic ores, oil, and metals, while exports lean heavily toward mineral products, fertilizer, drugs, and chemicals. Here, then, at Niihama has developed a unified complex of industry based originally upon copper mining and smelting, which in many ways resembles the Miike-Omuta complex in western Kyushu based upon local coal and developed by Mitsui capital.

In Imabari, a small, isolated center on the northeast coast of Takanawa Peninsula, 60–70 per cent of the value added by manufacturing derives from textiles and textile finished products. Unusual emphasis is upon the processing of cotton towels.

Somewhat isolated from the several centers of industry strung out along the shores of Hiuchi Nada in the north, is the Matsuyama district facing Iyo Nada to the west. This local metropolis owes its size to

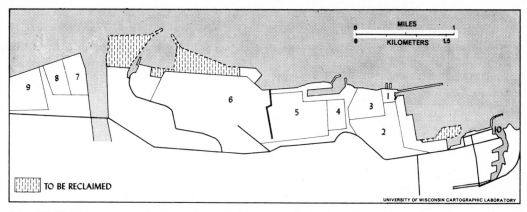

Fig. 15-17.—Matsuyama harbor industrial district, and the locations of its main factories: (1, 9) rayon, (2) oil refinery and petrochemicals, (3) oil storage, (4, 6, 8) chemicals, (5) synthetic fiber, (7) food, (10) general-cargo harbor. After map provided by the Matsuyama Municipal Office.

several functions, for it is a prefectural capital and university city, a ferry and domestic port, a market center serving the Matsuyama Plain and all of western Shikoku, and a new but rapidly growing center of factory industry which rivals Niihama in importance. The center of the old city, in the vicinity of the feudal castle, is somewhat removed from tidewater, but a new outer industrial port with water-front factories is in process of development which is bringing new life to what was a lagging city. Basic to Matsuyama's new tidewater industrial complex is oil refining and associated petrochemical industries, a repetition of the situation noted at several other manufacturing centers within Setouchi (Fig. 15-17). In addition to the Maruzen Company's large refinery and associated chemical and oil storage facilities, with piers that can accommodate tankers up to 40,000 tons, there are two rayon plants, a large synthetic-fiber establishment, two chemical, one food, and one soda factory. Within the city proper and removed from the water-front concentration, are scattered textile, food, machine, and lumber-processing plants. Noteworthy among them is a large establishment making farm machinery and one engaged in canning mandarin oranges and their juice. Chemicals represent 60 per cent of the total value added, followed by machinery with 13 per cent.

INTERIOR AND SOUTHERN SHIKOKU

This much larger part of Shikoku belongs to the Outer Zone of folded mountains and, like that region's other two subdivisions located in Kii

Peninsula and central and southern Kyushu, the Shikoku part is thinly populated and generally backward in economic development. Plains are strikingly limited in extent, the isolated Kochi lowland on the south side of the island, bordering Tosa Bay, being the largest. Because of its protected winter location in the lee of the Shikoku Mountains, the Kochi area has unusually mild winters, which have favored its development as a truck gardening center specializing in out-of-season vegetables. In recent years the growing of sensitive crops under protective vinyl covers has greatly expanded the truck-garden industry, with emphasis on such items as watermelons, tomatoes, and cucumbers.[22] Here in Kochi, also, is the only significant development of double rice cropping. Along the south coast of Kochi Prefecture is a rather remarkable concentration of *eta* people, the so-called untouchables, who in that region are engaged in both agriculture and fishing. Within Kochi they constitute about 6 per cent of the total population.[23]

The Shikoku Mountains have few important regional characteristics that are worthy of serious comment. Forest industries are important, as might be expected in such a region of high altitude and prevailing slope land. A special crop of the Shikoku Mountains, grown in rotation with food crops under a system of shifting cultivation, is *mitsumata* (*Edgeworthia papyrifera*), a perennial shrub from which Japanese paper is made.[24] About 74 per cent of the nation's harvested area of mitsumata is located in highland Shikoku. Along the deeply indented west coast of Shikoku fronting on Uwa-no Sea and Bungo Channel is to be observed one of the most spectacular developments of steep-slope terraced fields anywhere in Japan.[25] The importance of the Besshi ore mine, located in mountainous Shikoku, has already been commented upon in an earlier section dealing with the industrial district of Niihama, located on the north coast of Shikoku.[26]

16 · SOUTHWEST JAPAN: KYUSHU

Although differing from Chugoku and Shikoku in the directional trend of its long axis, Kyushu resembles them in the general northeast-southwest strike of its principal morphological and geological features. Like Shikoku, Kyushu belongs to both the Inner and the Outer Zones of Southwest Japan; a bold morphological fault-scarp which forms the steep northern face of the Kyushu Folded Mountains provides a distinct boundary between the two parts (Fig. 16-1). South of this fault-scarp, terrain in general resembles the folded mountains of Kii Peninsula and southern Shikoku, although intersection with the Ryukyu Arc in southern Kyushu has resulted in important modifications due to volcanic extrusions. North of the scarp, features resemble more closely the hill lands of Chugoku and northern Shikoku, granites being prevalent, although here, too, volcanics have added many modifications.

NORTHERN AND WESTERN KYUSHU (INNER ZONE)

Extremely complicated and varied morphological and lithic characteristics make a simple division of northern Kyushu unusually difficult. Three general divisions, each containing a considerable degree of lithic and terrain variety, are here recognized: (1) The Tsukushi Hills, and similar smaller but isolated areas in northwestern Kyushu, are essentially a

571

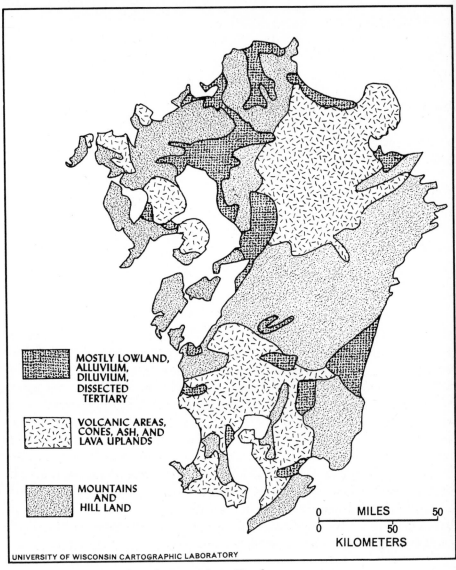

MOSTLY LOWLAND,
ALLUVIUM,
DILUVIUM,
DISSECTED
TERTIARY

VOLCANIC AREAS,
CONES, ASH, AND
LAVA UPLANDS

MOUNTAINS
AND
HILL LAND

0   MILES   50

0   50

KILOMETERS

UNIVERSITY OF WISCONSIN CARTOGRAPHIC LABORATORY

Fig. 16-1.—Terrain and earth materials in Kyushu.

separated fragment of Chugoku. Granite and metamorphic rocks are
most prominent, although Tertiary strata are not lacking, and fault-block
structure is widespread. (2) South of Tsukushi, and between it and the
Kyushu Folded Mountains, is a dissected lava-and-ash plateau with numerous large cones. This northern volcanic region occupies in Kyushu a
position comparable to the Inland Sea basin separating Chugoku and
Shikoku, although here what was the basin has been filled with recent

ash and lava. (3) Hizen Peninsula and the Amakusa Islands comprise the very irregular and loosely articulated region of western Kyushu. Geologically and morphologically this third subdivision combines features of both the other regions, for recent volcanic, Tertiary, and metamorphic rocks are all present.

Climatically this northern Kyushu region is intermediate in character between Sanin and Sanyo, having more rainfall and less sunshine, particularly in winter, than Sanyo, but more sunshine than Sanin. It is less well protected than Sanyo from the northwest monsoon, and the damp, windy, chilly winters, with considerable cloud, are disagreeable.

With the exception of the volcanic subdivision, northern Kyushu, like Kinki and the Inland Sea borderlands, is a part of Japan's ancient and most important culture zone. On alluvial lowlands, population is dense, cities are numerous, and manufacturing is relatively well developed, this sometimes being designated the fourth principal node in the country's manufactural belt.

Tsukushi hill lands and associated plains

Composed largely of uplifted and dissected tilted blocks with depressed intervening lowlands, the Tsukushi district gives the impression of being a hill country without much order or system in the arrangement of its various units. Elevations rarely exceed 1000 meters. In past geologic ages, particularly in the Tertiary epoch, some of the tectonic basins were estuaries which became partially filled with coal-bearing sediments. Since uplift these weak Tertiary strata have been thoroughly dissected, forming low hilly tracts with wide open valleys in the midst of more formidable granitic hills. What were formerly very irregular coasts with numerous islands, both along the Japan Sea and Setouchi, have been considerably smoothed as alluvium has been deposited at the heads of indentations behind crescentic beach ridges and bars. Numerous islands have been thereby tied to the mainland. The coastline is still far from smooth, however, although good natural harbors are not numerous.

Two large and unlike hill masses, separated by the fault-line Mikasa Valley, comprise most of Tsukushi proper. The western half, designated the Seburi (Seburu) Horst is a compact and relatively rugged block possessing typically bold granitic features. Occupance is relatively meager, although some cultivation is carried on, both in the narrow and intricate valleys, and on portions of the upland surface where slopes are not too steep.

In the Chikuho Block, east of the Mikasa fault-line valley, compactness of form gives way to greater variety, including detached clusters of granite hills, mostly fault-blocks, lower basin-like Tertiary areas of dissected hill country, and a modest amount of alluvial lowland as well. Sedimentary rocks are more common, and occupance is much more widespread than in Seburi. Extending from north to south through the middle of this area is a basin 40–50 kilometers long and 10–12 kilometers wide, drained northward by the Onga River and its tributaries. Low, rounded Tertiary hill masses, separated by alluvium-floored valleys of variable widths, characterize the basin. The supply of water from rivers and normal rainfall necessary to irrigate the paddies, which tend to monopolize the alluvium, is supplemented by numerous ponds situated among the hills.

The particular fame of this Onga River basin, however, is that it contains the Chikuho Coal Field, after Ishikari, in Hokkaido, Japan's greatest producer, its annual output of 12–13 million tons being about one-half of the total mined in northern Kyushu, and approximately one-quarter of the nation's total (Fig. 16-2). While much the larger part of the Chikuho Field lies within the basin of the Onga River and its tributaries, two or three very small and unimportant districts are located near, but outside, the basin.[1] Chikuho coal seams are embedded in Tertiary shale, sandstone, and conglomerate strata deposited in basins within the highlands composed of igneous and metamorphic rocks. The coal-bearing sedimentaries have suffered two or three different foldings, accompanied by numerous faults, resulting in the formation of a number of local coal basins of small extent. Both the interrupted character of the coal seams and their steep angles of dip make mining operations difficult and ex-

TABLE 16-1

Kyushu coal fields: number of mines and output, 1961

Field	Number of mines	Output (1000 T.)
Chikuho	160	12,732
Asakura		45
Fukuoka	19	1,095
Miike		3,859
Karatsu	32	2,324
Sasebo	84	3,763
Sakito-Takashima	9	2,963
Amakusa	11	412
TOTAL KYUSHU	315	27,194
TOTAL JAPAN	. . .	55,413

Source: Fukuoka Prefectural Office.

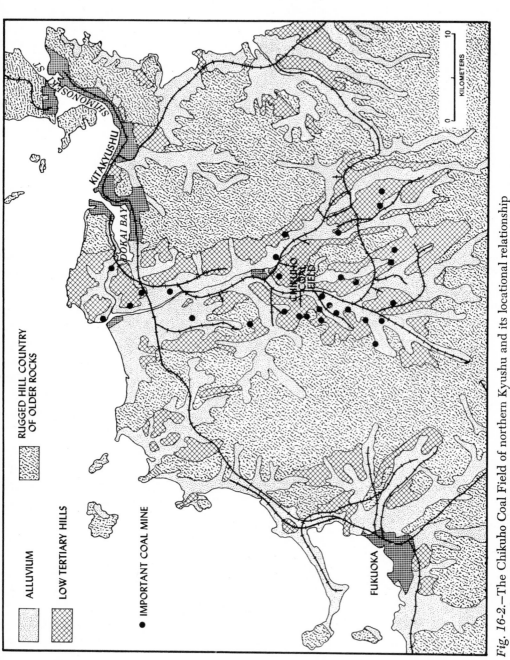

Fig. 16-2.—The Chikuho Coal Field of northern Kyushu and its locational relationship to the industrial centers of Kitakyushu and Fukuoka.

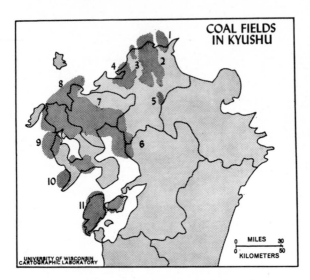

Fig. 16-3.—(1) Kokura, (2) Chikuho, (3) Munakata, (4) Fukuoka, (5) Asakura, (6) Miike, (7) Karatsu, (8) Sasebo, (9) Saketo-Matsushima, (10) Takashima, (11) Amakusa.

pensive. As many as fifteen seams occur whose thickness ranges from 0.6 to 1.5 meters. Some 160–170 mines of all sizes, many of them highly inefficient, are currently in operation, but only about twenty of these produce as much as 20,000 tons per month. Most of the larger mines are located in the upper or southern parts of the basin where exploitation began later than elsewhere. Larger mines are usually of the shaft type, whereas numerous smaller ones engage in working surface outcrops. Because of their low operating efficiency, there are plans for closing down a considerable number of the Chikuho mines. Most of the coal mined is bituminous and subbituminous in grade, and only small amounts are satisfactory for manufacturing even a low grade of coke. Those characteristic features of mining areas—conspicuous top works with extensive workmen's barracks, mine dumps, and long lines of coal cars in transit or on sidings—are all to be seen, although the mines are so scattered and hidden among the hills that in most parts the scene remains distinctly more agricultural than saxicultural. In recent years there has been a gradual shifting of mining operations southward up the valleys, as the older mines in the lower valley are worked out and abandoned.

Located only forty kilometers inland from Shimonoseki Strait, the Chikuho Field has a distinct advantage in situation. Within the basin a complicated rail system, dendritic in pattern, connects the coal field and its individual mines with the north coast, where the product is consumed locally in large quantities by the heavy industries of Kitakyushu or is exported to other parts of Japan through the ports of Kitakyushu and Fukuoka. Coal from northern Kyushu, chiefly Chikuho, dominates in the

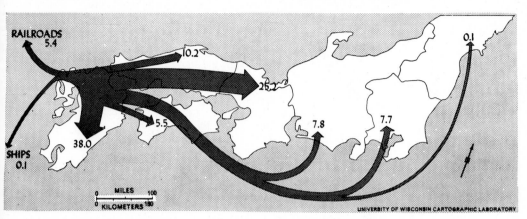

Fig. 16-4.—Proportions of Kyushu's coal output consumed in different parts of Japan. From *Nippon no Chiri,* Vol. 7.

markets of Japan as far north as Nagoya, beyond which latitude coal from Joban and Hokkaido tend increasingly to monopolize the market. Of the output of Kyushu coal, 79 per cent is used in southwestern Japan and 15.5 per cent in Central Japan. (See Table 16-2.) Most of the remainder is used in locomotives. Of the domestic coal used in Kyushu, 97 per cent originates in Kyushu, while comparable figures for other regions are 61.7 per cent for Chugoku, 90.2 per cent for Shikoku, 81.3 per cent for Kinki, 49.0 per cent for Tokai and western Central Japan, 15.2 per cent for Kanto and eastern Central Japan, and 0.5 per cent for northern Japan.

Except for a couple of cement plants, there are no important manufacturing developments within the Chikuho Field itself; instead these are concentrated at tidewater along the Shimonoseki Strait only thirty

T A B L E 16-2

Regional utilization of Kyushu coal

Region	Coal consumption	
	% of total Kyushu output	% furnished by Kyushu
Kyushu	38.0	96.9
Chugoku	10.2	61.7
Shikoku	5.5	90.2
Kinki	25.2	81.3
Tokai and Western Central Japan	7.8	49.0
Kanto and Eastern Central Japan	7.7	15.2
Tohoku and Hokkaido	0.1	0.5

Source: Fukuoka Prefectural Office.

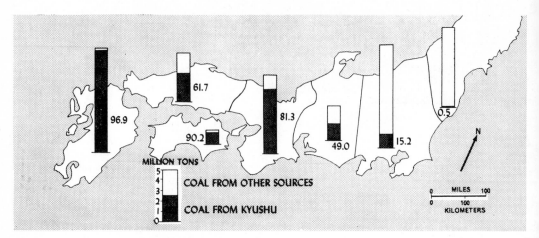

Fig. 16-5.—Relative dominance of Kyushu coal in the several main subdivisions of Japan. From *Nippon no Chiri,* Vol. 7.

or forty kilometers distant, to which location industrial raw materials can be brought by boat without transshipment.

The Tsukushi littoral

Along the indented, island-studded littoral of Tsukushi, with its numerous small and very irregular fragments of alluvium, agricultural utilization is intensive but has few unique or distinctive features. Double cropping of the paddies is common, which causes the transplanting of rice to be delayed until as late as mid-July. Not only the alluvial plains, but also the lower slopes of adjacent hills are cultivated, although not to the same degree as in Setouchi. Villages tend to concentrate along the drier inner margins of the lowlands adjacent to the hills or along the beach ridges. In the latter location they are commonly protected by a wall of conifers on their sea sides and surrounded by fields of unirrigated crops. The most extensive of the littoral lowlands of Tsukushi is the small Fukuoka alluvial plain (425 sq. km.) located at the north end of the Mikasa fault-line valley which separates the Tsukushi Highlands into an eastern and a western subdivision. In the low Tertiary hills bordering the plain, chiefly to the east of Fukuoka city, is the relatively unimportant Fukuoka Coal Field whose two or three main mines and several smaller ones have an annual output of slightly over one million tons.

Fukuoka city (522,000 population) comprises at present what were formerly two unlike urban units: (1) Fukuoka proper, an old castle town of great fame and a modern political and university center, and (2)

Hakata, the oldest port of Japan, with a history going back to the third century. The local port still goes by the name of Hakata. This large urban unit serves not only the Fukuoka Plain hinterland, but also, by way of the Mikasa Valley, the Tsukushi or Saga Plain, the largest in all Kyushu, lying south of Seburi Horst. Fukuoka's spit harbor is relatively well protected, but its shallow waters make it unsuitable for large ships. However, plans are under way for dredging a fairway and anchorage to a depth of ten meters, for constructing a breakwater, and for expanding wharf space—all of which when completed will make Hakata greatly more attractive as a port.[2] The port's present water-borne trade, amounting to about 3.4 million tons in 1961, is preponderantly domestic, the foreign trade being less than 300,000 tons, or only 8–9 per cent of the total. Hakata Port was first developed as an outlet point for locally mined coal, and even in 1961 coal comprised 63 per cent of the total domestic export tonnage, followed by ship equipment with 20 per cent. Domestic imports are slightly more varied, but petroleum comprises one-third of the tonnage, fish 13.5 per cent, vehicles 12 per cent, and metals 8 per cent. Most of the present port facilities occupy land reclaimed from the sea, and future reclamation is planned both for an expanded port and for a new seaside industrial area.

Considering its large size, Fukuoka is not a particularly important producer of manufactured goods, for while its workshops and factories employ over half as many as those of highly industrialized Yahata (now included in Kitakyushu), yet the value added by manufacturing is only one-fifth as great, not even equaling that of much smaller cities such as Niihama and Tokuyama. In contrast to Yahata, a center of heavy industry and large factories, manufacturing establishments in Fukuoka are characteristically of small or medium size, the average factory employing only seventeen workers. In addition to being smaller and hence less conspicuous than the steel mills and cement factories of Yahata, manufacturing plants in Fukuoka are less confined to exclusive factory areas, but instead are spread more widely, being intermingled with commercial and residential areas, and at the same time are more diversified in their products. Like Kyoto, Fukuoka is usually characterized as a center of craft industries, and indeed the wares of its artisans and workshops —including *Hakata-ori* brocade fabrics for neckties and kimono sashes, painted clay dolls, and special glassware—are distinctive, but they scarcely account for the employment of 36,000 workers in industry. A larger percentage of the labor force is employed in those more prosaic industries which manufacture paper, flour, sugar, electric and mining machinery, cotton textiles, agricultural equipment, and chemicals. Machinery of all kinds and foods are the foremost sectors of the industrial structure, followed by publishing, paper, ceramics, and textiles. But although Fukuoka is a processing center of only modest importance, it achieves greater distinction as a business, political, and educational center, not only for Fukuoka Prefecture, but for all of Kyushu.

Kitakyushu industrial center.—Although much of northern Kyushu could be considered as forming the southwestern end of Japan's manufacturing belt, within this general region factories characteristically are grouped in a number of scattered local centers, somewhat as they are along the margins of the Inland Sea and along the Tokai coast between Tokyo and Nagoya. In one area only, however, is there a genuinely first rank and relatively compact concentration of factories. This is along a narrow coastal strip thirty to forty kilometers long in northernmost Kyushu, reaching from Shimonoseki Strait on the east, which forms the western entrance to the Inland Sea, to Dokai Bay on the west, and including what were formerly the cities of Moji, Kokura, Tobata, Wakamatsu, and Yahata (or Yawata). Since the spring of 1963 these five cities have been amalgamated to form the single city of Kitakyushu (meaning north Kyushu) with a DID population of over 800,000 and a *shi* population of about a million. If Shimonoseki, across the Strait in southwesternmost Honshu, which is an integral part of this compact urban industrial area, is included, the DID population (1960) is about 960,000, and the *shi* population is over 1,230,000. Kitakyushu plus Shimonoseki is sometimes referred to as Kammon. Thus, new Kitakyushu becomes the seventh most populous city in Japan.

The Kitakyushu industrial center is sometimes designated as one of the principal nodes within the country's manufacturing belt. Actually, however, it can scarcely be compared with the other three nodes in either area or importance, for it is but a single industrial city, albeit a very important one, and not a hierarchy of contributing manufacturing centers such as characterizes Kanto, Kinki, and Chukyo. Still, among individual manufacturing cities of Japan it ranks seventh in number of employees, fifth in salaries paid, sixth in value of raw materials consumed, sixth in value of shipments, and fifth in value-added (ahead of both Kobe and Kyoto). So while it may seem unwise to speak of Kitakyushu city as the fourth principal manufactural node, this does not detract from its high industrial rank among individual cities.

What has probably given this city an industrial fame greater than that warranted by its size and output is that for many years it was the nation's greatest pig iron and steel center, for as late as 1926 it produced 79.4 per cent of the pig iron and 65.5 per cent of the crude steel made in

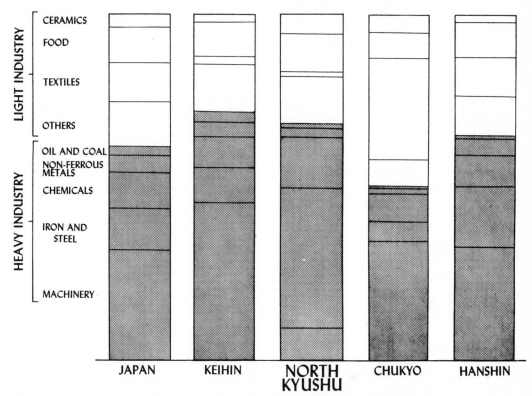

Fig. 16-6.—Composition of industry in North Kyushu, compared with Japan as a whole and with the three great manufacturing nodes, based upon value of shipments, 1960.

Japan. Even as late as 1960 its output of pig was one-third that of the entire country, which gave it first rank; and although it produced nearly one-quarter of the crude steel, both Hanshin and Keihin outranked it. Thus, Kitakyushu is essentially a single compact center, specialized in heavy industry, where large factories are unusually conspicuous and give to the region a thoroughly industrial appearance. In a sense, it is the Pittsburgh of Japan. And while pig iron and steel greatly predominate, chemicals, cement, and machinery are likewise important (Table 16-3). Still, fabricated metals and machinery are far less important relatively than they are in the country as a whole, indicating that much of Kitakyushu's pig iron and steel are sent in this crude form to the populous market areas of Japan where they are processed into machinery. Textiles are of negligible consequence. Within Kitakyushu, 48.1 per cent of the employees and 70.3 per cent of the output are associated with large factories having more than 300 employees. The comparable figures for all Japan are 29.6 per cent and 12.2 per cent.

Yahata, now included within Kitakyushu, was conceived and fostered

as a heavy-industry center by the central government at about the beginning of the present century. Several factors recommended it as a major iron and steel center: (1) There is easy access by rail to the nearby
Chikuho coal field. (2) The location on tidewater greatly facilitates
the import of bulky raw materials such as iron ore, coking coal, scrap iron,
pig iron, and petroleum, as well as the shipment of similar bulky products,
especially coal and steel. And not just ordinary tidewater location was
provided, but more particularly frontage on Shimonoseki Strait, which
even in 1900 was a focus of sea lanes. At this point also, rail and highway
lines connecting Honshu and Kyushu intersect the strait water route, so
that Kammon was, and still is, a crossroads of traffic. (3) Extensive deposits of limestone, sand, and clays are locally accessible. (4) It is in
an intermediate position between Asia and the great industrial domestic
markets of Kanto, Kinki, and Tokai. Thus, when in 1901 the Japanese
government established here the large Yawata Iron and Steel Works,
the total effect was to trigger the development nearby of a group of satellite plants that rounded out the Kammon heavy-industry complex. It
bears re-emphasizing that this North Kyushu steel center was oriented
to raw materials rather than to markets, for much of what it produced
had to be marketed in other more populous parts of Japan.

But while northern Kyushu and more particularly the Kammon District held a pre-eminent position in the early years of Japan's industrial
expansion, in more recent decades its rate of growth has fallen behind
those of the great market-oriented centers in central and southwestern
Honshu, so that its proportional share of the country's manufactural output has continued to dwindle. In 1938 the total value of its manufactures
was 8 or 9 per cent of the nation's total, but this was down to 5.3 per cent
in 1955 and 4.5 per cent in 1960. Major industrial firms in Kammon show
an increasing tendency to build their new factories outside of Kyushu.
This relative decline in the importance of northern Kyushu reflects the

TABLE 16-3

*Composition of manufacturing in Yahata, Wakamatsu, Kokura, Tobata, and Moji (now comprising
Kitakyushu), 1960, in percentage of value added*

Iron and steel	52
Chemicals	18
Machinery (all kinds)	8
Ceramic, stone, and clay products	7
Fabricated metals	4

Source: Ministry of International Trade and Industry, *Census of Manufactures, 1960: Report by
Cities, Towns, and Villages* (1962).

Fig. 16-7.—View of the large Yawata iron and steel plant fronting on Dokai Bay in Kitakyushu. Courtesy of Geographical Survey Institute, Japan.

operation of certain handicaps prevalent there. One of the most serious is the shortage of industrial fresh water, a feature that is difficult to correct without resorting to the expensive distillation of sea water, which the Yawata Steel Plant already has been obliged to do. There is likewise a serious deficiency of flattish land, especially with tidewater location, for future factory sites. The narrow seaside plain, part of it land reclaimed from the sea, is already in intensive use, and further reclamation is under way at present along the Tobata littoral which will permit the expansion of the Yawata Iron and Steel Plant in that location and the further improvement of its port facilities.[3]

Likewise handicapping production in present plants and discouraging the building of new ones is the congestion in sea, rail, and highway traffic. Attempts are being made to break this bottleneck by (1) construction of a highway-and-pedestrian suspension bridge across the neck of Dokai Bay between Wakamatsu and Tobata, to supplement the rail tunnels and the ferries traversing Shimonoseki Strait; (2) electrification of the Moji-Kurume National Rail Line; (3) modernization of National Highway No. 1 which traverses the area; and (4) deepening and general improvement of harbors. In addition, all northern Kyushu suffers perennially from a peripheral location as regards the great domestic markets of Japan, so that those industries that are in any degree market-oriented find in this region a less desirable location. North Kyushu, it must be recalled, was first industrialized by the national government for military purposes, so that the heavy industries originally established there, dependent as they were upon local coal and imported raw materials, had little relation to the country's markets.[4] Corollary to the above is the meager development in this region of the so-called "growth industries," such as oil refining, petrochemicals, electrical machines, equipment, and home appliances, motorcars, and the like.

Kitakyushu city, fairly coextensive with a narrow littoral plain, is some forty kilometers long but only a few kilometers wide at a maximum and, in its eastern part, barely more than a kilometer or two. Of the five cities which were amalgamated to form the new Kitakyushu, only one, Kokura, was an old castle town. All the others are industrial cities of relatively recent origin and owe their growth to modern manufacturing. Lacking roots in premodern Japan, Kitakyushu has little Oriental charm. On the contrary, most parts are grimy, smoke-stained, and wholly unbeautiful. In many Japanese industrial areas, where workshops rather than factories are the rule, the unsightly features common to some manufacturing districts can be partially hidden or camouflaged, but this is impossible in Kitakyushu where the very size and nature of the industries—blast furnaces, steel mills, machine shops, cement factories, flour mills, sugar refineries, and the like—preclude any softening of their starkness. With its massive, grimy buildings, tall chimneys belching black smoke, piles of coal and iron ore, huge waste dumps, scores of railroad tracks, and its canopy of murky sky, Kitakyushu is oppressively industrial.

An approximate zonation of the functional areas of the city shows the more exclusively residential district to lie on the land side, some of it on the bordering hill slopes. The retail-wholesale commercial parts of

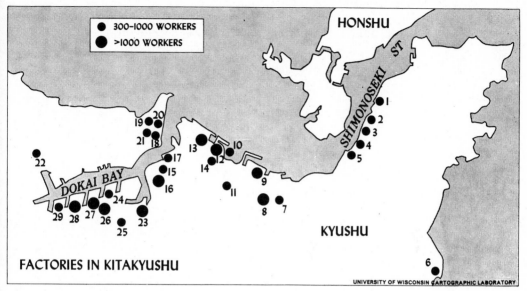

Fig. 16-8.—Kinds and locations of the main factories in Kitakyushu: (1, 6, 24, 29) cement, (2) electrical equipment, (3, 9, 13, 23) steel, (4) sugar refinery, (5) brewery, (7) pulp and paper, (8, 26) ceramic materials, (10, 11, 14, 27) machinery, (12, 28) chemicals, (15, 18, 19) non-ferrous metals, (16, 22) glass, (17) fish-canning, (20) ship-yard. After map in *Nippon no Chiri,* Vol. 7.   No scale given.

the new city, composed of several individual nodes which are a carry-over from the very recent past when there were five cities instead of the present one, chiefly occupy the inner parts of the coastal strip. Residences and some of the smaller manufacturing plants belonging to the ubiquitous industries are embedded within these commercial areas. Larger plants engaged in heavy manufacturing, together with warehouses, storage yards, docks, and piers, are concentrated along the sea margins of the coastal strip, part of it land reclaimed from the sea. Most large plants have sites on tidewater where they can be served directly by cargo ships and lighters. Other factories located slightly inland from the water front are commonly reached by canals.[5]

Three different ports—Dokai, Kokura, and Moji—formerly served the five cities now combined to form Kitakyushu. Dokai, serving what were Yahata, Tobata, and Wakamatsu, all fronting on Dokai Bay, is primarily an industrial port. Its tonnage of domestic trade greatly exceeds that of its foreign, and while domestic exports are several times the imports, the reverse is true of its foreign trade. Foreign imports are largely such bulky cargoes as iron and other ores, coal, salt, sugar, and wood, while similar bulky cargoes of pig iron, steel, and cement comprise the principal outgoing cargo. Farther east along the coast, beyond Dokai Bay, and consequently situated on the Kammon Fairway within Shimonoseki Strait, are located what were the individual ports of Kokura and Moji, now also serving the Kitakyushu industrial area. None of these three ports is of first rank, especially as general cargo commercial ports, although their combined tonnage in bulk materials is very large, exceeding that of Kobe, the first port of Japan in terms of value of foreign trade. From these ports large amounts of coal, pig iron, and steel move toward the industrial cities concentrated along the margins of the Inland Sea.

Dredged and marked Kammon Fairway within Shimonoseki Strait, an important international navigation channel which is 480 meters wide at its narrowest part, is characterized by rapid tidal currents that may reach a velocity of 8 knots an hour at the constriction. Two submarine tunnels, one for vehicular traffic and the other for rail, underlie the Strait and connect Honshu with Kyushu. Current improvements in the Kammon Fairway include deepening the channel to a minimum of 13 meters and widening several stretches, so that ships of up to 15,000 gross tons will be able to use the channel.[6]

The *Tsukushi-Kumamoto Plain* (2700 sq. km.) in northwestern Kyushu is the most extensive area of low relief on the island. It is frequently considered to be two plains, Tsukushi in the north and Kumamoto in the south, partially separated by a small area of hill country (Kumamoto Hill Land) in the vicinity of Omuta city, along whose seaside the coastal lowland is relatively narrow. The northern, or Tsukushi, section has connection with Fukuoka and Kitakyushu by way of the fault-line valley of the Mikasa River. Drained by the Chikugo River, this low, flat alluvial plain is covered by a network of small canals bordered by various kinds of aquatic plants whose violet, white, and scarlet flowers provide a brilliant display of color during the blossoming season. Occupance of the lowland goes back to prehistoric times, and evidence of the Jori system of land subdivision practiced here in the seventh and eighth centuries is still to be seen in the rectangular pattern of canals and fields.[7] Tsukushi is a lush region, advanced in its agricultural techniques and exhibiting a landscape dominated by rice paddy. Rice yields are among the highest in the nation. One might expect that in Kyushu, the southernmost island, where double cropping of paddies is climatically possible, rice planting characteristically would be late, following upon the harvesting of winter crops. Actually there is as much as two months difference in the planting time of rice in different parts of Kyushu, so that the distribution patterns both of planting seasons and of varieties sown are very complicated.[8] Along the southern margins of Tsukushi, where

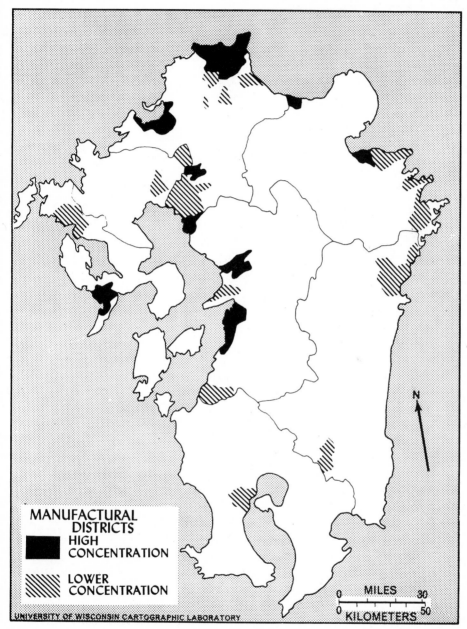

MANUFACTURAL
    DISTRICTS
    HIGH
    CONCENTRATION

    LOWER
    CONCENTRATION

UNIVERSITY OF WISCONSIN CARTOGRAPHIC LABORATORY

MILES   30

KILOMETERS   50

Fig. 16-9.—Main concentrations of manufacturing in Kyushu by *shi* and *gun*. After
map by Makoto Murakami.

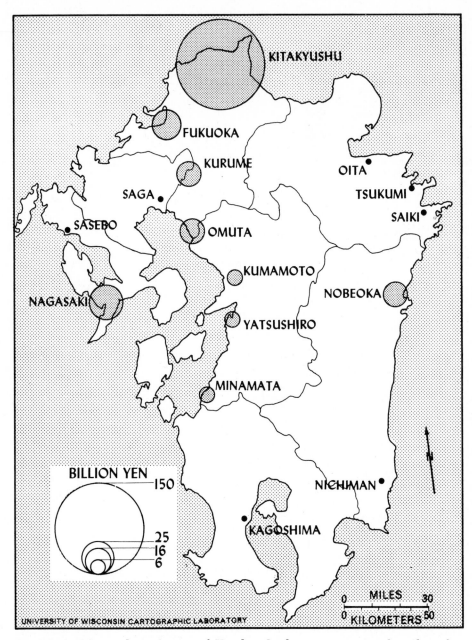

Fig. 16-10.—Main industrial cities of Kyushu. Circles are proportional to the value added by manufacturing, 1960.

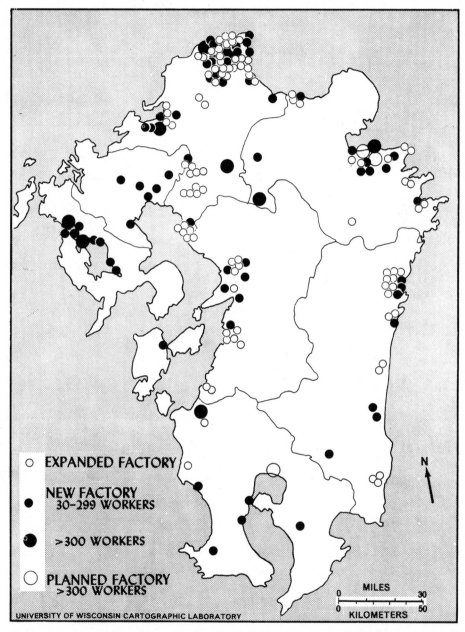

Fig. 16-11.—Distribution of new and expanded factories in Kyushu between 1955 and 1960. After map by Makoto Murakami.

it borders Ariake Bay, there has been accomplished one of the nation's two or three most extensive reclamations of seaside land for agriculture.[9] Kurume (113,000 population), an inland city and the plain's metropolis, is completely specialized in the manufacture of rubber tires and tubes, rubber products representing nearly 80 per cent of the value added by its factories. Smaller Saga is a minor machinery and textile center.

Kumamoto, the southern subdivision of the larger plain, is composed of two terrain types; for while its northern and southern extremities are low flat alluvial lowlands, its more extensive central part is a low upland surface which occupies the crest of a diluvial fan composed of volcanic ash deposited by the Shira River flowing westward from the *caldera* of Mt. Aso. Small hills, which are partially buried gneissic monadnocks, rise prominently above the diluvial upland surface. Paddy fields occupy much of the low new alluvium, but the extensive ash-covered diluvial upland, where irrigation is difficult, is largely given over to dry crops. Until about the 1930's this upland was specialized in mulberry, but as the foreign market for raw silk declined during that decade, there was a gradual shift to staple food crops such as sweet potatoes and wheat, a trend that was accelerated during World War II as the need for increased domestic food production became urgent. More recently the crop emphasis has been changing again, and this time toward commercial orchards and tobacco.[10] Kumamoto city (275,000 population), the third largest city of Kyushu, is an old castle town which in modern times has developed as a prefectural capital and a business center. For its large size, manufacturing is only very modestly developed, and the ubiquitous food industries hold first rank. Much smaller Yatsushiro, at the southern extremity of the Kumamoto Plain, with a population only one-seventh that of Kumamoto, has an equal value added by manufactures, but here there is a strong specialization in pulp and paper.

As noted earlier, the Tsukushi and Kumamoto Plains are given a certain amount of separation by the interposition of a small mass of granitic hill country which in lithic and terrain character appears to be a detached fragment of the Tsukushi Range farther north, although lower in elevation. In the tilted Tertiary shales, sandstones, and conglomerates which flank the granite upland on its sea side there exist modest coal measures which are designated as the Miike Coal Field.[11] Although only about ten square miles in area, and with usually only three important mines in operation, Miike in 1961 produced just slightly less than 4 million tons of coal and ranked next to Chikuho as the most important field in Kyushu. Normally about 20–25 per cent of the Miike product is used locally and the rest is exported, chiefly to domestic markets in southwestern Japan. The terrain of the coal field is that of hill country with low relief. One or two of the mines, located at tidewater, are engaged in mining submarine seams. Upwards of fifteen or twenty abandoned coal mines with their conspicuous dump heaps are to be seen in the vicinity of Omuta.

The *Omuta-Miike industrial district,* which owes its origins to Miike coal and which has Omuta city and its artificial harbor of Miike as its principal focus, lies seaward of the granite hills and occupies sites on both a dissected diluvial upland and a narrow coastal lowland. Omuta (157,000 population), together with a few small satellite towns, forms one of the more important local industrial centers within northern Kyushu.

Local coal is the *raison d'être* for Omuta's thoroughgoing specialization in chemicals. About three-fifths of the employees in manufacturing and over two-thirds of the value-added are attributable to the chemical sector. Fabricated metals and machinery are next in rank. Omuta is a product of Mitsui capital and management, this concern being responsible for the mining of the coal and the construction of Miike Harbor, as well as the establishing of upwards of ten large factories processing chiefly chemicals and metals. Among the great variety of products turned out, with commonly a single plant fabricating more than one product, are the following: sulfuric ammonia, liquid ammonia urea, sulfuric and nitric acids, dyestuffs, coke, coal gas, fertilizers, plastics, drugs and medicines, non-ferrous metals, firebrick, carbide, cement, salt, acetylene black, and machinery. Due to a common ownership, the work of the different plants is closely integrated. The district's grimy, smoky appearance and its conspicuously large factories give it a resemblance to Kitakyushu, but on a smaller scale. Unlike Kitakyushu, most of the manufacturing establishments are not located on tidewater, but instead are inland and located on the northern and northeastern outskirts of Omuta where they are served by a canal and also a belt-line railroad connecting with Miike Harbor (Fig. 16-12).

Miike Port, a part of Omuta city, serves an important function as the sea gateway to the local mining and industrial area, and it is the leading commercial port of middle and southern Kyushu. The shoal water of Ariake Bay has made necessary the dredging of a fairway, which is 1800 meters long and 6 meters deep at low tide. At the land end of this artificial channel, an inner and an outer harbor have been constructed with a lock gate between, made necessary by the large tidal range of 5.5 meters. The inner harbor is kept at a depth of 8.5 meters or more, while the outer harbor has a depth of 9.5 meters at low tide. Ships up to 10,000 tons may enter the inner harbor, while larger vessels are obliged

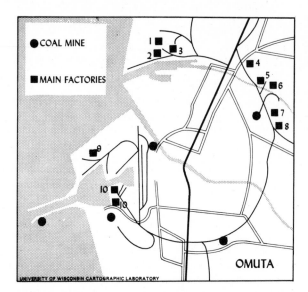

Fig. 16-12.—The Miike-Omuta industrial district and its coal mines: (1, 3, 4, 6, 7, 8, 9) chemicals, (2, 5) steel, (10) thermoelectric plant. After map in *Nippon no Chiri*, Vol. 7. No scale given.

to use the outer one.[12] Chief items of cargo loaded at Miike, destined for both domestic and foreign markets, are coal and fertilizer. Cargo unloaded at Miike consists of ores, industrial salt, and coal of special qualities.

The northern volcanic region

Between the Tsukushi Hills to the north and the Kyushu Folded Mountains to the south is an extensive area of ash-and-lava deposits which rises out of and completely buries the Setouchi depression. So diverse and complicated are the features of this volcanic area that any brief description must of necessity be inadequate.

In gross anatomy, the northern portion is composed of several relatively high strato, shield, and lava-dome volcanoes, with associated andesite plateaus, believed to be the product of fissure eruptions. The plateaus have been so altered by faulting and erosion that they present a great confusion of features. Mesa and butte forms with steep bordering cliffs are common. Where dissection has opened up relatively wide valleys or basins there is a considerable rural population, with isolated or loosely clustered farmsteads being more prevalent than compact rural villages. In places artificial terracing for rice as well as for dry crops is remarkably developed. Circular Kunisaki Peninsula, projecting into the Inland Sea, is composed of a maturely dissected conical volcano whose deep and wide radial valleys are devoted to paddy rice and to the Igusa reed from which tatami mats are made. Some of the nearby ash uplands are specialized in orchards and vegetable growing. Slight submergence along the northeastern side of Kunisaki has drowned the lower ends of the radial valleys, producing an irregular coastline. For the most part, however, the coast of the volcanic region is regular and bordered by low diluvial terraces whose smooth but sloping surfaces are planted in deciduous orchards and the usual dry crops.

Beppu city (86,000 population) on the coast of Beppu Bay just south of Kunisaki Peninsula is the center of a famous hot springs resort area which boasts of having over 1800 thermal springs. The daily flow of hot water amounting to 47,000 tons is used not only for pleasure and therapeutic bathing, but also for heating houses and vegetable hothouses and for cooking.[13] Beppu is a tourist city with several hundred hotels and inns and even more numerous bathhouses, together with restaurants, convalescent homes, and souvenir shops. It draws most of its clientele from Kyushu, especially the Kitakyushu area, making its strongest appeal to low-income groups such as students and school children.

The southern and southwestern part of this Northern Volcanic Region is largely composed of the mighty cone of active Mt. Aso[14] and its associated lava, mud, and ash uplands, extending across almost the entire width of Kyushu and covering an area of nearly 2000 square kilometers. While most of the lava is hidden under subaerially deposited ash, it is exposed along the sides of some valleys. Aso's crater, which measures 17 kilometers east-west by 24 kilometers north-south is one of the world's

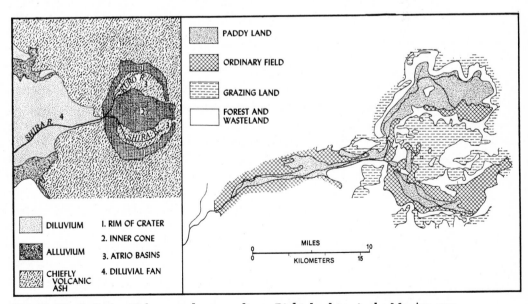

Fig. 16-13.—Left, Aso Volcano and surroundings. *Right,* land use in the Mt. Aso area. After map in *Regional Geography of Japan,* No. 5, *Inland Sea and Kyushu Guidebook.*

greatest *calderas*. From the crater floor rise five new volcanic cones, some of them active, with one reaching an altitude of nearly 1600 meters. Two crescentic crater basins, a north and a south, occupy positions between the crater walls and the central cones; both at one time contained crater lakes in which volcanic and water-borne sediments have been deposited. These elevated lacustrine plains within Aso's crater are now well occupied by a dense rural population growing rice in the lower, wetter portions and dry crops at the higher levels. Winds are strong at these elevations and as a consequence the villages are tree-enclosed. The intercoastal railroad line passes through the crater by way of the north basin, the ascent along the gorge-like valley which drains the crater to the west being both difficult and spectacular.

The outer slopes of Aso are relatively mild in gradient, nowhere more than 10°, and to the east often only 2° or 3°. Radial drainage is conspicuous. The original ash-mantled lava and mud surface was at first, no doubt, smoothly rolling, but rivers have incised relatively steep-walled valleys, between which are wide remnants of the slightly roughened original upland. Local relief is usually not much over one hundred meters. Higher portions of the ash uplands close to the crater, where showers of ash are frequent, are covered with coarse wild grasses upon which horses and cattle are grazed. At lower elevations, more accessible and less subject to current showers of ash, cultivated land is more abundant, although there is evidence that occupance has been relatively recent. The valley floors are covered with an intricate dendritic pattern of rice lands, whereas the cultivated ash uplands are devoted to dry crops. Dispersed settlement is common.

In spite of the handicaps of elevation and poor ash soils, some 15–20 per cent of Aso *gun*, or district, is under cultivation. Of the three-quarters in wild grassland (60 per cent) and forest, a majority of the latter is planted in needle trees such as cedar and pine.[15] Paddy fields occupy about one-half the cultivated area and are found both within the crater and in the radial valleys of the upland. Less than one-third of the cultivated area bears more than one crop, and production per unit area is low. On the ash uplands, where dry-field agriculture prevails, the commonest summer crops are maize, upland rice, and beans, while wheat and rapeseed are grown in winter. Cattle raising is an important adjunct of the upland farming, and grazing on the natural pastures is extensively practiced. As a consequence of overgrazing and a neglect of grass conservation, there has been a gradual decline in the quality of the range lands.

Insular and peninsular northwestern Kyushu

Fragmentation has been so complete in this transition region of both sedimentary and volcanic rocks in Saga and Nagasaki prefectures, and culture patterns are so complex and variable, that a simple and comprehensive synthesis of its geographic features is almost impossible. It is the southwesternmost extremity of Ogasawara's Peripheral Zone of land-use regions, where intensity of agricultural land use, as well as the development of manufacturing, is at a lower level than in the Kitakyushu district. Its extremely irregular outline, with deeply indented coasts and nearly enclosed arms of the sea, is the result of subsidence of a land surface made asymmetrically irregular by erosion, faulting, and volcanic activity. Natural harbors are numerous, and fishing is a major occupation of the coastal villagers.

North Hizen.—Matsuura Peninsula is a hilly region of weak Tertiary rocks, considerable areas of which are capped with basaltic lava flows. Where basalt is absent, typical Tertiary landscapes prevail, and a relatively dense population, living chiefly in dispersed settlements, cultivates not only the intricate maze of valleys, but also the adjacent slopes, whose artificially terraced fields remind one of the Setouchi borderlands. Where basalt caps the Tertiary rocks, rolling tableland surfaces, whose margins often are precipitous from the crest down to the geological unconformity, are typical. The resulting features are mesa-like in form. On top of the tablelands, smooth in parts and uneven in others, both rice and dry crops are raised. The terraced rice fields often have retaining walls built from the surrounding basaltic boulders.

Within the Tertiary rocks of the Matsuura Peninsula are two small coal fields, Sasebo on the west and Karatsu on the east. In 1961 the former produced about 3.8 million tons of coal and the latter 2.3 million. No generally important industrial development has grown up in the vicinity of these coal fields as is the case at Miike. Some coal is used locally in the Sasebo shipyards and engineering works and at Nagasaki, but most of it is exported to other parts of the country through the port cities of Sasebo and Karatsu.

Within this same Tertiary peninsula are two towns, Imari and Arita, which are nationally and internationally famous as centers of fine porcelain manufacture. Small and middle-sized plants predominate. Most of the product is not chinaware for common use, as is the case with Nagoya, but instead, specialization is in the crafting of fine art pieces. A good quality of local china clay, weathered from a feldspar found in the liparites of this district, first attracted a Korean potter in the early part of the seventeenth century, and from that small and early beginning the present industry has evolved.

South Hizen is composed of three sprawling peninsulas, Sonoki on the west, Nomo to the south, and Shimabara on the east, all joined to the northern Tertiary region by Tara Volcano (983 meters elevation), a dissected strato cone with associated lava domes. Radial drainage pattern approaches perfection on Tara, the lower portions of the diverging valleys being devoted to paddies, and the interstream ash-and-lava uplands sown to dry crops or left in woods. An almost continuous line of villages follows the volcano's shoreline; inland, there are scattered rural residences.

Shimabara, the eastern peninsula, is principally composed of Unzen Volcano.[16] The almost perfect elliptical curvature of the peninsula's north and east coasts is due to the conical elevation of the volcano. Physical and cultural patterns are rather similar to those previously described for Tara. Here the crescentic zone of cultivation on the lower, mildly sloping ash apron is as much as three miles wide in places. At an elevation of about 700 meters, in a gigantic explosion crater, there has developed the nationally famous Unzen hot-spring resort. Reduced summer temperatures, mountain scenery, and hot springs, together with golf links and splendid hotel accommodations, make Unzen an attractive hill-station for foreigners and Japanese who desire to escape the tropical summer heat of the lowlands.

Western Sonoki and southern Nomo peninsulas have earmarks of the Outer Folded Zone of southern Kyushu, for they are composed largely of old crystalline schists; on the other hand, they are joined to Hizen by volcanics and other rocks not characteristic of the Outer Folded Zone. In general the peninsulas are hilly regions and not well developed. In two separate areas—Sakito, on the northwest coast of Sonoki Peninsula, and Takashima, on Nomo Peninsula not far from Nagasaki—fewer than a dozen mines produce close to 3 million tons of coal annually. Together these are known as the Sakito-Takashima Coal Field.

Occupying the head of a deep and narrow ria indentation about three miles from the open sea, and therefore possessed of an excellent natural harbor, is the old port city of Nagasaki (261,000 population). Its development has never been closely associated with the immediate hinterland, however, which is prevailingly hilly, almost without alluvial plains, and meagerly settled. Furthermore, the city's isolation at the sea-end of a long and rugged peninsula has handicapped its modern development. During the more than two centuries of Japan's seclusion after 1625, Nagasaki was the sole gateway whereby Western and Chinese culture filtered into the Hermit Nation. Moreover, for years prior to the Russo-Japanese War it was the wintering port for the Russian Asiatic Fleet, so that Nagasaki bears numerous hallmarks of foreign influence.

The city is amphitheatre-like in form, occupying the steep slopes of the hills encircling the bay-head as well as the narrow coastal strip itself. Nagasaki's strong resemblances in site and form to many Mediterranean coastal cities and villages has been observed. Its atypical slope site helps to make it one of the most picturesque of Japanese cities. Many of the narrow and irregular streets of the residential sections on the slopes are so steep as to require steps and terraces. Business, commercial, and industrial forms tend to concentrate on the flatter land bordering the bay, and there streets are wider and more regular in pattern.

A single great manufacturing combination—the Mitsubishi Shipbuilding and Engineering Company, with its associated steel works, machine shops, engine works, motor air brake and electrical machinery plant— has given Nagasaki its principal industrial fame.[17] The Mitsubishi plant at Nagasaki is one of the world's foremost shipyards, and in 1961 it led all others by launching thirteen large ships totaling about 248,000 tons, nine of the large vessels being for export. Thus Nagasaki is a mono-industrial center where the metal-machine industry accounts for over 80 per cent of the total factory output and where the colossal Mitsubishi combination is responsible for 90 per cent of the metals and machines and 80 per cent of all manufactures. The mighty wharves and docks of the Mitsubishi Combination occupy several kilometers of water front along the west side of the bay. Because of the outstanding importance of the shipbuilding plants at Nagasaki and Sasebo, over 61 per cent of the total manufacturing output of Nagasaki Prefecture is classed as transportation equipment; electrical machinery and equipment processed in these same two centers add another 7 per cent.

As a port city Nagasaki is in a period of retrogression. In 1900 it was the third-ranking port of the country in foreign trade; in 1960 it had dropped to eleventh. Significant causes for the decline are (1) the development of competing manufactural, coal-exporting, and general port cities in northernmost Kyushu; (2) the substitution of oil for coal as ship fuel, thus reducing the importance of Nagasaki's coaling services; and (3) the loss of the China trade which had been of particular importance to Nagasaki. Tonnage of domestic trade is nearly ten times that of foreign trade, while in foreign trade the value of exports is more than four times that of imports. Since Nagasaki is one of the country's important fishing ports, it is not surprising that canned fish should be one of its most valuable outgoing cargoes. Petroleum, fertilizer, and rice are among its chief imports.

On the Amakusa Islands to the south of Hizen, hill country prevails, with numerous crests reaching elevations of 400–500 meters. Only very meager alluvial patches exist, for the hills come down to the sea margins in most parts. Agricultural-fishing settlements, both dispersed and compact, dot the indented coasts. A total of nearly a dozen mines produce some 400,000 tons of coal annually.

SOUTHERN AND EASTERN KYUSHU (OUTER ZONE)

Kyushu south and east of the Median Dislocation Line and its morphologic fault, which mark the northern margins of the Kyushu Folded Mountains, is one of the more isolated and backward parts of Japan. Along with southern Shikoku and Kii Peninsula, southern Kyushu comprises what is known morphologically as the Outer Zone of Southwest Japan. Like the other two segments, it is a region where population is relatively sparse, owing to the prevalence of slope lands and the very modest development of alluvial lowlands as well as the comparative isolation. Strongly oriented toward the primary economies, it has a low proportion of workers engaged in manufacturing. Cities are few and relatively small.

Southern Kyushu differs from its mountainous counterparts in southern Shikoku and in Kii, in that here the intersection of the Ryukyu Arc has resulted in important volcanic extrusions. Accordingly, southern and eastern Kyushu divides logically into two principal parts, a lower southern one composed chiefly of volcanic ash-lava upland and cones and a higher and more rugged northern one which is the folded Kyushu Mountains. (See Fig. 16-1.)

Volcanic South Kyushu

This much-fragmented region, complicated both geologically and morphologically, occupies a depressed area south and west of a crescentic fault-scarp which marks the southern boundary of the Kyushu Mountains. Essentially it consists of (1) several active or recently dormant volcanic cones, (2) areas of dissected andesite flow, irregular in outline, (3) an extensive lapilli-ash upland in a youthful stage of dissection, and (4) several steptoes, both sedimentary and granitic, which protrude well above the ash-plateau level (Fig. 16-14). At the extreme south, Kagoshima Bay, enclosed between Satsuma and Osumi peninsulas, is a tec-

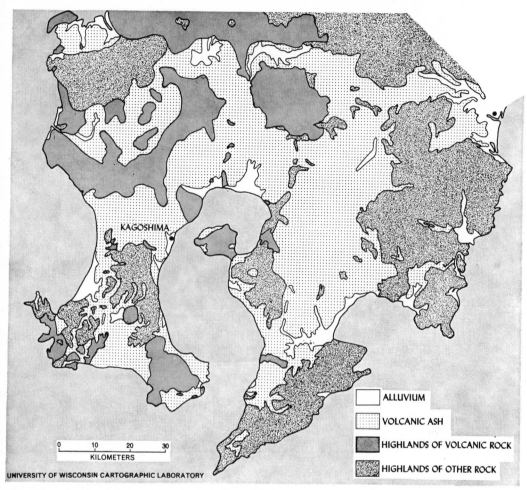

Fig. 16-14.–The southern volcanic region of Kyushu.

tonic depression in the plateau surface. Extensive alluvial lowlands are lacking and for this reason important population clusters of high density are absent. Settlement focuses chiefly on volcanic upland plains and a few scattered fragments of new alluvium.

Marking the highest elevations are the symmetrical volcanic cones, sometimes isolated, sometimes in groups. A number of them have slopes of such fresh lava or ash that they do not as yet support a forest cover and so appear stark and barren. On others forests are abundant, or, where cleared, the resulting moor-like areas provide pastures of low quality for a specialized horse industry. The andesite flows, confined to the northwestern part, have been so dissected as to produce areas of rugged hill country containing little settlement. A conspicuous feature of the andesite hills is the numerous loose boulders, some of great size, which cover both flanks and crests. In the andesite country the hills often have table-land crests, which though far from level are conspicuously less steep than the slopes of the intervening valleys. South of the main lava mass are smaller flows, less formidable in aspect, which are intricately intermingled with the ash upland.

The steptoes, about a half dozen in number, are simply higher remnants of the fractured sedimentary and igneous surface not yet submerged beneath the showers of ash or the lava flows. They vary somewhat in appearance, but in general are rugged hill country, usually less than 1000 meters in elevation, with few settlements.

Acting as a matrix binding together steptoes, volcanic cones, and andesite hills, the ash plateau is not continuous over extensive areas. Considerable portions of the original depositional surface still persist as flattish or rolling uplands, 150–350 meters above sea level. Over a large proportion of its area, however, stream erosion has reduced the region to slopes with a confused drainage pattern. It may be that part of the surface irregularity is due to the original unevenness of the hard-rock surface upon which the ash was deposited. A conspicuous feature of the region is the wide, flat-bottomed, steep-walled valleys, whose sides show excellent cross-sections of the gray ash. Where the valley sides are not sheer they are commonly mantled in dense subtropical evergreen forest, which gives an almost tropical aspect. Along the coasts where the ash plateau lies close to the sea are what appear to be old ash-covered fans or cones with incised stream valleys. The immediate sea margins are flanked with beach ridges, back of which lie partially filled lagoons.

Relatively isolated from northern Kyushu and the rest of Japan by the formidable barrier of the Kyushu Range, this Satsuma region of southern Kyushu is provincial in outlook and at least a quarter-century behind the rest of Japan. It remained a semi-independent state, relatively hostile to the new government in Tokyo, for some time after the Restoration. Even today the Satsuma dialect, spoken in the homes of this region, sounds strange to the ears of the native of Kanto. Cultural connections have been strong with the Ryukyu Islands to the south, and evidences of this affiliation are to be observed in house features and social customs. Settlement types are varied; though compact villages are by no means absent, semidispersed and isolated rural residences appear to predominate. Most settlements avoid the crests of the ash upland, where water is difficult to obtain and where the full force of the typhoons is felt in late summer and early fall. More commonly, rural inhabitants seek the shelter of the valleys, building their dwellings close

Fig. 16-15.—The somewhat uneven crest of the ash upland of southern Kyushu. The crops are unirrigated.

to the bases of valley walls, where there is the dual advantage of slight elevation, offering protection against floods, and the presence of spring water.[18] However, settlements so located are in some danger from landslides at times of heavy rains or earthquakes. Rural dwellings, many of them only crude huts with steep thatched roofs and overhanging eaves, are tropical in aspect. Characteristically they are set in the midst of a subtropical vegetation consisting of live oaks, bamboo trees, and, occasionally, banana trees.

Volcanic southernmost Kyushu is included within Ogasawara's agricultural Frontier Zone, where intensity of land use is low, subsistence farming is dominant, and commercial or cash crops are meagerly developed. Low quality ash soils, plus the handicap of relative isolation, have acted to retard development, so that the region has one of the lowest per capita income rates anywhere in Japan.

Rice is relatively less important in this region of ash uplands than where areas of new alluvium are more extensive. As might be expected, its cultivation is confined largely to the valley bottoms and the coastal lowlands. On the flattish to rolling ash uplands dry fields of sweet potatoes, vegetables, upland rice, winter grains, beans, tobacco, sugar cane, and fruit trees are characteristic. Kagoshima is Japan's ranking prefecture in rapeseed and sweet potatoes, produces almost all of the country's sugar cane, and is one of the leaders in tobacco culture. Recently there is some evidence of an increased emphasis on truck gardening, an attempt no doubt to capitalize on the mild climate. Since the upland surfaces are not level, farmers have resorted to crude terracing in laying out the fields. This has been one of the country's principal horse breeding areas, but that specialization is declining. By contrast, the number of swine has multiplied, and Kagoshima at present has a disproportionately large population of both swine and draft cattle.

Cities are not numerous in southern Kyushu, for those important commercial and industrial services which cities perform remain underdeveloped here. This is in marked contrast to the situation in northern Kyushu. Kagoshima (231,000 population), the metropolis of the Satsuma region, is located on a narrow coastal plain at the base of the ash plateau, fronting upon the quiet waters of Kagoshima Bay and situated almost under the shadow of imposing Sakurajima Volcano. Like scores of other cities and towns in southern Kyushu, it is an old castle town. Its functions are chiefly administrative and commercial, for it is a prefectural capital and a port of local importance. Manufacturing remains underdeveloped, and what little exists emphasizes chiefly foods and publishing. The port's trade, preponderantly domestic, is largely with Okinawa. Imports are mainly foodstuffs, while the more varied exports include lumber and wood, paper products, foodstuffs, and textiles.

Along the eastern margin of the ash plateau where it makes contact with a fragment of the Kyushu Mountains is the *Miyakonojo Basin*, containing the most important compact rural population cluster of the entire subdivision. It is probably a tectonic depression, partially filled with ash, which has been carved into a series of low terraces by stream erosion. Rice fields occupy extensive areas on the smooth terrace surfaces as well as on the floodplains.

Miyazaki Plain (800–900 sq. km.), second largest in southern Kyushu, is a wedge-shaped area along the southeast coast, terminated on its inland margins, where it meets the Kyushu Mountains, by a distinct morphologic fault. The plain consists of a base of weak Tertiary rocks, upon which rest diluvial sediments of variable thickness (Fig. 16-16). Haltings in the emergence of this coastal plain have resulted in several terrace levels which terminate in more or less abrupt wave-cut margins. In the narrow northern part where a mantle of diluvium covers most of the Tertiary rock, the upland surface is moderately extensive. But in the broader southern section where much more of the flattish diluvial surface has been removed by stream erosion, low rounded hill masses composed of Tertiary rock and volcanic ash are associated with a complicated system of broad, alluvial-floored valleys. A lagoon plain of poorly drained alluvium, bordered by beach ridges along its ocean front, occupies the seaward margins, the coastal strip widening at the mouths of the rivers where the alluvium extends inland along the broad, steep-sided valleys of several small streams. It likewise widens southward, and

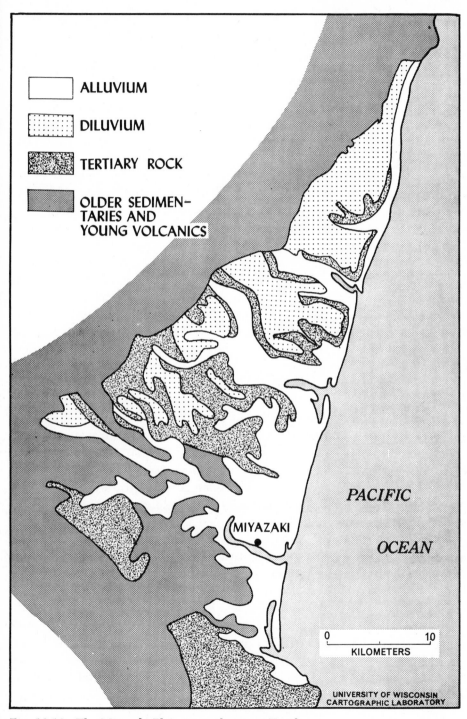

ALLUVIUM

DILUVIUM

TERTIARY ROCK

OLDER SEDIMEN-
TARIES AND
YOUNG VOLCANICS

PACIFIC

OCEAN

MIYAZAKI

0   10
KILOMETERS

UNIVERSITY OF WISCONSIN
CARTOGRAPHIC LABORATORY

Fig. 16-16.—The Miyazaki Plain in southeastern Kyushu.

in the vicinity of Miyazaki city it assumes the dimensions of a genuine alluvial lowland, whose surface is interrupted by numerous remnants of old beach ridges.

Half a century or more ago, before the eastern Kyushu rail line was put through, this somewhat isolated Miyazaki Plain was relatively undeveloped. But in more recent decades new settlers have reclaimed considerable areas of farm land, both on the diluvial upland and on the wet lagoon plain. On the upland surfaces, farms with isolated farmsteads, set in the midst of woodland and moor, are planted almost exclusively to unirrigated crops—soybeans, buckwheat, sweet potatoes, and vegetables as summer crops, followed by wheat and barley in winter. Lately there has come to be a degree of specialization in truck gardening. Similarly, the sandy beach ridges along the coast are sites for dry crops. By contrast, the poorly drained lagoon lowlands, where reclaimed for agriculture, are areas of rice monoculture. However, the more extensive lowland plain in the vicinity of Miyazaki city, with its numerous remnants of old beach ridges, has a mixture of dry crops and irrigated rice. Paddy fields, many of them irrigated from ponds, occupy the floors of the intricate valley systems back from the coast. Miyazaki city (95,000 population) is a local market and a prefectural capital, but is inconsequential as a manufacturing center.

Kyushu Mountains

The northern and eastern parts of southern Kyushu are largely occupied by the Kyushu Range. Highest toward its northern margins, where it terminates in a bold fault-scarp overlooking the Aso ash-and-lava plateau, this mountain mass resembles a tilted block with upthrust on the north. At its western end, the Kuma River flows northward against the general slope of the area, crossing it in a meandering antecedent gorge, a fact suggesting that the river once flowed upon an erosion surface, which was later tilted and folded to form the Kyushu Mountains. Slopes are usually steep and the valleys narrow, although the crests of the ridges still bear some evidence of the earlier erosion surface. The only lowlands of appreciable size are the tiny Hitoyoshi fault basin (224 sq. km.) and the still smaller Nobeoka Plain.

As in most mountain areas of rugged terrain, population is relatively sparse. Shifting cultivation is still practiced to some extent on the slope lands, but this is declining as drag roads are extended deeper into the mountains, making the forest resources more accessible.[19] Piles of logs and wood and bundles of charcoal at the railroad stations testify to the importance of the forest industries.

Where the folded structures project into Bungo Channel, which separates Kyushu from Shikoku, the rugged coast is of the deeply indented ria type. Here settlement is concentrated on the small areas of new alluvium which have accumulated at the heads of the long indentations. The meager area of level land has led to the artificial terracing and cropping of the adjacent hill slopes, sometimes to elevations of several hundred feet. Orange groves are numerous on these slope sites.

The advantages associated with protected natural harbors and relatively deep water in this section of ria coast in eastern Kyushu have recently begun to attract new industry. Already established for several decades is the copper, gold, and silver smelter and refinery located on the tip of Saganoseki Peninsula, where both local ores and imported domestic and foreign ores are processed. Additional plants include those making electric wire, iron products, and carbon, located in Oita city; paper and pulp, at Tsurusaki; and cement, rayon, and iron works at the smaller cities and towns of Usuki, Tsukumi, and Saeki located at the heads of coastal indentations somewhat farther south.

Most notable of the newer developments, and a forerunner of a probable expansion in manufacturing, is the reclamation along the seacoast between Oita and Tsurusaki of 8.8 square kilometers of new land for factory construction, which already is the site of active development of a steel and petroleum combine. A large oil refinery and a pig iron–steel mill of 3.9 million tons annual capacity are now under construction, and it is expected that these in turn will attract other chemical and heavy industries such as will cause this strip of coast to be one of the nation's important new local centers of heavy industry. This follows the overall national plan calling for a wider dissemination of industry, together with its location in areas that were previously backward in factory development. Deep water will permit the berthing of huge oil tankers of up to 100,000 tons capacity, while hills both to the north and to the south provide shipping with protection against strong winds. In addition the Ono River is able to provide a daily supply of industrial water equal to about 1.2 million cubic meters. Thus, the combined advantages of deep and protected harbors, available seaside land for factories, and a good supply of water are expected to make the region attractive for new industry. The lack, of course, is important local markets, for this region is well removed from the populous heartland of Japan.

Much farther to the south, and representing a striking exception along what is a very extensive stretch of industrially underdeveloped coastline, is the small and isolated city of Nobeoka (62,000 population) which, capitalizing on abundant hydroelectric power developed in the adjacent highlands, has acquired national fame as a local center completely specialized in the chemical and electrochemical industries. About on a par in importance with Omuta, the coal-based chemical center in western Kyushu, Nobeoka derives 90 per cent of its total value added by manufacturing from the chemical sector. Characteristic products are staple fiber, rayon, explosives, and resin.

REFERENCE MATTER

NOTES

CHAPTER 1

1 Akira Watanabe, "Geomorphology in Relation to Recent Crustal Movements in Japan," *Proceedings of the IGU Regional Conference in Japan, 1957* (Tokyo, 1959), p. 220.

2 Akira Watanabe, "Landform Divisions of Japan," *Bulletin of the Geographical Survey Institute* (Tokyo), 2, Part 1 (1950), 81–94.

3 Francis Ruellan, "La Vigueur de l'érosion normale au Japon," *Proceedings of the Third Pan-Pacific Science Congress* (Tokyo, 1926), 2, 1860–62.

4 Shuko Iwatsuka, "On Landslides and Related Phenomena in Mountainous Areas of Japan," *Proc. IGU Regional Conf. Japan, 1957*, pp. 154–57.

5 Hisashi Sato, "Distribution of Volcanoes in Japan," *Proc. IGU Regional Conf. Japan, 1957*, pp. 184–88.

6 For a fairly comprehensive treatment of this topic, in both its general and its regional aspects, see Ludwig Mecking, *Japans Häfen* (Hamburg, 1931), pp. 180–207.

7 Y. Otuka, "The Japanese Coastline" (in Japanese with English summary), *Geographical Review of Japan*, 9 (1933), 818–43.

8 Jörn Leo, "Küstengestaltung Japans nach neuesten Forschungen," *Petermanns Geographische Mitteilungen*, 80 (1934), 260.

9 Akira Watanabe, "Major Geomorphological Divisions of the Coast of Japan" (in Japanese with English summary), *Journal of Geography*, 66, No. 1 (1957), 1–16.

CHAPTER 2

1 On this general topic see B. W. Thompson, "An Essay on the General Circulation of the Atmosphere over South-East Asia and the West Pacific," *Quarterly Journal of Royal Meteorological Society* (London), 67 (1951), 569–97; Glenn

T. Trewartha, *The Earth's Problem Climates* (Madison, Wis., 1961), pp. 151–53, 180–86; and an article by Staff Members of the Section of Synoptic and Dynamic Meteorology, Institute of Geophysics and Meteorology, the Academia Sinica, Peking, "On the General Circulation over Eastern Asia," *Tellus*, 9 (1957), 432–36; *10* (1958), 58–75, 299–312.

2 H. Flohn and H. Oeckel, "Water Vapour Flux during the Summer Rains over Japan and Korea," *Geophysical Magazine*, 27 (1956), 527–32. See also T. Murakami, Y. Arai, and K. Tomatsu, "On the Rainy Season in the Early Autumn" (in Japanese with English summary), *Journal of the Meteorological Society of Japan*, Series 2, 40 (1962), 330.

3 T. Murakami, "The General Circulation and Water-Vapor Balance over the Far East during the Rainy Season," *Geophysical Magazine* (Tokyo), 29 (1959), 131–71.

4 *Ibid.*, pp. 142–43.

5 Ken Suda, "A Study of the Dynamic Climatology of Cold Outbreaks in the Far East," *Geophysical Magazine* (Tokyo), 29 (1959), 413–14.

6 *Ibid.*, p. 458.

7 *Ibid.*, p. 459.

8 James E. Miller and Homer T. Mantis, "Extratropical Cyclogenesis in the Pacific Coastal Regions of Asia," *Journal of Meteorology*, 4 (1947), 29–34.

9 Trewartha, *The Earth's Problem Climates*, pp. 196–97.

10 Hideo Suzuki, "Ueber die Bereiche des Winterlichen Niederschlags in Japan," *Geographical Review of Japan*, 34 (1961), 321–26.

11 Ren-ichi Saito, "The Climate of Japan and Her Meteorological Disasters," *Proceedings of the IGU Regional Conference in Japan, 1957* (Tokyo, 1959), p. 176.

12 Flohn and Oeckel, in *Geophysical Magazine*, 27, 527–32. See also Ken Suda and T. Asakura, "A Study of the Unusual 'Baiu' Season in 1954 by Means of Northern Hemisphere Upper Air Mean Charts," *Journal of the Meteorological Society of Japan*, 33 (1955), 233–34.

13 On this topic see Taiji Yazawa, "Der Jahreszeitliche Ablauf der Witterung in Japan," *Geographische Rundschau*, 9, No. 11 (1957), 407–11.

14 Takeshi Sekiguchi, "On the Year Climate in Japan" (in Japanese with English summary), *Geographical Review of Japan*, 24 (1951), 175–85.

CHAPTER 3

1 Koshi Nomoto, "Pulp-Wood Supply in Japan," *Geographical Review of Japan*, 33 (1960), 300–311.

2 General Headquarters, Supreme Commander for the Allied Powers, Natural Resources Section, "Reconnaissance Soil Survey of Japan," Report 110 (Tokyo, 1948–1951). Eight mimeographed regional reports (maps, 1:250,000; "Kanto Plain Area," 110-A, 1948; "Kyushu," 110-B, 1950; "Kyoto Area," 110-C, 1950; "Shikoku," 110-D, 1950; "Hiroshima Area," 110-E, 1950; "Nagoya Area," 110-F, 1951; "Northern Honshu Area," 110-G, 1951; "Hokkaido," 110-H, 1951); "Summary," 110-I, 1951 (maps, 1:800,000). See also Yutaka Kamoshita, *Soils in Japan, with General Map of Soil Types in Japan, 1:800,000*. Miscellaneous Publication B, No. 5 (National Institute of Agricultural Sciences, Tokyo, 1958).

3 Roy W. Simonson, U.S. Department of Agriculture, personal communication, Jan. 11, 1963.

4 Japanese Geological Survey, *Geology and Mineral Resources of Japan*, 2d ed. (1960), pp. 133–34.
5 *Ibid.*, p. 135.
6 Kiyoo Wadati and Takuzo Hirono, "Ground Subsidence in Industrial Districts of Japan and Disasters Subsequent to It," *Proceedings of the IGU Regional Conference in Japan, 1957* (Tokyo, 1959), pp. 202–9.
7 *Geology and Mineral Resources of Japan*, map on p. 131.

CHAPTER 4

1 Yasushi Oiwa, Carl G. Nordquist, and Oscar H. Lentz, "The Mineral Industries of Japan and Southeast Asia," *Quarterly of the Colorado School of Mines, 56*, No. 4 (1961), 183.

CHAPTER 5

1 J. E. Kidder, *Japan Before Buddhism* (London, 1959), pp. 87–89.
2 *Ibid.*, p. 26.
3 *Ibid.*, pp. 86–87.
4 *Ibid.*, pp. 89–91.
5 *Ibid.*, p. 91.
6 *Ibid.*
7 *The Kojiki,* trans, B. H. Chamberlain (Kobe, 1932); *The Nihongi,* trans. W. G. Aston (London, 1924).
8 Irene B. Taeuber, *The Population of Japan* (Princeton, 1958), pp. 9, 14.
9 *Ibid.*, p. 16; and Minoru Tachi, "Forecasting Manpower Resources: Population and Labor Force—Some Experiences in Japan," Institute of Population Problems, Tokyo, English Series, No. 55 (mimeographed reprint of the paper presented to the International Conference on the Middle Level Manpower, October, 1962).
10 Taeuber, *The Population of Japan*, pp. 23–24.
11 See the following map: "Population Density at the Beginning of the Meiji Era" (choropleth method), 1:2,000,000, Geographical Survey Institute (Tokyo, 1956).
12 Kingsley Davis, "The Theory of Change and Response in Modern Demographic History," *Population Index, 29* (1963), 349–52.
13 Tatsuo Honda, "Population Problem of Japan in the 1960's" (in Japanese with English summary), *Annual Reports of the Institute of Population Problems*, No. 7 (1962), pp. 1–4.
14 Kiichi Yamaguchi, "A Comparison of Standardized Birth, Death and Natural Increase Rates of Respective Prefectures in Japan: 1955–1960" (in Japanese with English summary), *ibid.*, pp. 41–44.
15 Masako (Sakamoto) Momiyama, "Geographical Variation of Death Rate in Postwar Japan," *Proceedings of the IGU Regional Conference in Japan, 1957* (Tokyo, 1959), pp. 405–12.
16 Minoru Tachi, Masao Ueda, and Hidehiko Hama, "Regional Characteristics of Population in Japan," *ibid.*, p. 482.
17 M. Tachi and M. Oyama, "Regional Distribution of Income and Population" (in Japanese with English summary), *Journal of Population Problems*, No. 82 (1961),

p. 17; also Minoru Tachi, "Regional Income Disparity and Internal Migration of Population in Japan," Committee for Translation of Japanese Economic Studies, No. 21, International House of Japan, p. 8 (mimeographed).

18 Masao Ueda, "Features of In-migrants by Age and Industry" (in Japanese with English summary), *Annual Reports of the Institute of Population Problems*, No. 7 (1962), pp. 25–30.

19 *Ibid.*, Fig. 1, p. 27.

20 *Ibid.*, p. 29.

21 Yoichi Okazaki, "Regional Migration of Male Labor Force: Measurement and Some Observations" (in Japanese with English summary), *ibid.*, pp. 21–22.

22 Tachi, "Regional Income Disparity and Internal Migration of Population in Japan"; Tachi and Oyama, in *Journal of Population Problems*, No. 82; Syoji Tsubouchi, "Population Pressure and Rural-Urban Migration in Japan," *Proc. IGU Regional Conf. Japan, 1957*, pp. 512–16; Okazaki, in *Annual Reports of the Institute of Population Problems*, No. 7, pp. 20–24, 91–92.

23 Tsubouchi, in *Proc. IGU Regional Conf. Japan, 1957*, p. 514.

24 Tachi, "Regional Income Disparity and Internal Migration of Population in Japan," pp. 17–18.

25 Bureau of Statistics, Office of the Prime Minister, *1960 Population Census, Densely Inhabited District: Its Population, Area, and Map* (Tokyo, 1961).

26 For a list and description of a series of relatively large-scale population maps of Japan based on 1950 and 1955 census data, compiled and constructed by the Geographical Survey Institute (Tokyo), see a review by Glenn T. Trewartha, in the *Geographical Review* (New York), 49 (1959), 283–86. See also Selected References, p. 635.

27 Tatsutaro Hidaka, "Population Density of Japan," *Bulletin of the Geographical Survey Institute* (Tokyo), 5, Parts 1–2 (1957), 31–64.

28 Geographical Section, "Population Density by Landform Division in Japan," *Bulletin of the Geographical Survey Institute* (Tokyo), 6, Parts 2–3 (1960). See detailed tables, pp. 160–65. For 1960 maps, see Selected References, p. 635.

29 Pierre Gourou, "China," *The Development of Upland Areas in the Far East* (Institute of Pacific Relations, New York, 1949), *1*, Part 1, 1–24; and Glenn T. Trewartha, "Japan," *ibid.*, Part 3, 59–82.

CHAPTER 6

1 Takeo Tanioka, "Différenciation Régional des Types de l'Habitat Rural au Japon," *Proceedings of the IGU Regional Conference in Japan, 1957* (Tokyo, 1959), pp. 503–12. See his Figs. 1 and 3 for distribution of dispersed settlement.

2 K. Kodera and H. Iwamoto, "On the Scattered Settlements (Strendorf) on the Alluvial Fan of Ooi-gawa River" (in Japanese with English summary), *Geographical Review of Japan*, 15 (1939), 686–710, 760–83.

3 Yukio Asaka and Masataro Nagai, "Settlement Originated in the Medieval Period (15th–16th Century)," *Distribution Maps on the Regional Geographical Study of Japan*, comp. Taro Tsujimura (Tokyo, 1952), pp. 20–21.

4 John F. Embree, *Suye Mura: A Japanese Village* (Chicago, 1939), pp. 12–35.

5 R. P. Dore, *Land Reform in Japan* (New York, 1959), pp. 201–40.

6 One of the best short descriptions of life in a small Japanese rural community is available in *Fortune*, September, 1936, pp. 87 ff. See also Embree, *Suye Mura:* and Richard K. Beardsley, John W. Hall, and Robert E. Ward, *Village Japan* (Chicago, 1959).

7 Tokuichi Asai, "On the Reorganization of *gun* and *shi,* or Countries and Cities in Japan," *Proc. IGU Regional Conf. Japan, 1957,* pp. 271–72; Shinzo Kiuchi, "Centrifugal and Centripetal Urbanization in Japan," *ibid.,* pp. 367–68.

8 Bureau of Statistics, Office of the Prime Minister, *1960 Population Census, Densely Inhabited District: Its Population, Area, and Map* (Tokyo, 1961), p. 40.

9 Takeo Tanioka, "Le Jôri dans le Japon ancien," *Annales: Économies, Sociétés, Civilisations,* No. 4 (Octobre–Décembre, 1959), pp. 631–32.

10 H. Sato, "Distribution of the 'Strassendorf' " (in Japanese), *Geographical Review of Japan, 6* (1930), 550–57.

11 Ryuziro Isida, *Geography of Japan* (Tokyo, 1961), pp. 50–55.

12 Tanioka, in *Annales* (Octobre–Décembre, 1959), pp. 631–32.

13 Toshio Kikuchi, "Geographical Function of the Reclamation Settlements During the Period of the Shogunate, 1603–1867," *Proc. IGU Regional Conf. Japan, 1957,* pp. 362–67.

14 *Ibid.,* p. 362.

15 International Geographical Union, Science Council of Japan, Regional Conference in Japan, *Regional Geography of Japan,* No. 1, *Hokkaido Guidebook* (Tokyo, 1957), pp. 10–12. For additional discussion of several regional forms of rural settlement in Japan see Robert Burnett Hall, "Some Rural Settlement Forms in Japan," *Geographical Review, 21* (1931), 93–125.

16 George B. Sansom, *Japan: A Short Cultural History* (New York, 1931), p. 193.

17 Keiichiro Yamaguchi, "Regional Differences in the Process of Urbanization in Japan" (abstract), *Proc. IGU Regional Conf. Japan, 1957,* pp. 528–29.

18 Yosoburo Takekoshi, *The Economic History of the Civilization of Japan* (New York, 1930), pp. 243–45, 258.

19 For a treatment of the structure and functional areas of castle towns, see Kenichi Tanabe, "Development of the Areal Structure of Japanese Cities in the Case of Castle Towns—As a Geographic Contribution to the Study of Urban Geography," *The Science Reports of the Tohoku University,* 7th Series (Geography), No. 8 (1959), pp. 88–105.

20 Kenjiro Fujioka, "An Explanation of Japanese Castle Towns," *Distribution Maps on the Regional Geographical Study of Japan,* comp. Taro Tsujimura (Tokyo, 1952), pp. 18–19; see also Kenjiro Fujioka, "Feudal Traditions in the Forms and Zone Structures in Japanese Cities," *Proc. IGU Regional Conf. Japan, 1957,* pp. 316–19.

21 Isida, *Geography of Japan,* p. 60.

22 Yukio Asaka, "Shukuba-machi," *Distribution Maps on the Regional Geographical Study of Japan,* comp. Taro Tsujimura (Tokyo, 1952), pp. 21–22.

23 Robert B. Hall, "The Cities of Japan: Notes on Distribution and Inherited Forms," *Annals of the Association of American Geographers, 24* (1934), 196.

24 Sachio Asaka, "Development of Towns and Villages in the Edo Period," *Proc. IGU Regional Conf. Japan, 1957,* p. 275.

25 Irene B. Taeuber, *The Population of Japan* (Princeton, 1958), p. 27. For descriptions of urban development in feudal Japan and particularly of Edo, Osaka,

and Kyoto, see Engelbert Kaempfer, *The History of Japan . . . 1690–92* (Glasgow, 1906), 3, 6–7, 20–24, 71–74, 306–7, 313–17.

26 Isida, *Geography of Japan*, pp. 40–41.

27 *Ibid.*, p. 41.

28 Minoru Tachi and Masao Uyeda, "A Statistical Study on the Variation of Basic Demographic Phenomena by the Size of Communities," *Archives of the Population Association of Japan*, No. 1 (1952), pp. 96–98.

29 Isida, *Geography of Japan*, p. 41.

30 Taeuber, *Population of Japan*, pp. 47, 49.

31 *Ibid.*, p. 71.

32 For an elaboration of this theme of recent changes in the six major cities of Japan, see Peter Schöller, "Wachstum und Wandlung japanischer Stadtregionen," *Die Erde* (formerly *Zeitschrift der Gesellschaft für Erdkunde zu Berlin*), 93 (1962), 202–34.

33 Yoshio Watanabe, "An Analysis of the Function of Urban Settlements Based on Statistical Data—A Functional Differentiation Vertical and Lateral," *Science Reports, Tohoku Univ.*, 7th Series, No. 10 (1961), pp. 72–76.

34 Thomas O. Wilkinson, "A Functional Classification of Japanese Cities: 1920–55," *Demography, 1*, No. 1 (1964), 179–83.

35 *Ibid.*, p. 182.

36 Watanabe, in *Science Reports, Tohoku Univ.*, 7th Series, No. 10 (1961), pp. 84–85, 93.

37 See Ludwig Mecking, "Kult und Landschaft in Japan," *Geographische Anzeiger,* 30 (1929), 137–46, for an analysis of landscape features as they are related to the spiritual life of the country.

38 A good description of the ordinary farmhouse appears in Embree's *Suye Mura,* pp. 89–94. See also Beardsley, Hall, and Ward, *Village Japan*, pp. 77–95.

39 *Ibid.*, pp. 90–92.

CHAPTER 7

1 For a treatment of the historical background of Japanese agriculture see Thomas C. Smith, *The Agrarian Origins of Modern Japan* (Stanford, Calif., 1959).

2 R. P. Dore, *Land Reform in Japan* (New York, 1959), p. 214. Other useful references on land reform and its effects are Lawrence I. Hewes, "Japanese Land Reform Program," Supreme Commander for the Allied Powers, Natural Resources Section, Report No. 127 (mimeographed, 1950); Owada Keiki, "Land Reform in Japan," in Kenneth H. Parsons, Raymond J. Penn, and Philip M. Raup, *Land Tenure* (Madison, Wis., 1956), pp. 219–29; Masaru Kajita, *Land Reform in Japan*, Agricultural Development Series, No. 2 (Agriculture, Forestry, and Fisheries Productivity Conference, Tokyo, 1959).

3 Smith, *The Agrarian Origins of Modern Japan*, Chaps. IX–X.

4 Japan FAO Association, *A Century of Technical Development in Japanese Agriculture* (Tokyo, 1959), pp. 36–37.

5 *Ibid.*, pp. 37 ff.

6 John D. Motz, "How Japan Has Expanded Its Agriculture," *Foreign Agriculture,* 24 (1960), 8–9.

7 *A Century of Technical Development in Japanese Agriculture*, p. 106.

8 Dore, *Land Reform in Japan,* pp. 255–56.

9 Japan FAO Association, *A Strategy for New Agriculture* (Tokyo, 1962), pp. 7–8.

10 Shiroshi Nasu, *Aspects of Japanese Agriculture,* Institute of Pacific Relations International Research Series (New York, 1941), p. 68.

11 "Of Many Men on Little Land," *Fortune,* September, 1936, pp. 92–93.

12 On this topic of land reclamation see the following: Edward A. Ackerman, *Japan's Natural Resources* (Chicago, 1953), pp. 376–431; Glenn T. Trewartha, "Japan," *The Development of Upland Areas in the Far East* (Institute of Pacific Relations, New York, 1949), *1*, Part 3, 59–82; Glenn T. Trewartha, "Land Reform and Land Reclamation in Japan," *Geographical Review,* 40 (1950), 376–96; Fukuo Ueno, *Land Utilization in Japan,* Agricultural Development Series, No. 12 (Agriculture, Forestry, and Fisheries Productivity Conference, Tokyo, 1960); and Nasu, *Aspects of Japanese Agriculture,* pp. 63–87.

13 Ueno, *Land Utilization in Japan,* p. 32.

14 *Ibid.*

15 Koji Iizuka, "Postwar Land Reform and Japanese Agriculture as Seen from the Standpoint of Human Geography," *Proceedings of the IGU Regional Conference in Japan, 1957* (Tokyo, 1959), p. 341.

16 Nasu, *Aspects of Japanese Agriculture,* p. 7.

17 For a discussion of these several types of physical land improvement and their estimated effects on production, see Ackerman, *Japan's Natural Resources,* pp. 388–99; also Ueno, *Land Utilization in Japan,* pp. 37–41; and Shiro Sasaki, *Land Development and Improvement Projects in Japan,* Agricultural Development Series, No. 3 (Agriculture, Forestry, and Fisheries Productivity Conference, Tokyo, 1959).

18 Ueno, *Land Utilization in Japan,* p. 37.

19 For a discussion of the feasibility of these various methods see Ackerman, *Japan's Natural Resources,* pp. 399–443.

20 T. Noh and K. Yosizaki, "Terrace Cultivation in Japan" (in Japanese with English summary), *Geographical Review of Japan,* 12 (1936), 352–68, 828–35; *14* (1938), 230–38.

21 Fukuo Ueno, "The Problems of the Utilization of Slope Land in Japan," *Proc. IGU Regional Conf. Japan, 1957,* p. 521.

22 *Ibid.* See also Fukuo Ueno, "The Problems of the Utilization of Slope Lands in Japan" (in Japanese with English summary), *Journal of Geography* (Tokyo), *62* (1953), 3. The latter reference contains detailed maps of slope-field distribution.

23 Robert B. Hall, "Revolution in Asian Agriculture," *Journal of Geography* (Tokyo), 69 (1960), 97–105; Robert B. Hall, Jr., "Hand Tractors in Japanese Paddy Fields," *Economic Geography,* 34 (1958), 312–20.

24 Mitsunori Saito, "Japanese Agricultural Regions with Special Reference to Part-Time Farm Households" (in Japanese with English summary), *Geographical Review of Japan,* 34 (1961), 200–221.

25 Yoshikatsu Ogasawara, "The Role of Rice and Rice Paddy Development in Japan," *Bulletin of the Geographical Survey Institute* (Tokyo), 5, Part 4 (1958), 1–2.

26 Y. Ichinose, "Distribution of Paddyland of High Elevations and in Cooler Climates of Japan," *Geographical Review of Japan,* 27 (1954), 108–16.

27 Ogasawara, in *Bulletin of the Geographical Survey Institute,* 5, Part 4 (1958), 6.

28 Y. Daigo and Y. Suzuki, "On the Distribution of Maximum Possible Cultivating

Period of Aquatic-Rice in Japan" (in Japanese with English summary), *Geographical Review of Japan, 24* (1951), 1–2.

29 Takane Matsuo, *Rice Culture in Japan* (Ministry of Agriculture and Forestry, Tokyo, 1954), pp. 47–48. See also Takane Matsuo, *Rice and Rice Cultivation in Japan* (The Institute of Asian Economic Affairs, Tokyo, 1961); Katsumi Amatatsu, *Growing Rice in Japan,* Agricultural Development Series, No. 11 (Agriculture, Forestry, and Fisheries Productivity Conference, Tokyo, 1959); and Chieko Aoki, "Regional Characteristics of Productivity of Paddy-Field Rice in Japan—Report II" (in Japanese with English summary), *Geographical Review of Japan, 36* (1963), 412–23.

30 Ogasawara, in *Bulletin of the Geographical Survey Institute,* 5, Part 4 (1958), 19–20.

31 Jiro Yonekura, "The Development of the Grid-Pattern Land Allotment System in East Asia," *Proc. IGU Regional Conf. Japan, 1957,* pp. 547–48. On this same topic see also Masatoma Ikeda, "Regionality of the Jori System," *ibid.,* p. 348; Takeo Oda and Takeo Tanioka, "The Jori System: A Rural Planning in Ancient Japan," *Distribution Maps on the Regional Geographical Study of Japan,* comp. Taro Tsujimura (Tokyo, 1952), pp. 16–17; Takeo Tanioka, "Le Jôri dans le Japon Ancien, *Annales: Économies, Sociétés, Civilisations,* No. 4 (Octobre–Décembre, 1959), pp. 625–39; Ryuziro Isida, *Geography of Japan* (Tokyo, 1961), pp. 50–51.

32 This topic is dealt with in some detail in Matsuo, *Rice Culture in Japan,* pp. 25–36; Matsuo, *Rice and Rice Cultivation in Japan,* pp. 62–80; and Amatatsu, *Growing Rice in Japan,* pp. 72–89.

33 John D. Eyre, "Japanese Inter-Prefectural Rice Movements," *Economic Geography, 38* (1962), 78–86; Matsuo, *Rice Culture in Japan,* pp. 10–11.

34 Yoshikatsu Ogasawara, "The Distribution Pattern of Cash Crop Fields in Japan," *Proc. IGU Regional Conf. Japan, 1957,* pp. 440–41.

35 John H. Thompson, "Urban Agriculture in Southern Japan," *Economic Geography, 33* (1957), 224–37.

36 Setsutaro Murakami, "The Formation of Citrus Growing Areas in Japan," *Proc. IGU Regional Conf. Japan, 1957,* p. 423.

37 Shozo Yamamoto, "The Preliminary Survey of the Tea Industry in Japan," *Human Geography, 10* (1958), 92–107.

38 Bureau of Statistics, Office of the Prime Minister, *Japan Statistical Yearbook, 1962* (Tokyo, 1962).

39 On the general topic of Japan's animal industries and the pastures and meadows that support them, see the following: H. Yasuta, "Regional Characteristics of the Landscape of Pasture and Meadows in Japan" (in Japanese), *Geographical Review of Japan, 29* (1956), 91–101; Hatsuo Yasuta, "Mountain Grazing in Japan" (in Japanese with English summary), *Journal of Geography* (Tokyo), *65* (1956), 12–21; Ken-ichi Tanabe, "Areal Analysis of the Milk Cow Keeping in Japan," *The Science Reports of the Tohoku University,* 7th Series (Geography), No. 4 (1955), pp. 1–24.

40 Takeshi Motooka, "The Changes in Agricultural Land Use in Japanese Agriculture after the War," *Proc. IGU Regional Conf. Japan, 1957,* pp. 413–14.

41 Yoshikatsu Ogasawara, "Land Use of Japan," *Bulletin of the Geographical Survey Institute* (Tokyo), 2, Part 1 (1950), 95–119.

CHAPTER 8

1 On the economic evolution of Japan see William W. Lockwood, *The Economic Development of Japan, Growth and Structural Change, 1863–1938* (Princeton, 1954).
2 Information Office, Consulate General of Japan, *Japan Report*, Jan. 15, 1961, p. 4.
3 Foreign Capital Research Society, *Japanese Industry* (1961), p. 12.
4 Peter Schöller, "Wandlungen der Industriestruktur Japans," *Deutscher Geographentag* (Cologne), May 22–26, 1961, p. 238.
5 See Ryuziro Isida, *Geography of Japan*, (Tokyo, 1961), pp. 91–93.
6 On this topic see Kikukazu Doi, "The Industrial Structure of Japanese Prefectures," *Proceedings of the IGU Regional Conference in Japan, 1957* (Tokyo, 1959), pp. 310–16. Kiyoki Koda, "Manufacturing Districts in Japan" (abstract), *ibid.*, pp. 372–74; Norio Hasegawa, "Distribution of Manufacturing in Japan: A Macroscopic Analysis of Localization of Manufacturing," *The Science Reports of the Tohoku University*, 7th Series (Geography), No. 8 (1959), pp. 39–48; Mutsuo Nishimura, "Industrial Distribution in Japan," second report (in Japanese with English summary), *Human Geography*, 4 (1952), 296–309.
7 John C. Weaver, "Crop Combination Regions in the Middle West," *Geographical Review*, 44 (1954), 175–200.
8 Doi in *Proc. IGU Regional Conf. Japan, 1957*, Table 2, p. 315.
9 One of the best maps showing in detail the distribution of manufacturing in Japan is to be found in John H. Thompson and Michihiro Miyazaki, "A Map of Japan's Manufacturing," *Geographical Review*, 49 (1959), 1–17 (facing p. 16).
10 Economic Planning Agency, *Economic Survey of Japan (1960–1961)*, Tokyo, 1961, p. 46.
11 *Ibid.*, p. 404. See also "Industrial Review of Japan, 1963," *Nihon Keizai Shimbun* [*Japanese Economic Journal*], Tokyo, 1963, p. 160.
12 Secretariat: The Conference of Hanshin Metropolitan Region and the Conference of Osaka Conurbated Area, *Synopsis of Various Planning Projects for Hanshin-Kinki Region* (December, 1961), pp. 1–5.
13 For a short sketch of the new industrial areas, see "Industrial Review of Japan, 1963," pp. 160–65.
14 Mutsuo Nishimura, "Industrial Distribution in Japan" (in Japanese with English summary), *Human Geography*, 3 (1951), 10–20, 75–76.
15 Minoru Beika, "Structure of Industrial Districts in Japan—the Cases of Tokyo, Osaka and Nagoya," *Kobe Economic and Business Review*, 3 (1956), 98.
16 On this topic of industrial zonation, see not only Beika's article cited in n. 15 but also his "Problems of Regional Industrialization in Japan," *ibid.*, 1 (1953), 127–38.
17 Alan D. Walton, "The Founding of the Japanese Synthetic Textile Industry," *Journal of the Textile Institute*, 51 (1960), 277–78. See also Alan D. Walton, "Japan's Synthetic Fiber Industry," *Geographical Review*, 3 (1961), 438–41.
18 Yasushi Oiwa, Carl G. Nordquist, and Oscar H. Lentz, "The Mineral Industries of Japan and Southeast Asia," *Quarterly of the Colorado School of Mines*, 56, No. 4 (1961), 34.
19 *Japanese Industry* (1961), p. 44.

20 *Japan Report,* 7 (April 15, 1961), 5.
21 Oiwa, Nordquist, and Lentz, in *Quarterly of the Colorado School of Mines, 56,* No. 4 (1961), 36.
22 *Ibid.,* p. 62.
23 *Ibid.,* pp. 183, 187.
24 Sadao Yamaguchi, "The Locational Changes of Iron and Steel Industry in Japan," *Proc. IGU Regional Conf. Japan, 1957,* p. 532.
25 Yoshiyuki Kitamura, "The Geographic Structure and Development of the Four-Wheeled Motor Vehicle Industry in Japan" (in Japanese with English summary), *Geographical Review of Japan, 34* (1961), 326–43.
26 *Japanese Industry* (1961), p. 133.
27 Japanese National Railways, *The New Tokaido Line* (1962), p. 1.
28 On Japanese rail lines and their traffic see the following: Koichi Aki *et al., The Economic Atlas of Japan* (Tokyo, 1954), Plates 45, 46, 47; Takeo Arisue, "A Regional Study on Passenger Traffic in Japan in Special Reference to a Traffic Community" (in Japanese with English summary), *Geographical Review of Japan, 30* (1957), 1016–30; and these publications of the Japanese National Railways: *Japanese National Railways: A General Description* (1962); *The New Tokaido Line* (1962); and *1962 Edition Facts and Figures.*
29 *Japanese Industry* (1961), pp. 133–34.
30 Robert B. Hall, "The Road in Old Japan," *Studies in the History of Culture* (Menasha, Wisconsin, 1942), pp. 122–55.
31 Road Bureau, Ministry of Construction, *Roads in Japan, 1963* (Tokyo, 1963), p. 26.
32 *Japanese Industry* (1961), p. 134.
33 *Ibid.* (1962), p. 153.
34 *Japan Report,* 9, No. 17 (Sept. 15, 1963), 8.
35 *Ibid.,* 8 (Aug. 15, 1962), 5.

CHAPTER 9

1 See K. Tanaka, "Geographical Units of Japan" (in Japanese), *Geographical Review of Japan, 3* (1927), 1–21.
2 For a synoptic treatment of winter precipitation in Hokkaido, see the following: W. C. Jacobs, "Wartime Development in Applied Climatology," *Meteorological Monographs, 1* (1948), 38–43; Takeshi Kawamura, "The Synoptic Climatological Considerations on the Winter Precipitation in Hokkaido" (in Japanese with English summary), *Geographical Review of Japan, 34* (1961), 583–95.
3 General Headquarters, Supreme Commander for the Allied Powers, Natural Resources Section, "Reconnaissance Soil Survey of Japan," Report No. 110–H, "Hokkaido" (Tokyo, 1951), p. 46.
4 International Geographical Union, Science Council of Japan, Regional Conference in Japan, *Regional Geography of Japan,* No. 1, *Hokkaido Guidebook* (Tokyo, 1957), p. 7. On the colonization of Hokkaido, see also Alfons Scheinpflug, "Die japanische Kolonisation in Hokkaido," *Mitteilungen der Gesellschaft für Erdkunde zu Leipzig, 53* (1935), 33–88; and F. C. Jones, *Hokkaido: Its Present State of Development and Future Prospects* (London, New York, 1958).
5 *Hokkaido Guidebook,* pp. 10–12.

6 John A. Harrison, *Japan's Northern Frontier* (Gainesville, Fla., 1953), p. 59.

7 *Ibid.,* p. 76.

8 "Die innere Kolonisation Japans," *Staats-und-Sozialwissenschaftliche Forschungen, 23,* Part 3 (1904), 37, 69.

9 See *American Influence upon the Agriculture of Hokkaido* (College of Agriculture, Imperial University, Sapporo, 1915); also *Hokkaido Guidebook,* pp. 76–77.

10 Irene B. Taeuber, *The Population of Japan* (Princeton, 1958), p. 174.

11 Hokkaido Planning Office, *Hokkaido Shorai Jinko No Suikei* [*Estimates of Future Population in Hokkaido*], 1961, pp. 8–11.

12 S. Kuwayawa, *An Outline of Hokkaido Agriculture* (Hokkaido National Agricultural Reseach Station, Sapporo, 1954), p. 5.

13 Office of Hokkaido Development Agency, *Hokkaido Kaitaku Chiku Gaikyo* [*Outline of Land Reclamation in Hokkaido*], Sapporo, 1957, p. 9.

14 Hokkaido Japan-American Rural Cooperation Society, *Agriculture of Hokkaido* (Sapporo, 1962), p. 38.

15 Jones, *Hokkaido,* p. 60.

16 *Ibid.,* pp. 67–70.

17 Hideo Fukui, "Recent Changes in the Distribution of Rice Farming in Hokkaido," *The Science Reports of the Tohoku University,* 7th Series (Geography), No. 10 (1961), p. 9.

18 *Ibid.,* pp. 10-11.

19 *Ibid.,* p. 11.

20 Hideo Fukui, "Rice Agricultural Region in Northeast Japan as Seen from the Distribution of Rice Varieties," *ibid.,* No. 11 (1962), pp. 3–7.

21 Yoshikatsu Ogasawara, "Land Use of Japan," *Bulletin of the Geographical Survey Institute* (Tokyo), 2, Part 1 (1950), 99–100.

22 Jones, *Hokkaido,* pp. 101–2; *Hokkaido Guidebook,* p. 61.

23 *Hokkaido Guidebook,* p. 64.

24 *Ibid.,* p. 66.

25 *Ibid.,* pp. 63, 67.

26 *Ibid.,* pp. 136–39.

27 Darrell Haug Davis, "Type Occupance Patterns in Hokkaido," *Annals of the Association of American Geographers,* 24 (1934), 206–9.

28 Norio Hasegawa, "Apple Culture Areas in Hokkaido, Northern Japan," *Science Reports, Tohoku Univ.,* 7th Series, No. 10 (1961), pp. 25–31.

29 *Hokkaido Guidebook,* p. 124.

30 Yoshio Okazaki, "Topographic Development of the Kushiro Moor and Its Surroundings, Hokkaido, Japan" (in Japanese with English summary), *Geographical Review of Japan,* 33 (1960), 462–73.

31 Hokkaido Prefectural Office, *Outline of Land Reclamation in Hokkaido* (1957), p. 8.

32 Fukui, in *Science Reports, Tohoku Univ.,* 7th Series, No. 10 (1961), p. 15.

33 *Hokkaido Guidebook,* pp. 49, 103.

34 *Outline of Land Reclamation in Hokkaido,* p. 8.

35 According to information obtained in an interview at Kushiro Shicho Government Office.

36 *Hokkaido Guidebook,* p. 108.

37 *Ibid.,* p. 71.

CHAPTER 10

1 International Geographical Union, Science Council of Japan, Regional Conference in Japan, *Regional Geography of Japan*, No. 2, *Tohoku (Northeastern Japan) Guidebook* (Tokyo, 1957), p. 22.

2 Norio Hasegawa, "Changing Features of Agriculture in Tohoku," *The Science Reports of the Tohoku University*, 7th Series (Geography), No. 11 (1962), pp. 13–14. See also Toshio Noh, "Agricultural Problems in Tohoku," *Papers of the Michigan Academy of Science, Arts, and Letters*, 47 (1962), 517–20; and "Patterns and Trends of Dry-Land Crops in Tohoku and Niigata," in *Hata Sakutsuke Hoshiki No Bunpu To Doko* (Nogyo Gijutsu Kyokai, 1958). Twelve folded maps, in color; scale 1:535,000.

3 Hasegawa, in *Science Reports, Tohoku Univ.*, 7th Series, No. 11 (1962), pp. 24–25.

4 Hideo Fukui, "Areal Difference and Its Yearly Change of Cold Disaster in the Rice Cultivation of Northeast Japan (1)," *ibid.*, No. 7 (1958), pp. 29–30.

5 Hideo Fukui, "Rice Agriculture of Tohoku District Seen from Yield Per Unit Area," *ibid.*, No. 9 (1960), pp. 12–13; Hideo Fukui, "Rice Agricultural Region in Northeast Japan Seen from the Distribution of Rice Varieties," *ibid.*, No. 11 (1962), pp. 7–11.

6 Fukui, "Rice Agriculture of Tohoku . . .," *ibid.*, No. 9 (1960), p. 15.

7 Ken-ichi Tanabe, "Differentiation of Livestock Keeping in the Northeastern Japan," *ibid.*, No. 3 (1954), pp. 97–100.

8 Norio Hasegawa, "Spatial Association of Manufacturing in the Aizu Basin," *ibid.*, No. 8 (1959), p. 52.

9 *Tohoku Guidebook*, p. 28.

10 International Geographical Union, Science Council of Japan, Regional Conference in Japan, *Regional Geography of Japan*, No. 1, *Hokkaido Guidebook* (Tokyo, 1957), pp. 29–30.

11 *Tohoku Guidebook*, p. 79.

12 *Ibid.*, p. 80.

13 *Ibid.*, p. 77.

14 Norio Hasegawa, "Apple Growing in Iwate Prefecture, Northern Japan," *Science Reports, Tohoku Univ.* 7th Series, No. 7 (1958), pp. 45–46.

15 Kenzo Fujiwara, "Geomorphological Development of the Kitakami Valley," *ibid.*, No. 8 (1959), pp. 8–38.

16 Hasegawa, "Apple Growing Areas . . . ," *ibid.*, No. 7 (1958), p. 48.

17 Kenzo Fujiwara, "Some Consideration of the Recent Faulting in the Western Fringe of the Fukushima Basin," *ibid.*, No. 7 (1958), pp. 2–7. A number of Japanese students of landform have concluded that not all of the piedmont fan-like surfaces in Tohoku are depositional in origin, but that some are pediment in structure and represent erosion surfaces covered only by a veneer of river deposits. See Tatsuo Wako, "Geomorphological Surfaces in the Koriyama Basin, Fukushima Prefecture," *ibid.*, No. 12 (1963), p. 71.

18 Hatsuo Yasuda, "The Physiognomy of Fukushima Basin, Fukushima Prefecture," *Journal of Geography* (Tokyo), 60 (1939), 381–87.

19 Kanokichi Saito, "The Decline of Mulberry Fields and the Expansion of Orchards in Fukushima Basin" (in Japanese with English summary), *Geographical Review of Japan*, 32 (1959), 432–42.

20 Details on the geography of the Mogami District, or Shinjo Basin, are to be found in the following ten papers, all in *The Science Reports of the Tohoku University*, 7th Series (Geography), No. 3 (1954): Ken-ichi Tanabe, "General Description of the Mogami District," pp. 1–4, and "Progression of Stock Keeping in Mogami Gun, Yamagata Prefecture," pp. 5–10; Hideo Fukui, "On the Adjustment of Rice Cultivation to the Physical Environment in the Backward District—an Example of the Shinjo Basin, Yamagata Prefecture," pp. 11–28; Hiroshi Shitara, "The Climate of the Shinjo Basin Viewed from Its Rice-Plant Seeding Season," pp. 29–40; Kiyofumi Takeuchi, "Their Floods and Their Damages on the Sake River, in Shinjo Basin, Yamagata Prefecture," pp. 41–50; Yoshiro Tomita, "Surface Geology and Correlation of River Terraces," pp. 51–58; Fujio Ouchi, "A Study of the Hydrological Geography of the Mogami Basins, (1) Obanazawa Basin," pp. 59–67; Ken-ichi Tanabe and Hideo Fukui, "Land Utilization of Some Reclaimed Lands in the Shinjo Basin, Yamagata Prefecture," pp. 68–76; Yoshio Watanabe, "The Service Pattern of the Shinjo Basin, Yamagata Prefecture—A Research in a Less Populated Basin in Japan," pp. 77–90; Reiko Fujimoto, "Shinjo City and Two Small Towns—A Study of the Inner Structure of Local Towns," pp. 91–96.

21 Yoshiro Tomita, "Structure of the Mamigasaki Fan in the Yamagata Basin, Yamagata Prefecture, with Regard to Its Facilities as a Groundwater Reservoir (A Preliminary Rept.)," *Science Reports, Tohoku Univ.*, 7th Series, No. 8 (1959), pp. 1–7.

22 Norio Hasegawa, "Apple Culture on the Midare Fan in the Yamagata Basin, Yamagata Prefecture," *ibid.*, No. 9 (1960), pp. 49–50; Kanokichi Saito, "Changes in the Distribution of Mulberry Field in Yamagata Prefecture" (in Japanese), *Geographical Review of Japan*, 30 (1957), 1030–42.

23 The Yonezawa Basin is treated in considerable detail in the *Tohoku Guidebook*, pp. 38–50; and in the following five papers, all in *The Science Reports of the Tohoku University*, 7th Series (Geography), No. 5 (1956): Kenzo Fujiwara, "Topography of the Yonezawa Basin Viewed from the Tectonic Movement in the Surrounding Hill-Lands," pp. 1–14; Yoshiro Tomita, Hydrography and Irrigation Systems in the Yonezawa Basin," pp. 15–26; Hideo Fukui, "Regional Approach on the Areal Differentiation of Agrar Land-Use Pattern in the Yonezawa Basin," pp. 27–58; Norio Hasegawa, "Grape Gardening Area in the Yonezawa Basin," pp. 59–86; Ken-ichi Tanabe, "Establishing Process of M_1 Milch-Cow-Keeping Region in Northeastern Part of Yonezawa Basin, According to Areal Differentiation of Land Utilization—Agricultural Geographic Description of Two Milch-Cow-Keeping Regions (1)," pp. 87–99. See also Ken-ichi Tanabe, "The City of Yonezawa and Its Subordinate Towns—as the Upper Structure of the Land Utilization of the Yonezawa Basin from the View Point of the Structure of a Region," *ibid.*, No. 6 (1957), pp. 1–41.

24 This subdivision of Owu has been described in some detail by Glenn T. Trewartha, "The Iwaki Basin: Reconnaissance Field Study of a Specialized Apple District in Northern Honshu, Japan," *Annals of the Association of American Geographers*, 20 (1930), 196–223.

25 *Tohoku Guidebook*, p. 90.

26 Norio Hasegawa, "Geographical Analysis of the Orchards with Decaying Aged Trees—Exemplified from the Case of the Apple Production in the Iwaki Basin," *Science Reports, Tohoku Univ.*, 7th Series, No. 4 (1955), pp. 50–55.

27 *Tohoku Guidebook,* p. 98.

28 *Ibid.,* pp. 105, 108–10.

29 *Ibid.,* pp. 100–102.

30 John D. Eyre, "Mountain Land Use in Northern Japan," *Geographical Review,* 52 (1962), 236–52. See also Ken-ichi Tanabe, "Formation Process of M_3 Milk-Cow-Keeping Region in Northeastern Part of Kitakami Mountainland," *Science Reports, Tohoku Univ.,* 7th Series, No. 5 (1956), pp. 100–128.

31 Yoko Okura, "Terrace Topography and Its Sediments at the Northeastern Foot of the Abukuma Plateau" (in Japanese with English summary), *Geographical Review of Japan, 31* (1958), 206–19.

32 Hiroshi Marui, "The Expansion of Coal Production in the Joban Coal Field" (in Japanese with English summary), *Geographical Review of Japan, 34* (1961), 22–36.

33 Hiroshi Marui, "Some Problems of the Labor Force in the Zyoban (Joban) Coal Field" (in Japanese with English summary), *Geographical Review of Japan, 33* (1960), 79–88.

CHAPTER 11

1 International Geographical Union, Science Council of Japan, Regional Conference in Japan, *Regional Geography of Japan,* No. 4, *Chubu (Central Japan) Guidebook* (Tokyo, 1957), p. 13.

2 For a more detailed study of a representative basin see Glenn T. Trewartha, "The Suwa Basin: A Specialized Sericulture District in the Japanese Alps," *Geographical Review, 20* (1930), 224–44.

3 Taizi Yazawa, "On the Tuka-dukuri (a Type of Land Utilization) on the Suwa Basin, Nagano Prefecture" (in Japanese with English summary), *Geographical Review of Japan, 14* (1938), 490–506.

4 Richard Fairchild Hough, "The Impact of the Drastic Decline in Raw Silk upon Land Use and Industry in Selected Areas of Sericultural Specialization in Japan" (unpublished Ph.D. dissertation, University of Wisconsin, 1963), pp. 162–64.

5 *Ibid.,* pp. 176–207.

6 Katsutaka Itakura, "The Industrial Change in Suwa Basin" (in Japanese with English summary), *Human Geography* (Japan), *11* (1959), 240–55, 290–91. See also John D. Eyre, "Industrial Growth in Suwa Basin, Japan," *Geographical Review, 53* (1963), 493–502.

7 *Ibid.,* pp. 495–96.

8 Tadao Yokota, "Horticulture in Kofu Basin, Yamanashi Prefecture" (in Japanese with English summary), *Geographical Review of Japan, 30* (1957), 1118–29.

9 Kanokichi Saito, "On the Relation Between Locations of Mulberry Fields and Orchards in Kofu Basin," *Human Geography* (Japan), *10* (1958), 107–19, 157–58. See also Paul Morrison, "Viticulture in the Kofu Basin, Japan," *Northwestern University Studies in Geography,* No. 6 (1962), pp. 113–34.

10 *Chubu Guidebook,* p. 111.

11 *Ibid.,* pp. 110–11.

12 Hough, "Impact of the Drastic Decline in Raw Silk . . . ," pp. 211–13.

13 Takeo Ichikawa, "The Formation of Apple-Growing Area in Nagano Basin" (in Japanese with English summary), *Geographical Review of Japan, 31* (1958), 147–52.

14 Hough, "Impact of the Drastic Decline in Raw Silk . . . ," pp. 241–44.
15 *Chubu Guidebook,* pp. 84–85.
16 *Ibid.,* p. 86.
17 *Ibid.,* pp. 87–88.
18 *Ibid.,* pp. 88–89.
19 Robert Burnett Hall, "Some Rural Settlement Forms in Japan," *Geographical Review, 21* (1931), 111–16.
20 Hatuo Yasuda, "The Landscape of the City of Takada, Niigata Prefecture" (in Japanese with English summary), *Geographical Review of Japan, 15* (1939), 509–23.
21 *Ibid.,* pp. 514, 523.
22 Hajime Kanasaki, "Migrant Workers from the Hokuriku District" (in Japanese with English summary), *Geographical Review of Japan, 35* (1962), 251–62.
23 Hatuo Yasuta, "Distribution of Hasagi in Hokuriku District" (in Japanese with English summary), *Geographical Review of Japan, 16* (1940), 657–72.
24 Haruo Inouye, "Retrogression of the Graded Sandy Coast of Niigata by Wave Erosion," *Proceedings of the IGU Regional Conference in Japan, 1957* (Tokyo, 1959), p. 145.
25 Hiroshi Sasaki, "Der Landschaftlichewandel der Agrarsiedlung in der Kanbaraebene, Japan" (in Japanese with German summary), *Geographical Review of Japan, 34* (1961), 650–62.
26 John H. Thompson and Michihiro Miyazaki, "A Map of Japan's Manufacturing," *Geographical Review, 49* (1959), 13–16.
27 Motoshige Sato, "Economic-Geographical Characters of the Three Ports, Sakata, Niigata, and Naoetsu," *Geographical Review of Japan, 31* (1958), 32–35.
28 A. Watanabe, "Consideration of the Elevated Delta-Fans of Japan" (in Japanese), *Geographical Review of Japan, 5* (1929), 1–13.
29 Saburo Fukai, "Toyama Plain and Its Geomorphological Developments" (in Japanese with English summary), *Geographical Review of Japan, 31* (1958), 416–29.
30 Isamu Matui, "Statistical Study of the Distribution of Scattered Villages in Two Regions of the Tonami Plain, Toyama Prefecture" (in Japanese), *Japanese Journal of Geology and Geography, 9* (1932), 251–66. See the valuable bibliography on p. 251.
31 *Nippon Chiri Taikei* [*Encyclopedia of Japanese Geography*] (Tokyo, 1930), *6B,* 267.
32 Thompson and Miyazaki, in *Geographical Review, 49,* 16.
33 "Ports of Japan," *Asian Affairs, 6* (1961), 105–6.

CHAPTER 12

1 On the terrain character of the Kanto Plain, see the following: Sohei Kaizuka, "The Physiographic Development of the Kanto Plain: Tephrochronological Study," *Proceedings of the IGU Regional Conference in Japan, 1957* (Tokyo, 1959), pp. 158–64; and Sohei Kaizuka, "Landform Evolution of the Kanto Plain" (in Japanese with English summary), *Geographical Review of Japan, 31* (1958), 59–85.
2 General Headquarters, Supreme Commander for the Allied Powers, Natural Re-

sources Section, "Reconnaissance Soil Survey of Japan," Report No. 110-A, "Kanto Plain Area" (Tokyo, 1951).

3 G. Imamura, N. Yazima, and Y. Tuzimoto, "Underground Waters and Rural Habitations," *Comptes Rendus du Congrès International de Géographie* (Amsterdam, 1938), *2*, Sec. A, 98–100.

4 Yoshikatsu Ogasawara, "Land Use of Japan," *Bulletin of the Geographical Survey Institute* (Tokyo), *2*, Part 1 (1950), 105.

5 Hyozo Shirahama, "On Upland Farming and Its Basic Structure of the Eastern Shimofusu Upland Region from the Viewpoints of the Agricultural Development" (in Japanese with English summary), *Geographical Review of Japan, 31* (1958), 362–78.

6 Hidezo Tanakadate, "Topographical Analysis of Density of Population in Kwanto" (in Japanese), *Journal of Geography* (Tokyo), *50* (1938), 541–55.

7 K. Okamoto, "Surface Configuration and Farm Dwellings in the Dispersed Settlements of the Kanto Plain" (in Japanese with English summary), *Geographical Review of Japan, 27* (1954), 96–107.

8 Imamura *et al.*, in *Comptes Rendus . . ., 2*, A, 300.

9 *Ibid.*, pp. 99–100.

10 Nikiti Yazima, "A Study of the Newer Settlements of the Musashino Upland" (in Japanese with English summary), *Geographical Review of Japan, 15* (1939), 807–21; T. Yamazaki, "Patterns of Settlement on Musashino Upland" (in Japanese), *Geographical Review of Japan, 9* (1933), 766–74.

11 Taiji Yazawa, "On the Distribution of Wind-Mantles in the Vicinity of Tokyo" (in Japanese with English summary), *Geographical Review of Japan, 12* (1936), 47–66, 248–68. See also Ryukiti Ito, "Types and Functions of *Yasiki-mori* (Forest in Farmyard) on the Musashino Upland, Western Suburb of Tokyo" (in Japanese with English summary), *Geographical Review of Japan, 15* (1939), 624–42, 672–85.

12 John D. Eyre, "Sources of Tokyo's Fresh Food Supply," *Geographical Review, 49* (1959), 458. See also Shohei Birukawa, "Vegetable Horticultural Regions in Relation to the Giant Urban Markets of Japan," *Tokyo Geography Papers* (Tokyo Kyoiku University), *6* (1962), 211–15.

13 On this topic see Richard Fairchild Hough, "The Impact of the Drastic Decline in Raw Silk upon Land Use and Industry in Selected Areas of Sericultural Specialization in Japan" (unpublished Ph.D. dissertation, University of Wisconsin, 1963); and K. Saito, "Changes in the Distribution of Mulberry Field in Gumma Prefecture" (in Japanese), *Geographical Review of Japan, 28* (1955), 449–59.

14 Jogyo Takeuchi, "The Recent Development of the Sagamihara Tableland," *Proc. IGU Regional Conf. Japan, 1957*, pp. 490–91.

15 Tokyo Metropolitan Government, *City Planning, Tokyo* (1962), p. 5.

16 *Ibid.* See also Tokyo Metropolitan Government, *Regional and City Planning for Tokyo—Basic Materials* (1961), p. 1–1.

17 Kiyoo Wadati and Takuzo Hirono, "Ground Subsidence in Industrial Districts of Japan and Disasters Subsequent to It," *Proc. IGU Regional Conf. Japan, 1957*, pp. 202–9.

18 International Geographical Union, Regional Conference in Japan, *Geography of Tokyo and Its Planning* (Tokyo, 1957), pp. 37–38. See also *Regional and City Planning for Tokyo*, pp. 3–2, 3–3, 3–4.

19 Yoshio Tsujimoto *et al.*, "Distribution of Industries in Tokyo" (in Japanese with English summary), *Geographical Review of Japan, 35* (1962), 491.

20 *Ibid.*, p. 503.

21 *Ibid.*

22 See *ibid.* for greater details on the distribution of various industries.

23 *City Planning, Tokyo*, pp. 48–49.

24 Takeo Arisue, "Road Transportation in the Tokyo Chiba Industrial District" (in Japanese with English summary), *Human Geography* (Japan), *14* (1962), 138.

25 *Ibid.*, p. 140.

26 Map (scale 1:50,000) published in November, 1962, supplied by the Chiba Prefecture Development Department, showing land reclamation and industrial land use.

27 "Ports of Japan," *Asian Affairs, 6* (1961), 45–46.

28 John D. Eyre, "Tokyo Influences in the Manufacturing Geography of Saitama Prefecture," *Economic Geography, 39* (1963), 283–98.

29 Yoshio Tsujimoto, "The Productive Structure of the Textile Industry in the Western Part of the Kanto District" (in Japanese with English summary), *Geographical Review of Japan, 28* (1955), 434–49.

CHAPTER 13

1 Most of this region has been treated in detail by Glenn T. Trewartha, "A Geographic Study in Shizuoka Prefecture, Japan," *Annals of the Association of American Geographers, 18* (1928), 127–259. See also Robert B. Hall, "Tokaido: Road and Region," *Geographical Review, 27* (1937), 353–77.

2 K. Kodera and H. Iwamoto, "On the Scattered Settlements (Strendorf) on the Alluvial Fan of Ooi-gawa River" (in Japanese with English summary), *Geographical Review of Japan, 15* (1939), 686–710, 760–83. See also Trewartha, in *Annals of the Association of American Geographers, 18*, 240–42.

3 *Ibid.*, pp. 251–52.

4 Shinkichi Yoshimura, Kenkichi Iwasaki, and Gohei Ito, "Microclimatic Observations on Strawberry Fields of Southern Foot of Mt. Kuno, Shizuoka Prefecture" (in Japanese), *Journal of Geography* (Tokyo), *47* (1935), 158–68; George H. Kakiuchi, "Stone Wall Strawberry Industry on Kuno Mountain, Japan," *Economic Geography, 36* (1960), 171–84.

5 Shozo Yamamoto, "On the Relation of the Tea Market in Shizuoka City to Its Market Region" (in Japanese with English summary), *Geographical Review of Japan, 26* (1953), 522–34. On the topic of Shizuoka tea, see also the following: Shozo Yamamoto, "The Tea Industry of Mountains in the Middle Shizuoka Prefecture" (in Japanese), *Tokyo Geography Papers* (Tokyo Kyoiku University), *4* (1960), 57–86; Shozo Yamamoto, "Some Observations on the Tea Industry of the Makinohara Upland in Shizuoka Prefecture" (in Japanese with English summary), *ibid., 5* (1961), 53–78.

6 Isamu Ota, "Industrialization of the Gakunan Pulp and Paper Industry District, Shizuoka Prefecture" (in Japanese with English summary), *Geographical Review of Japan, 35* (1962), 427–35.

7 Shozo Yamamoto *et al.*, "Land Use of the Shimizu Area, Central Japan—A Geographical Study on the Modernization of Region," *Tokyo Geography Papers* (Tokyo Kyoiku University), *6* (1962), 48, 50.

8 Fumio Tada and Misahiko Ohya, "The Flood-Type and the Classification of Topography," *Proceedings of the IGU Regional Conference in Japan, 1957* (Tokyo, 1959), p. 195.

9 International Geographical Union, Science Council of Japan, Regional Conference in Japan, *Regional Geography of Japan*, No. 4, *Chubu (Central Japan) Guidebook* (Tokyo, 1957), p. 31.

10 Atsuhiko Betsuki, "Waju or the Polder in Japan," *Comptes Rendus du Congrès International de Géographie* (Amsterdam, 1938), 2, Sec. E, 47–49.

11 *Chubu Guidebook*, p. 32.

12 *Ibid.*

13 Betsuki, in *Comptes Rendus . . .*, 2, E, 49.

14 *Chubu Guidebook*, p. 36. See, also Masuo Ando, "A Basin-Like Topography on the Nobi Alluvial Plain and the Sequence of Its Land Utilization" (in Japanese with English summary), *Geographical Review of Japan*, 25 (1952), 251–57.

15 *Chubu Guidebook*, pp. 43–44.

16 Kiei Ito, "On the Location of Woolen Yarn Spinning Industry in Japan Especially with Respect to the Accumulated Process in the Tokai District" (in Japanese with English summary), *Human Geography* (Japan), 12 (1960), 326–55.

17 *Chubu Guidebook*, p. 59.

18 Fusako Miura, "Pottery Industry in Nagoya City—Its Characteristics and Development" (in Japanese with English summary), *Human Geography* (Japan), 12 (1960), 31–49.

19 *Chubu Guidebook*, p. 57.

20 "Ports of Japan," *Asian Affairs*, 6 (1961), 47.

21 *Chubu Guidebook*, p. 25.

22 Terumichi Osako, "The Relation Between the Decrease of the Area of Mulberry Fields and the Agricultural Structure of the Regions in Northern Mie Province" (in Japanese with English summary), *Geographical Review of Japan*, 34 (1961), 68–82.

23 *Chubu Guidebook*, pp. 67–68.

24 *Ibid.*, p. 79.

CHAPTER 14

1 Akira Watanabe, "Landform Divisions of Japan," *Bulletin of the Geographical Survey Institute* (Tokyo), 2, Part 1 (1950), 91–92.

2 Tokuji Chiba, "On the Denuded Hillsides Surrounding the Kinki Basins" (in Japanese with English summary), *Human Geography* (Japan), 6 (1954–1955), 433–41.

3 This subdivision has been studied in detail by Robert B. Hall. See his published study, "The Yamato Basin, Japan," *Annals of the Association of American Geographers*, 22 (1932), 243–91.

4 International Geographical Union, Science Council of Japan, Regional Conference in Japan, *Regional Geography of Japan*, No. 3, *Kinki Guidebook* (Tokyo, 1957), pp. 13–14.

5 Hideharu Umezaki, "The Function and Distribution of Ukezutsumi or Protec-

tion Banks in the Yamato Basin" (in Japanese with English summary), *Geographical Review of Japan, 31* (1958), 590–601.

6 *Kinki Guidebook,* pp. 22–23.

7 Kyoto City Tourist Bureau, *Kankokyoto No Ugoki—Nyuraku Kankokyakusu To Kankoshunyu [Number of Visitors and the Income from Them],* in Japanese (Kyoto, 1960).

8 Bureau of Statisics, Office of the Prime Minister, *1960 Establishment Census of Japan* (1961).

9 *Kinki Guidebook,* pp. 42–43.

10 Takeo Tanioka, *Kyoto: Sono Chiritanbo [Geographic Survey of Kyoto],* in Japanese (Tokyo, 1961), p. 108.

11 *Guide-Book, Excursion D: Kyoto, Nara, Osaka, Kobe,* Pan-Pacific Science Congress (Tokyo, 1926); see especially the geological map of Kinki. On some geological maps the older diluvium is designated as Tertiary.

12 Yoshitaka Horiuchi, "A Geographical Study of Reverse Stream Irrigation on the Plain to the East of Lake Biwa" (in Japanese with English summary), *Geographical Review of Japan, 32* (1959), 70–82; Takeo Oda *et al.,* "A Regional Survey of the Southern Part of the Plain to the East of the Lake Biwa" (in Japanese with English summary), *Geographical Review of Japan, 26* (1953), 224–81.

13 Secretariat: The Conference of Hanshin Metropolitan Region and the Conference of Osaka Conurbated Area, *Synopsis of Various Planning Projects for Hanshin-Kinki Region* (December, 1961), p. 28.

14 Jogyo Takeuchi and Yoshitaka Horiuchi, "On the Irrigation Ponds of the Osaka Plain, Especially on the Kashii River Valley" (in Japanese with English summary), *Geographical Review of Japan, 32* (1959), 567–79.

15 "Basic Materials for Comprehensive Development Plan of the Hanshin Metropolitan Region," Report by Japanese Team, Part 1 (August, 1960), p. 19 (mimeographed).

16 *Ibid.,* pp. 118–20.

17 *Ibid.,* p. 119.

18 "Planning and Action Program for the Development of the Hanshin Metropolitan Region in Japan," Report by a Joint Japan–United Nations Team (Tokyo, June, 1962), Sec. 1, p. 2 (mimeographed).

19 "Basic Materials for Comprehensive Development Plan . . .," pp. 121–24.

20 Joint Japan–United Nations Team on Hanshin Metropolitan Region, "Reports of Committees of Special Areas of Work," Supplement, Review Report No. 1, "Report of Committee 1" (Sept. 10, 1960), pp. 11–14.

21 "Planning and Action Program for the Development of Hanshin . . .," Sec. 1, pp. 5–6.

22 *Ibid.,* pp. 7–8.

23 "Basic Materials for Comprehensive Development Plan . . .," Part 2 (August, 1960), p. 65. This same reference, pp. 58–89, contains an account, with detailed maps, of the several industrial zones and areas in Osaka. See also Institute of Research in Commerce and Industry of Osaka Prefecture, *Distribution Maps of Various Industries in Kinki and Keihanshin (Kinkichiho Oyobi Keihanshinwa Kogyochitai Kogyo Bunpuzu),* Osaka, 1958.

24 "Ports of Japan," *Asian Affairs, 6* (1961), 48.

25 "Basic Materials for Comprehensive Development Plan . . .," Part 2, 58–63.
26 "Outlines of Kobe Port's Plans for Creation of Seaside Zones," Reclamation Bureau, Kobe City (mimeographed).

CHAPTER 15

1 As general references, see the following: International Geographical Union, Science Council of Japan, Regional Conference in Japan, *Regional Geography of Japan*, Nos. 5 and 6, *Inland Sea and Kyushu Guidebook* and *Northern Shikoku Guidebook* (Tokyo, 1957); H. Schmitthenner, *Die Japanische Inlandsee* (Breslau, 1921); I. Watanuki, *Geography of Setouchi with One Hundred Illustrations* (in Japanese), Tokyo, 1932; Keiji Tanaka, "Some Geographical Notes on the Excursion to the Inland Sea (Setouchi) Region, Including Miyajima" (unpublished manuscript); N. Yamasaki, "Morphologische Betrachtung des japanischen Binnenmeers Setouchi," *Petermanns Geographische Mitteilungen*, 48 (1902), 245–53.

2 Singo Tanaka, "The Drainage-Density and Rocks (Granitic and Paleozoic) in the Setouchi Sea Coast Region, Western Japan" (in Japanese with English summary), *Geographical Review of Japan*, 30 (1957), 564–78; Singo Tanaka, "On the Relation Between the Drainage-Density and Lithology Along the Sea-Coast in Setouchi, Western Japan" (abstract), *Proceedings of the IGU Regional Conference in Japan, 1957* (Tokyo, 1959), p. 197.

3 T. Chiba, "The Bare Hills of the Okayama District" (in Japanese with English summary), a series of six articles in the *Geographical Review of Japan*, 27 (1954), 48–58, 158–66, 255–62, 374–84; 28 (1955), 121–32, 182–92.

4 Hiroshi Shitara, "On the Winter Weather Divide in the Chugoku Region" (in Japanese with English summary), *Geographical Review of Japan*, 31 (1958), 655–65.

5 Gihachi Tomioka, "Some Characteristic Features of 'Kihansen' (Steam-and-Sail-Boat) Traffic in the Seto Inland Sea" (in Japanese with English summary), *Geographical Review of Japan*, 33 (1960), 363–78.

6 *Northern Shikoku Guidebook*, pp. 3–4, 27–33.

7 Hikoichiro Sasaki, "The Cultural Landscape of Islands of the Inland Sea" (in Japanese with English summary), *Geographical Review of Japan*, 8 (1932), 38–47.

8 I. Watanuki, "Villages in Setouchi" (in Japanese), *Geographical Review of Japan*, 7 (1931), 1053–74.

9 Curtis A. Manchester, "Igusa: a Critical Cash Crop in the Rural Economy of Okayama Prefecture," *Economic Geography*, 34 (1958), 47–63.

10 *Guidebook of Excursion*, Regional Conference on Mineral Resources Development, Economic Commission for Asia and the Far East (Tokyo, April 20–May 10, 1953), pp. 1–5.

11 For a detailed account of reclamation and land use on the Okayama Plain, see John D. Eyre, "Japanese Land Development in Kojima Bay," *Economic Geography*, 32 (1956), 58–74; and *Inland Sea and Kyushu Guidebook*, pp. 19–31.

12 Tsutomu Fujimori, "Shipbuilding Industry and Its Surrounding Area" (in

Japanese with English summary), *Human Geography* (Japan), *12* (1960), 302–75, 378–79.

13 H. Nozawa, "The Development of the City of Hiroshima," *Geographical Bulletin* [*Chiri Ronso*] (Kyoto Imperial University), No. 5 (1934), pp. 47–93.

14 Hiroshi Morikawa and Kenji Kitagawa, "Hiroshima—Wandlungen der Inneren Struktur und Region," *Erdkunde, 17* (1963), 103.

15 *Guidebook of Excursion,* Regional Conference on Mineral Resources Development, Economic Commission for Asia and the Far East (Tokyo, April 20–May 10, 1953), pp. 17–28.

16 Kasuke Nishimura, "Chugoku Mountains as a Staircase Morphology," *The Science Reports of the Tohoku University,* 7th Series (Geography), No. 12 (1963), pp. 1–19.

17 John D. Eyre, "Cattle Raising in Japan," *Geographical Review, 52* (1962), 299–300.

18 Robert Burnett Hall, "The Hiinokawa Plain," *Proceedings of the Fifth Pacific Science Congress* (Toronto, 1933), *2,* 1360–61.

19 *Ibid.,* p. 1370.

20 *Northern Shikoku Guidebook,* p. 17.

21 *Ibid.,* p. 48.

22 John D. Eyre, "Market Gardening in Kochi Prefecture, Japan," *Papers of the Michigan Academy of Science, Arts, and Letters, 47* (1962), 485–505.

23 Masaki Yamaoka, "Social Outcasts Villages along the Shikoku Pacific Coast," *Proc. IGU Regional Conf. Japan, 1957,* pp. 537–43.

24 Masatane Soma, "The Cultivation of *Mitsumata* on Shifting Fields in Shikoku," *ibid.,* pp. 470–77.

25 Tokuji Chiba, "A Geographical Analysis of Terrace-Cultivated Landscape on the Coast of the Bungo Channel" (in Japanese with English summary), *Geographical Review of Japan, 33* (1960), 447–62.

26 *Guidebook of Excursion,* Regional Conference on Mineral Resources Development, Economic Commission for Asia and the Far East (Tokyo, April 20–May 10, 1953), pp. 7–16.

CHAPTER 16

1 International Geological Congress, *The Coal Resources of the World* (Toronto, 1913). For a map of the field see Map 16 in the accompanying atlas.

2 Port and Harbour Bureau, Ministry of Transportation, *Port and Harbour Construction in Japan, 1963,* pp. 95–101.

3 *Ibid.,* pp. 84–94.

4 Makoto Murakami, "Development and Accumulation of the Industrial Regions in Kyushu" (in Japanese with English summary), *Geographical Review of Japan, 35* (1962), 310–26.

5 Matsuo Higaki, "Industrial Location in the Heavy Industrial Cities in North Kyushu" (in Japanese), *Geographical Review of Japan, 29* (1956), 143–56; Matsuo Higaki, "The Forms of Sphere and the Types of Regional Conditions in the Industrial Zone in North Kyushu" (in Japanese with English summary), *Geographical Review of Japan, 31* (1958), 702–17; John H. Thompson, "Manu-

facturing in the Kita Kyushu Industrial Zone of Japan," *Annals of the Association of American Geographers, 49* (1959), 420–42.

6 *Port and Harbour Construction in Japan, 1963,* p. 79.

7 International Geographical Union, Science Council of Japan, Regional Conference in Japan, *Regional Geography of Japan,* No. 5, *Inland Sea and Kyushu Guidebook* (Tokyo, 1957), pp. 60–61.

8 Shoichi Chishaki, "Rice Cropping Season in Kyushu" (in Japanese with English summary), *Geographical Review of Japan, 27* (1954), 366–73.

9 Reiji Okazaki, "Ariake Bay Reclamation Works and Its Progress" (abstract), *Proceedings of the IGU Regional Conference in Japan, 1957* (Tokyo, 1959), pp. 446–47.

10 *Inland Sea and Kyushu Guidebook,* p. 77.

11 *Guide-Book, Excursion E-3: The Miike Coal Field,* Pan-Pacific Science Congress (Tokyo, 1926); *Guidebook of Excursion,* Regional Conference on Mineral Resources Development, Economic Commission for Asia and the Far East (Tokyo, April 20–May 10, 1953), pp. 29–38.

12 *Inland Sea and Kyushu Guidebook,* pp. 75–76.

13 *Ibid.,* pp. 93–102. See also *Guide-Book, Excursion E-1: Beppu, The Hot-Spring City,* Pan-Pacific Science Congress (Tokyo, 1926), p. 5.

14 *Guide-Book, Excursion E-4: Aso Volcano,* Pan-Pacific Science Congress (Tokyo, 1926).

15 For a detailed description of the Aso district see *Inland Sea and Kyushu Guidebook,* pp. 79–92.

16 *Guide-Book, Excursion E-1: Unzen Volcanoes,* Pan-Pacific Science Congress (Tokyo, 1926), pp. 3, 4.

17 Kan-ichi Kawaji, "A Study of Regional Structure and Forming Process of the Manufacturing Industries on Seaside Industrial City of Nagasaki" (in Japanese with English summary), *Human Geography* (Japan), *13* (1961), 16–33.

18 Robert Burnett Hall, "Some Rural Settlement Forms in Japan," *Geographical Review, 21* (1931), 102–10.

19 Hiromu Futagami, "Economic Structure of Gokanosho Village in the Kyushu Mountains," *Geographical Review of Japan, 31* (1958), 152–60.

SELECTED REFERENCES

CHAPTER 1: LAND-SURFACE FORM

Hall, Robert B., and Akira Watanabe. "Landforms of Japan." *Papers of the Michigan Academy of Science, Arts, and Letters, 18* (1932), 157–207.

Iwatsuka, Shuko. "On Landslides and Related Phenomena in Mountainous Areas of Japan." *Proceedings of the IGU Regional Conference in Japan, 1957* (Tokyo, 1959), pp. 154–57.

Mogi, Akio. "On the Shore Types of the Coasts of Japanese Islands" (in Japanese with English summary). *Geographical Review of Japan, 36* (1963), 245–66.

"Population Density by Landform Division in Japan." *Bulletin of the Geographical Survey Institute* (Tokyo), 6, Parts 2–3 (1960), 157–66.

Sato, Hisashi. "Distribution of Volcanoes in Japan." *Proceedings of the IGU Regional Conference in Japan, 1957* (Tokyo, 1959), pp. 184–88.

Shimomura, H. "On the Physiographic Provinces of Japan" (in Japanese). *Geographical Review of Japan, 2* (1926), No. 12, 1–14; 3 (1927), No. 4, 46–65.

Tokuda, S. "On the Echelon Structure of the Japanese Archipelagoes." *Japanese Journal of Geology and Geography, 5* (1926–1927), 41–76.

Watanabe, Akira. "Geomorphology in Relation to Recent Crustal Movements in Japan." *Proceedings of the IGU Regional Conference in Japan, 1957* (Tokyo, 1959), pp. 215–21.

———."Landform Divisions of Japan." *Bulletin of the Geographical Survey Institute* (Tokyo), 2, Part 1 (1950), 81–94.

———. "Major Geomorphological Divisions of the Coast of Japan" (in Japanese with English summary). *Journal of Geography, 66*, No. 1 (1957), 1–16.

CHAPTER 2: CLIMATE

Arakawa, H. "The Air Masses of Japan" (in Japanese with English summary).

Journal of the Meteorological Society of Japan, 13 (1935), 385–402; *14* (1936), 328–38; *15* (1937), 185–89.

———. "Die Luftmassen in den japanischen Gebieten." *Meteorologische Zeitschrift, 54* (1937), 169–74.

———, and J. Tawara. "Frequency of Air Mass Types in Japan." *Bulletin of the American Meteorological Society, 30* (1949), 104–5.

Asakura, T. "A Synoptic Study on the Seasonal Change of the General Circulation from Summer to 'Shurin' in 1951." *Journal of the Meteorological Society of Japan,* Series 2, *35* (1957), 278–87.

Flohn, H., and H. Oeckel. "Water Vapour Flux during the Summer Rains over Japan and Korea." *Geophysical Magazine, 27* (1956), 527–32.

Fukui, Eiichiro. "Climate of Japan," in *Regional Geography of Japan,* No. 7, *Geography of Japan* (International Geographical Union, Science Council of Japan, Regional Conference in Japan, Tokyo, 1957), pp. 12–20.

Industrial Meteorological Association. *Climatological Atlas of Japan.* Tokyo, 1948.

Kawamura, Takeshi. "The Synoptic Climatology of Winter Monsoon in Japan" (in Japanese with English summary). *Geographical Review of Japan, 37* (1964), 64–78.

Landsberg, H. *A Climatic Study of Cloudiness over Japan.* Institute of Meteorology, *Miscellaneous Reports,* No. 15. Chicago, 1944.

Miller, James E., and Homer T. Mantis. "Extratropical Cyclogenesis in the Pacific Coastal Regions of Asia." *Journal of Meteorology, 4* (1947), 29–34.

Murakami, T. "The General Circulation and Water-Vapor Balance over the Far East during the Rainy Season." *Geophysical Magazine* (Tokyo), *29* (1959), 131–71.

———, Y. Arai, and K. Tomatsu. "On the Rainy Season in the Early Autumn" (in Japanese with English summary). *Journal of the Meteorological Society of Japan,* Series 2, *40* (1962), 330–49.

Saito, Ren-ichi. "The Climate of Japan and Her Meteorological Disasters." *Proceedings of the IGU Regional Conference in Japan, 1957* (Tokyo, 1959), pp. 173–83.

Sekiguchi, Takeshi. "On the Year Climate in Japan" (in Japanese with English summary). *Geographical Review of Japan, 24* (1951), 175–85.

Snow Association of Japan. *Climatography of Snow in Japan.* Tokyo, 1949.

Suda, K., and T. Asakura. "A Study of the Unusual 'Baiu' Season in 1954 by means of Northern Hemisphere Upper Air Mean Charts." *Journal of the Meteorological Society of Japan, 33* (1955), 233–44.

Suda, Ken. "A Study of the Dynamic Climatology of Cold Outbreaks in the Far East." *Geophysical Magazine* (Tokyo), *29* (1959), 413–62.

Suzuki, H. "Klassifikation der Klimate von Japan in der Gegenwart und der Letzten Eiszeit." *Journal of Geology and Geography, 33* (1962), 221–34.

Takahasi, K. "On the Transformation of the Cold Dry Air Mass by Traveling ôver Warm Sea" (in Japanese with English summary). *Journal of the Meteorological Society of Japan, 18* (1940), 77–80.

Takahasi, Koichiro. *Dynamic Climatology: Especially the Weather of Japan* [*Dokikogaku*]. 2d ed. Tokyo, 1957.

Thornthwaite, C. W. "The Climates of Japan." *Geographical Review, 24* (1934), 494–96.

Trewartha, Glenn. *The Earth's Problem Climates.* Madison, Wis., 1961. See numerous references on Japan, pp. 321–22.

Wadachi, Kiyoo, ed. *Climate of Japan* [*Nihon no Kiko*]. Tokyo, 1958.
Yazawa, Taiji. "Der Jahreszeitliche Ablauf der Witterung in Japan." *Geographische Rundschau,* 9 (1957), 407–11.

CHAPTER 3: VEGETATION

Ackerman, Edward A. *Japan's Natural Resources and Their Relation to Japan's Economic Future.* Chicago, 1953. See especially pp. 250–73.
Cummings, Laurence J. "Forestry in Japan, 1945–1951." General Headquarters, Supreme Commander for the Allied Powers, Natural Resources Section, Report No. 153. Tokyo, 1951. (Mimeographed.)
———. "Japan." In Stephen Haden-Guest, John K. Wright, and Eileen M. Teclaff, eds., *A World Geography of Forest Resources.* New York, 1956, pp. 551–60.
———, Donald J. Haibach, and Harold F. Wise. "Forest Area, Volume, and Growth in Japan, Statistical Summary." General Headquarters, Supreme Commander for the Allied Powers, Natural Resources Section, Preliminary Study No. 37. Tokyo, 1950. (Mimeographed.)
Forestry Extension Associates of Japan, Forestry Agency. *Forestry in Japan, 1960.* Tokyo, 1960.
General Headquarters, Supreme Commander for the Allied Powers, Natural Resources Section. "Important Trees of Japan." Report No. 119. Tokyo, 1949. (Mimeographed.)
Kraebel, C. J. "Forestry and Flood Control in Japan." General Headquarters, Supreme Commander for the Allied Powers, Natural Resources Section, Preliminary Study No. 39. Tokyo, 1950. (Mimeographed.)

CHAPTER 3: SOIL

General Headquarters, Supreme Commander for the Allied Powers, Natural Resources Section. "Reconnaissance Soil Survey of Japan." Report No. 110. Tokyo, 1948–1951. (Mimeographed.) Eight regional reports and summary: "Kanto Plain Area," No. 110-A (1948). "Kyushu," No. 110-B (1950). "Kyoto Area," No. 110-C (1950). "Shikoku," No. 110-D (1950). "Hiroshima Area," No. 110-E (1950). "Nagoya Area," No. 110-F (1951). "Northern Honshu Area," No. 110-G (1951). "Hokkaido," No. 110-H (1951). "Summary," No. 110-I (1951).

Kamoshita, Yutaka. *Soils in Japan. With General Map of Soil Types in Japan, 1:800,000.* Miscellaneous Publication B, No. 5. National Institute of Agricultural Sciences. Tokyo, 1958.
Swanson, C. L. W. "Reconnanissance Soil Survey of Japan." Soil Science Society of America, *Proceedings, 1946, 11,* 493–507.

CHAPTER 3: WATER

Ackerman, Edward A. *Japan's Natural Resources and Their Relation to Japan's Economic Future.* Chicago, 1953. See especially pp. 41–47.
Bell, Francis, and Arvi Waananen. "Hydrology of Japan." General Headquarters,

Supreme Commander for the Allied Powers, Natural Resources Section, Report No. 43. Tokyo, 1946. (Mimeographed.)

Ishida, Ryuziro, ed. *Geography of Japan*. International Geographical Union, Science Council of Japan, Regional Conference in Japan, 1957. *Regional Geography of Japan*, No. 7. Tokyo, 1957. See pp. 20–23, 73–75.

Japanese Geological Survey. *Geology and Mineral Resources of Japan*. 2d ed. Tokyo, 1960. See pp. 129–35.

CHAPTER 4: MINERAL AND ENERGY RESOURCES

General Headquarters, Supreme Commander for the Allied Powers, Natural Resources Section. "Oil Fields of Hokkaido." Report No. 18. 1946.

———. "Dolomite Resources in Japan." Report No. 53. 1946.

———. "Nickel Deposits in Japan." Report No. 57. 1946.

———. "Zinc and Lead Resources of Japan." Report No. 65. 1947.

———. "Sulphur Resources of Japan." Report No. 66. 1947.

———. "Iron Ore Resources of Japan." Report No. 69. 1947.

———. "Pyrite Resources of Japan." Report No. 70. 1947.

———. "Petroleum Resources and Production in Japan." Report No. 80. 1947.

———. "Iron Sand Resources of Japan." Report No. 98. 1947.

———. "Ishikari Coal Field, Hokkaido." Report No. 99. 1947.

———. "The Coal Fields of Kyushu." Report No. 103. 1948.

———. "Copper in Japan." Report No. 106. 1948.

———. "Lignite in Japan." Report No. 112. 1948.

———. "Coal Fields of Eastern Honshu, Japan." Report No. 120. 1949.

———. "Gold and Silver in Japan." Report No. 128. 1950.

———. "Coal Fields of Western Honshu, Japan." Report No. 132. 1950.

———. "Coal Fields of Hokkaido, Japan." Report No. 133. 1950.

———. "Japanese Mineral Resources." Report No. 141. 1951.

Japanese Geological Survey. *Geology and Mineral Resources of Japan*. 2d ed. Tokyo, 1960.

———. *Guidebook of Excursion*. Regional Conference on Mineral Resources Development, Economic Commission for Asia and the Far East. Tokyo. April 20–May 10, 1953.

Oiwa, Yasushi, Carl G. Nordquist, and Oscar H. Lentz. "The Mineral Industries of Japan and Southeast Asia." *Quarterly of the Colorado School of Mines*, 56, No. 4 (1961).

CHAPTER 5: POPULATION

Bureau of Statistics, Office of the Prime Minister. *1960 Population Census, Densely Inhabited District* (Preliminary report, August, 1961).

———. *1960 Population Census, Densely Inhabited District: Its Population, Area, and Map*. Tokyo, December, 1961.

Hidaka, Tatsutaro. "Population Density of Japan." *Bulletin of the Geographical Survey Institute* (Tokyo), 5, Parts 1–2 (1957), 31–64 (colored map, scale 1:4,000,000).

Honda, Tatsuo. "Population Problem of Japan in the 1960's" (in Japanese with

English summary). *Annual Reports of the Institute of Population Problems,* No. 7 (1962), pp. 1–4.

Isida, Ryuziro. *Geography of Japan.* Tokyo, 1961. See especially pp. 37–48.

Kawabe, Hiroshi. "The Internal Migration of Japan: 1950–1955" (in Japanese with English summary). *Geographical Review of Japan, 34* (1961), 96–108.

———. "Migration to Cities in Japan: 1950–1955" (in Japanese with English summary). *Journal of Geography* (Tokyo), *70* (1961), 166–80.

Kawai, Reiko. "Population Density of Japan by Land Form Division." *Erdkunde, 15,* No. 3 (1961), 226–32.

Kidder, J. E. *Japan Before Buddhism.* London, 1959. See especially pp. 19–91.

Kishimoto, Minoru. "Studies on Population Geography of Japan." *Journal of Gakugei* (Tokushima University), *11* (1961), 19–38.

Kobayashi, Hiroyoshi. "The Population Changes in Prominent Sericultural Areas in Japan" (in Japanese with English summary). *Geographical Review of Japan, 31* (1958), 131–41.

Kornhauser, David H. "Urbanization and Population Pressure in Japan." *Pacific Affairs, 31* (1958), 275–85.

Kuroda, Toshio, *et al. Nippon no jinko ido* [*Internal Migration in Japan*]. Tokyo, 1961. (In Japanese.)

Miyakawa, Minoru. "Changes in the Distribution of Births by Prefectures" (in Japanese with English summary). *Annual Reports of the Institute of Population Problems,* No. 6 (1961), pp. 30–33.

Momiyama, Masako (Sakamoto). "Geographical Variation of Death Rate in Postwar Japan." *Proceedings of the IGU Regional Conference in Japan, 1957* (Tokyo, 1959), pp. 405–12.

Okazaki, Yoichi. "Regional Migration of Male Labor Force: Measurement and Some Observations" (in Japanese with English summary). *Annual Reports of the Institute of Population Problems,* No. 7 (1962), pp. 20–24.

"Population Density by Landform Division in Japan." *Bulletin of the Geographical Survey Institute* (Tokyo), 6, Parts 2–3 (1960), 155–66. See also "Population Density by Landform Division" (Part 3), 1955 Population Census of Japan, Population Maps (3 sheets; scale 1:800,000). Published by Bureau of Statistics, Office of the Prime Minister, March, 1958.

Population Maps of Japan, 1960 Population Census. Part 1: "Population Distribution by *Shi, Ku, Machi* and *Mura.*" Part 2: "Population Change by *Shi, Ku, Machi* and *Mura.*" Part 3: "Population Density and Densely Inhabited Districts by *Shi, Ku, Machi* and *Mura.*" Scale 1:1,000,000; 3 sheets per part. Published by Bureau of Statistics, Office of the Prime Minister, 1963.

Population Problems Research Council. *Facts about Japan's Population.* Population Problem Series No. 17. The Mainichi Newspapers, Tokyo, 1960.

Tachi, Minoru. "Regional Income Disparity and Internal Migration of Population in Japan." Committee for Translation of Japanese Economic Studies, No. 21. The International House of Japan. (Mimeographed, no date.)

———, and Masao Ueda. "A Statistical Study on the Variation of Basic Demographic Phenomena by the Size of Communities." *Archives of the Population Association of Japan,* No. 1 (1952), pp. 94–112.

———, Masao Ueda, and Hidehiko Hama. "Regional Characteristics of Population in Japan." *Proceedings of the IGU Regional Conference in Japan, 1957* (Tokyo, 1959), pp. 480–84.

————, and Misako Oyama. "Regional Distribution of Income and Population" (in Japanese with English summary). *Journal of Population Problems,* No. 82 (1961), pp. 1–17.

————. "Potential of Internal Migration in Japan—from the Viewpoint of the Interrelationship between Regional Distribution of Income and That of Population" (in Japanese with English summary). *Annual Reports of the Institute of Population Problems,* No. 5 (1960), pp. 38–42.

Taeuber, Irene B. "Japan's Population: Miracle, Model, or Case Study." *Foreign Affairs, 40* (1962), 595–604.

————. *The Population of Japan.* Princeton, 1958.

Tsubouchi, Syoji. "Population Pressure and Rural-Urban Migration in Japan." *Proceedings of the IGU Regional Conference in Japan, 1957* (Tokyo, 1959), pp. 512–16.

Ueda, Masao. "Differential Net Migration by Age and Sex in Prefectures, 1920–1935 and 1950–1955" (in Japanese with English summary). *Annual Reports of the Institute of Population Problems,* No. 6 (1961), pp. 24–29.

Yamaguchi, Kiichi. "A Comparison of Standardized Birth, Death and Natural Increase Rates of Respective Prefectures in Japan: 1955–1960" (in Japanese with English summary). *Annual Reports of the Institute of Population Problems,* No. 7 (1962), pp. 41–44.

CHAPTER 6: SETTLEMENTS

Asai, Tokuichi. "On the Reorganization of *gun* and *shi,* or Counties and Cities in Japan." *Proceedings of the IGU Regional Conference in Japan, 1957* (Tokyo, 1959), pp. 271–72.

Asaka, Sachio. "Development of Towns and Villages in the Edo Period." *Proceedings of the IGU Regional Conference in Japan, 1957* (Tokyo, 1959), pp. 273–75.

Asaka, Yukio. "Shukuba-machi." *Distribution Maps on the Regional Geographical Study of Japan,* comp. Taro Tsujimura (Tokyo, 1952), pp. 21–22.

Beardsley, Richard King, John W. Hall, and Robert E. Ward. *Village Japan.* Chicago, 1959.

Embree, John H. *Suye Mura: A Japanese Village.* Chicago, 1939.

Fujioka, Kenjiro. "An Explanation of Japanese Castle Towns." *Distribution Maps on the Regional Geographical Study of Japan,* comp. Taro Tsujimura (Tokyo, 1952), pp. 18–19.

————. "Feudal Traditions in the Forms and Zone Structures in Japanese Cities." *Proceedings of the IGU Regional Conference in Japan, 1957* (Tokyo, 1959), pp. 317–19.

General Headquarters, Supreme Commander for the Allied Powers, Natural Resources Section. "The Japanese Village in Transition." Report No. 136. Tokyo, 1950. (Mimeographed.)

Harasawa, Bunya. "A Study in the Relation Between Post Towns and the By-Way Traffic in the Edo Period" (in Japanese with English summary). *Geographical Review of Japan, 31* (1958), 277–91.

Inami, Etsuji. "Correlation Between War Damage and the Rate of Population Growth in Japanese Cities Before and After World War II" (in Japanese with English summary). *Geographical Review of Japan, 26* (1953), 495–504.

Isida, Ryuziro. *Geography of Japan.* Tokyo, 1961. See especially pp. 49–62.

Kikuchi, Toshio. "Geographical Function of the Reclamation Settlements During the Period of the Shogunate, 1603–1867." *Proceedings of the IGU Regional Conference in Japan, 1957* (Tokyo, 1959), pp. 362–67.

Kiuchi, Shinzo. "Centrifugal and Centripetal Urbanization in Japan." *Proceedings of the IGU Regional Conference in Japan, 1957* (Tokyo, 1959), pp. 367–71.

Schöller, Peter. "Wachstum und Wandlung japanischer Stadtregionen." *Die Erde* (formerly *Zeitschrift der Gesellschaft für Erdkunde zu Berlin*), 93 (1962), 202–34.

Tanabe, Ken-ichi. "Development of the Areal Structure of Japanese Cities in the Case of Castle Towns—As a Geographic Contribution to the Study of Urban Geography." *Science Reports of the Tohoku University*, 7th Series (Geography), No. 8 (1959), pp. 88–105.

Tanioka, Takeo. "Différenciation Régional des Types de l'Habitat Rural au Japon." *Proceedings of the IGU Regional Conference in Japan, 1957* (Tokyo, 1959), pp. 503–12.

Watanabe, Yoshio. "An Analysis of the Function of Urban Settlements Based on Statistical Data—A Functional Differentiation Vertical and Lateral." *Science Reports of the Tohoku University*, 7th Series (Geography), No. 10 (1961), pp. 63–94.

Yamaguchi, Keiichiro. "Regional Differences in the Process of Urbanization in Japan" (abstract). *Proceedings of the IGU Regional Conference in Japan, 1957* (Tokyo, 1959), pp. 528–29.

CHAPTER 7: AGRICULTURE

Ackerman, Edward A. *Japan's Natural Resources and Their Relation to Japan's Economic Future.* Chicago, 1953. See especially pp. 55–108, 153–61, 224–49, 359–64, 376–443.

Amatatsu, Katsumi. *Growing Rice in Japan.* Agricultural Development Series, No. 11. Agriculture, Forestry, and Fisheries Productivity Conference. Tokyo, 1959.

Dore, R. P. *Land Reform in Japan.* New York, 1959.

Eyre, John D. "Japanese Inter-Prefectural Rice Movements." *Economic Geography,* 38 (1962), 78–86.

———. "Water Controls in a Japanese Irrigation System." *Geographical Review,* 45 (1955), 197–216.

Fielding, Gordon J. "Dairy Industry of Japan" (33 pages, mimeographed). University of Auckland, New Zealand, 1962.

Geographical Survey Institute, Staff Members, Map Division. *Land Use in Japan.* 6th ed. Tokyo, 1961.

Hall, Robert B. "Revolution in Asian Agriculture." *Journal of Geography* (Tokyo), 69 (1960), 97–105.

Hall, Robert B., and Toshio Noh. "Yakihata, Burned Field Agriculture in Japan, with Its Special Characteristics in Shikoku." *Papers of the Michigan Academy of Science, Arts, and Letters,* 38 (1952), 315–22.

Hall, Robert B., Jr. "Hand Tractors in Japanese Paddy Fields." *Economic Geography,* 34 (1958), 312–20.

Japan FAO Association. *Agriculture at the Crossroads: What Are Japanese Farmers Thinking of Tomorrow?* Tokyo, 1961.

———. *Agriculture in Japan.* Tokyo, 1958.

————. *A Century of Technical Development in Japanese Agriculture.* Tokyo, 1959.

————. *A Strategy for New Agriculture: A Supplement to Agriculture at the Cross-roads.* Tokyo, 1962.

Kajita, Masaru. *Land Reform in Japan.* Agricultural Development Series, No. 2. Agriculture, Forestry, and Fisheries Productivity Conference. Tokyo, 1959.

Kakiuchi, George H. "Toji: A Study of a Semiagricultural Community Exploiting Marine Resources." *Papers of the Michigan Academy of Science, Arts, and Letters,* 47 (1961), 507–15.

————, and Setsutaro Murakami. "Satsuma Oranges in Ocho-mura: A Study of Specialized Cash Cropping in Southwestern Japan." *Geographical Review,* 51 (1961), 500–518.

Mathiesen, R. S. "The Japanese Salmon Fisheries." *Economic Geography,* 34 (1958), 352–61.

Matsuo, Takane. *Rice and Rice Cultivation in Japan.* Institute of Asian Economic Affairs. Tokyo, 1961.

————. *Rice Culture in Japan.* Ministry of Agriculture and Forestry. Tokyo, 1954.

Mukumoto, Tsutomu. *Agricultural Machinery and Implements in Japan.* Agricultural Development Series, No. 10. Agriculture, Forestry, and Fisheries Productivity Conference. Tokyo, 1959.

Murakami, Setsutaro. "The Formation of Citrus Growing Areas in Japan." *Proceedings of the IGU Regional Conference in Japan, 1957* (Tokyo, 1959), pp. 420–28.

Nasu, Shiroshi. *Aspects of Japanese Agriculture.* Institute of Pacific Relations International Research Series. New York, 1941.

Ogasawara, Yoshikatsu. "The Distribution Pattern of Cash Crop Fields in Japan." *Proceedings of the IGU Regional Conference in Japan, 1957* (Tokyo, 1959), pp. 440–45.

————. "Land Use of Japan." *Bulletin of the Geographical Survey Institute* (Tokyo), 2, Part 1 (1950), 95–119.

————. "The Role of Rice and Rice Paddy Development in Japan." *Bulletin of the Geographical Survey Institute* (Tokyo), 5, Part 4 (1958), 1–23.

Ozaki, Chujiro. *Farm Household Economy Survey in Japan.* Agricultural Development Series, No. 13. Agriculture, Forestry, and Fisheries Productivity Conference. Tokyo, 1960.

Sasaki, Shiro. *Land Development and Improvement Projects in Japan.* Agricultural Development Series, No. 3. Agriculture, Forestry, and Fisheries Productivity Conference. Tokyo, 1959.

Smith, Thomas C. *The Agrarian Origins of Modern Japan.* Stanford, Calif., 1959.

Tanioka, Takeo. "Le Jôri dans le Japon ancien." *Annales: Économies, Sociétés, Civilisations,* No. 4 (1959), pp. 625–39.

Thompson, John H. "Urban Agriculture in Southern Japan." *Economic Geography,* 33 (1957), 224–37.

Tobata, Seuchi. *An Introduction to Agriculture of Japan.* Agriculture, Forestry, and Fisheries Productivity Conference. Tokyo, 1958.

Ueno, Fukuo. *Land Utilization in Japan.* Agricultural Development Series, No. 12. Agriculture, Forestry, and Fisheries Productivity Conference. Tokyo, 1960.

————. "The Problems of the Utilization of Slope Land in Japan." *Proceedings of the IGU Regional Conference in Japan, 1957* (Tokyo, 1959), pp. 520–24.

CHAPTER 8: MANUFACTURING

Aki, Koichi, Shigehito Tsuru, Shinzo Kiuchi, and Kaoru Tanaka. *The Economic Atlas of Japan* [*Nihon Keizai Chizu*]. Tokyo, 1954.

Beika, Minoru. "Problems of Regional Industrialization in Japan." *Kobe Economic and Business Review*, 1 (1953), 127–38.

———. "Structure of Industrial Districts in Japan." *Kobe Economic and Business Review*, 3 (1956), 97–124.

Cohen, J. B. *Japan's Economy in War and Reconstruction*. Minneapolis, 1949.

Crawcour, Sydney. "Progress and Structural Change in the Japanese Economy." *Asian Survey*, 1 (1961), 3–9.

Doi, Kikukazu. "The Industrial Structure of Japanese Prefectures." *Proceedings of the IGU Regional Conference in Japan, 1957* (Tokyo, 1959), pp. 310–16.

Foreign Capital Research Society. *Japanese Industry*, vols. *1962* and *1963*. Tokyo.

Ginsburg, Norton, and John D. Eyre. *A Translation of the Map Legends in the Economic Atlas of Japan*. Chicago, 1959.

Hasegawa, Norio. "Distribution of Manufacturing in Japan: A Macroscopic Analysis of Localization of Manufacturing." *Science Reports of the Tohoku University*, 7th Series (Geography), No. 8 (1959), pp. 39–48.

Industrial Review of Japan, vol. 8 (1963) and vol. 9 (1964). Nihon Keizai Shimbun, Tokyo.

Isida, R. "Geography of the Industrialization of Japan." *Proceedings of the IGU Regional Conference in Japan, 1957* (Tokyo, 1959), pp. 349–54.

———. "The Industrialization of Japan: A Geographical Analysis." *Annals of the Hitotsubashi Academy*, 7, No. 1 (Tokyo, 1959), 61–80.

Itakura, K. "Distribution of Chemical Plants in Japan" (in Japanese with English summary). *Geographical Review of Japan*, 32 (1959), 351–64.

Itakura, Katsutaka. "The Distribution of Major Industries" (in Japanese with English summary). *Geographical Review of Japan*, 31 (1958), 95–105.

Ito, Kiei. "On the Location of Woolen Yarn Spinning Industry in Japan" (in Japanese with English summary). *Human Geography*, 12 (1960), 326–46.

———. Regional Characteristics of Minor Textile Industries of Japan" (in Japanese with English summary). *Human Geography*, 9 (1957), 339–57.

Kikuchi, Ichiro. "Cement Plant Location in Japan" (in Japanese with English summary). *Geographical Review of Japan*, 34 (1961), 361–74.

Kitamura, Yoshiyuki. "The Geographic Structure and Development of Four-Wheeled Vehicle Industry in Japan" (in Japanese with English summary). *Geographical Review of Japan*, 34 (1961), 326–42.

Kiuchi, S. "The Technical Factors in the Location of Industrial Regions of Japan" (in Japanese with English summary). *Bulletin of the Geographical Institute of Tokyo University* (1952), pp. 54–66.

Kiuchi, Shinzo. "Centrifugal and Centripetal Urbanization in Japan." *Proceedings of the IGU Regional Conference in Japan, 1957* (Tokyo, 1959), pp. 367–71

Koda, Kiyoki. "Manufacturing Districts in Japan." *Proceedings of the IGU Regional Conference in Japan, 1957* (Tokyo, 1959), pp. 372–74.

———. "Regional Development of Manufacturing" (in Japanese with English summary). *Geographical Review of Japan*, 31 (1958), 8–14.

Lockwood, William W. *The Economic Development of Japan: Growth and Structural Change, 1863–1938.* Princeton, 1954.

Nishimura, Mutsuo. "Industrial Distribution in Japan" (in Japanese with English summary). *Human Geography, 3* (1951), 10–20; *4* (1952), 296–311.

Schöller, Peter. "Wandlungen der Industriestructure Japans." *Deutscher Geographentag* (Cologne), May 22–26, 1961, pp. 238–54.

Schwind, Martin. "Die regionale Verteilung der japanischen Industrie." *Petermanns geographische Mitteilungen, 98* (1954), 280–88.

Takeuchi, Atsuhiko. "Location of Bicycle-Manufacturing Industry in Japan" (in Japanese with English summary). *Geographical Review of Japan, 33* (1960), 412–24.

Thompson, John H., and Michihiro Miyazaki. "A Map of Japan's Manufacturing." *Geographical Review, 49* (1959), 1–17.

Walton, Alan D. "The Founding of the Japanese Synthetic-Textile Industry." *Journal of the Textile Institute, 51* (1960), 275–86.

Yamaguchi, Sadao. "The Locational Changes of Iron and Steel Industry in Japan." *Proceedings of the IGU Regional Conference in Japan, 1957* (Tokyo, 1959), pp. 529–35.

Yamaguti, Sadao. "Die regionale Struktur der japanischen Industrie." *Geographische Rundschau, 9* (1957), 438–43.

CHAPTER 8: TRANSPORTATION

Arisue, Takeo. "A Regional Study on Passenger Traffic in Japan in Special Reference to a Traffic Community" (in Japanese with English summary). *Geographical Review of Japan, 30* (1957), 1016–30.

Japanese National Railways. *Japanese National Railways: A General Description.* Tokyo, 1962.

———. *The New Tokaido Line.* Tokyo, 1962.

———. *1962 Edition Facts and Figures.* Tokyo, 1962.

Mecking, Ludwig. "Japans Häfen, ihre Beziehungen zur Landesnatur und Wirtschaft." *Mitteilungen der Geographischen Gesellschaft in Hamburg, 42* (1931).

Moriwaki, Ryoji. "Some Characteristics of Foreign Trade Ports in Japan." *Science Reports of the Tohoku University,* 7th Series (Geography), No. 13 (1964), pp. 95–113.

"Ports of Japan." *Asian Affairs, 6,* No. 1 (1961), 1–168.

Principal Ports in Japan, 1960. Japan Port and Harbor Association and Central Secretariat of the International Association of Ports and Harbors. 1960.

Road Bureau, Ministry of Construction. *Roads in Japan.* 1963.

CHAPTER 9: HOKKAIDO

Davis, Darrell Haug. "Agricultural Occupation of Hokkaido." *Economic Geography, 10* (1934), 348–67.

———. "Present Status of Settlement in Hokkaido." *Geographical Review, 24* (1934), 386–99.

———. "Type Occupance Patterns in Hokkaido." *Annals of the Association of American Geographers, 24* (1934), 201–23.

Fukui, Hideo. "Recent Changes in the Distribution of Rice Farming in Hokkaido." *Science Reports of the Tohoku University,* 7th Series (Geography), No. 10 (1961), pp. 9–22.

———. "Rice Agricultural Region in Northeast Japan as Seen from the Distribution of Rice Varieties." *Science Reports of the Tohoku University,* 7th Series (Geography), No. 11 (1962), pp. 1–11.

General Headquarters, Supreme Commander for the Allied Powers, Natural Resources Section. "Reconnaissance Soil Survey of Japan," Report No. 110-H, "Hokkaido." Tokyo, 1951. (Mimeographed.)

Harrison, John A. *Japan's Northern Frontier.* Gainesville, Fla., 1953.

Hasegawa, Norio. "Apple Culture Areas in Hokkaido, Northern Japan." *Science Reports of the Tohoku University,* 7th Series (Geography), No. 10 (1961), pp. 23–30.

Hokkaido Japan-American Rural Cooperation Society. *Agriculture of Hokkaido.* Sapporo, 1962.

International Geographical Union, Science Council of Japan, Regional Conference in Japan, 1957. *Regional Geography of Japan,* No. 1, *Hokkaido Guidebook.* Tokyo, 1957.

Ito, Hisao. "Geographical Division of the Fishing Areas Around Hokkaido" (in Japanese with English summary). *Geographical Review of Japan,* 37 (1964), 377–86.

Jones, F. C. *Hokkaido: Its Present State of Development and Future Prospects.* London and New York, 1958.

Jones, Wellington D. "Hokkaido, the Northland of Japan." *Geographical Review,* 11 (1921), 16–30.

Kawaguti, J. "The Cultivated Region in Hokkaido as Related to Density of Population" (in Japanese with English summary). *Geographical Review of Japan,* 13 (1937), 649–68, 689–705.

Kawamura, Takeshi. "The Synoptic Climatological Considerations on the Winter Precipitation in Hokkaido" (in Japanese with English summary). *Geographical Review of Japan,* 34 (1961), 583–95.

Mecking, Ludwig. "Japans Häfen, ihre Beziehungen zur Landesnatur und Wirtschaft." *Mitteilungen der Geographischen Gesellschaft in Hamburg, 42* (1931).

Nippon no Chiri [*Geography of Japan*], Vol. 1, *Hokkaido.* Tokyo, 1961.

Okazaki, Yoshio. "Topographic Development of the Kushiro Moor and Its Surroundings, Hokkaido, Japan" (in Japanese with English summary). *Geographical Review of Japan, 33* (1960), 462–73.

Scheinpflug, Alfons. "Die japanische Kolonisation in Hokkaido." *Mitteilungen der Gesellschaft für Erdkunde zu Leipzig, 53* (1935), 1–132.

Yasuta, Hatsuo. "On the Dairy Regions in Hokkaido" (in Japanese with English summary). *Human Geography* (Tokyo), *16* (1964), 1–18.

JAPAN, ELEVATIONS

□ 0–100 METERS

▨ 100–1000

▨ 1000–2000

■ OVER 2000

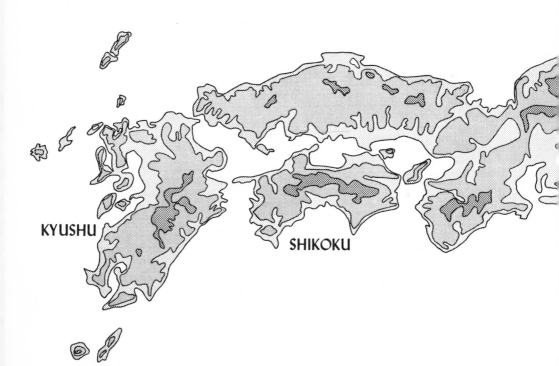

KYUSHU

SHIKOKU